NANO-HYPE

NANO-HYPE

THE TRUTH BEHIND
THE NANOTECHNOLOGY BUZZ

DAVID M. BERUBE

Foreword by Mihail C. Roco

 Prometheus Books

59 John Glenn Drive
Amherst, New York 14228-2197

Published 2006 by Prometheus Books

Inquiries should be addressed to
Prometheus Books
59 John Glenn Drive
Amherst, New York 14228–2197
VOICE: 716–691–0133, ext. 207
FAX: 716–564–2711
WWW.PROMETHEUSBOOKS.COM

10 09 08 07 06   5 4 3 2 1

Library of Congress Cataloging-in-Publication Data

Berube, David M.
    Nano-hype : the truth behind the nanotechnology buzz / David M. Berube.
        p. cm.
    Includes bibliographical references and index.
    ISBN 1–59102–351–3 (hc. : alk. paper)
    1. Nanotechnology—Social aspects. 2. Nanotechnology—Government policy.
3. Nanotechnology. I. Title: Nanohype, the truth behind the nanotechnology buzz.
II. Title: Truth behind the nanotechnology buzz. III. Title.

T174.7.B376 2005
620'.5—dc22

                                                                2005017490

Printed in the United States of America on acid-free paper

CONTENTS

ACKNOWLEDGMENTS   **11**

FOREWORD   **15**

INTRODUCTION   **19**

**CHAPTER 1: EXAGGERATION,
HYPERBOLE, AND HYPE-STERIA**   **29**

Hyperbole as a Variable   29
Hyperboles and Goo   37
    Nanotechnology in Popular Culture   40
    Nanotechnology and the Media   44
    Nanotechnology in Academe   45
Conclusion   47

CHAPTER 2: SPECULATION AND CRITICISM ABOUT NANOTECHNOLOGY

Advocates   49
   Richard P. Feynman   49
   K. Eric Drexler   54
      Early Years—*Engines* and Goo   55
      Later Years—*Nanosystems* and Denialism   56
      Direction of the Nanotechnology Movement   57
Critics   65
   Technical Critics   65
      George Whitesides, Harvard   66
      Richard Smalley, Rice   69
   Popular Critics   73
      Bill Joy, Sun Microsystems   73
      Zac Goldsmith, the *Ecologist*   76
      HRH Prince Charles   78
Conclusion   80

CHAPTER 3: GOVERNMENT ACTORS IN NANOTECHNOLOGY

Individuals   82
   George Allen and Ron Wyden, the Senate   82
   Neal Lane, from the NSF to Clinton to Rice   85
   Mihail (Mike) Roco, NSET   87
   Thomas Kalil, Former Clinton Adviser   89
   Philip Bond, Department of Commerce   90
Government Science Promotion   94
   National Science Foundation   94
   National Institute of Standards and Technology   100
   Advanced Technology Program   101
Executive Branch   104
   Executive Departments   106
      Department of Energy   109
      Department of Defense and DARPA   113
      Other Departments and Agencies   119
Conclusion   121

CHAPTER 4: GOVERNMENT INITIATIVES IN NANOTECHNOLOGY 123

Initiatives and Spin 124
National Nanofabrication Users Network (NNUN) 124
National Nanotechnology Initiative (NNI) 125
President William Clinton 129
President George W. Bush 130
National Nanotechnology Infrastructure Network (NNIN) 130
Twenty-First Century Nanotechnology Research
and Development Act 131
International Actors 135
The United Kingdom and the European Union 137
Japan 144
China 151
Conclusion 153

CHAPTER 5: PROMOTIONAL REPORTS ON NANOTECHNOLOGY 155

United States 155
United Kingdom 165
Economic and Social Research Council 165
UK Royal Societies 167
UK House of Commons 170
European Union 172
Business Stakeholders 176
Credit Suisse/First Boston 176
Forbes/Wolfe Nanotech Report 178
Swiss Reinsurance Company 181
Conclusion 183

CHAPTER 6: APPLICATIONS OF NANOSCIENCE 185

Instruments and Apparatuses 186
Manufacturing and Materials 188
Agriculture and Food Production 193
Electronics and Computing 195
Healthcare 200

Energy   207
Luxury Products   210
Conclusion   211

CHAPTER 7: NANO-INDUSTRY AND NANO-ENTREPRENEURS   213

Economics of Nanotechnology   216
Business of Nanotechnology   218
Established Transnational Firms   220
Start-ups and Venture Capital   222
    Punk Ziegel and Company Index   228
    Merrill Lynch's Nanotech Index   229
    Lux Nanotech Index   229
The Nanosys IPO Story   230
Individuals   234
    Josh Wolfe—Lux Capital   234
    Steve Jurvetson—Draper, Fisher, and Jurvetson   236
    Charlie Harris—Harris and Harris Group   238
Nanotechnology Business Alliance   239
Conclusion   242

CHAPTER 8: NONGOVERNMENTAL ORGANIZATIONS AND NANO   245

Proponents   247
    The Foresight Institute   247
        FI: Institute for Molecular Manufacturing   251
        FI: Center for Constitutional Issues in Technology   254
        FI: The New Foresight Nanotech Institute   255
    Center for Responsible Nanotechnology   256
Opponents   263
    ETC Group   263
    Greenpeace Environmental Trust   270
Conclusion   273

CHAPTER 9: NANOHAZARDS AND NANOTOXICOLOGY   275

Time Frames and Ethical Calculi among the Frames   276
Fear and Trepidations   279
Investments in Nanotoxicology   280
A Primer on Nanotoxicology Research   281
    Range of Concerns   283
    Are Nanoparticles Biodegradable?   285
    Are Nanoparticles Toxic?   287
Risk Analysis   300
Environmental Concerns and Their Ethics   303
Conclusion   304

CHAPTER 10: SOCIETAL AND ETHICAL IMPLICATIONS OF NANOTECHNOLOGY RESEARCH   305

Diagnosis   310
Defining SEIN   310
    Who Benefits from SEIN Research?   314
    The Warning   315
    Teasing Motives   317
SEIN as Symbology   319
Ongoing Research   321
    American Universities   322
        UCLA—Nanobank   324
        University of South Carolina—nanoSTS   324
        Michigan State University   325
        The Nanotechnology Business Alliance—HEITF   325
State of SEIN   329
Conclusion   333

CHAPTER 11: A PUBLIC SPHERE IN NANOSCIENCE AND TECHNOLOGY POLICY MAKING   335

The Call   336
The State of the Public   337

The Challenge   339
Defining the Public Sphere   341
    Architecture of the Public Sphere   342
    Status of the Public Sphere   342
Science and the Public Sphere   344
Publics and Counterpublics   348
About Movements   350
    The Anti-GMO Movement   353
    The Anti-Nanotechnology Movement   354
Experiments   356
Problem Solved?   359
Conclusion   360

NOTES   **361**

BIBLIOGRAPHY   **469**

INDEX   **507**

ACKNOWLEDGMENTS

This work was made possible because of the financial, intellectual, and emotional assistance I received from dozens of people. The list of these colleagues comes from many corners of the nanoworld, scientists and humanists, investors and pitchmen, professors and undergraduates, and enthusiasts and curmudgeons.

I am thankful to the resources provided by the University of South Carolina and its libraries; our university president, Andrew Sorensen, who has been a strong supporter of the type of work we do; the Office of the Vice President of Research under the leadership of Dr. Harris Pastides; provosts Jerry Odom and Mark Becker; the College of Arts and Sciences, under the management of Dean Mary Ann Fitzpatrick; and the support from USC's NanoCenter, in which I serve as an active member with cohorts from biology, chemistry, physics, and engineering. Thanks go to the NanoCenter for the offices that gave me the physical space to complete this work and Tom Vogt who is forging a space for the NanoScience and Technology Studies program in the center.

I am also appreciative of the support received from friends and colleagues, including K. Eric Drexler, for whom I have great respect and who started me on my journey and kept me informed of happenings with his e-mail messages; Rachelle Hollander from the National Science Foundation (NSF), who more than anyone marshaled support for societal and ethical implications research; Jeff Karoub, formerly of *Small Times*, who always has something cooking and is a good friend, Kristin Kulinowski of Rice's CBEN, for her honesty and patience in dealing with a social scientist and for reading an early version of chapter 8; Sonia

Miller of the Converging Technologies Bar Association, for her passion and zeal to speak what is on her mind; Christine Peterson of the Foresight Institute, who dedicated most of her life to nanotechnology and whose role is underappreciated; Mike Roco from the NTSC, who never ceased to amaze me with his enthusiasm— no one seems more committed to nanotechnology as a movement; Jim von Ehr from Zyvex, for his vision and commitment and risk taking; Josh Wolfe from Lux Capital, who provided me with an education in the venture capital industry; and a host of others. I also need to thank my colleagues in the Nanotechnology Inter-disciplinary Research Team (NIRT) at South Carolina, especially Professor Mickey Myrick from chemistry and biochemistry and Bob Best from the School of Medicine, who answered questions about chemistry and physiology.

Some sources were invaluable in completing this project. Josh Wolfe's *Forbes/Wolfe Nanotech Report* demonstrated that it is well worth its steep subscrip-tion price. The *NanoBusiness News*, a publication of the NanoBusiness Alliance, is another remarkable source of information. *Small Times* is one of the few nano-specific publications I read from cover to cover and find both interesting and useful. NanoInvestor News and NanoTechWire.com are additional Web sources worth perusing. The Web resources on nanotechnology are expansive. While most reprint or link to articles and other Web resources, a few are worth mentioning here: the ETC Group, the Foresight Institute, and the Center for Responsible Nanotechnology. All these sources are well endnoted and are in the bibliography. *Nanotechnology Law & Business Journal* is the new publication in town. It is highly readable and has offered some of the best articles on nanotechnology yet.

Finally, I am thankful for the emotional and personal encouragement from family and friends, especially the small group of undergraduate yahoos who worked on the NIRT abstract project at the University of South Carolina. They tracked down research and references in a body of literature that is plagued with less than impeccable scholarship. My student research assistants were an endless source of entertainment as they plodded through dozens of books and hundreds of articles in preparing this work, ending up some of the better-informed people in the country on the nanotechnology debate. In addition, I need to thank a few of my intercollegiate debaters who helped me investigate the commercial, rhetorical, and scientific nature of nanotechnology: Katie Dennis, Lane McFadden, M. Glenn Prince, Mark Smith, Shawn Starkey, Michael Yehl, and especially Tim Wilkins, who introduced me to Drexler's *Engines of Creation.*

Four of my students deserve special mention. Each had a major impact on this project and one changed my life. They helped enormously as sounding boards for some of the more outrageous assumptions and conclusions.

When John (J. D.) Shipman decided to write his undergraduate honors thesis on the business of nanotechnology, I turned over early versions of chapters 4, 6, and 7 to him, and he made significant contributions to them. In addition, he had a profound and enduring impact on my life. Not only is he one of the most complex people I have ever known, but he is also one of the most intelligent and clever. His immunity to being duped by hyperbole made him a powerful colleague. He is exceptionally gifted and sophisticated far beyond his chronological age. He remains a friend, and I will never forget how much I took pleasure in the years we worked together. I aspire to work with him again. I know he will find what he is looking for in law, and this book is unconditionally dedicated to him.

M. Kevin McCarrell was hired during 2004 to help with the bibliography and stayed on for over a year. He tracked down source materials and worked long hours soliciting and acquiring permissions and releases for tables and graphic materials. He is a bright young man. I anticipate he may have a rewarding future in law.

When a graphic designer became no longer a luxury, T. K. Wickham joined the project and saved the day. His help was greatly appreciated.

Finally, Christopher Dickson joined me in the summer of 2005 and did a tremendous amount of work. He edited much more material than he was paid to do. He is diligent, sharp witted, and willing to take on the unknown. I turned over to him many small components of the final editing process, and he never let me down. He is not only one of the nicest persons you will ever meet, but he will also have a great future ahead of him in law and politics.

In addition, I had the luxury of four understanding colleagues who never turned away when my demands for their attention and resources would have tired others. Without them, this book might never have been completed.

Davis Baird negotiated my working relationship with the folks at our NanoCenter, and he was a leading force in efforts to secure funding from the NSF. He is personally responsible for creating nanoSTS (NanoScience and Technology Studies). He took the lead in a proposal for the NSF-supported National Science and Engineering Center called the Center for Nanotechnology in Society. Though it went largely unfunded, Baird did manage to get a portion of it funded. Finally, he was instrumental in our focus on societal and ethical issues in nano-technology even as he moved on to become the dean of our honors college. He authorized the hire of research assistants who helped me enormously, and he is a powerful intellect and a respected friend and colleague.

Steve Lynn chairs a very large English department with many fields of con-centration, including a small and diverse communication studies program. In addi-tion to being a member of the NIRT grant, Steve provided me support, financial and release time, whenever I needed some. His decision to hire an instructor to take over some of my duties allowed me the time I needed to complete this.

Special thanks go to John Skvoretz and Gordon Baylis, who were acting dean and associate dean of liberal arts, the college in which humanities and social sci-ences were housed up to 2005. A sociologist and a psychologist, respectively, by trade, they were honest and straightforward toward me and my pursuits, providing as much help as they were able to muster. Their support of my debate and argu-mentation program, including hires and graduate assistantships, provided me with administrative and instructional release to write this book.

A different type of special thanks goes to my friends, especially J. D., and my family, especially my mother, Aurore, for putting up with me. I am not easy to live with and work with, and I know it. I also want to thank my colleagues in compet-itive policy and parliamentary debating who kept me honest and USC's Debbie Kassianos who kept my computers up and running.

Chapter 7, especially the section on venture capital and Nanosys, owes much to Ann Thayer's excellent article on investment in *Chemical and Engineering News* from May 2, 2005. In addition, David Forman's articles in *Small Times* were an important source of my technology investment education.

Thanks go out to BBC, Inc.; Cientifica; Community Research and Develop-ment Information Service (CORDIS); Credit Suisse Equity Research/First Boston (Michael Mauboussin and Kristin Bartholdson); Crown Business Pub-lishing, a division of Random House (Jack Uldrich and Deb Newberry); the Euro-pean Nanobusiness Alliance (Cristina Roman); the European Society for Preci-sion Engineering and Nanotechnology (EUSPEN); the *Forbes/Wolfe Nanotech Report*, Gartner Financial Services (research director David Schehr); NanoApex Corp. (NanoInvestor News); and John Wiley and Sons, Inc. (Glenn Fishbine) for letting me use graphs, tables, figures, and PowerPoint images in this book.

Portions of chapter 2 began as a paper at a conference in South Carolina in 2003 and were published in a book of selected articles from the same conference and in the Winter 1995 issue of *IEEE Technology and Science*. Parts of chapter 10 started as a lecture at Rice's Nanodays 2003 and at various conferences, and most of chapter 9 appeared as a commentary at a 2005 NAS meeting and a 2005 EPA Public Hearing on the regulation of nanoscale materials.

Finally, thanks go to Prometheus Books and its editorial staff, especially Steven L. Mitchell, Jeremy Sauer, and Chris Kramer; Jill Maxick in publicity; and Nicole Lecht in the art department for their attention and patience. They were incredibly helpful in editing, designing, and promoting a book on time-sensitive material.

FOREWORD

M. C. ROCO

Senior Advisor for Nanotechnology—National Science Foundation, and
US National Science and Technology Council's Subcommittee
on Nanoscale Science, Engineering, and Technology (NSET)

M ajor advances in technology are often marked by the process of convergence and divergence of the fields of science and engineering. One of the most significant developments in modern times is the convergence at the nanoscale. This convergence allowed us to "see and touch" at the first level of organization both atoms and molecules. This is where the fundamental principles of life can be found. Convergence on the nanoscale promises an understanding of the basic properties and functions of materials and systems which may lead to their economical use in the manufacturing process.

At the nanoscale there is the transition from the individual to the collective behavior of atoms and molecules. This opens up multiple pathways for assembling atomic and molecular structures with significantly different results. For example, once we have the basic understanding and tools at the nanoscale we can transform many technologies, manufacturing processes, electronics, and medicine, and create new procedures, for example, to filter water more efficiently and convert energy more effectively.[1]

Scientific convergence and technological integration at the nanoscale were captured by the National Nanotechnology Initiative (NNI) and announced in the year 2000. I had the privilege to propose the NNI in March 1999 at a US National Science and Technology Council Committee on Technology meeting. I recall at that time that a major concern of the White House and Congress was the hype surrounding nanotechnology that was present on several Web sites. However, the NNI approval advanced rapidly.

Professor Samuel Stupp, chair of the first National Research Council (NRC) review of the NNI in 2002, told the second NRC review panel on August 26, 2005:

"Nano is not hype; the nanoscale and its integration with larger-length scales is where science wants to be today. It must be protected at all costs." This is confirmed by the robust industrial, biomedical, and other investments in nanotechnology around the world. While nanotechnology may be oversold in the short-term in some areas, its overall implications seem to be underestimated in the long-term.

The NNI adopted a specific definition for nanotechnology and a long-term view for research and development. Nanotechnology is the ability to understand, control, and manipulate matter at the level of individual atoms and molecules, as well as at the "supramolecular" level involving clusters of molecules. Its goal is to create materials, devices, and systems with essentially new properties, functions, and principles of operation that arise because of their exceedingly small structure. The NNI was motivated by a long-term scientific and technological vision: The ability to systematically control matter at the nanoscale will lead to a technological and industrial revolution. The new technology will be developed progressively, from passive nanostructures that are currently in production to complex molecular nanosystems by the period 2015 to 2020. Because the environmental, health, and societal implications are long-term, and some are unknown, we need to build the capacity to address those implications at the beginning of each large research and development project. All NSF-funded nanocenters are required to address societal and environmental aspects, and additional program solicitations have provided funding for societal-implications studies since 2001. The Nanotechnology Informal Science and Engineering Network established by the NSF in 2005 will bring a more realistic view of nanotechnology to the public at large and K–12 students in the United States.

After earlier research breakthroughs in nanoscale science and engineering, public interest has grown. And so have exaggerations about the benefits and potential unexpected consequences. After scientific discoveries of molecular self-assembly at the end of the 1970s and the creation of self-assembled nanostructures (by Jean Marie Lehn and others), several groups interpreted the results as leading in the short-term to few nanometer-sized assemblers with extraordinary functions, such as self-reproduction. After two decades we are still wondering if the required degree of complexity can be achieved in such nanometer-sized assemblers.

Nanotechnology hype has its parallels in the beginnings of most discoveries. Hype may motivate risky research that is not necessarily clear at the outset. As such, some hype can be highly productive. Hype also has various origins. One source is science fiction literature. Another source may be aggressive public relations outreach and lobbying. Media coverage tends to be dominated by extreme points of view because such opinions are more sensational ("It makes news") and find more dedicated supporters. Utopian speculations, such as "anything that we need is possible" may be taken as goals by various lobbying organizations or individuals. Hype also may be used by some groups as a rallying point for new investments.

A special feature of this book is that it was written by a leading social scientist from the University of South Carolina (which was one of the first institutions to address the societal and environmental aspects of nanotechnology, as well as one of the first institutions to receive NSF support for societal-implications studies), who has a broad view of the nanotechnology community based on existing public reports and personal interviews. Professor Berube is an objective observer of the physical sciences and engineering nanotech community, so he is not tainted by any special interest in supporting respective research and development initiatives. His presentation is fluent and easily accessible, reminiscent of a respected journalist.

The book does not feature much disagreement between the experts who are developing nanotechnology. What controversy that does exist can be found in the opinions, some grounded and many not, of some outliers whose conception of what is happening is built from perceptions sometimes wholly at odds with science and engineering principles and with what is happening in labs and production facilities. While those voices do not constitute an objective rebuttal to what science is up to, they need to be heard, addressed, and vetted.

Besides the science and engineering that spurs discovery in nanotechnology, it is important to understand the impact of public perception, public involvement, and various societal forces. This volume addresses public perception of nanotechnology, and whets the reader's appetite for more realistic assessment of the potential in this field. It includes a great deal of information from direct and indirect sources, such as media as well as nanotechnology experts and observers. Many points of view therefore are expressed. The diversity of ideas is evident, and arose from the breadth of nonscientific sources.

For instance, the science and engineering community would clearly identify George Whitesides and Richard Smalley as major innovators and proponents of nanotechnology. In this book they are presented as critics in the Drexler debate. In another example, the Credit Suisse/First Boston and Swiss RE reports are listed as indications of industry interest, while there are many other reports from the electronics, chemical, and biomedical industries that probably deserve to be mentioned as well.

Nanotechnology is a revolutionary technology, growing progressively from rudimentary nanostructures toward molecular nanosystems. This presents the potential to create a broad technology platform for industry, medicine, and other fields. There is a definite need for scientific overviews, information on trends and forecasts, studies on future scenarios, societal-implications studies, and indeed public involvement.

This book is a diverse collection of opinions and ideas expressed by leading scientists and engineers as well as nonspecialists with an interest in visionary ideas, public interactions, as well as economic and policy matters. Whether the accounts Professor Berube has given here can be determined to be "the truth about nanotechnology," only the reader can judge. I commend him for his drive to uncover the multidimensional image of the nanotechnology buzz and for wading through such a vast amount of hyperbole to achieve a significant degree of clarity.

INTRODUCTION

Nanomania is everywhere. On television, Hewlett Packard claimed to have brought it into our homes. Dr. Nathan Parker, a professor of nanotechnology, not only married a supermodel but also equipped his home with GE's Profile Series of Appliances. According to GE spokesman Jim Healy, "One of the pieces of evidence that something is becoming more real is that in enters into the lexicon of society. And these days the commercial is the ultimate form of popular expression."[1]

It's found on our TV serials and in our movies. The soldiers in *Stargate ST-1* have been fighting the Replicators for a few seasons. *Star Trek*'s Borg is a species that uses nanoprobes as a means for conscription. Recently, Agent Cody Banks's training is put to the test when he's sent to pose as a prep school student and befriend fellow teen Natalie Connors in order to gain access to her father—a scientist unknowingly developing a fleet of deadly nanobots for an evil organization. "Eric Bana's character in the Ang Lee film *The Hulk* is a doctor researching the use of nanotechnology as a means to repair damaged tissue. An accident with the nanotechnology combined with the young man's unique physiology to produce the Hulk."[2] On UPN, Jake 2.0 was a simple technician in a government lab before he got infected with nanobots and became the National Security Agency's new superhero. The children haven't been ignored. "Nano of the North" was a *PowerPuff Girls* episode. There is even an online game called "Dib's Nano Chase" from *Invader Zim*. PC and Xbox have *Deus Ex: Invisible War*.[3] *Duckboy in Nanoland* recorded fifteen hundred players within its first three weeks, and the video game in the Science Museum of London's gallery chalked up even more.[4] And the list continues.

Of all the popular-culture artifacts, one has attracted the attention of government policy makers—Michael Crichton's *Prey*: An unemployed stay-at-home computer programmer is recruited by his wife to a government research installation in the Nevada desert. Julia Forman is on a team that designs military nanobots and nesting intelligence, which leads to some gruesome encounters between them and both indigenous fauna and the researchers. Jack Forman comes to the rescue and liberates both the company doing the research and some friends and family from the beasties with his Holmesian deductive skills and knowledge about software. Except for intriguing "feeding" scenes and the similarity between the fictional company Xymos and the not-so-fictional Zyvex, the book is substandard pulp fiction written as a novelized screenplay about parasitism and symbiosis. You can pick up a copy in most Wal-Mart discount bins.

To legitimize this whimsy, Crichton begins with an introduction with footnotes on artificial evolution, quoting some of the people you will be introduced to in the following chapters. Expert citation and footnotes notwithstanding, Crichton can pray that Hollywood produces his book into a blockbuster; whether *Prey* will have an impact on popular consciousness sufficient to sway public sentiments about nanotechnology is unknown.[5]

Twentieth Century Fox bought the rights to the book for about $5 million and wanted to get going on *Prey* quickly.[6] What Crichton did for dinosaurs in *Jurassic Park* he threatens to do with nanobots. In a few years from now, tweens, teens, and young adults will be drawn to the summer release of this SFX movie. Of course, the recent release of *I, Robot*, which has only a passing commonality with Isaac Asimov's work, features nanites as the technology of choice to halt a robot revolution directed by an out-of-control artificial intelligence, VIKI. Government and industry hope that America's first introduction to nanotechnology does not involve murder and mayhem.

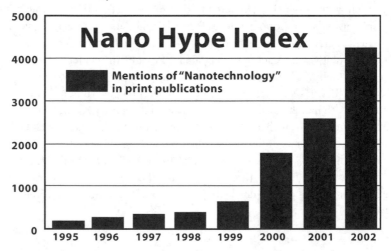

The pervasiveness of the term *nanotechnology* in print publications of all sorts appeared graphically in an early issue of the *Nanotech Report.* Josh Wolfe detailed the rising number of mentions of the word *nanotechnology* in print publications, calling it the *Nano Hype Index* (see previous page). He confidently added that "during 2003, a wave of concern and pessimism would fall by the wayside as the topic is thoroughly discussed and the public is educated about nanotech. Such fears will ultimately by superceded by a series of unexpected scientific breakthroughs and investor momentum resulting from returns reaped by early investors."[7]

Particles are manufactured on the nanoscale every day, and businesses have sprung up everywhere claiming a foothold in the nanophenomenon. While we may not have built nanomachines per se, we have produced an array of nanoparticles. "Worldwide governments, corporations and venture capitalists invested greater than $3 billion into nanotechnology in 2003."[8]

On yet another level, university laboratories have relabeled their micro-electro-mechanical-systems (MEMS) research as nanoscience in an effort to acquire lucrative government research funds. While lithography may be occurring on a scale approaching the nanolevel, most of this work is not taking full advantage of the nanolevel, the unique quantum world.

Is the technology only about chemosynthesis, catalysis on the nanoscale? Or is the technology about nanobots working together to construct macroscopic products? If the former interpretation is accurate, then we need to examine the consequences of nanoparticles in terms of its interaction with the environment and its impact on life and world values. If the latter interpretation is accurate, then we may need to consider whether a world with nanobots doing our bidding is such a good idea. Or maybe we are approaching something between these two interpretations. Or even further, we may need to consider the interaction between nanotechnology and other technologies like biotechnology and computerization. The emergent nature of these convergences may introduce a new series of questions and issues.

In an effort to simplify an incredibly discrete field of science so that policy makers and the public might better understand what our tax dollars are supporting, the proponents have painted themselves into a corner. The exaggeration and hyperbole infecting the discussion of nanoscience and nanotechnology have led to false expectations and apprehensions. This has led to worries that we may be fearful of a bogey.

"In a move that industry supporters blame on a conflation of facts with popular fiction, activists have begun to organize against the science."[9] However, it is irresponsible to suggest their environmental concerns are fictional. "While some concerns have already been raised that seem more in the realm of science fiction, there are also some very real issues with the potential health and environmental effects of nanosized particles."[10]

As media, businesses, and universities make overclaims associated with their works, they have muddied the debate over the societal and ethical implications of science and technology on the nanoscale. Unless we rewrite the story of nano-technology, we will not be able to evaluate what science and technology are doing on the nanoscale. If there is a reasonable probability that the implications of nanotechnology might be problematic, then we may need to engage some fore-sight to decide whether we can afford to continue blithely supporting research and applications without a parallel and comparatively vigorous vetting of its societal and ethical implications. "An effort must be made by the research community to open lines of communication with the public to make clear that potential safety risks are being explored and not ignored."[11]

We know technologies are neither wholly desirable nor undesirable. By their nature, they require a balancing of values and benefits in order to make a prudent decision to continue on the course set for us and often by us. We cannot assume that science will police itself. We cannot defer to our policy makers and elected representatives when it comes to issues of right and wrong because technology is what we allow it to become. "There is no Law of Nature that guarantees that each new technological introduction will be able to safely walk the tight rope above dis-aster. Every time, the technology becomes more powerful and the potential for ruin becomes that much greater. Technology is nothing more than a manifestation of accumulated human genius—nasty or noble. So, as always, it is not technology we need fear or trust—it is we."[12]

What is interesting about nanotechnology is that it functions as a technolog-ical multiplier. It moves other technology to new levels. Credit Suisse Equity Research labels it a general-purpose technology (GPT). Other GPTs, including the steam engine, electricity, and railroads, have been the basis for major economic revolutions. Some have argued that nanotechnology's impact could be akin to plastics.[13] The term *general-purpose technology* was investigated by Elhanan Helpman.[14] GPTs have four characteristics:

- scope for improvement (the technology evolves and over time it progresses while costs fall)
- wide variety of uses (inherent in GPTs are major opportunities for improvements, adaptations, and modifications)
- wide range of uses (over time the proportion of an economy's productive activities that uses the technology increases)
- strong technological complementarities with existing or potential tech-nologies

All prior GPTs have led directly to major upheavals in the economy. Nanotech-nology may be larger than any of the GPTs that preceded it. "The questions around nano are no longer whether it's coming or if it's real but just how big it is.

Some see nano simply as a new material revolution, akin to the dawn of petro-chemical-based products. Others herald a transition as dramatic as mankind's advance from stone to metal tools."[15] This could lead to profoundly *creative destruction*. Creative destruction is the process by which a new technology or product provides an entirely new and better solution,[16] resulting in the complete replacement of the original technology or product.[17]

For example, the ETC Group makes a dire prediction about the convergence of biotechnology and nanotechnology: "If biotech and nanotech merge, the two great sources of productive power—minerals and microbes—also come together. The coming together of nano- and biotechnologies does not merely spell the death of distance; it foretells the *death of dissent*."[18] They advocate a moratorium on some nanoresearch, but to purloin an adage from *SciFi* magazine, "When nanotech is outlawed, only outlaws will have nanotech."[19]

First of all, nanotechnology is about size, not about artifacts per se. At one billionth of a meter, we are at the nanometer. *Nanotechnology* was a term coined by Norio Taniguchi,[20] while Arthur R. Von Hippel may lay claim to the term *molecular engineering*.[21] The credit for the functionalized concept coupled with the term is usually associated with Richard P. Feynman, Nobel Prize–winning Cal-tech physicist who spoke about it at an after-dinner lecture in 1959.[22] However, it didn't make a splash until the mid-eighties with the publication of *Engines of Creation* by K. Eric Drexler, "an astonishing original work of futurism."[23]

"What passes for nanotechnology today is really just materials science. Mainstream nanotechnology, as practiced by hundreds of companies, is merely the intellectual offspring of conventional chemical engineering and our new nanoscale powers."[24] Materials nanoscience is nothing new. "Renaissance artists used paints and glazes that got their appealing color and iridescence from nanoparticles. The ancients too found uses of nanoparticles in soot."[25]

The materials science class of nanotechnology is very different from the Drexlerian version of nanotechnology. "While mainstream nanotechnology gives you better eye shadow, Drexler's nanotechnology gives you a whole new face."[26] Drexler's version involves incredibly tiny machines building macroscopic products; to accomplish this task requires an army of nanomachines. Many experts deride his vision as pseudoscience, science fiction, and unrealistic utopianism.

Why the virulence remains is unclear, though Adam Keiper offers two explanations. First, "Drexler's talk of the great boon and bane of nanotechnology may cast a pall over their own modest research." Second, "they simply don't know what they're talking about.... They are the wrong experts—they are excellent in their discipline but have little expertise in systems engineering."[27]

As such, my goal is to provide the reader with a better understanding of how nanotechnology has been communicated to the many audiences willing, sometimes even anxious, to listen. While this may not be the first time (or the last) that a technological phenomenon has been hyped, it represents a wonderful case study, for it has come to dominate many forums in science and technology policy and decision making. It has become associated with a want and need and occupies our dreams and nightmares. Why individuals and organizations selected to use hyperbole as the trope from communicating nanotechnology will be examined in detail. Nanotechnology is another in a long list of media- and government-sanctioned fears. It should be unsurprising that America feels totally disempowered when it comes to functioning as a legitimate public sphere to both direct and check the agenda of the purveyor's science and technology, often funded by the tax dollars paid by its citizens.

Chapter 1 reviews the role of hyperbole in nanoscience and nanotechnology, especially as instigated by the new culture of public scientists and perpetuated by the media. Nanostorytelling is examined against the role of hyperboles in advocacy, especially the profoundly disturbing "grey goo" scenario and other bizarre fictional depictions in popular culture, especially science fiction.

Chapter 2 concentrates on the two scientists who have been labeled as "fathers" of nanotechnology: Richard Feynman and K. Eric Drexler. Feynman's 1959 speech is compared with a second on the same subject given a few decades later. The modification of his original thesis is criticized. Drexler is studied from his earliest works in the 1980s through his rhetoric in the first half of the first decade of the twenty-first century. Examined is what happened to Drexler's vision of molecular manufacturing and why it didn't play a significant role in the set of initiatives that grew out of President Clinton's Caltech speech, which is usually associated with the first rumblings of policy of any sort regarding nanoscience and nanotechnology. The scent of some conspiracy hangs over the subject as fueled by Drexler and some of his supporters. After avoiding the limelight of controversy and character assassination, Drexler has returned with a pocketful of existence proofs and an axe to grind. The focus shifts to five critics. The first two address science and the second set, science fiction. They are George Whitesides of Harvard; Richard Smalley of Rice and Carbon Nanotechnologies, Inc.; Bill Joy from MicroSystems; Zac Goldsmith, the editor of the *Ecologist*; and HRH Prince Charles. While Whitesides and Smalley have significant problems with molecular manufacturing as envisioned by Drexler and his followers, Bill Joy, Zac Goldsmith, and Prince Charles rant and rave about health, the environment, and uncontrolled replicating assemblers with fantastic scenarios of gloom and doom.

Chapter 3 reviews some important government actors in nanoscience and nanotechnology. The congressional call for nanotechnology has come from the Senate Committee on Commerce, Science, and Technology led by George Allen (R-VA) and Ron Wyden (D-OR). Within the executive branch, four leaders are examined: Neal Lane, former NSF director and adviser to President William Clinton, helped write the speech Clinton delivered that inaugurated the National

Nanotechnology Initiative; Mihail Roco chairs the Interagency Working Group on Nanoscience Engineering and Technology and is the senior NSF advisor on nanoscience and technology; Thomas Kalil was one of Clinton's technology and economic advisers; and Philip Bond was the former principal adviser to Commerce Secretary Don Evans on science and technology policy, which included import/export controls and nanotechnology, and outspoken proponent and frontperson in the Department of Commerce. Contemporary agenda setting in American science and technology policy is reviewed within three different entities: the National Science Foundation, the National Institute of Standards and Technology (NIST), and the Advanced Technology Program (ATP) in NIST. In addition, institutions with their own nanocultures—the Department of Energy; the Department of Defense, especially DARPA; and departments of agriculture, justice, and homeland security—are studied as major players in designing the trajectory of America's nanoscience and technology policy.

In chapter 4, government initiatives and their history are reviewed including the National Nanotechnology Initiative (NNI), the National Nanofabrication Users Network (NNUN), and the National Nanotechnology Infrastructure Network (NNIN). Finally, the chapter concludes with some explication how these initiatives on nanoscience and technology are being matched by non-US government actors: the European Union, Japan, and the People's Republic of China.

Chapter 5 reviews a set of promotional reports on nanoscience and nanotechnology. It starts with a review of some of the more important publications coming from the National Science Foundation in the United States, comparing them with the recent publications from the United Kingdom's Royal Academy of Sciences and Engineering. While both sets of reports discuss fundamental efforts to support basic research, health and environmental considerations, and commercial applications, they are contrasted with three major reports coming out of business and industry: one done by First Boston and Credit Suisse Equity and another from Forbes written by Josh Wolfe, who traces the companies and associated business issues. The final report was recently released by the Swiss Reinsurance Company. It is included because of its different sense of risk and is an object lesson in the cascading impact of nanoscience through a variety of commercial entities. Choices had to be made, and some other reports were excluded. However, many of them are cited elsewhere. More reports will come. These seem to be the best, most important, and received the most notoriety.

In chapter 6, the beginnings of nanoscience applied in terms of technological processes and products are examined in instruments and apparatuses, manufacturing and materials, agriculture and food production, electronics and computing, healthcare, energy, and luxury products. Most of this material was gleaned from professional journals, business and industry publications, and industry reports.

In chapter 7, the beginning of nanobusinesses and the markets of nanotools and commercial application of nanoscience are studied. Three categories of businesses are examined, including transnational corporation in-house programs, start-ups that are dependent on venture capitalists, and start-ups that moved into public offerings (the case of Nanosys is discussed in detail). Next, three of the most prominent venture capitalists and their strategies and tactics are detailed: Josh Wolfe of Lux Capital and the *Forbes Nanotechnology Report*; Steve Jurvetson, from Draper, Fisher, and Jurvetson; and Charlie Harris from Harris and Harris. Finally, the chapter concludes with an analysis of the NanoBusiness Alliance as directed by Mark Modzelewski, Sean Murdock, and others.

Chapter 8 considers the roles played by nongovernmental organizations. On the proponent side, we begin with the Foresight Institute (FI) with Christine Peterson at the helm. FI not only was one of the first promoters of nanotechnology research but most recently proposed that the National Academy of Science complete a study to determine the feasibility of molecular manufacturing. FI is joined by the Center for Responsible Nanotechnology with Chris Phoenix and Mike Treder, both heavy supporters of Drexler; they publish a large amount of online documents favoring a tempered approach to molecular manufacturing. On the opposite side, there is Pat Mooney and the ETC Group (Action Group on Erosion, Technology, and Concentration, formerly Rural Advancement Foundation International—RAFI), an international group that supports socially responsible developments of technologies useful to the poor and marginalized societies, addresses international governance issues and corporate power, and wants a moratorium on some nanoresearch. The second opposing organization is the Greenpeace Environmental Trust, which, while more moderate than the ETC Group, also calls for a limited moratorium on commercialization.

In chapter 9, research associated with nanoparticles and their effect on environmental health and safety is summarized. While somewhat technical in nature, the chapter treats the reader respectfully, and while some terms may appear unfamiliar, they were not dumbed down when doing so would have affected the summaries. It begins with a review of the actual investments made in learning more about nanotoxicology. Next, a primer on nanotoxicology research covers the most important studies that are being cited in meetings on environmental health and safety. Then, the range of concerns is covered within two areas: biodegradability and toxicology. The chapter concludes with a basic review about how risk assessment and analysis occurs within this area and how it intersects with ethical issues.

In chapter 10, an attempt is made to ferret out the rationale for the recent emphasis on societal and ethical implications of nanotechnology (SEIN) by the NSF and other government agencies.[28] Rhetorically, the case has been made that a second genetically engineered seeds-and-food problem like what happened in Europe might emerge around commercialized nanoscience. The SEIN issue is reexamined as a political phenomenon. It begins with teasing out the motives government agencies may have in supporting this type of research leading to speculations that it may be little more than perception and fear management. Cynical in tone, the chapter attempts to forewarn that it is not unlikely that this research will function as mere window dressing. There is tremendous trepidation among SEIN researchers—they may serve a public-relations function to rationalize policy initiatives already in the works, acting as apologists should things go awry. To better evidence this interpretation, current efforts undertaken at a handful of American universities, such as UCLA's commercialization studies under Lynn Zucker, the University of South Carolina's nanoScience and Technology Studies program directed by Davis Baird, and a recent program developed by Paul Thompson at Michigan State University. These efforts are evaluated against a moribund initiative by the NanoBusiness Alliance called the Health and Environmental Issues Task Force (HEITF).

In the final chapter, we observe the reaction to nanohype by citizens. Traditionally, a discussion of checks and balances associated with public policy has included the "public sphere." However, while it may be prudent to examine the role of the public sphere when addressing traditional political considerations, such as those found in platforms candidates build when they run for office, it is much more problematic in science. Nanotechnology provides a special setting to evaluate the public sphere. Many policy makers and scientists have indicated a concern for an antinanotechnology consensus growing in the West. As a social movement, nanotechnology may find opponents from among those protesting against globalization and genetically modified organisms, especially food products. In response, government regulators have demanded that societal and ethical implications should be examined against the rush to fund research into nanoscience and its applications in an effort to defuse the situation.

It is worth making a few comments about style. The endnotes were used extensively for a few reasons. First, it is important to know that this book was heavily researched and involved over a decade of reading and evaluating government, industrial, journalistic, and Web resources. Second, the decision to overresearch the material is to build a bulwark against criticism that someone in the field of communication studies, in a department of English—an outlier—may be underqualified to speak on the subject. My location in the department of English is a quirky happenstance peculiar to the University of South Carolina and reflects choices made long before I joined the faculty. For some years, the communications faculty members were in a department of theater, speech, and dance. Nonetheless, I have degrees in biology, psychology, and communications, and I am a tenured full professor. I teach graduate courses in the rhetoric of science and technology and risk communication. I am a member of the USC NanoCenter and have studied this phenomenon since 1986.

While it is true that I penned some articles on nanosocialism for an online journal, they were written as arguments—not as some sort of manifesto—unlike some of the transhumanists out there. As an argumentation scholar on some level, I wrote them to generate rebuttal. Over the years, I have received responses to these articles, and many critics were surprised when I refused to advocate the overthrow of capitalism. For anyone who really knows me, that should come as absolutely no surprise.

This work treats scientific experts preferentially when they make scientific claims, especially in chapter 9 on toxicology. However, the full range of hyperbole involves many voices, almost all of which are mediated by the press in one form or another. When an NGO makes a claim that a scientist disagrees with, it is not enough for science to speak from authority to silence criticism. If we are truly interested in hearing from stakeholders, then we need to listen to them even when we disagree with what they are saying. To privilege one voice over another would move public participation downstream when the call for stakeholder involvement needs to move them upstream in the decision-making process.

The conclusions drawn in this book may not be all things to all disputants in the debate over nanotechnology. There was a bona fide effort, however, to include many different points of view to unpack the hyperbole pervading the field (notice the Stupp quote in Roco's foreword). If you disagree with me, that is fine. I respect your opinion and think healthy and vigorous debate is only good for the burgeoning world of nanoscience and nanotechnology.

This work has been made possible by a grant from the National Science Foundation, NSF 01–157, NIRT (Nanotechnology Interdisciplinary Research Team): Philosophical and Social Dimensions of Nanoscale Research—From Laboratory to Society; Developing an Informed Approach to Nanoscale Science and Technology. All opinions expressed within are the author's and do not necessarily reflect those of the University of South Carolina or the National Science Foundation.

Finally and most importantly, this work will remain a work in progress as long as government agencies, like the NSF, and universities, such as the University of South Carolina, continue to support the scholarship by humanists and social scientists in science and technology studies. Science and technology studies, while a young discipline, is a very important one and will prove its worth during this millennium.

While the story is ongoing, your introduction to it is about to start.

CHAPTER 1

EXAGGERATION, HYPERBOLE, AND HYPE-STERIA

Nanotechnology, according to N. Katherine Hayles, "has become a potent cultural signifier. It represents not so much a theoretical breakthrough as a concatenation of previously known theories, new instrumentation, discoveries of new phenomena at the nano-level, and synergistic overlaps between disciplines that appear to be converging into a new transdisciplinary research front."[1] To justify the government spending and media interest in this area, people on the nanofront have engaged in exaggeration and hyperbole to repackage this idea as something new and exciting. As Tim Harper put it, "Nobody makes nanotechnology. It will not be an industry unto itself."[2] Mostly, nanoscience and nanotechnology have been an evolutionary product of where science, especially biology, and chemistry, is at the turn of the millennium. Nonetheless, this new research front has produced a set of diverse narratives about nano that share a commonality: hyperbole.

HYPERBOLE AS A VARIABLE

While emerging technologies have always been a challenge to communicate to the public, sometimes even to technically competent audiences, nanotechnology has some unique features. Though biotechnology shares some of them, it had the advantage of being directly linked with health and medicine, at least until it got into the seeds-and-food business. On the other hand, nanotechnology has all the negatives associated with biotechnology and is plagued with starry-eyed expectations and equally perverse exaggerations of gloom and doom. It is the inherent

29

linkage with hyperbole of all sorts that has made nano such an incredibly difficult sell. This does not mean nanotechnology has uniquely attracted hyperbole. For example, Pat Mooney of ETC claimed that "over-hype is always there. There's never been a technology in the last century or so that hasn't been over-hyped."[3] Yes, but there is the matter of degree and the intense mediation. In addition, nano-technology has become associated with major national initiatives in nearly all developed and some developing countries. In addition, the hyperbole is being used by proponents and opponents, including Mr. Mooney.

"Hyperbole involves the conveying of a proposition that so distorts the obvious truth that the hearer recognizes the non-literal intention on the speaker's part."[4] While many hyperboles are apparent because they are absurd, others are much less apparent. More importantly, hyperbole, like irony, demands the hearer perceives the intention. If and once the audience accepts an example of hyperbole as literal, it becomes increasingly problematic to debunk the mismeaning, the mis-understanding. Denial is insufficient. The entire reasoning matrix that led to the misunderstanding needs to be revised. The hearer must decide that the misun-derstanding must be corrected because the warrants originally used are incom-mensurable with more salient and exigent issues of personal and group impor-tance. Simply put, rebutting a hyperbole taken as literal requires rewiring of the process that led to the misunderstanding. Saying "You're wrong" doesn't work.

Don Eigler of IBM tells a story about sea monsters cogent to the following discussion. Prior to the voyages of Columbus, there were stories about sea mon-sters and serpents out there in the Atlantic Ocean that could destroy whole ships, do all kinds of havoc. Where was the proof of the sea monsters? You see, it didn't matter that there was no empirical proof.

Misunderstanding predicated on the improper decoding of hyperbole is not all bad. There are opportunists hyping the hype and trying to capitalize on the momentum, and this is especially true with nanotech. For example, "journalists, talking heads and misinformed investors could be heard recommending 'nanotech' companies that in fact had little to do with nanotechnology, boasted weak funda-mentals or traded at unsightly valuations."[5]

Who's using nanohype? It's used by those who for whatever reason serve to gain from a nanotechnology initiative. This includes visionaries, bureaucrats, elected officials, industrial leaders, recipients of government grants, and investment bankers. Indirectly, there is a second group that also benefits from nanohype, such as the media that use it to generate interest and controversy, nongovernmental organiza-tions that use it to create points of interest for fund-raising purposes, and various and sundry others who crave attention of all sorts, whether the undercompensated academic, the marginalized environmentalist, or the oddball (generally an outlier).

Hyperbole is associated with the full range of effects nanotechnology could produce. Its capabilities are described as *dramatic*. Its effect on reducing energy consumption is *drastic*. Its impact on the advancement of medicine is *fundamental*. Its result on the precision and effectiveness of military weaponry is *significant*.[6] The concern: "Experts in nanotechnologies across the world identified hype (misguided promises that nanotechnology can fix everything) as the factor mostly likely to result in a backlash against it."[7] There are costs involved when we accept low-probability, high-positive valence events in deciding policy. "Such rhetoric may encourage people to drop their guard with prudential actions that are important today. The promise of electricity too cheap to meter does not encourage energy conservation in the present. The promise of the end of resource scarcity can do nothing but foster the profligate use of currently available resources. Promises to end all pollution and clean up all toxic waste dissuade people from worrying about the messes they are creating today."[8]

What rationale that does exist for preferring exaggeration seems to revolve around the general observation that "research is progressing at a speed that outpaces the predictions of the most optimistic prognosticators."[9] In other words, we do not have the communication skills to describe the phenomenon, so we default to tropes. Similar to a professor using a simile or a metaphor in describing something novel, exaggeration and hyperbole are used when adjectives and adverbs elude us. As such, we have simply run out of language tools to explain what is happening in the field.

If we divide nanotechnology into time periods, we immediately notice that the hype mostly surrounds the long-term, the ten-years-and-beyond category. Furthermore, the most egregious hype is associated with molecular manufacturing involving mechanosynthesis popularized by K. Eric Drexler. Nanotechnology in the two immediate time periods, the next five years and five-to-ten years, is not without some risks, but hardly as revolutionary or apocalyptic as those associated with universal assemblers.

Two people can play the shell game. While it may behoove industrialists to misdirect our attention from current applications, it is quizzical they chose the hyperbole of molecular manufacturing to accomplish that. We are simple creatures; generalization is a fall-back strategy we learned in our early years, and many of us find it incredibly difficult to surrender. The other player uses hyperbole to misdirect attention as well but this time toward the promises of nanotechnology drawing with it investment and government subsidy.

Recently, we've noticed the prefix "*nano* has become a surefire market cap turbo-charger for many unscrupulous stock promoters. The number of low priced stocks waving a nanotech banner grows by the day."[10] "As nanotechnology has become a buzzword on the street, it has attracted a load of scam artists some of whom have nothing to do with it have actually added *nano* to their name."[11] For example, US Global Nanospace "is in the business of making armor and other defense products. But the name isn't all that doesn't make sense. Since March 2002, this $85 million market cap company brought in a paltry $125,679 in revenue with a net loss of over $7 million."[12] NanoPierce Technologies "was called Sunlight Systems and was also known as Mendell-Denver. They've got nothing to do with nanotechnology."[13] It has gotten so bad that Asensio and Co., an investment firm, approached New York attorney general Eliot Spitzer "charging that misuse of the nano label has become a favorite tactic for fraudulent stock promotion."[14] Manuel Asensio, chief executive of the firm, claimed, "Investors are being harmed on a daily basis."[15] Ansenio and Co.'s main trading strategy is short-selling, hence they have a financial stake in the definition.

While *Crains Detroit* writer Andrew Dietderich cautions we should "forget all the hype surrounding nanotechnology," he admits, "it's here to say."[16] And it's being perpetuated by the dozens of conferences "mostly about hype and rallying up investors, potential alliance partners, clients and journalists about the prospects for the industry."[17]

Rick Snyder of Ardesta identified some of the pros and cons of hype: "It's bad from the perspective that it could attract investors who shouldn't be investing." But some good does come from the hype: "In many cases, it is spawning a responsible discussion of health and societal issues."[18]

On the other hand, NanoTex CEO Donn Tice insists his company is "undergoing a marketing retrofit in which we're moving away from these nano prefixes for individual products. Nano-Care, that treatment that helps fabrics stand up to spills, is being redubbed Resist Spills. Nano-Touch is being named Cotton Touch. Coolest Comfort is the new name for our moisture wicking technology, wrinkle-free cotton garments and resin-treated knits. You get the picture."[19] According to Thiemo Lang, founder of the first-ever nanotechnology mutual fund for Activist, "It could be advisable for companies to underemphasize their nano profile. Investors could become oversaturated by the *nano* buzz—especially if more companies employ the nano label just to capitalize on the financial markets. The slogan 'We are a nanotechnology company' emblazoned on a Web site might suddenly be perceived negatively."[20]

Everyone involved should have a consistent message. If investors are told a technology will change the world, someone who is concerned about the risks cannot then be told that the same technology is no big deal. It strikes a false note to say that something can be both revolutionary and nothing to worry about, Lang says. "Such inconsistencies will breed public mistrust and fear."[21] "Nanotechnology is being hyped and over-hyped today," said Thomas Cellucci of Zyvex. An odd remark, since Cellucci is not unwilling to hype it up on occasion. Others, like John Fan, CEO of Kopin, claim otherwise. "In many cases the products touch everybody—chances are, people are really using products which use nanotechnology already, they are in your hands. It's not just talk."[22]

Hype serves an amplifying function. Accordingly, "the fears push technologies to improve and get society to look at consequences and decide what trade-offs are acceptable."[23]

The problem is not isolated to the entrepreneurial and investment communities. Even "some scientists contend that much of its *hype* may not match the reality of present scientific speculation."[24] This disconnect is particularly troubling because while scientists—especially those dissociated from business, industry, and government and associated with universities—still garner some trust, the message of nanotechnology is rife with challenges of jargon, cross-disciplinary static, and other problems.

It doesn't stop there either. Universities across the entire country have opened nanocenters mostly populated by faculty from well-established departments who have been relocated to a new building or a few rooms in a wing. Other than a new sign out front and an administrative assistant who answers the phones part-time, not much has changed. Senator Wyden during early hearings on the Twenty-First Century Nanotechnology R&D Act admitted this same problem: "The joke these days in the world of science is that everyone is doing nano work. Just as the '90s saw everyone putting *Dot.com* after titles, everyone if putting *nano* before their science."[25]

Here's the rub:

> There are really two different Nanotechnology Movements in the world today. One—based in industrial corporations, university laboratories, and government research-funding agencies—remains closely tied to chemistry, physics, and material science, working to create the actual technical breakthroughs of the coming decade or two. The other—based largely in science fiction literatures—postulates a future century in which nanotechnology revolutionizes human capabilities, based more on metaphor than on careful calculation, but having a profound influence on perspectives of people who are not professional scientists or engineers.[26]

Hyperbole is a trope that is meant to draw attention to an unnoticed or obscured referent by exaggerating one or more of its features. Like most exaggerations, hyperbole develops a life of its own as it is replicated in story after story. Left unchecked, it is a lot like a conspiracy theory. However, once the products of applied nanoscience reach the market and they fail to meet expectations, consumers will balk and citizens will question the price tag associated with this failed experiment. The National Nanotechnology Initiative (NNI) may appear simply a pork barrel initiative. As such, affirming hyperbole leads to heightened expectations. When unmet, it leads to resentment. When nanotechnology, especially the NNI, is dependent on the goodwill of taxpayers to foot the bill, it becomes vulnerable. A vulnerable national initiative is a loser's game.

For example, Rustum Roy has described the term *nano* as a *halo-regime*—a term that is sold to budget managers in order to increase funding.[27] Roy believes that by association, appropriations are drawn to government programs, universities and industries that rename their micro-electro-mechanical systems (MEMS) programs to dine on the gravy train:

> The labeling is a standard marketing device. Find a *halo* word that most people, including the decision maker, do not understand. The word should be new, different, euphonious, and connected somehow, however tenuously, to science. This is rarely difficult, with appropriate embellishments of connections and exaggerations of the potential benefits. This is the contemporary manifestation of the human need for mythopoesis: creating iconic ideas to which we attach special, even *magical* meanings.[28]

Rhetorical theorists have examined the significance of *god* and *devil* terms in many different settings.[29] Generally, as language becomes increasingly polarized, god and devil terms emerge. Companioning a person, artifact, phenomenon, or concept with another conception, even a word or term, with high positive or negative valence, the person or phenomenon absorbs the valence and by connection is deemed positive or negative as well. For example, associating the political campaign of Arnold Schwarzenegger for the recall gubernatorial election in California in 2003 with his roles as the Terminator transferred his fictional depiction of a savior of sorts with his real-world persona. Rightly or wrongly, voters recalled Gray Davis and selected the Governator as a substitute state executive officer.

In terms of science and technology policy, the associated variable can be positive or negative in its depiction as well. For example, describing a line of research as "another cold fusion" transfers the negative valence associated with the work of Stanley Pons and Martin Fleischman to another's work, while referring to a pharmaceutical as "the next penicillin" transfers positive valence to it.

It would seem that Roy's halo category works similarly. Government grants and, where possible, venture capital may be attracted to projects that are haloed. Just as the figure in a Renaissance painting wearing a halo must be blessed, so it may be for a company or product carrying the nano prefix.

For the time being, nano seems to be a halo or god term attracting interest, attention, and capital. This would explain on at least one level the proliferation of the term and the power that seems to be associated with it. It might also help to explain the media's fascination with anything and everything nano.

There has been some hand-wringing over the state of nanohype. "There is some understanding among industry that the level of hype surrounding nanotechnology has, to some extent, damaged investment potential."[30] Others fear more insidious consequences:

> Many communications and public policy experts say it is essential that nanotech researchers and supporters get on board with efforts to dampen unquestioning enthusiasm for nanotechnology. They argue that without discussion of the potential pitfalls, future nanotechnology research could be subjected to such extreme pressures that funding is jeopardized and research progress is slowed, perhaps halted altogether in some cases.[31]

The Hype Cycle Defined

"Usually at the early stages of hype and hope, actors start moving, taking positions, building alliances—this is how networks and industry structures emerge,"[32] and that is what we are seeing today. Gartner is a research and advisory firm with over eleven thousand clients. It consists of over forty-three hundred associates, including twelve hundred research analysts and consultants. In FY 2001 it had revenues of $952 million.[33] This Stamford, Connecticut–based performance-management company publishes the *Gartner Hype Cycle Special Report*,[34] which traces the progression of an emerging technology.

This cycle begins with an event that generates significant interest such as with the NNI. Next, it works through inflated expectations when the only enterprises making money are conference organizers and publishers, such as InfoCast and *Small Times*, and that seems to be the point we are at today. What follows is disillusionment, which some say we are nearly at since we have been able only to point to nanopants and the like as major accomplishments. The process ends with a shaking-out period of enlightenment, which the NSF is trying to nurture along with its SEIN-related grants. Hopefully, this will all end at some plateau of productivity.

Gartner has used this model[35] to discuss information technology,[36] Web services,[37] intrusion detection systems,[38] and other hyped products and services.

Government and industry are committed to diffuse some of the wilder claims surrounding the field. Presumably, once the hype recedes, a more serious discussion about the true nature of nanoscience and nanotechnology can proceed, though this may be an incorrect assumption. There is no evidence that informed debate supplants unsubstantiated overclaims and hyperbole. R. Stanley Williams

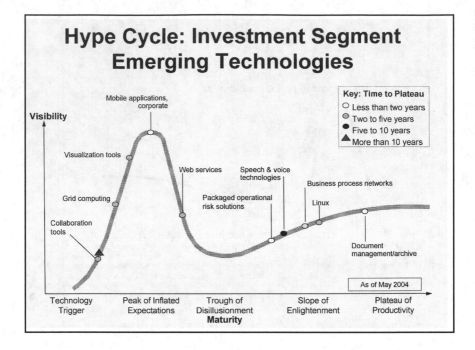

from Hewlett-Packard Labs said just that: "My greatest fear about all of this nano-technology going forward is that expectations will be raised too high, too fast, and the consequences will be that the field will lose credibility, and we'll lose a lot of the momentum that we've been trying to get together for a long time."[39]

The previous is not meant to suggest nanotechnology is a sham. Truth be told, "there's plenty of smoke," said Cliff Detz, who invests for Chevron Texaco Technology Ventures, "but also at least an ember in there."[40] Actually, there are some who feel the level of hype is just right much like Baby Bear's porridge in the Goldilocks story. Representative Mike Honda (D-CA) doesn't think the hype is that much of a problem. "There is always a danger of too much hype, and it is true that nanotechnology has received a lot of hype. But I think it is warranted in this case. It won't be the next dot-com bust because with nanotech, there is actual *stuff* there that we can touch and see and make things out of. At first, nanotech's impact probably won't be completely revolutionary, and some people might be tempted to prematurely say that it failed to live up to expectations."[41] So, maybe, it is not just right.

There are others who think nanotechnology is underhyped. While they seem to be in the minority, their voices are heard loudly and clearly. There are two promoters who are probably the most positive advocates of nanotechnology. Take James von Ehr. He's the founder of Zyvex and speaks about nanotechnology everywhere. You can see him in the photo opportunity taken in the White House

Oval Office when Bush signed the Twenty-First Century Nanotechnology R&D bill into law. Von Ehr said:

> There's a lot of hype these days. There's a lot of promises being made, primarily by funders, funding agencies, venture capitalists, the press. A lot of people are saying what they need to say to get funding so they can proceed. In the short-term, the field is overcommitted and over-hyped. In the long-term though, certainly the field is under-hyped. We should be investing a lot more in the field than we are because it is so vital to our country and the world.[42]

Another example was Philip Bond from the Department of Commerce, who appeared at nearly every conference and meeting on the commercialization of nano-technology. His nano-enthusiasm was unquenchable:

> In all of this, at least as a policy person, I try to separate hype from hope. But the more I thought about that, the more I determined that in this political town, maybe the separation isn't all that important, because hype and hope end up fueling the social passion that forms our politics. It gets budgets passed. It makes things possible for all of you. Without some passion in the public square, we will not achieve many of our goals. These goals are mind-boggling—what we used to think of as miraculous—the deaf to hear, the blind to see, every child to be fed. And that's just for starters.[43]

Unfortunately, there is a difference between emphasis and overzealousness and hyperbole. Hype is not intended to be understood literally. Since so few of us are sufficiently competent about nanotechnology to separate literal from figurative, and since what constitutes the literal is masked by wildly divergent definitions of what nanotechnology is or even could become, the problem of unfulfilled heightened expectations remains.

Hyperbole is not limited to figurative positive claims. Indeed, some of the more memorable ones are highly negative, even apocalyptic. The quintessential illustration of negative hype as it applies to nanotechnology, nanohype if you will, is "goo."

HYPERBOLES AND GOO

Hyperbole has haunted serious discussions of nanoscience and nanotechnology. "Highly improbable scenarios have become tiresome pieces of apocalyptic baggage."[44] The scenarios of doom have been dubbed *goo* because the result is some sort of uncontrolled replication of nanobots that devour all competing life on the planet. What remains is a mass of nanobots writhing in some slurry—hence *goo*. The reason they are powerful scenarios and highly persuasive is their risk calculus. "Even a slight risk of such a catastrophe is best avoided."[45]

According to the ETC Group, "an Internet search produced sixty entries referring to the threat of grey goo."[46] Drexler started the gray goo debate in *Engines* when he wrote a short passage, "The Threat from the Machines." As Chris Phoenix puts it: "Drexler spent about three paragraphs or a page in *Engines* mentioning this possibility. Unfortunately, he gave it a name that alliterates, also unfortunately it tapped into primal human fears about bugs and germs."[47]

Drexer wrote: "Tough, omnivorous 'bacteria' could out-compete real bacterial: they could spread like blowing pollen, replicate swiftly, and reduce the biosphere to dust in a matter of days. Among the cognescenti of nanotechnology, this threat has become known as the gray goo problem."[48] Drexler retells "science fiction authors and journalists focused on this scenario, and runaway replicators became closely associated with molecular nanotechnologies."[49]

Two major goo scenarios have dominated the discussion of the societal and ethical implications of long-term nanotechnology—actually, the nanotechnology envisioned by Drexler, which is called *mechanosynthesis*. The first is called grey or gray goo and the second green goo. Gray goo is directly associated with the universal assembler, and it is assumed the veracious beasties will devour the planet in some uncontrolled process of replication. Green goo is much like gray goo, but it results from the convergence of nanotechnology and biotechnology and the creation of new forms of life that will behave malevolently, threatening natural habitats, biodiversity, and human health.[50]

Gray goo is an absurd construction and an excellent example of fearmongering. "But the idea has taken on a life of its own," says Peter Singer, director of the University of Toronto's Joint Center on Bioethics.[51]

The problem with these goo stories is that it is not sufficiently removed from how Drexler's universal assemblers would work. When Drexler's supporters debunk goo they also tend to delegitimize the assembler. Take, for example, the following remark by a Foresight Institute associate. Robert Freitas, the nanomedicine expert, tore through the scenario with a technical analysis he titled *Some Limits to Global Ecophagy by Biovorous Nanoreplicators, with Public Policy Recommendations*:

> Such a high-speed attack would instantly cause a massive spike in temperature, alerting authorities to the situation and allowing them to respond. Conversely, it is theoretically possible for biosphere-eating self-replicators to create an almost undetectably small increase in temperature—but then it would take them twenty months to complete their task, leaving plenty of time to observe the destruction and organize a defense.[52]

The time it takes for the biosphere to become goo challenges the time its takes for assemblers, universal or otherwise, to build macroproducts at the molecular level. Another illustration comes from a team of attack dogs referred to as the Center for Responsible Nanotechnology. They function much like the vice presidential candidate in national elections. Their rhetoric is sufficiently distant from Drexler for plausible deniability, and they are the first to attack anything challenging molecular manufacturing as a potential route to nanotechnology:

> A gray goo robot would face a much harder task than merely replicating itself. It would have to survive in the environment, move around, and convert what it finds into raw materials and power. This would require sophisticated chemistry. A gray goo robot would also require a relatively large computer to sort and process the full blueprint of such a complex device. A nanobot or nanomachines missing any part of this functionality could not function as gray goo.[53]

These claims would also apply to the assembler problematizing its utility as an engine of any sorts. As a consequence of these incoherencies, it is understandable why the debate has become so muddy.

Bill Joy punctuated the goo scenario in his *Wired* article. In the wake of predictions of Judgment Day at the mechanical arms of nanobots, Drexler and his supporters have attempted to reject the scenario by denying it unequivocally in nothing short of an apologia: "Although biosphere-eating goo is a gripping story, current molecular manufacturing proposals contain nothing even similar to gray goo. The idea that nanotechnology manufacturing systems could run amok is based on outdated information."[54]

2004 was the year of the gray goo retrenchment. ETC now calls the goo scenario simply a red herring. Prince Charles joined the reversals, denying he ever addressed it. Amanda Armstrong attests: "Drexler must rue the day he coined the term to describe nanobots replicating out of control and consuming the biosphere."[55]

Drexler has revisited the issue of self-exponential manufacturing and "slayed the myth that molecular manufacturers must use dangerous self-replicating machines"[56] and repeated "that the easiest and most efficient systems will not have the capabilities required for autonomous runaway manufacturing.... The development and use of molecular manufacturing need not at any step involve systems that could run amok as a result of accident or faulty engineering."[57] He added, "A nanofactory simply would not have the functionality required." Also, "There appears to be no technological or economic motive for producing a self-contained manufacturing system with mobility, or a built-in self-description, or the chemical processing system that would be required to convert naturally occurring materials into feedstocks suitable for molecular manufacturing systems."[58] That's denial.

According to the BBC, Drexler commented: "What I did not expect was that efforts to quiet concerns over grey goo would lead to false scientific denials of feasible technologies. I also underestimated the popularity of depictions of swarms of tiny nanobugs in science fiction and popular culture."[59] Singer believes "it's theoretically possible to stifle the notion of gray goo because the term has yet to become widely used by the general public. The battle against the concept of gray goo will be lost when the movie *Prey* comes out. *Prey* is the thing that will make the link between the potentially inaccurate word that policy makers use today and a galvanizing fear on the part of the general public."[60] Singer also adds, "There are just some phrases that capture the public imagination—*Frankenfood, gray goo, Jurassic Park*. Once they are out there, they electrify and galvanize the public imagination completely unrelated to how accurate they are."[61] There are lot of reasons why gray goo is seductive. It's alliterative and falls trippingly. It has elements of drama, mystery, and conspiracy. Finally, it's one of those terms that characterize a conceptual narrative that can be substituted for a legitimate understanding of nanotechnology. The story of goo becomes a substitute proposition on nanotechnology, and both Drexler and Singer have come to realize that.

The term has even found its way into debates by policy makers on nanotechnology. The following statement began congressional hearings on the Hill: "As many people here know, the most extravagant fear about nanotechnology is that it will yield nanobots that will turn the world into gray goo. That's not a fear I share, but I do worry that the debate about nanotechnology could turn into gray goo—with its own deleterious consequences."[62] Representative Sherwood Boehlert's (R-NY) concern is shared by the media. The worry is "these futuristic visions will overshadow more practical concerns, such as the potential new forms of pollution and the safety of workers that handle nanomaterials."[63] An official UK report by the Economic and Social Research Council corroborated the misdirection: "The public debate focuses on long-term possibilities of radical nanotechnology rather than the rather mundane applications that have arrived so far."[64]

Nanotechnology in Popular Culture

Beyond the examples in the introduction to this book, fictional depictions about nanotechnology have taken many forms, from Bill Spence's cartoonish *Nanoquest* and Drexler's animated nanofactories, even to artistic renditions of views through a scanning electron microscope. Undoubtedly the vision of nanotechnology most embraced in our popular culture are depictions of Armageddon instigated by the nanobot replication image of gray goo first discussed in Drexler's *Engines of Creation* and popularized by Bill Joy's article "The Future Doesn't Need Us."

In addition, its "potential and perception as a threat stems from its multiple causalities: so many people and institutions have their hands in it....As with the technology itself, the lines of agency are blurred, and risk as such emerges as the most powerful force in determining the nature of the nanotechnology represented in the popular forum."[65]

This is especially true of science fiction writers "who focused on this idea, and gray goo became closely associated with nanotechnology spreading a serious misconception about molecular manufacturing systems and diverting attention from more pressing concerns."[66] This complaint seems genuinely unfair. It is clearly not the duty or obligation of fiction writers to garner support for a science initiative. Sci-fi writers are not writing nonfiction. The duty to accurately frame the issues of nanotechnology rests with others.

Nanotechnology is an ideal subject for sci-fi writers. First, it involves the type of time frames in which much science fiction is situated. Second, it crosses over to engage futurists. "For many years, futurists steeped in the culture of science fiction and prone to thinking in time frames that reach decades ahead have been dreaming up a fantastic future built using nanotechnologies."[67] Third, it "connects to numerous preexisting themes of science fiction and offers writers an extraordinarily broad palette of capabilities, all imbued with the appearance of scientific plausibility."[68] Fourth, nanotechnology is not only something we can't see but also something that isn't here yet. As such, it "gives science fiction writers a chance to engage in the art of predicting and warning about possible futures."[69] Fifth and finally, it taps into two foundational themes in science fiction literature: superhumanism, "the dream that *homo sapiens* can achieve complete mastery over nature and has utter freedom to shape its own destiny,"[70] and human arrogance—"the dark vision of nanobots running amok is a new wrinkle on the old golem/ Frankenstein myth, the dangers of meddling with godlike powers or bringing too much hubris to science."[71]

Converging Technology Bar Association founder Sonia Miller wrote that "the public's reliance on science fiction as a primariy source of information sadly hinders the effective advancement of nanotechnology."[72] The real challenge is hype management. "It doesn't help that many nanotechnologists tout nanotechnology as having the potential to virtually change or reengineer everything we know today that is man-made."[73] The public is hearing hyperbole from both sides of the fence. As such, when a highly exaggerated interpretation is offered in the public culture, whether in print or on film, and it is sufficiently like the hype of the proponents and opponents, we get a problematic linkage. Differentiating fiction from reality becomes incredibly difficult when this occurs and the Hollywood blockbuster seems as real as anything found in *Science* or *Nature* to the rhetorically challenged.

In 2002 Michael Crichton wrote *Prey,* in which he raised fears associated with military research into nanotechnology. The book described "how biologically synthesized nanorobots wreak havoc in the U.S. The book is being made into a film. Within weeks of its release, tens of millions will know something about nanotechnology,"[74] says David Rajeski, director of the Foresight and Governance program at the Woodrow Wilson International Center for Scholars in Washington, DC. Rice's Vicki Colvin warned, "This is science fiction, not science fact, however, the public relations nightmare it could spawn is just as frightening to me, a nano-

technology researcher, as nanorobots might be to some people."[75] Steve Crosby of
Small Times chimed in as well: "Michael Crichton, the author who gave us killer
pathogens from space and the return of killer dinosaurs, is back on the pop culture radar with *Prey*, the fictional account of killer nanoswarms. There's nothing
wrong with escapist films and fiction, but we might ask ourselves if we're offering
the public an informed reality."[76] Kelly Kordzik of Winstead and head of the
Texas Nanotechnology Initiative believes an aggressive response might be prudent. "In response to Michael Crichton's new book, *Prey*, I would like to encourage
those of us involved in the development of this emerging science to take responsibility for defining what nanotechnology is and not rely on Hollywood or the
news media to do so."[77] A book that is not yet a movie has become the subject of
broad speculation from many venues.

A more level-headed remark came from Mike Roco, a fan of the book and a
movie buff. "The fact that the book speaks about nanotech, it brings in the attention of the general public to this field."[78] Roco may be correct. There is no such
thing as bad publicity.

On the positive side, it is worth noting the book was on the best-seller list for
only eight weeks and never reached number one; Twentieth Century Fox has yet to
find a director, a producer, or any actors for the movie.[79] The ETC Group claims
Prey "would have been relegated to flash-in-the-pan status if it weren't for the nanotech boosters keeping it alive."[80] How ironic if our rhetorical misgivings about *Prey*
actually increase its visibility and pertinence, especially since *Prey* is no *Silent Spring*.

Josh Wolfe cautions investors from overinvesting Crichton's *Prey* (which he
called "a modern version of the classic story we know as *The Blob*"[81]) with significance, but it was Mark Modezelewski, as the business and industry leader, who
placed this novel in perspective for the commercial community and the investing
public: "Crichton is a great fantasy writer and great writers make you believe they
are on to something real and true—like *Star Wars*, *Harry Potter* or *Jurassic Park*.
You'll no more see nano-robots destroying things in the backyard than you'll see
your child put on Harry Potter's cape and fly around the neighborhood on a
broom. It's all good harmless fun, but people shouldn't confuse fantasy with
reality."[82] While his lack of concern is mostly premised on "nobody reads," he
admits that "the movie version could be significant."[83]

Oddly enough, *Prey* has been beaten by *Paths of Destruction*, a made-for-TV
movie that debuted on the SciFi Channel in late September 2005. Whether a big-
screen or a small-screen release is important, we await to gauge the public reaction to this out-of-control-nanobot disaster film.

Are we worrying about the real danger? Americans have a dismal track record
when it comes to what they select to fear.[84] Children are safer in school than at
home. Strangers aren't waiting to steal your children—family members are. Airplanes are incredibly safe compared with automobiles. And the list goes on.

How can so many people be so concerned about a novel currently in discount sale bins? Will a movie mobilize people to boycott nanoproducts? We didn't need a killer asbestos movie to do that. *The China Syndrome* was a scary movie, but it was Three Mile Island and Chernobyl, the events themselves, that buried the nuclear energy industry. Two unpublished studies have both concluded that *Prey* is likely to play a small role in public opinions about nanotechnology, nevertheless the nanocommunity simply can't leave it alone.[85] We can probably draw the same conclusion for any *Prey*-like film.

However, if our goal is to move the public into a sphere that participates as a legitimate stakeholder in science and technology decision making, then we *may* have a legitimate issue here:

> The vision of a few nanometer intelligent robots mentioned in science fiction literature leads to immediate criticism by some groups that are concerned that such robots would take over the world and damage the environment. This dialogue is carried out, ignoring input from researchers who note that basic laws of mass and energy conservation may not lead to infinitely multiplying material objects, and that only a system of already known living systems may multiply and be intelligent.[86]

Without a tradition of meaningful scientific dialogue, construction of *the real* drawn from the world of implausible and impossible science and engineering would make building a communication model for a truly functional and empowered public sphere in science and technology decision making much more challenging. Also, "what is unfortunate is that the nanotechnology aspect of the novel is receiving the most press, especially in view of recent opinion pieces warning us of the potential unknown dangers of nanoparticles."[87]

There is a palpable concern that Crichton's book, once it becomes a movie, might prove difficult to handle. The wrong-headed science in the movie might establish expectations that will need to be corrected. Correcting misinformation is incredibly difficult since it involves unlearning and reeducation. *Forbes'* Josh Wolfe predicted, "There will be a backlash against nanotechnology, largely in the media, spurred on by Crichton's new book and movie to come. Expect to see ethical debates, like those on cloning, on what regulations should be put in place to limit the manipulation of molecules."[88] This might actually provide venues for enlightening debate. Preparing a model for that moment might be more productive than hand-wringing.

While the release of his book seemed to have left nanoscience and nanotechnology relatively unscathed, we can only hope the movie will do for nanotechnology what *The Day After Tomorrow* did for global warming: very little. "The general public seems humdrum about the fear of the great nanotech unknown." Stephen Herrera references the recent studies that likened carbon nanotubes to asbestos: "And yet there is nary a shiver." He offers a few reasons, one of which has legs. Maybe "the nation is too distracted with fears about terrorism, war, keeping their job and the moribund stock market." He feels Americans have inoculated themselves against fear and concluded: "Fortunately for nanoscience, and the generations of people who stand to gain more that they stand to lose, elected officials, corporate R&D directors, and investors seem to have developed immunity and at least thus far have resisted the temptation to let themselves be scared off by nano fearmongers."[89] If this is true, then the odds may be in favor of informed discourse. Regardless, designing a response strategy would seem to be the order of the day rather than a Chicken Little initiative.

Nanotechnology and the Media

Traditionally, citizen-consumers have learned about science and technology through an interpretive medium, a college of scientific journalism. Though many journalists lay claim to membership, very few carry the experience, credentials, or both. One of the main reasons the debate about nanotechnology has been so misinformed is the science media. News media is an essential variable in every conception of a functioning public sphere. Houston, we have a problem!

A major criticism of science journalists is simply that they lack zeal. Steve Fuller argues "except in cases of scientific misbehavior sufficiently grave to worry Congress, journalists will often print watered down or mystified versions of a scientist's own press release, which ends up only increasing the public confidence in science without increasing its comprehension."[90] When things go awry, the public is left with insufficient information to explain what may have occurred, and they default to simplistic generalizations and conspiratorial narratives.

Journalists are not only underzealous, but their publishers are also underconcerned about accurate reportage. Citing John Burnham,[91] Fuller reported "the supermarket tabloids [remain] the public's primary source of information about the latest developments in science."[92] Tabloids are more likely to report scares and disasters than recent breakthroughs from a university lab. Journalists and publishers are driven by market considerations: selling copy. Though their motives may not be universally impeachable and suspect, they hardly advance the best interests of accurately informing citizen-consumers.

While some journalists make a genuine effort to report accurately scientific claims, fighting for column space, they must give their editors what they can sell. For example, when coverage is given to science, the sensational is often accentuated. Too often many scientific claims are reported before definitive burdens and standards of proof are met (cold fusion, for instance). Highly impatient readers tend to blame inconclusive results on bad science rather than premature reporting and outrageous overclaims.

Not only do science journalists devalue the time frame between theorizing and verification, but they also present issues in "winner-take-all contexts that turn on some crucial fact or event,"[93] an overly simplistic model of causation. Furthermore, science journalists do not appreciate proof obligations associated with scientific claim making. "The more provocative the theory under dispute, the more likely journalists will champion it, which often serves to shift the burden of proof onto the opponents."[94]

Moreover, trying to balance their reporting, reporters solicit respondents from a local college. These experts express opinions on claims when they are often unprepared to make truly informed comments. This often leads to attacks on credibility, sometimes personality assaults, which leave readers with a view of scientific discourse as a schoolyard brawl.

The result: the majority of those writing, attempting to communicate to a mostly scientifically unsophisticated audience, write articles with flash, sparkle, and pizzazz, but with little information and insight.

Nanotechnology in Academe

Traditional intellectual media plays a vertical game: journalists write up and professors write down. This is an activity referred to as popularization. Today *third-culture* thinkers avoid the middleman and write their own books, much to the consternation of those with a vested interest in preserving the status quo. Some scientists have seen that the best way to present their deepest and more serious thoughts to their most sophisticated colleagues is to express these thoughts in a manner that is accessible to the generally intelligent reading public.[95]

C. P. Snow portrayed twentieth-century British and, by filiality, American intelligentsia, stratified into two "cultures": literary and scientified.[96] Snow blamed the resultant "gulf of mutual incomprehension between scientists and humanists largely on the refusal of humanists to integrate the scientific culture into their understanding."[97] John Brockman took Snow's second essay on culture, "A Second Look," and suggested that recently a third culture, somewhat unlike Snow's vision, has begun to emerge. Whereby Snow felt the third culture would involve "literary intellectuals...on speaking terms with the scientists,"[98] Brockman says this is not the case and "[s]cientists are communicating directly to the general public"[99] and "the traditional intellectual has become increasingly marginalized."[100]

While traditional intellectuals bemoan this trend, it suggests a very intriguing phenomenon. "The emergence of this third-culture activity is evidence that many people have a great intellectual hunger for new and important ideas and are willing to make the effort to educate themselves."[101] Rebutting elitists' claims that the public is naive and disinterested, we continue to see "scientific topics receiving prominent play in newspapers and magazines over the past several years including ...nanotechnology."[102]

Third-culture intellectuals have begun to avail their writing to the more general-reader markets. Luckily for them, readers have begun demanding more science-related literature. According to W. Daniel Hills, "People and things are moving so fast that you can't really imagine the life your child is going to lead. That's never been true before, and it's clear the course of that change and that discontinuity is science."[103] The citizen-consumers recognize a compelling need to learn to survive, and "one way to do it is to read books by scientists."[104]

The new *public intellectuals* are motivated to publish directly to citizen-consumers for two additional reasons. First, they are often interdisciplinarians. Their fields of theorizing and reasoning are insufficiently distinct. Their ideas and claims "don't fit within the neat structures of their internal disciplines. Many of the scientists who write popular books do so because there are certain kinds of ideas that have absolutely no way of getting published within the scientific community."[105] Their work seems outside a publication's usual fare—partially pertinent but not wholly so.

The second reason: Scientists have begun to understand that consensus building and outright support for their interests and fields are necessary corequisites to their theories and findings in order to procure and sustain third-party interest and backing for their research agenda. Because science exists in a dialectical relationship within the broader society and culture, scientists must justify their pursuits to the political leaders and other persons who control essential resources.[106] This may be truer in a post-9/11 world.

Popular support can move a government as well as create and sustain demand for industrial products and services. Third-culture scientists are marketing their ideas directly to citizen-consumers engendering support to help secure patronage on many different levels: public-interest groups, foundations, university and college administrators, government agencies, and policy makers. Science has its lobbyists and third-culture scientists contribute in their own way toward popularization of their projects. Brockman's observations are particularly true regarding nanotechnology.

The next question: Are citizen-consumers sufficiently versed in science to separate fact from fiction? There will always be quasi-altruistic watchdog individuals and groups who help restrain overly ambitious corporations. Nonetheless, the overly interrelated world has made singular-focus interest groups much less useful. Life issues, including especially those of science and engineering, have become interwoven into political, social, and personal concerns.

Brockman includes Stephen Jay Gould, Freeman Dyson, Stephen Hawking, Richard Leakey, and others on his roster of third-culture intellectuals, a group he also calls *new public intellectuals*. These are experts in science and technology who take their cases directly to citizen-consumers through their popular writing. K. Eric Drexler needs to be added to this list and to some lesser extent a host of others, including debunkers and critics, bureaucrats and elected officials, industrialists, business leaders, and some investment bankers, nongovernmental organizations and academic research centers, and the people who write government and business reports on nanotechnology.

CONCLUSION

Stories are the vehicles for communicating adventures of all sorts, and that is true for nanotechnology. Feynman may have introduced tiny technology in a talk to Caltech researchers. Drexler popularized it and entertained us with his *Engines of Creation*. Joy told tales of disaster. George Whitesides and Richard Smalley derided Drexler's version provoking a debate. Roco attests nanotechnology will change everything we do and will lead America's economy into the millennium. Crichton built a story around a misguided defense industrial complex. They have been joined by hundred of voices taking sides and spinning yarns about feasibility, costs, and benefits.

Right or wrong, these voices are writing the narrative about nanotechnology. They are moving the issues and controversies into the public venue and citizen-consumers are attempting to decide what they think is fact from fiction. Unprepared to do so, leaders in the nanotechnology movement are concerned that hyperboles might squeeze out reasoned discourse. As third-culture intellectuals ourselves, we need to address the public. For that to happen, the story needs to be deconstructed and vetted, especially those stories told outside the venues of science and by the underinformed and misguided.

"It is time to convert the passive approach of studies and research to active consumer protection education and review of the laws on the books. The media, books and movies are feeding the consumer while advocacy groups are providing unsubstantiated studies of the adverse effect of nanotechnology. Fears must be property assuaged, now. This is critical for nanoscience and nanotechnology to realize its true potential."[107]

Fears might affect development and investment, forgoing the United States' role in nanotechnology to other national actors. The net effect might be significant. Even if we shave back some of the exaggerations, there does remain some fascinating benefits to whoever captures the ember that powers the engines of nanotechnology.

The following chapters are written as a guide toward breaking down the narrative being constructed. We want to help build a better story that will speak to as many of the relevant stakeholders as possible.

CHAPTER 2

SPECULATION AND CRITICISM ABOUT NANOTECHNOLOGY

W e associate the term *character* with fictional depictions of people found in novels and plays. While oftentimes realistic, they tend to have other-worldly characteristics. This story, while grounded in reality, has characters flamboyant and larger than life. Their larger-than-life nature and penchant for exaggeration and hyperbole sometimes are their own construction while at other times the product of rumor and speculation. The following examines two groups of characters: advocates and critics.

ADVOCATES

The following two individuals classified as proponents proposed the possibility of nanotechnology. Neither of them offered highly articulated programs to get from nanoscience research to full-scale development per se. Those people are studied in the next chapter. These men are associated with the genesis of a nanotechnology movement of sorts. The roles they played and continue to play in the history of the nanotechnology movement remain significant even though they are separated by decades.

Richard P. Feynman

One of the first persons to articulate a future with nanotechnology was Richard Feynman, a drum-playing, jester-spirited, irreverent Nobel laureate who died in 1988. Freeman Dyson, who worked with him at Cornell, called him "half genius and half buffoon." He smoked pot, picked up women in bars, gave massages, learned to draw nudes and persuade women to pose for him, and became skilled at picking Yale locks. He worked on the Manhattan Project where most of what was to be done, was to be done for the first time.[1]

He has been described as a magician of the highest caliber. Originality was his obsession. He was the enemy of pomp, convention, quackery, and hypocrisy. He was generally embarrassed about having won the Nobel Prize. His lectures were theatrical and some were recorded and available for study.

He despised philosophy as soft and unverifiable and disliked people who called themselves philosophers even more. "It struck him as an industry built by incompetent logicians."[2] For him, philosophers and their brothers and sisters, the ethicists, were parasitic, and when they theorized they spoke in scholasticism. They generated hypotheses without any interest in validating them. They created little to nothing. "The philosophers were always a tempo behind, like tourists moving in after the explorers have left."[3] We can only assume that he would be apprehensive with the current NSF preoccupation with involving social scientists, including philosophers, in SEIN research.

While he had a reputation as a great teacher, he created barriers against students who searched him out as a thesis advisor because he didn't have the patience to guide a student through a research problem.[4] His isolation from his students may have allowed him to become very reflexive. This may have given him the time and opportunity to examine humanity's role in ecosystem Earth.

Feynman's brilliance was validated not only by his Nobel Prize but also by his participation in major scientific projects. There is a universal conclusion drawn by the people who knew him. For example, Paul Olum, John Wheeler's assistant at Princeton, said, "He was such an extraordinary special person in the universe."[5] Hans Bethe from the Los Alamos Project observed that "if Feynman says it three times, it was right."[6]

While Feynman's theory on superfluidity, the strange, frictionless behavior of liquid helium; a theory of weak interactions, the force at work in radioactive decay; and a theory of partons, hypothetical hard particles inside the atom's nucleus, that helped produce the modern understanding of quarks might have won him a Nobel Prize, he won one for his work on the interaction between light and matter.[7]

The role Feynman played in the development of nanotechnology is incredibly unclear, though Neal Lane suggested that it was Feynman's vision that was communicated to President Clinton, whose Caltech address led to the birth of the NNI. Even Drexler attributes Feynman's vision to the launch of the global nanotechnology race,[8] though he stops short of associating the current trend of research to his influence. James Gleick, one of Feynman's biographers, refers to him as the intellectual father of a legion of self-described nanotechnologists;[9] their spiritual father seems much more accurate.[10]

There's Plenty of Room at the Bottom was an address delivered on December 29, 1959, to the American Physical Society at Caltech in Pasadena. This was the speech in which Feynman offered a financial prize of "$1,000 to the first person to make a working electric motor that was no bigger than one 1/64 of an inch and another $1,000 to the first person to shrink a page of text 1/25,000 its size."[11] Feynman confronted "the problem of manipulating and controlling things on a small scale."[12] He intriguingly discussed miniaturizing text and predicted much of the nanograffiti that clogged many of the science and popular news monthlies and biweeklies in the 1990s, heralding the entry of the basic instruments of nanotechnology: the scanning tunneling electron microscope (STM). His speech, however, was mostly about computation and how in nanotechnology we may see developments that are profound in terms of memory storage and other computational capabilities. The speech ended with a series of rebuttals to some of the more obvious limitations associated with working on that scale.

On February 23, 1983, he delivered a sequel to the 1960 speech. He presented it to a large audience of scientists and engineers at the Jet Propulsion Lab (JPL) in Pasadena. A tape of the speech can be found in Caltech's video archives. He used slides, hand gestures, and sketches on a blackboard to supplement his speaking. It was informal. He titled his talk *There's Plenty of Room at the Bottom, Revisited*, and not *Infinitesimal Machinery* as reported in the *Journal of Microelectromechanical Systems* in 1993. In it, Feynman admitted that the small machinery he discussed in 1960 had "no particular use" and "there has been no progress in that respect."[13] He categorized his prediction "that soon we would have small machines" was certainly misguided.[14] He even labeled his interest in small machines "a misguided one.... There is no use for these machines, so I still don't understand why I'm fascinated by the question of making small machines with movable and controllable parts."[15] In usual Feynman fashion, he speculated on the use of small machines as drills, grinders, and adjustable masks. He then debunked some of the problems associated with construction at that level such as its energy source, control, movement, and so on. He developed theoretical solutions to some difficulties, like friction and sticking. He fantasized about "swallowing the surgeon," a device controlled by an attachable wire tail. He ended his talk with an in-depth discussion of quantum computing and the role little machines might play in its development.[16]

What led to Feynman's interest in microtechnology? According to Dyson, Feynman struggled "to understand the workings of nature by rebuilding physics from the bottom up."[17] Hence, it might not be much of a stretch for Feynman to transpose his bottom-up view of the discipline of physics to fabrication strategies of the very small.

Jeffrey Robbins commented in a collection of Feynman's short works that he was fascinated by the *kick* of discovery, the sudden feeling that you've grasped a wonderful new idea, and "he did physics for the sheer pleasure of finding out how the world works, what makes it tick."[18] He seemed particularly fascinated with realities on the microscopic scale. For example, in a 1981 BBC television program, he discussed what an artist sees and what he saw in a flower. The artist looked at the flower and expressed disdain for the scientists who would take it apart, but Feynman insisted he saw much more because he could imagine the cells in there, the complicated actions inside, which also have a beauty. He said, "I mean it's not just beauty at this dimension of one centimeter, there is also beauty at a smaller dimension."[19] Consequently, he demonstrated some interest in the building blocks of the macroscopic world.

In addition, Feynman regaled his biographers with many stories on how he learned a lot of science through his father's storytelling, and he admitted doing the same with his son. Occasionally, his anecdotes turned to a theme underlying stories he told his son, one of which was "a story about little people that were about so high [who] would walk along and they would go on picnics and so on and they lived in the ventilator."[20] Fascinating his son with "little people" living on his bedroom rug, he introduced him to scientific principles through fairy tales that involved interacting macro-objects on the microlevel.

Feynman selected nanotechnology for the *Plenty of Room* talk. Gleick had his own reason:

> He had several reasons for thinking about the mechanics of the atomic world. Although he did not say so, he had been pondering the second law of thermodynamics and the relationship between entropy and information; at atomic scales came the threshold where his calculations and thought experiments took place. The new genetics also brought such issues to the surface. He talked about DNA (fifty atoms per bit of information) and about the capacity of living organisms to build tiny machinery, not just for information storage but for manipulation and manufacturing.[21]

For Gleick, it wasn't too far a stretch then to move from living organisms to material sciences.

The entire process of ideation was Feynman's desktop. "He wasn't so much worried about whether something was a real idea or not but was it a way of generating new ideas? He found flaky people stimulating, because the people who are doing the ideas that really change the world are—or appear—flaky."[22] Being associated with K. Eric Drexler, for example, would not have been as problematic to him as it seems to be with some contemporary scientists, such as George Whitesides and Richard Smalley.

Feynman was intrigued by the limits of things. He "had always been interested in the limits that physics puts on the things one would like to do. When he theoretically probed the possibilities of what kind of *room was there at the bottom*, he conceived of a realm that is rapidly being realized in laboratories thirty years later."[23]

Feynman understood science was "a key to the gates of heaven, and the same key opens the gates of hell and we do not have any instructions as to which is which gate. Shall we throw away the key and never have a way to enter the gates of heaven? Or shall we struggle with the problem of which is the best way to use the key?"[24] He added, "You cannot understand science and its relation to anything else unless you understand and appreciate the great adventure of our time."[25] He was comfortable with uncertainty. He understood that many of the most important discoveries were accidental and extrapolative.

He said there is "imagination in science."[26] "Extrapolations are the only things that have real value."[27] "Knowledge is of no real value if all you can tell me is what happened yesterday. It is necessary to tell what will happen tomorrow if you do something—not only necessary, but fun."[28] "There is no harm in being uncertain. It is better to say something and not be sure than not to say anything at all."[29] Speculating about nanoscience before the discovery of the tools that enabled work at that scale was acceptable and even preferred.

It might have been interesting to conjecture about what Feynman would have said about the burgeoning fields of science and technology studies and college courses in the philosophy of science and technology. Assuredly, he would have found some distinction between the speculations in science versus those in the humanities that would warrant his despair about philosophers.

In the 1974 Caltech Commencement Address he remarked, "We ought to look into theories that don't work, and science that isn't science."[30] This remark began to make sense only when compared to two earlier remarks he made while touring a Buddhist temple in Hawaii: "In fact, it is from the history of the enormous monstrosities created by false belief that philosophers have realized the apparently infinite and wondrous capabilities of human beings."[31] Feynman understood

that to solve any problem that has never been solved before, you have to leave the door to the unknown ajar. You have to permit the possibility that you do not have it right.... The rate of development of science is not the rate at which you make observations alone but, much more important, the rate at which you create new things to test.... So what we call scientific knowledge today is a body of statements of varying degrees of certainty. Some of them are most unsure; some of them are nearly sure; but none is absolutely certain.[32]

While at Cornell, he worked with Wheeler and together they investigated some of the challenges of quantum physics. Wheeler once quoted the White Queen's remark to Alice: "It's a poor memory that only works backwards" to describe their absorber theory.[33] On another level, this simple quote may encapsulate Feynman's view of knowledge. It may be necessary to speculate in order to arrive at some ground to which science may need to be dragged for its own good.

How influential was Feynman? Commenting on his 1960 talk, he admitted, "I pointed out what everybody knew."[34] While he did produce a modicum of interest with the two contests he sponsored, there is little, if any, strong evidence that he instigated research agenda anywhere. "Feynman's talk was so visionary that it didn't really connect with people until the technology caught up with it. Many of those who manipulated atoms using the scanning tunneling microscope (STM) did not know about Feynman's talk until after they got into the atom-moving business."[35] While this observation isn't meant to belittle his vision, it does contextualize his importance.

Feynman commented why science and the humanities must find some common ground, especially when science has a strong social agenda:

> It seems to be generally believed that if the scientists would only look at these very difficult social problems and not spend so much time fooling with the less vital scientific ones, great success would come of it. It seems to me that we do think about these problems from time to time, but we don't put full-time effort into them, the reason being that we know we don't have any magic formula for solving problems, that social problems are very much harder than scientific ones, and that we usually don't get anywhere when we do think about them. I believe that a scientist looking at nonscientific problems is just as dumb as the next guy.[36]

Did Feynman's vision instigate the NNI? If so, has it been supplanted by nanolithography and other top-down models? Have leaders of *the* funding coalition attempted to narrow nanotechnology to exclude Feynman's vision? If so, why? Is it intentional? If so, what are the motivations? Feynman might applaud the SEIN agenda of the NNI or, at least, he would tolerate it. How he would feel about SEIN including philosophers and ethicists is another story altogether. Even federal bureaucrats are not quite certain that it is such a good idea after all.

K. Eric Drexler

Though Richard Feynman may have discussed and written about small machines before K. Eric Drexler,[37] his view, like Norio Taniguchi's, was different from Drexler's. Feynman discussed a path of miniaturization in which large machines could be used to make smaller machines, which would then build smaller machines, and so on toward the molecular level.[38] For Feynman, we might start big and build smaller. For Drexler, we simply build small. As Christine Peterson of the Foresight Institute has observed, Feynman's view of nanotechnology is substantially different from Drexler's nano. Drexler's model involves materials and devices that are eutectic.[39] Feynman's model does not. Feynman's tools would be sculpted rather than built. "The resulting tools, while small, would not be made with molecular-scale control."[40] Drexler's would.

Early Years—Engines and Goo

Josh Wolfe's characterization of Drexler may be unfair:

> Drexler credits himself with introducing the term *nanotechnology* in the mid 1980s and was the author of the futurist book *Engines of Creation* that caught the science community's imagination. A lightning rod for controversy, Drexler's critics say his futuristic visions of nanotech are science fiction and litter the popular media, detracting from nanotechnology's more practical scientific progress.[41]

Burdening anyone with the full range of uncertainties associated with nanotechnology is simply overkill.

Engines of Creation is an extraordinary exercise in prolepsis. The rhetorical trope of prolepsis examines a future event by hindsight. The writer situates himself, in this case, at some point in the future and examines an event in a futuristic past. As such, treating *Engines* entirely as a work of nonfiction demeans its rhetorical function. Also, treating *Engines* as a work of nonfiction makes it incredibly vulnerable to criticism. While Drexler may not elect to have *Engines* described as fiction per se, it is clearly uncomfortably situated between imagination and reality.

In the 1970s K. Eric Drexler was a student in MIT's Department of Interdisciplinary Sciences. His studies centered on the design of manufacturing systems for use in space. His undergraduate thesis was on light sails using reflective sheets of aluminum as thin as twenty to one hundred nanometers (nm). In spring 1977 a core concept of molecular nanotechnology was envisioned by him: "Consideration of self-assembled systems of molecular machines based on biological models led to the realization that such machines could be used to position reactive molecules, guiding chemical reactions under programmable control so as to build complex structures, not unlike the operation of a ribosome."[42] Despite the existence proof, the concept of directed self-assembly has haunted him ever since.

Drexler received a research affiliate appointment first at MIT's Space System Laboratory and later with MIT's Artificial Intelligence Laboratory under the sponsorship of Marvin Minsky. The first scientific paper on Drexler's model of nanotechnology appeared soon thereafter in 1981.

A series of articles followed. They addressed technical advances, model differentiation, and applications. In 1986 *Engines* was published, and the Foresight Institute, a nonprofit organization for the study of nanotechnology, was started by Drexler and associates. It will be examined in chapter 8.

In spring 1988 Drexler taught the first university class in nanotechnology at Stanford.[43] Presumably, Steve Jurvetson, the venture capitalist and a strong advocate of nanotechnology as an economic driver, took the course (see chapter 7). In 1991 Drexler received his doctorate from MIT in molecular nanotechnology.

Minsky, his advisor, reported that he "had to change his department in order to [put *nanotechnology* into his doctoral title]. This involved, according to Minsky, "getting control of the dean."[44] Regardless, Drexler seemed to hold the only extant doctorate in nanotechnology for over a decade.

Later Years—Nanosystems *and Denialism*

Drexler's dissertation is titled "Molecular Machinery and Manufacturing with Applications to Computation (Nanotechnology)." It is not much unlike his 1993 book *Nanosystems: Molecular Machinery, Manufacturing and Computation.* The opening paragraph of his dissertation abstract reads: "Studies were conducted to assemble the analytical tools necessary for the design and modeling of mechanical systems and molecular-precision moving parts of nanometer scale. These analytical tools were then applied to the design of systems capable of computation and of molecular-precision manufacturing."

Nanosystems was better received than *Engines.* Leo Paquette, a chemist from Ohio State; Roald Hoffman, a chemist from Cornell; Clark Still, a chemist from Columbia; and others found many of Drexler's ideas palatable. While the bulk of Drexler's stuff prior to *Nanosystems* was on the popular level, *Nanosystems* was a departure, an extension of his *Proceedings* work. *Nanosystems* is science and engineering.[45]

Nanosystems received its share of criticism as well. "To hard-nosed engineers, the juxtaposition of *theoretical* and *applied* quickly becomes an oxymoron. Their response to the author of *Nanosystems*: Come back when you can tell me how to make those things."[46] Philip Barth, an engineer at Hewlett-Packard said, "The holes are bigger than the substance.... There's a plausible argument for everything, but there is no detailed answers to anything."[47]

Ralph Merkle of Xerox PARC, who was a member of Drexler's inner circle at the Foresight Institute, responded, "Over 10,000 copies are now in print. No technical errors of significance have been found, despite several years of exposure to the entire scientific and technical community, including both critics and supporters."[48]

In addition to teaching and leading his own foundation, Drexler spoke at many seminars and conferences.[49] He inadvertently fathered a host of followers. Ed Regis said that at one of Drexler's lectures "there was a veiled feeling of being one of the Elect, the Select, the Knowledgeable, the Chosen."[50] Drexler has been called the Einstein of nanotechnology.[51] While Drexler shakes off the term *Drexlerian*,[52] he "will likely be considered the Father of Nanotechnology."[53]

After writing *Nanosystems*, Drexler participated in a few vocal debates within the pages of the Foresight publications and elsewhere. His most contentious debates were over practicality and gray goo. He has spoken at a wide array of conferences and conventions, including a keynote address, *Caution and Conservation*, at EXTRO 3 in 1997,[54] *Openness* at the Cato Institute– and Forbes-sponsored Annual

Conference on Technology and Society in 1998, and at nearly all of the Foresight-sponsored conferences.

Drexler grew weary of the vituperative nature of the debate over his brand of molecular manufacturing. He explained his absence from the fray over the last five years at a conference we hosted in South Carolina in early 2004:

> I had been away from this field since the late 1990s because there was so much noise about nanotechnology that I found it offensive and I thought maybe if I go away people will settle down and they'll focus on the technology and the science and they'll make some progress, and besides I was interested in computer science so I went off and did that.[55]

Evidently, absence does *not* make the heart grow fonder. Drexler discovered he was only an incidental variable in the mean spiritedness of the debates: "Last Spring [2003], I finally got fed up with this kind of thing, came back in, and started trying to change things." In March 2004, he was invited to speak at the South Carolina conference, which was partially funded by an NSF grant to study SEIN. We invited him, and some members of the sanctioned and funded nanocommunity were not pleased at our decision to provide him a dais of any sorts, but as academics that is what we do. He said:

> My presence here is a sign of the extent to which the current regime is crumbling. They've lost control of a conference finally. This is the first NNI event of any sort—from the initial conversations of scientists saying we should have a program, through all the meetings on all the different aspects that you can think of—to which I've been invited. So apparently something I have to say the leadership doesn't want people to hear.[56]

It would be more correct to say there is a conceptual conflict over Drexler. Some think that providing a microphone to an outlier awards value to his claims. While that may be true in some instances, we felt that by hearing what he had to say and vetting his remarks in the ballroom in which he spoke, in the halls and lobbies of the Adam's Mark that weekend, and in dozens of discussions and seminars since then, would be a better way to test Drexler's arguments, as we shall see below.

Direction of the Nanotechnology Movement

After an in-depth study of the current players and their tactics, the concept of nanotechnology that was sold to Clinton and others used hyperboles of nanohype. When it came to cataloging discoveries, developments, and advancements, more traditional nanoscience, such a chemosynthesis, crowded out the more hypothetical mechanosynthesis propounded by Drexler et al. There are many possible reasons, but primarily it seems to be one of funding. Distrusting its promise and fearing the public concern regarding its dangers simply might interfere with research funding.

There are two primary debates over the direction of government-sponsored nanotechnology research and development. They are over the focus of government initiatives. One debate involves Drexler's role and the second, why molecular manufacturing was extracted from the new enabling legislation for the NNI. While some nanotechnology proponents argue that molecular manufacturing should play a role in the drive to the illusive breakthroughs, the powers that be may believe otherwise. In the bout over direction, the majority of the science community rejects advanced or radical molecular manufacturing as a vision, instead opting for a less revolutionary and more evolutionary view that is grounded in applied nanoscience. University of Tennessee law professor Glenn Reynolds provides the thumbnail summary of the second debate:

> A sometimes bitter war has been waged within the nanotechnology community itself, between the scientists and visionaries on one hand and the business people on the other. The business community is afraid that advanced nanotechnology just seems too, well, spooky—and worse, that discussions of potentially spooky implications will lead to public fears that might get into the way of bringing products to market.[57]

As Howard Lovy, formerly of *Small Times*, put it, "Business leaders and policy makers did this by carefully selecting which theories are the ones the general public is supposed to believe, then marginalizing the rest."[58] Both debates conflate on the resolutional level.

What ended up happening may have less to do with the science than with the politics of science funding. As the editors of *Chemical and Engineering News* put it,

> Richard Smalley's objections [see below] go beyond the scientific. They are a strategy—if so-called dangerous nanotech can be relegated to summer sci-fi movies and forgotten after Labor Day, then serious work can continue, supported by billion-dollar funding and uninhibited by the idiocy that buries, for example, stem cell research. Given the politics of science, this strategy is understandable. Yet it is a strategy inspired not by the laws of nature but by the perverse nature of how we make laws. We dissemble rather than reason, because we can't imagine rational government policy addressing these reasonable fears.[59]

What happened to Drexler's vision? Why have he and other proponents of molecular manufacturing been closed out of the nano debate? Why have we convinced ourselves the only way to experience the nanofuture is through the production and use of passive materials, like nanoparticles and carbon nanotubes? What about molecular manufacturing convincing the government agencies funding research and, in some cases, development that molecular manufacturing is not a route worth investigating? Why did the Twenty-First Century Nanotechnology R&D Act exclude a mandate to test molecular manufacturing feasibilities?

Christine Peterson attempted to include a mandate into the Twenty-First Century Nanotechnology R&D Act that would have funded feasibility studies on Drexler's version of mechanosynthesis.[60] She advocated "a one-time study to determine the technical feasibility of molecular self-assembly for the manufacture of materials and devices on the molecular scale."[61] It never made it into the final version of the act, and Drexler blames Mark Modzelewski (former executive director of the NanoBusiness Alliance, the NbA) for its exclusion. Modzelewski categorized this allegation as "an elaborate fantasy about how molecular manufacturing research work was pulled from the bill."[62] He continued, however, to malign molecular manufacturing advocates as "bloggers, Drexlerians, pseudo-pundits, panderers, and other denizens of their mom's basement."[63]

When mechanosynthesis advocates challenged Modzelewski's characterizations, Modzelewski added, "The industry is not hiding from any real problems by ignoring your [referring to Reynolds] delusional fantasies and rantings, any more than one truly ignores a wino's claims on skid row that bugs are crawling under this skin.... We avoid mixing in the comic relief of the writings of Drexler and yourself on the subject."[64] The animosity is clearly palpable.

One rationale was suggested by *Betterhumans* commentator Simon Smith: "Nanotechnology business leaders are trying to avoid validating the fears in an effort to avoid potentially stringent regulations."[65] However, this categorization may be overly simplistic and merely a less creative form of name-calling. While Modzelewski may not be polite, it doesn't make him wrong. He is reacting to repeated claims and counterclaims about nanoproducts that are incredibly hyperbolic. Business and industry fear that claims about nanobots and uncontrolled replication might spur ethicists and environmentalists and cranks to frustrate a powerful economic force in the next few decades. In addition, business and industry are concerned that claims of an end to the limits of growth might set up the field to many of the dot-com problems.

"Drexler envisaged a particular way of achieving radical nanotechnology, which involved using hard materials like diamond to fabricate complex nano-scale structures by moving reactive molecular fragments into position."[66] Scientists claim the study itself would have been impractical given the fundamental discoveries that are a requisite for any organized effort to complete feasibility studies. While there may be little chemical or physical reasons why mechanosynthesis is impossible, there are many affirmative steps that are missing. While it is true that cells are proof positive that molecular manufacturing can occur, how to go about programming ribosomes to build macrostructures remains outside the realms of current chemistry and biotechnology.

Drexler extended an argument that the original Feynman vision that he popularized has been used by advocates of the NNI to sell the budget, but when it
comes to funding, grants go to solution-based chemistry, nanolithography, and
more conventional routes. In other words, his rhetoric was used to spark interest
but trashed when it came to actual research initiatives.

There is a peculiar assumption found in the claim: Feynman and Drexler's
versions are sufficiently similar to be conflated. The fallacy here is one of the basic
subset fallacies. While Feynman's view includes Drexler's, it does not mean all of
Feynman's remarks can be attributed to Drexler.

As mentioned above to some degree, it's not just the leadership of the NNI
that rejected Drexler's version: "business leaders, particularly in the U.S., are characterizing molecular manufacturing as a fanciful delusion—an approach that
could be very bad for business, and possibly bad period."[67]

Strangely, the debate over nanotechnology and molecular manufacturing is
becoming less academic and more personal, and the angst is almost palpable. On
March 3, 2004, I had the opportunity to listen to a talk by Drexler at a conference
titled "Imaging and Imagining," hosted by our Nanotechnology Interdisciplinary
Research Team. In addition, I also had the opportunity for some interpersonal discussions that shed a great deal of light on the direction and overall tone of the
debate between the two emerging schools of nanotechnology.

For the record, Mike Roco, who chairs the President's Council of Advisors on
Science and Technology, questioned our decision to invite Drexler to speak at the
conference, something that Drexler alluded to several times during his presentation. Although Roco made no effort to pressure us into disinviting Drexler, another
invitee did refuse to speak on the same dais as Drexler, and she chose not to attend.

Hayles tells this story: "Many scientists look on him with suspicion and even disdain; one researcher working in the field told a colleague he was so upset with Drexler
that, when seated with him at a conference, he challenged him to a fistfight."[68]

Strangely, in 1998, the NSF sponsored a forum in conjunction with the Sixth
Foresight Conference and James Murday and Mike Roco were session chairper-
sons. What happened over the last half decade is a little unclear, though the business and funding issue seems to explain why Drexler's version was marginalized.

Quite simply, Drexler and some of his associates seem anguished by the current state of affairs in the field of nanotechnology. Much of the debate over nanotechnology has become less focused on academic and scientific reason and has
digressed to insult-slinging. Since the 1986 publication of *Engines*, Drexler has
drawn a great deal of criticism for his extreme views. While his 1992 book,
Nanosystems, was less controversial, many in the scientific community remained
convinced his vision remains technically science fiction.

Today, Drexler makes at least two major allegations regarding his role in
nanotechnology. First, his vision of nanotechnology was used to get Congress

onboard to fund the NNI. However, once the funding was garnered, his ideas were largely discarded in favor of more conservative views of the technology's potential. And second, the inclusion of a feasibility study of molecular manufacturing in the Twenty-First Century Nanotechnology R&D Act was skillfully excised at the last minute by the NbA and is likely to further his marginalization in the field.

On the first claim, Drexler contends misrepresentation has led to a policy of denialism in the field, where the government and industry leaders have taken the "excitement about nanotechnology that came from the molecular manufacturing vision...and involved a host of researchers from other disciplines...including everything from particles to nano-pants...to define nanotechnology as everything but the original vision."[69]

Even the staunchest opponents of Drexler's vision of molecular manufacturing have difficulty denying that he has played an immensely important role in the popularization of nanotechnology. In many ways, Feynman's talk was so visionary that it wasn't fully able to bring nanotechnology into the mainstream of the scientific community. And while he may have created the concept of nanotechnology, it was Drexler who popularized it. Cynthia Selin from the Copenhagen Business School believes that Drexler's "book marks the origins...of the term and the formulation of what is said to be a new industrial revolution,"[70] hence the basis of some of the offending nanohype.

However, Clinton's science advisor, Neal Lane, who helped fashion the speech that Clinton delivered at Caltech, indicates that the vision sold to Clinton was not Drexler's at all. "President Clinton was not given any extreme version of where nano might go. He is a very smart person who understood that to the extent that we can control and manipulate atoms and molecules we can expect advances in technology, maybe revolutionary ones."[71] Clinton's speech references Feynman, not Drexler, and his most exaggerated claims dealt with data storage. The report put out by the interagency group in September 1999, *Nanotechnology—Shaping the World Atom by Atom*, describes what they were talking about at the time. It is much more Feynman than Drexler. Actually, Feynman is mentioned thirteen times, and Drexler isn't mentioned at all. Even despite Lane's counterclaims, Drexler continues to contend the NNI has become a game of bait and switch, where the original vision for nanotechnology is being replaced by a host of new-age products, which, while fascinating, are not really nanotechnology in its truest sense.

Undoubtedly, Drexler has been displaced as the key professional *spokesman* of the field because *serious* scientists are simply *burdened* by his visions. Gary Stix of *Scientific American* wrote that for nanotechnology to have any chance whatsoever, we need to discard nanobots and the idea of reanimating cadavers. "The field's bid for respectability is colored by the association of the word with a cabal of futurists who foresee nano as a pathway to techno-utopia: unparalleled prosperity, pollution-free industry, even something resembling eternal life."[72]

Furthermore, venture capitalists such as Forbe's Josh Wolfe warn that the Drexlerian hype surrounding nanotechnology may very well stifle new invest-ment. Scientists were weighed down with a sexy term that they didn't want. Stix warned that the overheated rhetoric of Drexler could derail the funding effort and situated his views as one of the greatest risks to nanotechnology. "If the public expects robots in their bloodstream, they are going to be left largely unimpressed by ultra-fine dust." And while this reinterpretation of nanotechnology may make goals easier to achieve, it clearly goes awry of Drexler's original vision of molec-ular manufacturing.

Thus far, the business community has joined the government in transitioning to a more conservative approach to nanotechnology, focusing on computer chips and sunscreen rather than Drexler's tiny robots. Even companies such as the Dallas-based Zyvex Corporation, which originally focused on molecular manu-facturing, have ventured into the more conservative applications of nanotech-nology. Many believe that in order for this technology to move forward, the field needs to be de-nanobotted,[73] which inherently involves relegating Drexler to the sidelines. Consequently, Roco and others have been dismissive of molecular man-ufacturing and nanobots, as has the NbA, the foremost business-development organization in nanotechnology.

According to some people, government and business fret that any talk of nanobots might conjure up the *Magician's Apprentice*. With doomsday scenarios such as Bill Joy's vision of a gray goo Armageddon, it comes as little surprise that business and government leaders have swept both Drexler's visions and his omi-nous warnings under the rug. In short, many just don't want to see nanotechnology derailed by groups who would likely attack nanotechnology just as they have attacked genetically modified seeds and foods. By casting aside Drexler's opti-mistic vision, as well as his warnings, leaders in both industry and government are finding it easier to bring nanotechnology out of the fringe and into the main-stream, whetting the public's appetite with rudimentary commercial applications.

Drexler's view of molecular manufacturing has been almost uniformly rejected by the scientific community, and he has been outed. With Drexler and his followers advocating a type of molecular manufacturing often associated with images of very tiny machines, it comes as no surprise that his claims are met with serious skepticism within the scientific world. Roco and his camp of scientists view nanotechnology as more than the next step in MEMS and the natural product of chemical synthesis and protein engineering. They anticipate remark-able advances once we are able to master the quantum characteristics of the nanoscale. While Drexler and most of his followers understand and appreciate the Roco version, the mainstream treats Drexler like a coot, as you will see below.

Second, in the final version of the Twenty-First Century Nanotechnology R&D Act, there was a feasibility study of molecular manufacturing. It was changed to molecular assembly. "This change in wording, from manufacturing to assembly, made at a very late stage in the legislative process, may seem insignificant, but its actual effect was to gut the intended feasibility study of all usefulness: molecular self-assembly is not merely feasible, it has actually already been achieved."[74]

Drexler claims that fears led to the removal of a feasibility study of molecular manufacturing in the final version of the Twenty-First Century Nanotechnology R&D Act, and he blamed the NbA. Reportedly, NbA official Nathan Tinker admitted to a reporter that the alliance approached the staff of Senator John McCain to have the study removed from the legislation.[75] Presumably, "Modzelewski and the NbA do not want to be associated with the Drexler version of nanotechnology."[76] While Modzelewski's role might have been minor to nonexistent, NbA's role seems to be evident.

Some regret this decision by the Senate Committee. As Lawrence Lessig put it in his 2004 *Wired* article:

> The world of federal funding would only be safe, critics believed, if the idea of bottom-up nanotech could be erased. Molecular manufacturing, Smalley asserted, was just a dream and facts of nature [would] prevent it from ever becoming a reality. In an ideal world, such scientific controversy would be settled by science. But not this time: Without public debate, funding for such *fantasy* was cut from the NNI-authorizing statute. Thanks to Senator John McCain, not a single research proposal for molecular manufacturing is eligible for federal dollars.[77]

Whether McCain is wholly blameworthy seems less important than the claim that no research proposal would ever be funded. As Roco told me on two separate occasions when we discussed a draft of this work, all grant proposals are peer-reviewed and never rejected beforehand. If a proposal were favorably reviewed, it would be eligible for funding.

Drexler's supporters at CRN make the same argument: "The scientists working on nanoscale technologies and the administrators funding them had several incentives to try to discredit molecular nanotechnology, including justifying the current funding decisions and avoiding anything associated with *gray goo* and other doomsday scenarios."[78] Later in 2004, Mike Treder from the CRN and Drexler attempted to divest themselves from the gray goo scenario. This reversal notwithstanding, Drexler remains adamant in his claim that the NNI was built on his vision.

According to his supporters, "the real reason the government is willing to shell out for nanotech is because our leaders in Washington believe in the more revolutionary version of nanotech espoused by Drexler, with all its great promise and grave perils." Keiper of the *New Atlantis* continues, "were it not for Drexler and his ambitious vision of molecular manufacturing, no one would have heard of nanotechnology today—and the federal government would certainly not be investing billions of dollars in nanotech research if they know only of Modzelewski's modest mainstream aims."[79] While Keiper's passion is commendable, his understanding of the construction of the NNI is not.

In short, Drexler seems to want it all, to be both visionary and scientist. However, the first must be shared, at least, with Feynman and the second is being denied by most of the scientific community. In addition, "scientists worry that he promises far too much with far too little experimental work to back up his claims. They fear he is squandering the cultural capital that science accumulates by patient and often laborious laboratory work."[80] The word *nanotechnology* can enclose many concepts of the very small. Drexler is a futurist, prophet, avatar, seer, guru, sage, messiah, and so on. He's a lot of things to a lot of people. The metaphors he helped craft have served as catalysts for much creative thinking. Today, the case for evolutionary rather than revolutionary nanotechnology is the case that may need to be made.

Watching Drexler speak about denialism, it is difficult not to notice that he is emotionally upset at the tone and direction of the debate over the NNI. He concluded his March speech at South Carolina contending that

> if the policy of denialism continues within the field of nanotechnology, the U.S. would end up being on the wrong side of a technology gulf comparable to hand-crafted spears and mass-produced machine weapons. Terrorism cannot destroy the United States of America or conquer it. Current NNI policy, if continued, has an excellent prospect of doing so.[81]

While I may not agree with Drexler, it might behoove the field if the debate became more civil.

According to biographer Ed Regis, "Drexler is barely solvent. He recently moved from his three-bedroom ranch house in silicon Valley into a modest apartment."[82] There is some legitimate concern that creating a martyr of Drexler might backfire. Melody Haller of the Antenna Group, a public-relations firm, indicated that "marginalizing people such as Drexler and others who believe in the feasibility of molecular manufacturing might create heroic martyrs for nanotech opponents to exploit."[83] There is a parallel concern expressed by British MP Ian Gibson:

> Attempts to calm public fears by simply denying the feasibility of molecular manufacturing will inevitably fail. A better course would be to show that its consequences are manageable and still distant. Of course, this approach doesn't work so well if its chief spokesman has to make his case over a chorus of false denials and attacks on his reputation.[84]

We probably should avoid martyring Drexler if our overall goal is to reduce unreasonable arguments against applied nanoscience. Conspiracy arguments are incredibly difficult to rebut. Gibson is correct. Simple denial is no longer a viable response to Drexler's version of molecular manufacturing because the script to the contrary is too well written at this point. Whether or not scientifically valid, there is a plausible version of nanotechnology that includes molecular manufacturing. As such, a more sophisticated and detailed strategy may be in order.

CRITICS

Drexler is clearly a critic of the NNI's current direction and its rejection of his version of molecular manufacturing. Others challenge the wisdom of sinking so much government money into nanotechnology research given the state of our economy, domestic needs, health and environmental concerns, and international geopolitics. However, the following examines two classes of complainants who look elsewhere for grist. By and large, the first set is pragmatic. It argues against the feasibility of molecular assemblers and Drexler's brand of nanotechnology, and it does so with a noticeable degree of disdain. The second set is idealistic and draws its criticisms from the health and environmental implications of applied nanoscience and the long-term implications of mature nanotechnology, especially the renowned gray goo scenario.

Technical Critics

Some writers argue simply that many of the manifestations of nanoscience, especially those articulated by Drexler and his followers, are impossible in the foreseeable future. These critiques are claimed to be anchored in scientific reality. While their denials have not gone unrebutted, they still seem to retain some credibility among the scientific and technical establishment. Two outstanding members of this sect are George Whitesides of Harvard and Richard Smalley of Rice University. Both these men are powerfully knowledgeable proponents of applied nanoscience. At the same time, they are critics of Drexler's brand of molecular nanotechnology. Both Whitesides and Smalley could easily have joined Feynman as advocates in a book with a different organizational scheme.

Most chemists see fundamental barriers that preclude any type of Drexlerian molecular manufacturing, regardless of time frame. George Whitesides of Harvard doesn't believe Drexler's engines can be powered and is unsure how molecular bonds would be broken by brute force. He thinks Drexler's intravascular minisubs would need to be so large they'd seriously impede blood flow.

But it's been the Nobel Laureate and Rice chemist Richard Smalley who had led the charge against Drexlerian nanotechnology. He was certainly the most vocal of the opposition. He also in many ways lowered the debate to a point of character assassination with broad rhetorical flourishes. Though ironically, Smalley claimed reading *Engines* was the trigger event that started his own journey in nanotechnology, today, Smalley derides Drexler. The recent debate between Drexler and Smalley addressed fat and sticky fingers, enzymes, and catalysis in the December 1, 2003, issue of *Chemical and Engineering News*. It hardly settled anything, especially when Smalley ended the debate condemning Drexler for scaring our children.

Whitesides and Smalley have challenged Drexler and led him to complain: "By falsely declaring molecular assembly technology to be impossible, detractors have associated it with warp drives in official circles and relegated it to fringe status."[85] They have engaged in a dialogue of sorts with him, although it has been unproductive, as we will see below.

George Whitesides, Harvard

George M. Whitesides received an AB degree from Harvard University in 1960 and a PhD from Caltech in 1964. He was a member of the faculty of MIT from 1963 to 1982. He joined the Department of Chemistry of Harvard University in 1982 and was department chairman from 1986 to 1989. He is now Mallinckrodt Professor of Chemistry at Harvard.[86] Whitesides is among the most highly cited high-impact researchers in nanotechnology, not to mention the number-one-ranked chemist in the ISI Web-based evaluation tool, Essential Science Indicators, with more than ten thousand citations to his credit since 1991. He is the author of over seventy papers with more than one hundred citations each. He has won multiple honors and is a renowned scientific figure.[87]

Whitesides's arguments can be distilled to this statement: not likely anytime soon, if ever. He observed that breakthroughs have been modest as researchers struggle to come to grips with the challenge of creating three-dimensional structures and functional tools on a nanoscale. It's because of this gap between reality and promise that the most influential research has been on technologies that make it possible to create simple nanostructures and to study those structures once they've been laid down on surfaces. Whitesides invented a microfabrication technology known as soft lithography and developed a fabrication technique for self-assembled monolayers. "Surface Logix, one of the many start-ups he helped to launch, is marketing test platforms to the pharmaceutical industry that incorporate his soft lithography techniques."[88] Therefore, his point of view seems motivated by many factors including self-interest.

Whitesides has rejected the Drexler concept of molecular machinery with stinging remarks: "It's complete nonsense. If there were new kinds of self-replicators, there might be a problem. But there are not. The level of hard science in these ideas is really very low."[89]

Whitesides's article "The Once and Future Nanomachine" established the primary body of disagreements he has had with Drexler's mature nanotechnology. His criticism is merciless. He begins by simplifying Drexler's point of view:

The Drexlerian vision imagines society transformed forever by small machines that could create a television set or a computer in a few hours at essentially no cost. It also has a dark side. The potential for self-replication of the assembler has raised the prospects of what has come to be called *gray goo*: a myriad of self-replicating nanoassemblers making uncountable copies of themselves and ravaging the Earth while doing so.[90]

His first set of complaints relates to practical problems with nanobots. Many of his concerns have been rebutted by Drexler and his followers, though Whitesides seems oblivious and unconcerned in continuing a tit-for-tat discussion. First, he articulated the stiction issue: "Small devices have very large ratios of surface to volume; surface effects—both good and bad—become much more important. Some of these types of problems will eventually be resolved if it is worthwhile to do so, but they provide difficult challenges now."[91] Next, he asks: "Where is the power to come from for an autonomous nanomachine? There are no electrical sockets at the nanoscale."[92] Next, he posits: "How would self-replicating nanomachines store and use information?"[93]

He continues. Beyond some general claims that "we have a long path to travel before we can produce nanomechanical devices in quantity of any practical purpose,"[94] Whitesides reiterates Smalley's complaint about *fat* and *sticky* fingers (see below). "First, is the pincers, or jaws, of the assembler. If they are to pick up atoms with any dexterity, they should be smaller than the atoms. But the jaws must be built of atoms and are thus larger than the atom they must pick and place."[95] "Second is the nature of atoms. Atoms, especially carbon atoms, bond strongly to their neighbors. Substantial energy will be needed to pull an atom from its place and substantial energy would be released when it is put in place. More important, the carbon atom forms bonds with almost everything. It is difficult to imagine how the jaws of the assembler would be built so that in pulling the atoms away from their starting materials they would not stick."[96] Finally, Whitesides argues random battering by water molecules would "make a nanoscale object bounce about rapidly" and "the tiny crafts would probably be impossible to steer."[97]

Beyond these beleaguering doubts, he tackled two profoundly troubling examples of hyperbole. First, he tackles the nanosubmarine, a version of which not only appeared in the 1966 movie *Fantastic Voyage* but also was alluded to in *Engines*.[98] Nanotechnology applications in health and medicine have been the source of wildly curious exaggerations for a handful of reasons. Technological developments that improve medicine and medicalization have been much less resisted than technological adjuncts in other fields. Everyone knows someone who died from some incredibly debilitating and dehumanizing illness, often a close loved one. Also, both the medical care and pharmaceuticals research industries have the resources for the expensive R&D needed for mature or advanced nanotechnology. As such, Whitesides offers three powerful reservations: information density, navigation, and power utilization.[99] These hunter-killer nanosubmarines "would have to carry on board a little diagnostic laboratory that would require sampling devices and reagents and reaction chambers and analytic devices: it would cease to be little."[100] Whitesides added,

> You've got to have these nano devices powered, and they have to have a certain size to talk to one another. It is not an accident that bacteria are one micron to three microns, not nanometers. It takes about that much space, at a molecular level, to store the information in the machinery to make them do the things that they do. It's also not an accident that they eat glucose—they need power.[101]

His final reservation gravitates around self-replicating assemblers and gray goo. "For the foreseeable future we have nothing to fear from gray goo.... To be self-replicating, a system must contain all the information it needs to make itself and must be able to collect from its environment all the materials necessary both for energy and for fabrication. It must also be able to make a copy of itself."[102]

These reservations have not been relinquished. In December 2003 at an NSF SEIN Workshop, Whitesides described these "apocalyptic visions as unfounded and irrational."[103] In two separate interviews, Whitesides maintained his objections. In 1998 he characterized the entire field of nanotechnology, especially Drexler's version, as "an area prone to overblown promises, with speculation and nanomachines that are more likely found in *Star Trek* than in a laboratory."[104] In 2003 he maintained his incredulity, claiming as he did two years earlier: "The dream of an assembler holds seductive charm in that it appears to circumvent these myriad difficulties. This charm is illusory: it is more appealing as metaphor than as reality, and less the solution of a problem than the hope of a miracle."[105]

In 2003 Drexler wrote an open letter to Whitesides:

> This atom-picking, sticky-jawed, universal-but-impossible *assembler* is an absurdity unlike anything proposed in the research literature. When you use it to dismiss my work, you attack a straw man. My actual proposal is, and always has been, to guide chemical synthesis by mechanically positioning reactive molecules.[106]

Drexler's letter ended with a paragraph of condemnation:

> By creating the illusion of a scientific argument against this concept, your misdirected remarks have needlessly confused discussion of both research objectives and long-term security concerns. If you have found a genuine scientific argument to support your speculation that molecular manufacturing is impossible, I believe you are under some obligation to state it. Otherwise, I urge you to avoid placing your considerable authority behind a position for which you offer no legitimate argument.[107]

However, Whitesides was not sufficiently baited to pen a response.

Richard Smalley, Rice

Richard Smalley was one of the winners of the 1996 Nobel Prize in Chemistry for discovering buckminsterfullerenes. He was a professor of chemistry, physics, and astronomy at Rice University. For the past decade, "he [had] been a leading proponent of a coordinated national research effort in nanoscale science and technology."[108] Anyone who saw Smalley speak understands Wolfe's comment that "he's been aggressive on issues of policy and championing using nanotech for alternative energy."[109] Recently, he has been touting carbon nanotubes as the answer to energy supplies. He served on the Scientific Advisory Board of the biotech start-up CSixty and NanoSpectra Biosciences.

Smalley has an enormous financial interest in nanotechnology. "Smalley's knack for building novel research equipment paved the way for the start-up Carbon Nanotechnologies, Inc. CNI manufactures single-wall carbon nanotubes based on a high-pressure reactor created by Smalley."[110] Like most university-budded start-ups, CNI drew from venture capital funding; "with $15 million in venture financing, Smalley spun his Rice research [into] CNI."[111] CNI is becoming a major player in the commercialization of carbon nanotubes. For example, it "has inked a deal with Sumitomo Corp. to market CNI's nanotubes in Japan, partnered with start-up NanoInk and joined MIT's Institute for Soldier Nanotechnologies."[112] Also, "DuPont licenses CNI's nanotube production process for flat panel display TVs."[113] Recently, CNI "teamed up with Kostat, Korea, to develop conducting packaging components for the electronics industry."[114]

Ray McLaughlin, its executive vice president, describes CNI as "the world's leading producer of carbon nanotubes. In 2005 or 2006, we could be up to 1,000 to 1,500 pounds a day."[115] In July 2004 CNI announced a patent on a composition of "single-walled carbon nanotubes with a nanometer-scale coating of another material that can include polymers and metals."[116] CNI also holds patents, issued or allowed, on a variety of technologies involving coverage for cutting, removing end caps, composition of matter coverage with other substituents covalently bonded to them, and others. Accordingly, its portfolio of application-enabling intellectual property is substantial:

> CNI has over 100 patent and patent applications with a total of about 5,000 claims in various stages of prosecution. Twenty-five of these with a total of about 900 claims have been issued or allowed. The portfolio of 100 patents and applications includes about 650 composition of matter claims, over 40 of which have been issues or allowed so far.[117]

Unsurprisingly, Smalley simply had little patience for the science fiction depictions of nanotechnology:

The principle fear is that it may be possible to create a new life form, a self-replicating nanoscale robot, a nanobot. For fundamental reasons, I am convinced that these nanobots are an impossible, childish fantasy. The assembly of complex molecular structures is vastly more subtle and complex than is appreciated by the dreamers of these tiny mechanical robots. We should not let this fuzzy-minded nightmare dream scare us away from nanotechnology. Nanobots are not real. Let's turn on the lights and talk about it.[118]

He was reluctant to waste any more time conjecturing about nanobots and finds those responsible for introducing the concept tiresome, especially in the form of Drexler. Disdain is expressed in the other direction as well: "Drexler and his followers say Smalley is a dinosaur and he sounds very much like a professor scolding a student."[119] Howard Lovy framed the disagreement very well: "The tenor of the debate is about personal pride, reputation, and a place in the pantheon."[120]

Smalley may have been best known in the nanotechnology policy debate for challenging Drexler's molecular-manufacturing thesis. His arguments appeared in the September 2001 issue of *Scientific American*. Initially, Smalley's primary complaint had to do with "*fingers* that are too fat and sticky to manipulate single atoms since the fingers themselves would be the same size or bigger." He also "calculated the barriers facing a single nanobot, capable of building at a speed of one billion atoms per second, would take 19 million years to build 30 grams (less than an ounce) of product."[121]

Smalley's conception of nanotechnology was profoundly different from Drexler's and doesn't include molecular assemblers. "Smalley does not think molecular assemblers as envisioned by Drexler are physically possible."[122]

In response, Drexler's defenders claim "Smalley's highly publicized objection in *Scientific American* was based on a distorted description of the Drexler proposal: it ignored the extensive body of work in the past decade.... Drexler's proposal, and most of those that have followed, have a single probe, or 'finger.'"[123]

Drexler goaded Smalley into responding to his rebuttals of Smalley's claims when he posted an open letter to Smalley in April 2003. Drexler followed up with a second open letter in July. In the *Chemical and Engineering News* debate in December 2003, Drexler claimed Smalley had "attempted to dismiss [his] work in [the] field by misrepresenting it.... [Drexler claims] a 20 year history of technical publications in this area and consistently describes systems quite unlike the straw man [Smalley] attacks."[124] Drexler responded in April 2003:

You have described molecular assemblers as having multiple *fingers* that manipulate individual atoms and suffer from so-called *fat finger* and *sticky finger* problems and you have dismissed their feasibility on that basis. I find this puzzling because, like enzymes and ribosomes, proposed assemblers neither have nor need these *Smalley fingers*.[125]

Smalley retorted with a series of questions:

> And now that we're thinking about it, how is it that the nanobot picks just the right enzyme molecule it needs out of this cell, and how does it know just how to hold it and make sure it joins with the local region where the assembly is being done, in just the right fashion? How does the nanobot know when the enzyme is damaged and needs to be replaced? How does the nanobot do error detection and error correction?[126]

For some reason, he included that enzymes and ribosomes work only in water solutions and preemptively added:

> If the nanobot is restricted to be a water-based life-form, since this is the only way its molecular assembly tools will work, then there is a long list of vulnerabilities and limitations to what it can do. If it is a non-water-based life-form, then there is a vast area of chemistry that has eluded us for centuries.[127]

Smalley's claim that enzymes require water has been rebutted elsewhere and does not need to be repeated here.[128] It is safe to say that Drexler and his supporters responded with a defense of mechanical positioning and added, "U.S. progress in molecular manufacturing has been impeded by the dangerous illusion that it is infeasible."[129]

Smalley countered: "You cannot make precise chemistry occur as desired between two molecular objects with simple mechanical motion along a few degrees of freedom in the assembler-fixed frame of reference. Chemistry, like love, is more subtle than that."[130]

Ray Kurzweil, author of *The Age of Spiritual Machines*, got into the debate:

> Smalley is ignoring the past decade of research on alternative means of positioning molecular fragments using precisely guided molecular reactions. Precisely controlled synthesis of diamondoid [diamondlike material formed into precise patterns] has been extensively studied, including the ability to remove a single atom from a hydrogenated diamond surface. Related research supporting the feasibility of hydrogen abstraction and precisely guided diamondoid synthesis has been conducted at the Materials and Process Simulation Center at Caltech, the Department of Materials Sciences and Engineering at North Carolina State University, the Institute of Molecular Manufacturing, the University of Kentucky, the United States Naval Academy, and the Xerox Park Alto Research Center.[131]

Chris Phoenix of the CRN offered his summary of the debate:

> Smalley began by inventing molecular *fingers* and describing why they don't work. Then, he invented a nanofactory based on wet chemistry, and described why it

cannot produce many useful products. Not until the end did he address Drexler's actual proposals, and his arguments at that point depended heavily on a clearly incorrect understanding of enzymes. In addition to being largely off-topic, and apparently contradicting his own statements of 1999 and 2003 that were references in Drexler's open letter, Smalley's arguments are sprinkled with factual error about chemistry.[132]

Smalley also argued that "speculation about the potential dangers of nanotechnology threatens public support for it,"[133] hinting at another motivation. Drexler responded that "your attempt to calm the public through false claims of impossibility will inevitably fail, placing your colleagues at risk of a destructive backlash. Your misdirected arguments have needlessly confused public discussion of genuine long-term security concerns."[134] Smalley countered:

> You and people around you have scared our children. I don't expect you to stop, but I hope others in the chemical community will join me in turning on the light, and showing our children that, while our future in the real world will be challenging and there are real risks, there will be no such monster as the self-replicating mechanical nanobot of your dreams.[135]

In response, Howard Lovy chimed in that Drexler's vision actually inspires and challenges children:

> Kids do not get excited about new nanotech companies and products. But they do enjoy the challenge of proving their elders wrong and achieving what was once thought *impossible*. If some old scientist says self-replicating nanomachines are out of the question, I'll bet there are a few bright kids out there plotting ways to send comets raining down on that dinosaur.[136]

Phoenix jumped in with his two cents: "The best antidote to fear is knowledge, not denial.... Hasty arguments against the possibility of mechanically guided surface chemistry will not, in the long run, be either comforting or productive. If accepted, these arguments could lead to bad policy that cannot respond to real risks and benefits."[137] Next, the Foresight Institute, hardly an objective source, circulated a press release: "Smalley is just not addressing the issues. Instead, he veers off into metaphors about boys and girls in love. He describes mechanosynthesis as simply *mushing two molecular objects together* in *a pretend world where atoms go where you want.*"

Finally Drexler concluded that

> [Smalley] offers vehement opinions and colorful metaphors but no relevant, defensible scientific arguments, hence no basis for crucial policy. Smalley has struggled for years to dispel public concerns by issuing false denials for the capa-

bilities of advanced nanotechnologies. That campaign has failed. It should be abandoned.[138]

Smalley had to wonder why he ever bothered to respond in the first place. The response machine was akin to political party war machines, springing responses to speeches at national political conventions. This debate was two people talking over each other. The categorizations and the overall volume and pitch of the responses were not conducive to reasonable rebuttal. The undertone of the claims and counterclaims hint at hidden motives and allegations of general incompetence, which merely damages any hope of consensus making.

Smalley simply stopped arguing with Drexler. He was much too busy doing other things in academia and business, as an advocate and a proponent of nanoparticles. The spat will need to continue without Smalley, who passed away in late October 2005 after a seven-year bout with leukemia.

Popular Critics

This second category of critics is dubbed *popular critics* not because they are particularly popular or critical, but rather since their concerns are much less technical and much more extrapolative in nature. Only three were selected, though there are many more. They are Bill Joy for his role in popularizing the gray goo scenario and Zac Goldsmith and HRH Prince Charles for their overall health and environmental concerns, some of which are difficult to ignore.

Bill Joy, Sun Microsystems

According to Jon Gertner, there are two Bill Joys. One is the Silicon Valley deity. He was the principal designer of the Berkeley version of Unix and one of the lead developers of Java. The other Joy is the subject of the following, though his twenty-one-year career at Sun Microsystems added credentials to his protests. According to Zac Goldsmith, "Bill Joy is our Gates-Kaczynski hybrid, and his vision of the future is worth listening to, because he knows, better than anyone else, exactly what he is talking about."[139]

In reality, there are more than two Bill Joys. The third Joy was the one interviewed on NPR in 2000. After a brief discussion on Kaczynski and successor species to humanity, he told Noah Adams:

> I'm an optimist about these technologies. I believe that wonderful things are possible with genetic technology, with nanotech, with robotics, . . . Nanotechnology will reduce the cost of manufacturing and make it . . . almost just unthinkable for people to have material poverty because things will be so inexpensive to manufacture.[140]

For purposes here, we will examine the Bill Joy responsible for popularizing the gray goo scenario. "In the hype and horror that has surrounded nanotechnology, the scare about nanobots running amok has figured large."[141] While Drexler might have been the first person to hypothesize about uncontrolled replication, Bill Joy is responsible for publicizing it. Joy thinks the probability of a civilization-changing event is most likely in the double digits, perhaps as high as 50 percent. His article and his defense of it has many detractors. For example, Virginia Postrel wrote:

> If Joy's article isn't really about coping with dangerous technologies, what is it? Read carefully, it is exactly what the man-bites-dog press accounts promised: a screed against unpredictable change, a call for a static world, and an assault on commerce and the individual desires it serves. It is the same old attack on the open society, just wrapped in cool clothes.[142]

Bill Joy warned in his *Wire* article, "we are at the cusp of extreme evil."[143] He worried we were nearing the birth of a robot species. "We are designing technologies that might literally consume ecosystems."[144] The wonders of nanotechnology might produce "plants with leaves no more efficient than today's solar cells that could out-compete real plants crowding the biosphere with inedible foliage. Tough omnivorous bacteria could outcompete real bacteria; they could spread like blowing pollen, replicate swiftly, and reduce the biosphere to dust in a matter of days."[145] In other words, uncontrolled replication of nanobots would turn our *green* world into a *gray* one, hence gray goo.

Joy taps into collateral issues surrounding nature and robust molecular manufacturing:

> The self-replicating nature of these dangers leaves an inordinate amount of responsibility in the hands of a single scientist or corporation. In Joy's view, the legal system is now powerless to stop a rogue scientist or a negligent technologist.... He says he believes that businesses doing research in areas deemed risky by their peers should be forced to take out insurance against catastrophes. He also said that science guilds should have the authority to limit access to potentially dangerous ideas.[146]

This may be the true undercoating of his argument, similar in many ways to the ETC Group's (see chapter 8).

He added that "the mad rush for patents and market share and money will trump caution. Regulatory agencies are structured to catch CFOs, not reckless private-sector technologists. And markets are ill-equipped to play traffic cops."[147] Joy singled out James Watson's belief that it makes little sense to stop research on account of an unspecified risk or evil. "That position has to be wrong." According to Joy, we are ethically bound to slow down.[148] In some ways, Joy's concerns are some of the same that instigated the National Academy of Sciences to request a SEIN component in federal grants under the NNI, though gray goo wasn't the warrant for their argument.

Joy advocates backing off, stopping research that might animate the nanobots of nightmarish scenarios and vesting the decision process in the hands of the enlightened few. "Joy's policy prescriptions are breathtakingly naive, blinkered, and totalitarian. They are contradicted by his own arguments, and by the uncomfortable historical realities Joy skirts even as he invokes them. Though it strains to seem wise, Joy's manifesto is at once childlike and childish."[149] Responding to a comparison between Joy and Einstein made by Joy in an interview with the *Washington Post*'s Joel Garreau,[150] Postrel countered: "Einstein did not advise Roosevelt to renounce development of the bomb, or to end research in atomic physics. In 1939, even pacifist-leaning physicists couldn't pretend that you could shut down science and count on evildoers to leave it, and you, alone. Facing the Nazis, Joy blinks. Facing Stalin, he closes his eyes altogether."[151] Postrel is especially at issue with the verification regime underlying Joy's solution:

> That unprecedented regime would require frequent inspection and constant surveillance of every garage, every basement, every bathroom, every mountain hideaway on the entire planet—a planet inhabited by six billion potential technology developers, not a small clique of a few dozen world-class physicists. Does Joy have any idea how big the world is?[152]

Joy's solution would be the Patriot Act on steroids. It's infeasible, and no one wants a world like that, but how easy it would be for Homeland Security to move from gathering information on individuals to wholesale surveillance of everyone doing everything.

Presumably, Joy has a "50,000 word manuscript written between projects at Sun and never published, that he said he hoped would awaken the lay public to the potential dangers of genetic engineering, robotics, and nanotechnology." It's been through a few drafts. He admits "the book got away from me," and he has finished paying back a six-figure contract advance.[153] Not unlike Drexler, Joy has experienced a lot of criticism. Called a "neo-Luddite" and "an outlier, a software engineer unqualified to speak authoritatively," he admits that "such insults and marginalization have been difficult."[154]

John Markoff reported a phone interview in which

> Joy said he doubted the development of advances could be reined in in the commercial world, and he criticized scientists as being largely silent on the inherently destructive potential of rapidly evolving technologies. Asked if he thought a technological species could expect to survive the ever-accelerating evolution of

its market-driven technologies, Joy said: "The answer is yes, but not without additional care. I think it's possible—but it's not a given. Survival won't come free."[155]

Zac Goldsmith, the Ecologist

While the ETC Group has its own spokepersons, one of the more colorful propagators of the faith has been a wealthy British editor and organic-food enthusiast, Zac Goldsmith, editor of the *Ecologist* and allegedly the unofficial advisor to HRH Prince Charles on nanotechnology. Mr. Goldsmith also led much of the media criticism in the United Kingdom over genetically modified foods and seems to have put his sights on nanotechnology. He says he will not be happy until he and others have radically changed the global, political, corporate, and social systems, so there's a Neo-Marxist agenda as well. "There's not much difference between capitalism today and communism. It's just different ways of slicing up the cake."[156] His detractors say he's a spoiled rich kid indulging romantic personal obsessions that have little to offer the real world.[157] Fortunately, that depiction is much too simplistic and uncharitable.

Goldsmith, born in 1975, may be the golden boy of British environmentalism and anticapitalism.[158] He is the son of the anti-Euro campaigner Sir James Goldsmith, founder of the Eurosceptic Referendum Party. His father made his fortune from the sale of sugar mice, Slimcea, and Ambrosia Creamed Rice. Ironically, Zac would undertake a career to discourage people from eating the very food his father helped to produce. (Sir James died in 1997 of cancer with a fortune estimated to be £1.5 billion, a portion of which Zac inherited.) Zac's mother, Lady Annabel Goldsmith, is a member of a landed, moneyed Londonderry family. Lady Annabel was Sir James's third wife and is the family matriarch moving in royal circles. Zac has an older sister, Jemima, married to Imran Kahn, and has a younger brother, Ben, in Ormeley Lodge, his mother's Queen Anne house on the edge of Richmond Park. Also at Ormeley, there are three older half brothers and sisters from Annabel's earlier marriage. "As a child, Zac liked animals and cared what they thought, read Gerald Currell and watched Attenborough films and became aware that the natural world was slowly dying."[159] He is worth nearly £300 million, has ruffled good looks, wears crewneck jumpers, drives an old green Land Rover, loves cricket, and has two children. He was Eton educated until he was expelled, and instead of attending university, he went to Ladakh in the Himalayas. He plays poker once a week at Aspinalls and hangs with Etonian socialites Ben Elliot and Tom Parker Bowles. He is an organic farmer and wants us all to eat organic food, locally grown, seasonally produced, and genetically unmodified, hence his interest in the genetically enhanced food issue.

Zac speaks very softly and very fast, is polite, and is "laced with all the political contradictions and irony of youth, wealth and ideology: the pin-up who dresses down, the chain smoker who fights cancer, the green who stubs out his fag in the organic apple juice bottle, the rich man who says he fights for the poor, the man with power who seeks to give it away."[160] Oddly enough, he chews Nicorette gum to reduce his tobacco addiction, and Nicorette is owned indirectly by Monsanto. He is using his wealth and stature as a platform, which is not inherently wrong, and "with both a family and personal foundation to draw on, he funds groups taking on genetically enhanced crops, industrial agriculture, nuclear power or whatever is the flavor of the month."[161]

The *Ecologist* was started in 1970 by Zac's uncle Teddy Goldsmith. After Nicholas Hildyard left the editorship twenty-seven years later, the International Society for Ecology and Culture took over the magazine. Zac was a researcher with them. He became the letters editor. As the staff decreased, he overtook editorship and moved the magazine to London. While partly funded by Zac Goldsmith's foundation, it began to pay for itself because in no small part of the roles undertaken by unpaid interns. His philosophy for the magazine involves editing stories so they can be understood by laymen, a process he shared with deputy editor Jeremy Smith. Goldsmith's editorship of the *Ecologist* has been turbid. In 1998 his printers, Penwells of Saltash, pulped a fourteen thousand print run without notice. They were concerned about facing libel action if they printed a highly negative issue described as a diatribe against Monsanto.[162] Recently, he has nested the magazine at the heart of the ever-burgeoning antiglobalization movement.[163]

Goldsmith has argued that "genetically modified crops are now interbreeding to produce a chemically resistant super-weed."[164] Goldsmith cites two English organizations—English Nature and the Royal Society—both of which announced that "unintended breeding between different modified varieties is leading to super-weeds that could dominate agricultural systems and require a new generation of toxic chemicals to deal with them."[165] In the same 2002 article, he cited an EPA-approved test at Oregon State University of a genetically modified (GM) bacterium that "could survive and replicate...and could have ended all plant life on this continent."[166] Of course, that did not happen because the Oregon students did their research properly. In general, Goldsmith's campaign involved serious fear-mongering, warning readers that "Britain hosts more than 100 GM sites, most of which are for crops that are resistant to pesticides and designed therefore to increase, not decrease, the use of chemicals in agriculture."[167] He opposes nanotechnology for many reasons but mostly because of the GM-nanotechnology connection.

As an opponent to nanotechnology, he has debated and defended his position. For example, he debated chemist Sir Harry Kroto on BBC's Radio 4's *Today* program on April 28, 2003. He argued, "It's mad that we're charging ahead without any debate. People are nervous because scientists have made a lot of mistakes—DDT, CFCs, thalidomide. A mistake with nanotechnology could be very much more serious than anything we've seen before."[168] He added, "There are labs all over the world trying to create self-replicating machines and for them to do this without proper legislation is wrong.... We should put the brakes on and have a proper debate before it's too late."[169] As such, Goldsmith has become associated with two claims: gray goo and health and environmental concerns. He would have a minor postscript but for his connection with Prince Charles.

HRH Prince Charles

Indeed, Goldsmith, more than any other, is held responsible for Charles's decision to convene a nanosummit at his country home, Highgrove House in Gloucestershire. Presumably Goldsmith sent Charles a copy of ETC's *The Big Down,* which warns of the nanonightmare.[170] As Goldsmith put it, the prince "can play an enormously important role in the debate."[171] Hardly a fool, Goldsmith found in Charles a high-visibility celebrity that the tabloids would cover for almost any reason, especially if he started spouting warnings about a bizarre technology.

Prince Charles, however, has attempted to distance himself from the gray goo debate:

> My first gentle attempt to draw the subject to wider attention resulted in *Prince fears grey goo nightmare* headlines. So, for the record, I have never used the expression and I do not believe that self-replicating robots, smaller than viruses, will one day multiple uncontrollably and devour our planet. Such beliefs should be left where they belong, in the realms of science fiction.[172]

Regardless, the linkage remains to some degree.

Charles has added the counsel of John Carroll, retired professor of engineering at Cambridge University, to Goldsmith's. "Referring to the thalidomide disaster, it would be surprising if nanotechnology did not offer similar upsets unless appropriate care and humility is observed."[173] According to Charles, "those are my sentiments too."[174]

Prince Charles's remarks have been generally discounted and unsurprisingly so. "Professor Steve Jones of University College London called Charles a *classic woolly thinker,* and Lord Winston, said he had raised *specters* and *science scares.* Mark Welland, professor of nanotechnology at Cambridge University, said the reference to thalidomide was *inappropriate and irrelevant."*[175] This is not the first time Prince Charles has been rebuked for some of his views on science. Earlier remarks on cancer led "Michael Baum, professor of surgery at University College London to say Charles may have overstepped the mark by promoting unproven therapies for cancer such as coffee enemas and carrot juice."[176]

In reality, Prince Charles voiced some legitimate concerns:

> First, it is known that chemical reaction rates often increase with increase in surface area.... Second, it is believed that small particles can behave as carcinogens

when larger particles of the same substance are benign.... Third, I do not believe that we know how nanoparticles diffuse through the skin and the body.[177]

How unfortunate that Prince Charles's remarks about nanotechnology are clumped with some other less sensible things he has said about architecture and health because these statements could have been released by any of the hundreds of scientists currently researching nanotoxicity. On another level, his entry into the debate may demonstrate "failure adequately to consider and understand how nanotechnology can bring benefits to the 5 billion people in developing countries. [The Toronto Bioethics group adds this comment about Prince Charles and the ETC Group:] The significant nanotechnology activity in developing countries may be derailed by a debate that fails to take adequate account of developing countries."[178]

Charles, a committed environmentalist, generally responded: "There will also have to be significantly greater social awareness, humility and openness on the part of the proponents of emerging nanotechnologies than we have seen with other so-called technological advances of recent years."[179] This position has placed him on a collision course with Tony Blair, who backs cutting-edge science. Blair believes nanotechnology could be worth billions to Britain and the Department of Trade and Industry, and he has committed more than £50 million to support the UK Strategy for Nanotechnology. While no one seems to takes Charles seriously, his notoriety has kept the issues salient.

Importantly, Zac Goldsmith does not like Tony Blair. It's also difficult to determine what level of concern Goldsmith has for Blair's technology policy per se since he seems to dislike Blair and nearly everything about his government:

> The current British government's one of the worst we've had.... On every important issue, the government has either lied or U-turned. Behind each and every decision his government makes you can smell the greasy hand of a large corporation. Biotech campaigners are seeing laws change that makes it almost impossible to combat biotechnology legally. And the government is getting away with it. Why? Because it has a huge mandate and has mastered the art of fooling the public.... We've never faced a crisis like that confronting us today.[180]

Goldsmith calls for a new party, maybe an invigorated Green Party with a worldview that would challenge large-scale development policies. In nanotechnology, he only sees more of the same. Science Minister Lord Sainsbury recently announced £90 million in grants on top of the £50 million already allocated to nanotechnology research annually. In turn, Goldsmith calls for a whole new world order. As he put it, "the globally integrated economy is both undesirable and unachievable. It's time to change direction. It's time for a new agenda—a new system of world leadership."[181]

Goldsmith is all over the map in his protestations, while Prince Charles seems to be calling for level-headed deliberation or revolution. It is awfully unfortunate they are associated because while Goldsmith wants devolution, Prince Charles seems to have voiced some legitimate concerns about health and environmental safety. It's difficult to determine who is to blame for this conflation; maybe we can blame the tabloids. Everyone else does.

CONCLUSION

Any good story needs characters: protagonists, foils, and so forth. The people we met above are characters and some of the loudest voices on the sidelines of the debate over nanotechnology. With limited resources at their disposal, they speculate about what the engine might accomplish but lack the immediacy to meaningfully affect its direction. While their testimonies have helped inform and animate argument between the major players, it is the government actors who direct nanotechnology. In the next three chapters, we examine the rhetorical functions served by legislators, bureaucrats, and government institutions; initiatives in the United States and abroad; and industry and capital markets. These three major variables are currently powering the engine and driving the nanotechnology movement.

CHAPTER 3

GOVERNMENT ACTORS IN NANOTECHNOLOGY

T he role of government in science and technology decision making, according to the government itself, is to provide financing to support basic research in areas where the market cannot. In addition, officials have touted certain scientific efforts as important to reaching national goals. John F. Kennedy's space race was set in motion to compete with the Soviet Union's *Sputnik* program. Nanotechnology has been hyped by government officials as a means to create jobs at home. "Government officials have called nanotechnology the foundation for the *next industrial revolution*."[1]

Hyperbole produces expectations—and expectations can pull a program along, but it can also condemn it. "To skeptics, such hyperbole is a sure sign that the science of manipulating individual molecules will ultimately fail to meet expectations, as happened with industrial ceramics, superconductors, and other scientific innovations."[2] Nonetheless, hyperbole by government barkers of nanotechnology has been and continues to be extensive.

The government's role in the nanotechnology movement is broad. "The U.S. government is catalyzing change by promoting convergence and cooperation in engineering, biology, physics, chemical, electronic, and materials science disciplines by directing funds to new dedicated nanotech centers."[3] All these funds are expended to meet the various goals and missions articulated by government promoters. The problems may arise as studies and commentaries are published implicating nanotechnology with a host of troubles. William G. Schulz writes:

Many communications and public policy experts say it is essential that nanotech researchers and supporters get on board with efforts to dampen unquestioning enthusiasm for nanotechnology. They argue that without discussion of the potential pitfalls, future nanotechnology research could be subjected to such extreme pressures that funding is jeopardized and research progress is slowed, perhaps halted altogether in some cases.[4]

Unless some ventilating of the claims and counterclaims in the public arena occurs, the nanotechnology movement may be at risk. A first step might include tempering the hyperboles, though for right now that doesn't seem very likely.

INDIVIDUALS

There are many individuals who serve and have served formative roles in promoting nanotechnology, and they cannot all be detailed here. As such, representative examples were selected to cover the breadth of the participants. Two are legislators. Three are professionals; two are scientists, and the other is an economist, who became advisors and bureaucrats. The last is a bureaucratic spokesperson.

George Allen and Ron Wyden, the Senate

Senators George Allen (R-VA) and Ron Wyden (D-OR) championed the Twenty-First Century Nanotechnology R&D Act in Congress. When the leadership in the US Senate changed in 2002, the chairmanship of the Senate Committee of Science changed as well. Allen became chair and Wyden, the ranking Democratic member. Wyden first introduced the Twenty-First Century Nanotechnology R&D Bill, but it was under Allen's leadership that it became an act and was signed by President George W. Bush.

We begin with Senator Wyden. Wyden claims he was a friend of nano before it was fashionable: "There's no question today that if you walk into a senior center or a Safeway store or something, people are not exactly buzzing about nanotechnology. There's not much awareness of the potential for good-paying jobs that this field promises."[5] Wyden understood his role as promoter as well as steward. Back in 2002, at the first Senate nanotechnology hearings, Wyden began his testimony with his sense of the importance of federal-government participation in nanotechnology:

> My own judgment is that the nanotechnology revolution has the potential to change America on a scale equal to, if not greater than, the computer revolution.
> ... If the federal government fails to get behind nanotechnology now with organized, goal-oriented support, this nation runs the risk of falling behind others in the world which recognizes the potential of this discipline.[6]

While his rhetoric may appear excessive, it is in line with most promotional statements coming out of the federal government. Wyden views his role as a tech-evangelist of sorts. "A big part of what we have to do is use this committee as a sort of a bully pulpit to walk through technology questions, show the practical applications and relate them to the goals that often sound strange at the beginning, but could become part of people's daily conversation before too long."[7] Wyden articulated some concern about overhyping nanotechnology: "We're not going to over-hype it; we're going to say the government can't afford not to make these early investments." Of course, in the same interview, he crows: "There isn't a scientific field that can't be shown to be ripe for revolutionary changes in the way it operates. This is about a fundamental restructuring of a set of technologies with near limitless potential."[8]

According to Wyden, the NNI existed at the whim of this and future administrations, and a more permanent structure for nanotechnology research and development was needed, hence his support for the Twenty-First Century Nanotechnology R&D Act. As he attested, "the bill doesn't spend that much more than before."[9] It's mostly about organizing and institutionalizing the entire process.

Wyden promoted three main goals of the act:

First, a National Nanotechnology Research Program should be established to superintend long-term fundamental nanoscience and engineering research.... Second, the federal government should support nanoscience through a program of research grants, and also through the establishment of nanotechnology research centers.... Third, the government should create connections across its agencies to aid in the meshing of various nanotechnology offices.[10]

In addition, Wyden saw the bill as significant beyond itself: "A lot of what we can do here will have implications for the country beyond nanotechnology, and can be very positive. If we really get this right, we're going to have a [funding] model that is going to be of enormous benefit for the country in a variety of applications."[11]

When leadership on the Senate Committee changed, Senator Allen became the chairman. Senator Allen's rhetoric was more nationalistic in tone than Wyden's: "It is important for health care, for communications, for commerce, for manufacturing, for aeronautics, and indeed for our national security that the United States is a leader in this nanotechnology or nanoscience revolution.... The potential economic and societal benefits are far too great to be overlooked."[12]

Allen observed a real need for improvements in management and promotion so costs and benefits could be communicated to stakeholders and evaluated reasonably. Allen, in an April 2, 2004, speech noted that "researchers and supporters of nano-tech need to do a better job of making sure [science and engineering] advances are understood and judged based on their actual applications and merits rather than disregarded due to unbounded fears and misguided perceptions."[13] One of his concerns: "People in the general public are still susceptible to misinformation." Allen cited *Prey* as a common source for misguided myths relating to nanotechnology.[14]

Senator Allen was a strong voice supporting the bill and the creation of a nanotech preparedness center as well. "Getting the nanotech bill passed was Senator Allen's biggest priority in 2003," said Allen's spokesperson.[15] Allen, like Wyden, had an interest in nanotechnology for many reasons, especially as projects were cropping up in his own state. Allen's alma mater, the University of Virginia, recently broke ground on a $28 million nanotech center,[16] and "Virginia has a burgeoning nanotechnology industry."[17] In addition, Allen crows about Luna Innovation, Inc., in Blacksburg, which has initiated six new companies over the last five years in Virginia. Allen sees high technology, especially nanotechnology, as a opportunity to rejuvenate the tobacco belt, which runs right through his state. "[Luna] is pursuing a wide range of übertechnology, including manufacturing process control, next-generation cancer drug treatment, analytical instrumentation, novel nanomaterials, advanced petroleum monitoring systems, and wireless remote asset management."[18]

In March 2004 Allen announced he would be "organizing a new congressional caucus aimed at promoting nanotechnology and trying to educate other lawmakers and their constituents about the new and emerging industry."[19] He perceived the caucus as the industry's portal to Congress. His spokesperson, John Reid, admits it remains in the organizational stage.

For Allen and Wyden, and a handful of other government leaders, "The notion of nanotechnology is emblematic for U.S. economic competitiveness."[20] According to Wyden, "The United States will not *miss*, but will *mine* the opportunities of nanotechnology.... America can't afford to miss the nanotechnology revolution. The potential not just for direct revenue, but also for jobs and the growth of related industries, is too large."[21] The linkage between nanotechnology and employment remains a recurrent theme in governmental rhetoric. Wyden saw nanotechnology as a digital equalizer. "This is a technology that can close the digital divide, and give opportunity to kids who haven't had very many options."[22]

Wyden repeated claims that nanotechnology would be an employment engine over and over again. Citing an NSF projection, he "estimated that over the next decade, America will need 800,000 to 1 million nanotechnology workers."[23] In April 2003, he remarked, "There is no question that the NNI has a big price tag but this is an investment we can't afford to pass up."[24]

Wyden also promoted his own state as a potential center for nanotechnology research and development. He spent his time attracting more nanotechnology funding to Oregon and lobbied President Bush to name Oregon's Nanoscience and Microtechnologies Institute one of the nation's nanotech research centers under the new law.[25] In 2004 he managed to get the Senate Appropriations Committee to approve at least $103.5 million in funding for defense-related projects in Oregon. "This funding for Oregon nanotech and other cutting edge technologies will set the table for significant job creation in our state, while also making Oregon a key player in our national defense efforts," he said.[26]

Senator Wyden believes that "nanotechnology, eventually, is just going to be a matter of life and death. The medicines that are going to better target cancers [will] kill them with little or no damage to surrounding tissue." He worried about the release of studies implicating nanoparticles with health and other environmental concerns.

On April 2, 2004, Wyden said that

> he would like to see further study into the potential health and environmental effect of nanomaterials before lawmakers react with new laws. He pointed to the nanotech bill's creation of a nanotech *preparedness center*, which will examine environmental, health and societal issues relating to nanotechnology as a way to help policymakers sift through some of the concerns. "We're not ready yet to put in place a battery of new regulations in nanotech."[27]

For Wyden, this new center might help vet health and environmental questions to preclude preemptively threatening the nanotechnology movement with onerous regulation.

Recently, the president's Export Council subcommittee on export administration began scrutinizing nanoproducts that could be used to further or counter chemical and biological warfare and other weapons that would make them susceptible to export controls. Wyden wants to keep nanotechnology in front of Washington's agenda and to protect it from measures such as export controls.[28]

David Golston, the House Senate Committee chief of staff, explained in April 2004, "We need to study the problems and come up with solutions, not rhetorical arguments. This is a prime time to intervene and address emerging concerns. People are still willing to listen."[29]

Neal Lane, from the NSF to Clinton to Rice

Neal Lane is a man of principle. In 2004 he, along with more than sixty other prominent scientists, signed a statement condemning the Bush administration's policies toward science. Outspoken against the Office of Management and Budget's proposal to centralize peer review, he recently commented, "The peer review situation at the OMB is frightening on many levels. The integrity of information is going to be seriously undermined in a process that requires political approval."[30]

Dr. Lane served as director of the National Science Foundation (NSF) beginning in October 1993. He was sworn in as director of the Office of Science and Technology Policy (OSTP) in August 1998.

Testifying before the Senate Subcommittee on Veterans Administration, Housing and Urban Development, and Independent Agencies, he offered his projections about nanotechnology on May 7, 1998: "If I were asked for an area of science and engineering that will most likely produce the breakthroughs of tomorrow, I would point to nanoscale science and technology."[31] Lane was one of the first policy wonks to hyperbolize the nanotechnology landscape. He testified, "Some nanoscale scientists and engineers even envision nanomanufactured objects that could change their properties automatically or repair them. When you think about it, this idea is not so outlandish—DNA molecules in our own bodies can replicate themselves with incredibly small rates of error."[32]

He lauded the NSF in the same hearings: "NSF support over the years has allowed nanoscale science and engineering to go from the realm of science fiction to science fact."[33] He felt that

> through these and other investments, NSF's portfolio sets the stage for a twenty-first century research and education enterprise that continues to lead and shape the information revolution, addresses key national priorities in such areas as the environment and nanotechnology, improving teaching and learning at all levels of education, and commits itself to reaching out and advancing public understanding of science and technology.[34]

One element that Lane added to the nanotechnology portfolio was Knowledge and Distributed Intelligence (KDI). He helped direct the NSF investments toward turning the deluge of information out there into learning and progress. Lane advocated the interlinkages between a variety of disciplines in approaching challenging studies, especially as they related to nanoscience and nanotechnology. A variation of this belief is found in NSF support during his tenure for experimental computer and communication networks. "The use of high speed networks to enable distributed groups of scientists and engineers to work together as one—in almost real time—is transforming the way discoveries and innovations are occurring. Their use of these cutting edge experimental systems will also lead to more powerful communication tools for society."[35] This model would be used in the network or Web configuration of many of the NSF's major initiatives in nanotechnology.

Later that year, Neal Lane became the assistant to the president for science and technology. As director of the OSTP, he played a formidable role in crafting the NNI. Lane returned to Rice University in January 2001. He said that

> I have had the privilege to serve in the Clinton-Gore Administration for over seven years and now am excited to be coming home to Rice, where Joni and I have so many friends. I look forward to teaching again and working with Rice's outstanding students and faculty on physics research and science and technology policy.[36]

At Rice, Lane became a university professor in the Department of Physics and Astronomy, a special appointment entitling him to teach in any department in the university. This was the first such appointment at Rice. Lane served as a senior fellow at the Baker Institute, worked with the Clinton library on the administration's accomplishments in science and technology, and was instrumental in procuring the NSF Nanoscale Science and Engineering Center (NSEC) grant that established the national research gem known as the Center for Biological and Environmental Nanotechnology (CBEN) at Rice.

Mihail (Mike) Roco, NSET

President Clinton established the National Science and Technology Council (NSTC) by executive order on November 23, 1993. It acts as a *virtual* agency for science and technology to coordinate the diverse parts of the federal research and development enterprise. The NSTC is chaired by the president. Membership consists of the vice president, assistant to the president for science and technology, cabinet secretaries and agency heads with significant science and technology responsibilities, and other White House officials.

The NSTC created the Interagency Working Group on Nanoscience, Engineering, and Technology (IWGN) in 1998 with members from eight federal agencies interacting closely with the academic and industrial communities. The IWGN's charge was to access the potential of nanotechnology and to formulate a national research and development plan. The NSTC's subcommittee on Nanoscale Science, Engineering, and Technology (NSET) is chaired by Mihail C. Roco. He is the NSF's senior advisor for nanotechnology, and he was "instrumental in the launch of the National Nanotechnology Initiative (NNI) in January 2001."[37]

Roco "has been a professor at Caltech, Johns Hopkins, Tohuku University in Japan, Delft University in the Netherlands and is a member of the Swiss Academy of Engineering Sciences. He has been the author on 250 publications and has patents for 13 inventions."[38]

Roco is the United States' leading nanobooster. He claims to have begun thinking about nanoscience in the '80s, especially in terms of bringing together all the sciences and developing a science-based vision for the field. He is the "U.S. government's original voice on nanotech."[39] He "has the ears of Beltway power players, corporations, and academic researchers."[40] He never ceases proclamations about the future with nanotechnology. For example, he made this claim in an introduction to an NSF document on SEIN in 2001:

> Nanotechnology has the potential to realign society, change business, and affect economics at the structural level. New business models, design tools, and manufacturing strategies may emerge at price points much reduced and highly effi-

cient. Nanotechnology will touch all aspects of economics: wages, employment, purchasing, pricing, and capital, exchange rates, currencies, markets, supply and demand. Nanotechnology may well drive economic prosperity or at least be an enabling factor in shaping productivity and global competitiveness.[41]

"In ten to fifteen years, nanotechnology will enter our lives in a big way. Early payoffs [he predicts] will come in computing and pharmaceuticals, where powerful new tools and methods will benefit industries that already work at, or near, the molecular level."[42]

Roco claims three primary goals as chair of the NNI: "My main goal is to maintain a consistent vision and keep new ideas continuously coming to bring the benefits for nanotech sooner.... Secondly, it's crucial to maintain strong interaction between members of the NNI and to expand the membership further.... Thirdly, we must address social implications, including maintaining U.S. competitiveness."[43] On societal consideration, he wears two hats. First, he unabashedly proposed to increase the nanotechnology workforce. He claims "the U.S. alone will need 800,000 workers trained in some area of nanotech . . . to reach a $1 trillion global industry by 2010–2015." As such, he is pushing a five-year plan to give 50 percent of undergraduate and graduate students access to nanotech courses and labs, and "in the longer term, training has to start before high school."[44] Second, "part of the strategy is to have an open interaction with the public and to keep in mind the main purpose is to serve society as a whole while ensuring the U.S. has a competitive lead." He clarified, "The U.S. still has a lead position but it is a weak lead." He warned, "We provide lots of info to other countries, but don't get much back." With over thirty countries with their own versions of the NNI, "it's important to get more formal agreements to get a balanced exchange of information."[45]

Roco's advocacy goes back a decade, and he is one of the stronger advocates of aggressive research arguing that there are benefits foregone if we fail to accept some levels of risk associated with research and commercialization. "Halting research carries its own risks," Roco says, giving the example of diseases that are beginning to resist conventional antibiotics. "We don't want to find after twenty years that our drugs don't work and we don't know what to do."[46] Under Roco's leadership, "nanotech research funding has grown from $116 million in 1997 to $3.7 billion over the next four years."[47]

His hyperbolic rhetoric is worth noting because it taps into nationalistic fervor more often than not. For example, he says nanotechnology is the latest of three megatrends that have emerged in the past fifteen years—the other two being information technology and biotechnology.[48] In addition, he seems to revel in fear appeals and nationalistic rhetorical flourishes. And, he adds, just because the United States stops nanotech research does not mean our competitors—and enemies—will follow suit. "There is a risk," he says, "that someone else will develop these technologies and we won't know how to counter them."[49]

On the other hand, Roco has his detractors. Some feel he "appears to live in a world of pure science, where truth is truth only if it's peer-reviewed. Yet, the rational approach does not always work in an irrational world." This led Howard Lovy to write: "The science of nanotech needs to develop some social skills. That's why it's probably a good thing that [Clayton] Teague [director of the U.S. National Nanotechnology Coordination Office] is the new public face of the U.S. nanotech program, rather than the NNI architect Mike Roco."[50] Actually, it has been suggested that Teague is the front man, while Roco remains in charge. Regardless about how anyone may feel about Roco, there's seldom a conference on nanotechnology at which he is not a featured speaker, and his name seems to appear in every article in print that covers the promotional aspects of the NNI.

Thomas Kalil, Former Clinton Adviser

In the early days of nanotechnology research, funding from the federal government was sparse and inconsistent. Scientists and research labs were struggling to find an ear in the executive branch willing and able to champion the cause for increased funding. Enter Thomas Kalil, the man whom Neil Lane referred to as "the most influential person in the Clinton Administration in getting the National Nanotechnology Initiative off the ground." As Clinton's deputy assistant for technology and economic policy, Kalil served as the bridge that connected the laboratories to the Oval Office. From this position, Kalil helped spur and orchestrate funding for nanotechnology and the beginnings of the NNI.

Prior to working with the NNI, Kalil had an extensive background with other tech-related initiatives such as the Next Generation Internet Initiative and another called Information Technology for the Twenty-First Century. His experience with research initiatives led him to believe that a similar model would help guide nanotechnology during its infancy. Kalil explains further: "I thought the initiative model had been one way in which I'd been successful in getting high-level support, interest and visibility for a particular area of research. I was confident enough that nanoscale science was such a broad area that it was appropriate for the government to emphasize that in its investment strategy."[51] It also didn't hurt that Kalil possessed a common characteristic among government men in nanotechnology at the time: a strong tendency toward controlled hyperbole and an appreciation for the federal government's unique role in encouraging technology. "Long term, nanotech can be as significant as the steam engine, the transistor, and the Internet. There is a critical role for government in areas of science and technology that are risky, long term, and initially difficult to justify to shareholders."[52]

Kalil's connections across a variety of institutions really granted him access to the wide world of applied nanoscience and helped it get off the ground. His knowledge of the field was not limited to any particular aspect. Kalil made all the right contacts at all the right times. He "communicated directly with the leaders of science agencies, and told them that [he] would fight for any increase they proposed, and also push for overall increases in their budget."[53] In 1997 Kalil contacted Roco, then program director of the National Science Foundation. "After six months of back-and-forth between them, Kalil helped create and lead interagency working groups devoted to nanotechnology."[54] Kalil's ability to balance Clinton's political desire to see nanotechnology funding increase against the economic benefits of the technology kept the issue fairly nonpartisan. "That was very important for getting Office of Management and Budget (OMB) to believe this was a very well-thought-through proposal, as opposed to sort of the whim of some political appointee in the White House,"[55] Kalil said. The pragmatic approach embodied by Kalil explains his ability to attract supporters from the other side of the aisle, such as former Speaker of the House Republican Newt Gingrich.

Success quickly mounted for Kalil. His proposal for the NNI passed an OMB evaluation. Next, the President's Council of Advisors on Science and Technology gave it a thumbs-up. By January 2000, Clinton had unveiled the plans to the public, and that following October, Congress approved the $465 million package.

Prior to the official start of the program, Kalil was hard at work behind the scenes trying to break down barriers between the scientists and the bureaucrats, the lab and the Oval Office. "In August of 2000, the White House boosted Kalil's working group to the level of subcommittee of the National Science and Technology Council. The group's mission: to implement the NNI, engage the public and create the National Nanotechnology Coordinating Office to provide information on nano-related topics, such as unintended environmental effects."[56]

While Kalil's term in the executive branch coincided with Clinton's departure in 2000, he still remains active in the field and concerned over the future of his baby, the NNI.[57] He is currently the Special Assistant to the Chancellor for Science and Technology at UC-Berkeley, where he is charged with "developing major new multi-disciplinary research and education initiatives at the intersection of information technology, nanotechnology, microsystems, and biology."[58] The interdisciplinary nature of his work in pushing for the NNI has come in handy for him at Berkeley as well. "Solving many of these problems will require the tools, technology, and insights from multiple disciplines," Kalil says.[59] His departure from the federal government may have been untimely as nanotechnology has been a staple of the Bush agenda, but we have not heard the last from Thomas Kalil on nanotechnology and the government's nanopolicy.

Philip Bond, Department of Commerce

Why include a departmental undersecretary? It seems to be obvious given the frequency that Philip Bond speaks to public gatherings about the promises of nanotechnology. With NIST and the Advanced Technology Program (ATP) under commerce, his enthusiasm for hyperbole is forgivable.

In Realis distributed an investor's guide. In it, it touted nanotechnology. In Realis claimed "for a recession-plagued economy, nanotech material advances offer opportunities to reduce factor costs and contribute toward renewing growth in profitability, and for a booming economy; it offers an attractive direction for growth investment."[60] This is the commerce department mantra.

"This technology is coming, and it won't be stopped," said Philip Bond.[61] "The nanotechnology race is on around the world,"[62] he portends. He sees a growth driver in nanotechnology. "Across the world, nations are investing in the research, development, technology, infrastructure, education, and training that will enable them to compete and win in a technology-driven global economy.... In today's highly competitive global economy, technology leadership delayed is technology leadership denied."[63]

"Bond promotes nanotechnology publicly as a means for expanding the national economy."[64] Given the current weak economic recovery, this argument rests well with investors, industrialists, and the general public. At a meeting in early April 2004 in Washington, he told this anecdote: "Perhaps most striking to me is the news this week that nanomanufacturing is arriving in Danville, Virginia." He was referring to an earlier remark made by Senator Allen at the same meeting. He continued that if nanotechnology has made it to Danville, "it is fair to say that nanotechnology—and nanotechnology-related jobs and economic growth—are no longer science fiction but economic reality."[65]

Philip Bond was the front man for commerce on nanotechnology. Previously, he was Hewlett-Packard's director of federal public policy. Forbes described him as "the most outspoken advocate of nanotechnology in Washington, D.C."[66] He was brought onboard by President Bush in 2001. He served as the principal advisor to Commerce Secretary Don Evans on science and technology policy, which includes import/export controls and nanotechnology. There was seldom a conference on commercialization or investment that did not include Mr. Bond among its featured speakers. He said:

> The NNI will bring three challenges to businesses. First, business must engage with research institutions to reap the benefits of nanotechnology innovations. Second, business must guard against an environment of fear of nanotechnology by engaging and addressing the legitimate societal and ethical issues of nano- technology. Lastly, the technology from NNI must be able to excite and inspire the next generation of scientists and engineers in the United States.[67]

For Bond, this hefty agenda is not implausible if business can marshal nano- technology through the misunderstandings promulgated by uninformed and mis- guided critics and miscreants.

When asked what he wished to accomplish for commerce, he responded, "First, maintaining the consensus around nanotech.... Secondly, engaging the scientists, particularly in this societal and ethical discussion. And, thirdly, really using nano where I can fire up the next generation."[68] He advocated educational outreach from the beginning of a child's education. "I think we have to find ways to excite kids [about science and technology] probably as early as elementary school adding that girls especially should be encouraged to follow career paths in those fields. We have to keep them going through the process—we're losing half the talent pool." In the same interview with FOXNews, he noted: "Nanotechnology—the art of manipulating materials on an atomic or molecular scale to build microscopic devices, such as robots—could be the next big thing that gets kids' attention. I'm hopeful that it's going to fire up kids."[69]

His tale usually portrays nanotechnology as evolutionary and revolutionary. Improved golf balls would be an evolutionary impact, while self-assembled nanowires for microprocessors would be a revolutionary impact. "Some knowledge of what nano means will creep into the public lexicon. You have to move the whole country towards a more innovation mindset. You have to understand that, yes, change is going to accelerate as knowledge accumulation is accelerating."[70] He feels the benefits lost from failing to aggressively promote nanotechnology outweigh precautionary concerns over risks. He suggested that the best way to deal with environmental issues is straightforwardly but worries that "our companies could be hung up waiting for the public debate to play out. In the meantime, their businesses are going down the tubes."[71]

At the April 2004 meeting, he cited Eva Oberdörster's fish study (see chapter 9) that alarmed many gathered in DC for the event. While there was a general feeling the media had overstated the results, Bond offered a spin of which an election warroom chief would have been proud: "The research is a good example of the government's commitment to ensuring public health by studying potential negative implications of nanotechnology. This study proves that our system of checks and balances is working to protect public health and the environment."[72] Of course, the study also suggested that carbon nanotubes were responsible for retarding fish.

Bond also felt that it's absolutely unnecessary to adopt more regulations of nanoproducts because all federal rules that control the safety of drugs, products, and materials should equally apply to nanoproducts.[73] While there is serious concern whether the current regime might be sufficient, Bond's optimism emanated.

His advice to the nanotechnology research and business community was to engage the debate aggressively: "You immediately establish your credibility because you've acknowledged the concern and have thought about it. There are always fringe groups who raise money based on a fringe message and the moment you engage them on the concept that any part of that is legitimate, you drive them further to the fringe."[74] He offered this ominous sound bite: "The body politic is susceptible to the virus of fear."[75] His orientation of the SEIN debate couldn't be much clearer.

"The time for action is now," said Bond. "We need your help to tell the positive story of nanotechnology and to prevent the flame of fear."[76] Though researchers may pursue their work with more objectivity, Bond seemed so committed to promoting nanotechnology that he was unable to contemplate that research results about its social effect could be anything but positive.

Bond was very cognizant about the level of misinformation in current thinking. "Fear is the product of ignorance."[77] While some warn that "aggressively battling public perception would invite counterattack in the form of louder protests,"[78] in an interview with Lovy, Bond attested that is not his philosophy. "I don't really buy into the old Hollywood maxim, 'Just spell my name right and you can say what you want.' I think that kind of publicity does matter and that's why today I wanted to urge people to be very active and aggressive at the local level when they see a premature report or a misleading report."[79] Bond seemed to be advocating perception management—a public relations function—and this has led to much unease and consternation with the SEIN research community (see chapter 10).

Bond, at a 2003 NanoBusiness Alliance conference, "showed products like Nano-Tex's wrinkle-free khakis to CNBC viewers before his keynote address."[80] In his address, he remarked that, "Nanotechnology has reached a tipping point."[81] Now is the moment when nanotechnology takes off or moves offshore. At a conference hosted by Swiss RE, he nearly silenced the international audience when he bluntly told them the United States would be in the lead.

Bond left commerce in 2005 for a job with MonsterWorldwide.com, where he is in charge of government communication. As of the printing deadline for this book, the Bush administration had not named his successor. Bond remains active in the industry through an advisory affiliation with NanoDynamics in Buffalo, New York.

The next promoter is an institution and within it are subsets of bureaucracies that encourage nanotechnology in their own fashion. While Wyden, Allen, Lane, Roco, Kalil, and Bond were or are outspoken pitchmen for nanotechnology, they all work or worked for the US federal government.

Recall that according to Kalil, "Long-term, nanotech can be as significant as the steam engine, the transistor and the Internet. There is a critical role for government in areas of science and technology that are risky, long-term and initially difficult to justify to shareholders."[82] The role of government in science and technology policy is changing, especially as it applies to the physical sciences. While basic research policy has changed very little, applied research, in terms of encouraging applications and early commercialization, has become part of American science and technology policy. As Kalil suggests, some areas that just don't quite fit into the business management paradigm, like nanotechnology, are special cases.

Early government support of applied science stimulates commercialization. According to Lux Capital's Josh Wolfe, editor of the *Nanotech Report*, "in nanotechnology there are early signs of a promising pattern. We know from past experience that government funding for science and technology eventually creates a pattern wherein new ventures spring up and some of them prosper."[83] Wolfe and others have begun to track companies that started up and budded from government-grant-supported laboratories in academia. Often, these start-ups are headed by the same scientific researchers who received the seed grants in the first place.

What follows is a discussion of some of the lead institutions and programs promoting basic and applied research in the physical sciences, specifically nano-technology. Following that brief review, how nanotechnology is being pitched and the effect of that strategy on policy making will be detailed.

GOVERNMENT SCIENCE PROMOTION

Why does government promote physical science? For decades, the bulk of funding from the federal government was associated with the National Institutes of Health (NIH). Quite simply, this research enabled improvements in health and well-being. However, government support for the physical sciences, especially when apparently unrelated to health, has traditionally been underfunded except during wartime (e.g., the Manhattan Project) or after a major initiative was answered (e.g., Kennedy's Space Initiative).

The US government must make a strong case to authorize and appropriate major investments in science. The public may support science as a concept, but when asked to fund it the public's reaction is very different. For example, an unpublished study found that less than 30 percent of respondents would vote for a politician who supported federal funding of nanotechnology. Oddly enough, another 30 percent would vote for the politician for the very same reason.

In the case of nanotechnology, two major arguments were advanced. First, the time to move research to commercialization exceeded the period of time private markets are willing to wait for equitable returns on investment. Second, tapping national-istic and exceptionalistic fervors, nanotechnology became associated with national economic competitiveness, employment goals, and national pride. Secondary arguments involving support for education, from "K to gray," national security, especially in the war on terrorism, and tertiary arguments from food safety to renewable energy supplies were also made. However, along the way the messages were sidetracked by both well-meaning and self-serving stakeholders. America was not prepared for an initiative that promised both revolutionary and evolutionary technologies on the level of the unseen.

National Science Foundation

Vannevar Bush has been credited with conceptualizing a national science program. In 1945, at the request of President Franklin D. Roosevelt, he recommended that a foundation be established by Congress to serve as a focal point for the federal government's support and encouragement of research and education in science and technology and for the development of a national science policy. Created by Congress in the National Science Foundation Act of 1950 and signed into law by President Harry S. Truman, the National Science Foundation was created to develop a broad scientific program for the nation.[84]

The NSF faced tumultuous times in the beginning. Immediately after the NSF bill was made law, North Korea invaded South Korea. US forces were committed to a peacekeeping mission that was expensive and distracted from the new science initiative.

At the same time, William Golden, a Wall Street investment banker, lobbied the administration to create the position of a presidential science advisor. However, that proposal was set back, and the position was relegated to the status of a mere advisory committee to the Office of Defense Mobilization within the executive.[85] The focus linked issues of science to national security, not unlike today.

The first formal appropriation for the NSF occurred in May 1951 when $13.5 million was authorized. However, in October Congress approved only $3.5 million, citing the Korean conflict as a reason for pared-back domestic spending.[86]

The NSF remained a relatively underfunded and unrecognizable feature of policy making until the *Sputniks* were launched by the Soviet Union. Congress, reacting to the Soviet buildup in scientific knowledge and personnel, responded by appropriating $136 million in fiscal year 1959, a big increase from $49.75 million from the previous fiscal year.[87] Notably, Congress saw the NSF as a source for basic research that might help the United States win the cold war.[88]

In recent times, the NSF has seen a change both in its prioritization and its levels of appropriations. Before turning to its political history, an examination of its goals, both stated and implicit, and its structure and organizational history is necessary to understand its role. In essence, the NSF has a broad mission of providing resources and infrastructure to science and engineering and funding basic research projects across and between disciplines.[89] The NSF "provides support for investigator-initiated, merit-reviewed, competitively selected awards, state-of-the-art tools, instrumentation and facilities."[90] It also has an explicit goal in maintaining a healthy supply of scientists, engineers, and educators. In total, the NSF has nearly twelve hundred employees and is divided into seven directorates. It is an independent agency in the executive branch, under the direction of a presidentially appointed director and a National Science Board composed of twenty-four scientists, engineers, and university and industry officials who specialize in research and education. This twenty-four-member board and the director are responsible for all decisions and policy made by the NSF.[91]

The NSF is remarkably efficient. Around 95 percent of the agency's total budget goes directly to support its research and education efforts. Less than 5 percent of its budget is spent on administrative and personnel costs. Before the fiscal year 2004's budget requests, the NSF was the only agency in the entire federal government to "receive *green lights* for its implementation of the President's management agenda, garnering high marks for both its financial management and E-government."[92]

The NSF is the lynchpin to most non-life-science basic research at colleges and universities in the United States. Although the NSF is responsible for less than 4 percent of the total research and development appropriations, it funds or contributes to approximately 50 percent of all basic research in higher education that is not related to medicine, which is the province of the NIH.[93] In different terms, the NSF funds nearly twenty thousand research and education projects through grants, contracts, and other agreements with over two thousand colleges, universities, and other partnerships. In excess of two hundred thousand people are directly involved in NSF-grant-related research and education programs. "In FY 2004, these include [approximately] 43,000 senior researchers and other professionals, 69,000 postdoctoral, graduate and undergraduate students, 15,000 K–12 students and 87,000 K–12 teachers."[94]

Through the 1990s until now, the NSF has benefited from gradual increases in funding except between FY 1995 and FY 1996.[95] From FY 1989 to FY 1998, the total appropriations for the NSF almost doubled.[96] Even accounting for inflation, it was an increase of 44.7 percent.[97] Beginning with FY 1999, Congress provided the NSF with increases greater than any other scientific research agency except the NIH.[98] In FY 2001 Congress passed the single largest budget increase for the NSF, an increase of roughly 13 percent.[99] The congressional fervor culminated in the passage of H.R. 4664, the NSF Authorization Act of 2002, a bill that sought to double the NSF's budget over five years.[100] The act, also known as P.L. 107–368, had bipartisan support from the ranking Democrat and the chairman of the Senate VA, HUD, the Independent Agencies Subcommittee, Christopher Bond (R-MO), and Barbara Mikulski (D-MD). James Walsh (R-NY), the chair of the House VA, HUD, and Independent Agencies Subcommittee also named the NSF his top priority within the VA-HUD bill.[101]

FY 2005 is an interesting case study in nanotechnology, especially in relation to the NSF. The 2005 budget proposed the highest amount ever requested for an NSF operating budget. The NSF wanted to strengthen its main research initiatives, one of which was nanotechnology.[102] In touting the NSF for increased funding, maintaining economic strength and homeland-security issues were raised as deliverables that the NSF could provide.[103] In relation to nanotechnology, the NSF would receive a 20 percent increase over FY 2004 for the National Nanotechnology Initiative, which would be funded at $305 million.

The Office of Management and Budget's report on the NSF's budget used nanotechnology as the example of "NSF investment in areas that will link discovery to innovation and learning, to maximize the likely benefit to society."[104]

This funding for nanotechnological research was a 104 percent increase, from $150 million to $305 million as requested in 2005.[105] Roco claims the final figure was $338 million.

Nonetheless, the NSF has been one of the few agencies that received increased funding since 9/11. The requested overall operating budget for the NSF was $5.745 billion, a 3 percent increase for the agency.[106]

However, unlike specified funding for nanotechnology, budgets for the NSF have not received the proposed increases as envisioned in the NSF Authorization Act of 2002. As Representative Eddie Johnson (D-TX) noted, "The President has sent us two consecutive budgets that fall short of reaching that goal [doubling the NSF's budget]. With this budget submission we stand $3 billion below the doubling path."[107] Even with this situation, nanotechnology still received an increase within the NSF's budget. Nanoscale Science and Engineering will see an increase of 20.3 percent, to $51.6 million.[108] The question becomes: how did nanotechnology become a protected category within the NSF's budget?

In FY 2001 the Clinton administration created the NNI. In its first year, the act sought to increase government investment in nanotechnology and research by $227 million. The initiative would require doubling of federal spending on nanotechnology within five years.[109]

The NSF was there at the beginning of the NNI. The NSF was not satisfied to be second fiddle behind the NIH in funding within the federal government. "From 1970 through 2000, federal support for life sciences more than tripled in constant dollars, whereas money for the physical sciences and engineering had by comparison remained flat."[110] When the NNI was proposed, its leaders saw an opportunity to seize a whole new level of funding. Rita Colwell, former head of the NSF, had already been posturing for more money in comparison to the NIH. In 2000, before the House Science Committee, Colwell was asked if there were any *unmet needs* in the NSF's proposed budget of $4.6 billion.[111] Her response: "I'd like to bring the size of our grants at least to the level of the average [NIH] grant." That would have quadrupled the average NSF grant allocation.[112] She further indicated that if the grant size were to be increased, scientists would increase their efficiency by ameliorating the time spent submitting applications and reviewing said proposals.[113] Regarding a time frame in which she would like to see the goal achieved, Colwell offered a four-year window.[114] Colwell had an ally in Senator Ron Wyden (D-OR).

In 2002 Senator Wyden and others noted that the physical sciences should not be left by the wayside, while the biological sciences receive unequal budgetary treatment.[115] By the year 2004, the average grant size had increased from $114,000 per year in 2001 to an estimated $139,000 per year.[116] This was still far short of even the average grant size for a participant in an NIH grant: $291,502.[117] However, it is clear from looking at the numbers that the NSF was able to alleviate some of its NIH envy with its placement within the NNI budget. NIH envy became one of the arguments for increasing the NSF's budget. The NNI would fortuitously be the program around which the lobby effort was coordinated.

From the beginning, the NSF struck its claim for NNI-related funding. In FY 2001, the NSF had almost two times as much funding as any of the other executive agencies in the NNI budget, at $217 million.[118] In FY 2005, the NNI would be funded holistically at nearly $1 billion, with the NSF "continuing to have the largest share of federal nanotechnology funding."[119] Unlike other agencies who only lamented about the woes of non-biological-science funding, the NSF actively pursued and changed its agenda to meet the goals of researching nanotechnology. The NSF created a massive theme area to deal with it: nanoscale science and engineering.

Interestingly enough, the NSF may now be poised in a more competitive position than the NIH in the coming years, at least in terms of new infusions of government money. Initially, the size of the average new grant request is set to fall in 2005.[120] In the FY 2005 budget outline, there was a proposed budget for the NIH through FY 2009. In FY 2006, the NIH budget would decline to $28.2 billion and is set to increase only to $28.7 billion in FY 2009. If one adjusts for inflation, the projected NIH budget would actually put the NIH's 2009 budget 6 percent below that of FY 2005.[121] This is in spite of the fact that "the NIH engages in research that offers the hope of breakthrough advances in the treatment of disease... it has long been popular both in Congress and with presidential administrations of both parties, and benefits from strong support not only in the scientific community but also from an extensive network of disease-oriented patient advocacy groups."[122] Although it is unlikely that the NSF will surpass the bottom-line budget of the NIH, its percent increases may surpass that of the NIH. Even if the NSF faces budgetary cuts in FY 2006,[123] it is unlikely it will dent its major research initiative, the NNI.

There are other budgetary hurdles for the NSF, and that is "one of the difficulties always faced by Congressional appropriators in trying to increase funding for the NSF. It receives its funding from the same appropriations bill that funds other research agencies."[124] However, a glance at the FY 2005 budget indicates that Congress may be able to set aside some of the competitive concerns and increase funding for some initiatives under the NSF, principally the NNI.

Pitting agency against agency in a stand for equity may not be enough to get money from the federal government. The NSF went a step further. It sold itself as the organization that could handle the massive interagency coordination to make the NNI a success.[125] The NSF was positioned well at the time of the NNI's inception. Roco made the case himself: "NSF has been a pioneer in the field . . . and is currently making the largest investment among federal agencies in fostering the development of nanoscale science and engineering."[126] In addition, the first nanotechnology-related program funded by the NSF in 1991 was followed by the Nanofabrication User Network in 1994.[127] At the time of the NNI's proposal, the NSF had over six hundred fifty projects, twelve centers, and twenty-seven hundred faculty members and students working on nanotechnology.[128] Thus, it had a strong claim to the technical expertise to administer NNI.

Some critics questioned the competence of the NSF to handle the parameters of the NNI. According to the Congressional Research Service, "Senator Bond indicated he had concerns about whether the NSF would be able to handle such a rapid infusion of money if the proposed increase was approved."[129] Others in Congress questioned whether or not the NSF had the staffing and personnel required to handle the task.[130] These considerations never translated into sufficiently serious reservations to threaten the NSF's leadership. The NSF received the lion's share of the NNI's budget; it remains the chief component of its basic operation functions.

Since the NNI began, the NSF has asserted a myriad reasons why it in coordination with (and sometimes without) the NNI should receive an increased funding level. One of the strongest was to strap its coordinating expertise to the linkage between the NNI and economic growth and prosperity and a host of other net benefits. Newt Gingrich was an early supporter of nanotechnology. Accordingly, he attested nanotechnology would drive a plethora of national goals. "Approaches to health, the environment, productivity and national security will all be profoundly shaped by this emerging revolution in knowledge."[131] One of the most salient arguments had to do with the association between the NNI and economic growth and security, especially in a postindustrial economy when manufacturing jobs were moving abroad and white-collar employment rates were mostly flat. John Podesta, the former Clinton chief of staff, labeled the NSF as "a principal foundation for U.S. economic strength."[132] Rita Colwell added, as did others, that global competitiveness would be a key factor in increasing funding to the NSF and the NNI because a "commitment to [these] disciplines needs to remain constant if the U.S. is to maintain its international science superiority."[133]

In this setting, international superiority in science is code for international economic leadership. Irrespective of the reasons articulated by the NSF, the simple fact remains that the NSF budget continues to balloon, especially in terms of the NNI. No matter what tactic deployed by the NSF, it remains clear that at worst it may face an overall budget cut; at best, it will continue to reign as the most important part of the NNI and soar above other highly lauded departments and agencies in terms of its funding from the federal government.

While executive departments impacted by the NNI will be reviewed later, the Department of Commerce is a major independent player in the NNI. Commerce has the National Institute of Standards and Technology, which managed the Advanced Technology Program. ATP, always controversial, helped fund the transfer of basic research to applied research and the early steps toward commercialization.

National Institute of Standards and Technology

NIST (originally the National Bureau of Standards) is in the Department of Commerce. The bureau was founded in 1901 to work specifically with industry, the only such agency with that authority.[134] It provided the venue to establish nationwide standards for products and processes. It also worked with industry to determine safety protocols and appropriate test environments for products and procedures.[135] The bureau was renamed in 1988 under the Omnibus Trade and Competitiveness Act of 1988, the same act that christened the arrival of the Advanced Technology Program. Recently, the budget of NIST has faced the same woes of that of its daughter program.

NIST funds a host of projects, but its budget has gone through a tumultuous period. In the FY 2005 budget, NIST saw a 26 percent increase in funding to its laboratories, receiving $417 million, on the heels of a discovery of a new state of matter.[136] However, on the whole, its funding has fallen by nearly 15 percent to $521.5 million since then.

Specifically in relation to nanotechnology, FY 2002 NIST funding for nanotechnology-related studies neared $40 million. For FY 2004, the president requested an increase of $5.2 million to bolster nanotechnology support within NIST.[137] In general, NIST focuses on nanomagnetics, "nanocharacterization research to produce standards and tools for visualization and characterization at the nanoscale, which are in high demand by a broad base of U.S. industries," and research to yield fundamental measurements needed to establish standards and allow future research into the field.[138] NIST anticipated stronger ties to industries, commerce, and university systems in order to achieve the best possible research. In the FY 2005 request, NIST expected half of its funds would be allocated to external organizations to achieve the research goals outlined in its request, which would circumvent costly one-time expenditures within the NIST program itself.[139]

NIST would like to dedicate the Advanced Measurement Laboratory to nanotechnology experiments to generate standards for industry. In essence, "the new facility would allow NIST to provide the sophisticated measurements and standards needed by U.S. industry and the scientific community for key twenty-first century technologies such as nanotechnology."[140] NIST also plans a litany of new research projects for the center, from nanoscale measurement tools to measuring biological molecules.[141] Another research and development user facility is the NIST Center for Neutron Research (NCNR). The NCNR uses beams of neutrons, which are nondestructive, highly penetrating probes integral to studying the "structure, properties, and dynamics of materials of many types—from proteins to nanocomposite coatings."[142] Because of the unique nature of neutrons, they make an exemplary ruler to establish uniform measurement standards.

Recent hearings on the NIST budget lauded the value of NIST and a hyped case for funding. Thomas Cellucci, president of Zyvex, testified NIST "is responsible for developing the measurements, standards, and data critical to private industry's development of products for a potential market that is estimated to exceed a trillion dollars in the next decade."[143] The nature of NIST's work has been important in fueling economic growth, and its applications were outlined by Cellucci. In discussing the track record of NIST, he noted the rather astonishing statistic that NIST's weights and measurements systems underpin $5 trillion in sales, which is half of the US economy.[144] He added these standards are especially critical given that nanotechnology is an emerging field void of standards at the present moment. He grounded the empirical example of in vitro devices (IVD), which but for NIST would not have met the new EU regulations, and the United States would have lost 60 percent of a $7 billion market.[145]

NIST seems important enough, but Cellucci, a master of entrepreneurial spin, stepped forward with correlative claims. "We're not only at war with terrorism, we are in the midst of a significant world-wide battle for technical prowess to sustain and increase our technological leadership in the world—the greatest economic battle of our lifetime."[146] The linkage between NNI and economic growth is hardly unique to the NSF, it seems. The spin and exaggeration come a few sentences later: "Anything but increasing NIST funding is surrendering our economic prosperity and giving up on our promise to our children—a promise for a higher quality of life."[147]

Associated with commerce and NIST is the Advanced Technology Program. The ATP has always been the hot potato of the appropriation process for two reasons: first, it looks like welfare for business and industry, and, second, it seems to place the US government in a game that supports one firm over another such that an infusion of grant money is seen as economic nepotism.

Advanced Technology Program

While the federal government has tended to shy away from directly participating in the marketplace, it invested into moving products into the commercial market through the Advanced Technology Program to bridge the gap between the research lab and the marketplace, stimulating prosperity through innovation. Through partnerships with the private sector, the ATP's early-stage investment accelerated the development of innovative technologies that promise significant commercial payoffs and widespread benefits for the nation. As part of the highly regarded NIST, the ATP changed the way industry approaches R&D, providing a mechanism for industry to extend its technological reach and push the envelope of what can be attempted.

The ATP was created in 1988 as part of the Omnibus Trade and Competitiveness Act under the direction of NIST. The creation of both ATP and NIST was actually done through a rider to the otherwise cumbersome act passed under the

Reagan administration. The goal of ATP was to focus on "fostering the development in the private sector of innovative, high-risk enabling technologies."[148] During the early years, fiscal years 1990 through 1993, the ATP functioned mostly as an ad hoc agency within NIST. It became a line item in the NIST budget only in FY 1994.

In theory, the ATP was visionary because it sought explicitly to "accelerate the development of high-risk technologies that promise significant commercial pay-offs and widespread benefits for the economy."[149] Thus, the explicit goal of the ATP was to increase economic growth through the development of enabling technologies. The awards given by ATP were determined by competition and based on three main criteria: technological ideas, potential economic benefits to the nation, and the strength of the plan for eventual commercialization of the results.[150] However, the recipient of the grant had to pay at least half of the projected costs of the project, including all the indirect costs associated with the grant work. Between 1990 and 2002, the ATP received 5,451 proposals and funded 642 at nearly $2 billion,[151] with the private sector matching approximately at that level at $1.9 billion.[152]

With regard to nanotechnology, since 1992, the ATP made thirty-nine awards, totaling nearly $142.5 million, to US industries wishing to invest in nano-technology.[153] Twenty-three awards, or almost 60 percent, were made since the beginning of FY 2000. Unsurprisingly, the number of nanotechnology-related projects as a percent of the total number of proposals submitted to the ATP was also increasing.[154] The ATP classified nanotechnological projects under the following theme areas: "nanostructured materials, nanofabrication processes and tools, nanobiotechnology, nanoelectronics, and nanometrology."[155]

The political history of the ATP is an interesting case study within science and politics. From its inception, the ATP has faced immense political opposition, especially from the Republican Party. In FY 1995, the ATP was funded at $430.6 million, which was only a slight cut from the proposed $451 million. Interestingly enough, this would be the last time that Democrats were the majority in the House of Representatives. At that time, the incoming House Science Committee chairperson Robert S. Walker (R-PA) favored the abolition of the ATP. Adjusting to meet the rhetoric, President Clinton further scaled down the funding to $340.7 million.[156] Following the Republican takeover of the Congress, the House and Senate called for not only the elimination of the ATP but also the entire Department of Commerce. While the executive department survived, the Clinton administration asked for $490.9 million for ATP, but received only $221 million for the year, and this funding came in the middle of the fiscal year in an omnibus bill.[157] FY 1997 was also a treacherous year for the ATP. The House and Senate wanted simply to fund it at a level that would end all current commitments that the ATP had begun. Fortunately, the ATP survived and was actually funded at a modest $225 million. In the FY 1998, Congressman Hal Rogers (R-KY) threatened to zero out the ATP, but it survived again with $192.5 million for that year. In FY

1999, the ATP's $259.9 million request was actually cut at the signing of a rescission bill to $197.5 million. In FY 2000, the House committee once again attempted to zero it out, but it was ultimately funded at $142.6 million for FY 2000.

From FY 2000 to the present, there has been a relentless attempt by the Bush administration to eliminate the ATP. Arguing for the administration, the FY 2004 budget noted "the administration believes that other federal R&D programs have a clearer federal role and are of higher priority.... Large shares of ATP funding have gone to major corporations and projects often have been similar to those being carried out by firms not receiving such subsidies."[158]

However, even though the Bush administration had tirelessly campaigned for the elimination of the ATP, in FY 2004 it was funded at $179.2 million, including over $60 million for new R&D ventures.[159] In FY 2005, it is once again on the chopping block. While once only rhetoric within subcommittee hearings, now the ATP was targeted for elimination by the president himself. This disdain for the ATP is understandable. The "program was conceived as a pseudo-venture capital (VC) firm for the federal government, focusing on emerging technologies that were too risky for private investors."[160] The arguments were the following: The rate of return for some emerging technologies is so far from the point of initial investment that the market cannot sustain their activities and be responsible to investors and shareholders. Where in the past, wealthy angels would drop fortunes into start-ups, it occurs too infrequently to be taken seriously as a variable in science and technology policy making. After the dot-com bubble burst, venture capitalists, especially some of the major VC firms, have grown overly conservative, though some exceptions are noted in chapter 7, to be counted on to usher in the new economy riding the wave of emerging technologies, like nanotechnology, which if you believe the rhetoric may be the *mother* of them all.

In response, the Bush administration believed that the private sector could do a better job at this research than the federal government. Moreover, the administration argued that in times of war and fiscal responsibility, the government simply did not have the flexibility to fund such initiatives.[161]

The ATP, as a percentage of the total federal budget, does not trade off in any significant way with another program within NIST or beyond. Moreover, Joe Lichtenhan, the president of Hybrid Plastics, noted "it's one of the few programs that really tried to transition technology to the nonmilitary sector."[162] He continued testifying that his company could not have completed its research into Polyhedral Oligomeric Sil Sesquioxanes (POSS) nanomaterials without the ATP because "we talked with numerous VCs and corporations and there was way too much risk for them to get involved."[163] Nevertheless, it has fewer and fewer advocates.

In the past, the ATP was funded largely by the concerted efforts of Senator Fritz Hollings (D-SC) and others because the ATP is no longer an explicit category in the federal budget. It seemed likely that as long as Bush or the Republican Party were in power, the ATP remained a tenuous program, one whereby grant recipients can never be sure whether they will be funded from year to year. With Hollings retiring from the Senate, it is unclear who will rise to take on the mantle, if anyone at all.

In 2004 a failed amendment to restore funding for the ATP on the House side by Representative Mike Honda (D-CA), a nanotechnology advocate in the House, motivated him to substitute HR 3598, the Manufacturing Technology Competitiveness Act (MTCA). Another nanotechnology supporter, Sherwood Boehlert (R-NY), expressed concern that 3598 would hamper any reauthorization of funding for the ATP and viewed it as a dangerous trade-off.

While the MTCA would be a one-time appropriation ($750 million) that required matching from the private sector of one-third the amount, it confronted a crowded calendar and the lack of support from the Republican leadership in Congress. The Nanomanufacturing Investment Partnership it would have created is an idea that may need to be revisited. In the meantime, ATP funding remained the only real source for support to help bring nanotechnology advances to the market.[164] Reportedly, Honda planned to reintroduce the legislation in the 109th Congress.

The political climate for the ATP has never been more precarious. The House Appropriations Committee zeroed out funding for the ATP for FY 2005. In 2006 the budget provides no funding for it. At a recent meeting of an NAS committee to evaluate the NNI, it became evident that the Small Business Innovation Research (SBIR) and Small Business Technology Transfer (STTR) programs were the only funded programs that could help fill the need for funds to help commercialize laboratory discoveries. SBIR helps small businesses commercialize laboratory research. STTRs allow start-ups to partner with nonprofits research institutions. Federal departments set aside a portion of their R&D budgets to fund them, especially SBIRs that purportedly have the same goal as the ATP.[165]

The remaining pertinent government players will be visited below and in the following chapters. It's enough at this point to recognize that even if nanotechnology may not be truly revolutionary, the government's decision to adopt an initiative to encourage it may have been. There would be a new gorilla in town for which the science-and-technology-policy-making establishment would need to make room. Moreover, this was not just another player at the table. This player came to the table with a lot of promises and a load of baggage.

EXECUTIVE BRANCH

Believe it or not, the federal government has been the primary patron of technology and emerging industry throughout much of the twentieth century. Revolutionary and world-altering achievements have been made possible by government-sponsored advancements.

With nanotechnology, the stakes may be significantly greater. Government research scientists seem confident that manipulating matter on an atomic scale may revolutionize society and forever change the world in which we live. The question that lies with most government bureaucrats is not *if* nanotechnology will affect society, but *how* it will affect it.

Realizing that in order for private-sector investment in some new technologies to flourish, the government must help subsidize much of the basic fundamental research, the government has led the way in placing nanotechnology on the economic radar screen in the United States. On occasion, the government has had more than an economic incentive to lead the way in investing in new technologies.

President John F. Kennedy recognized that the world's most important battles were not being fought in fields or jungles, with combat engagements between opposing armies with conventional arms. In order to adjust to a new world where the battle lines were not always clear and situations were growing increasingly unstable, Kennedy was ready to initiate aggressive government programs to increase research and development in areas such as space exploration. Ensuring that America was on the cutting edge of technology was perhaps Kennedy's most undervalued yet important contribution to our nation's future.

Today, America may be facing battles brought onto our homeland for the first time since Pearl Harbor, and the armies of yesterday are being replaced by small groups of insurgents, extremists, and terrorists with a unified hatred of American democracy.

In response, nanotechnology has been forwarded as one of the means to build our defense in these difficult times. Roco offered this assessment: "A keystone of U.S. defense posture includes maintaining a strong science and technology R&D program in order to have leading edge technologies available for timely weapons development as required. Nanotechnology represents one of these emerging technologies that can provide much needed enhanced capabilities."[166]

As the table on page 106 indicates, even prior to 9/11 almost half of all government-sponsored research and development was dedicated to national defense. Now that number is even higher, with anywhere between 62 percent and 66 percent of all funding for new technologies being directed toward defense-related programs.[167] With such a large bulk of the funding for nanotechnology research and development pouring from government coffers, it comes as no surprise that the major entities subsidizing nanotechnology research outside of private investment are university laboratories and research programs and various government agencies, most notably of which are the Department of Defense and the Department of Energy.[168]

Total Federal R&D Funding Based on Category

Budget funding	FY 2000 actual	FY 2001 actual	FY 2002 actual	FY 2003 preliminary	FY 2004 proposed	Percent change FYs 2003-04
		Billions of current dollars				
Total	78.644	86.756	97.624	111.593	117.967	5.7%
National Defense	42.580	45.713	53.016	62.463	66.835	7.0%
Nondefense	36.084	41.043	44.608	49.129	51.132	4.1%
Health	17.869	20.758	23.560	26.358	28.059	6.5%
Space research & technology	5.363	6.126	6.270	7.215	7.550	4.6%
General science	4.977	5.468	5.753	6.165	6.441	4.5%
Natural resources & environment	1.999	2.096	2.160	2.234	2.195	-1.8%
Transportation	1.636	1.640	1.838	1.867	1.860	-0.4%
Agriculture	1.426	1.657	1.606	1.710	1.564	-8.5%
Other functions*	2.814	3.298	3.421	3.581	3.463	-3.3%
		Billions of constant FY 1996 dollars				
Total	73.614	79.251	88.516	99.985	104.073	4.2%
National Defense	39.847	41.768	48.070	55.915	58.963	5.5%
Nondefense	33.768	37.492	40.446	43.979	45.11	2.6%
Health	16.722	18.962	21.362	23.595	58.963	4.9%
Space research & technology	5.019	5.596	5.685	6.459	6.661	3.1%
General science	4.657	4.995	5.216	5.519	5.682	3.0%
Natrual resources & environment	1.871	1.915	1.958	2.000	1.936	-3.2%
Transportation	1.531	1.498	1.667	1.671	1.641	-1.8%
Agriculture	1.334	1.514	1.456	1.531	1.38	-9.9%
Other functions*	2.633	3.013	3.102	3.206	3.055	-4.7%

*Other functions include energy, veterans benefits and services; education, training, employment and social services; commerce and housing credit, international affairs, admniistration of justice, community and regional development, income security and general government.

Executive Departments

All organizations have cultures. The confluence of government, industry, and academe produces both a dense and a loose form of culture. They share commonalities imposed by initiatives and deliver unique capacities fundamental to their identities. This is no less true for the departments and agencies vying for NNI-related budgets.

In testimony before Congress in summer 1999, Nobel laureate Richard Smalley told the Senate Committee on Science and Technology that "someone needs to go out, put a flag in the ground and say, Nanotechnology: This is where we are going."[169] As far as federal funding goes, the government has done just that with the creation of the NNI. Established in 1996, it was designed to "accelerate the pace of fundamental research in nanoscale science and engineering, creating the knowledge needed to enable technological innovation, training the workforce needed to exploit that knowledge, and providing the manufacturing science base needed for future commercial production."[170] Almost a decade after its inception, the NNI now helps to fund sixteen executive agencies and remains the cornerstone of US nanotechnology policy. The table on this page illustrates the trend in nanotechnology government spending. It evidences the government's commitment to integrate nanotechnology into the US economy.[171]

Since 1999, more than $3 billion has been appropriated for nanotechnology research—putting it on track to becoming the largest government-funded science initiative since the space program (see graph on page 108).[172] Note the growth in millions of dollars. (The 2004 bar is based on requested rather than appropriated budget.)

With funding for nanotechnology R&D pouring in from government coffers, unsurprisingly, the major entities subsidizing nanotechnology research outside of private investment are university laboratories and research programs.[173]

Prior to 1980, "the U.S. government was pouring billions into research and getting nothing quantifiable to show for it," says William Hoskins, director of technology licensing at Berkeley.[174] The wave of nanotechnology research brought about by the NNI has led to amendments and changes to the 1980 Patent and Trademark Law, more popularly known as the Bayh-Dole Act, which was meant to "provide commercial incentives to develop federally funded technologies."[175] Given the major role that the act played in the creation of many commercial applications in the biotechnology sector in the late 1990s, most experts agree that the Bayh-Dole Act will likely be the guiding force for the commercialization and popularization of much of the academic research associated with nanotechnology in the United States. In 1999 Congress passed the Technology Transfer Commercialization Act, which combined Bayh-Dole with many of its subsequent amendments in order to form a coherent government policy on the commercialization of federally funded research. This served as a major formative

NNI Appropriations Within Federal Departments
(Millions of Dollars)

Federal Department or Agency	FY 2000 Actual	FY 2001 Actual	FY 2002 Actual	FY2003 Actual	FY 2004 Request
Dept of Defense	70	125	224	243	222
Natl Science Found	97	150	204	221	249
Dept of Energy	58	88	89	133	197
Natl Inst Health	32	40	59	65	70
Natl Inst of Tech	8	33	77	66	62
NASA	5	22	35	33	31
EPA	-	6	6	5	5
Homeland Security	-	-	2	2	2
Dept of Agriculture	-	1.3	0	1	10
Dept of Justice	-	1.4	1	1	1
TOTAL	270	465	697	770	849
(% of 2000)	(100%)	(172%)	(258%)	(287%)	(314%)

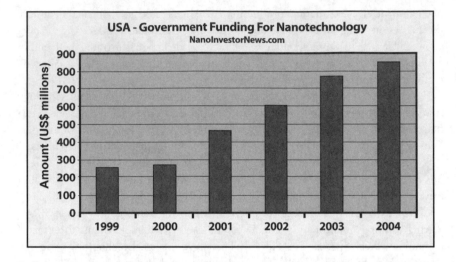

legislative action that has allowed nanotechnology to emerge from research and enter into commerce.[176] The key provisions of the 1999 legislation have greatly influenced the way in which nanotechnology research is commercialized in an academic setting:[177]

- Universities were encouraged to collaborate with businesses to promote the utilization of inventions arising from federally funded research.
- University guidelines for granting licenses were provided, and existing regulations were reformed.
- Universities could elect to retain title to inventions arising from government funding.
- Preference in licensing had to be given to small businesses.
- Regulations and restrictions on those companies that seek to commercialize research being done at academic centers outside of the United States were increased.

There have been some noteworthy improvements in the number of technology transfers and patents coming out of major research institutions,[178] evidenced by the wide variety of new research programs that have a special focus on commercial applications.[179] However, opponents are concerned that the act's provisions simply don't go far enough in creating an atmosphere conducive to commercializing research. In testimony before Congress, Hewlett-Packard's development director Stan Williams complained that the congressional legislation on technology transfer was in fact having a negative effect on industry-academic collaboration because of the lengthy transfer process that was involved in commercializing federal research.[180] As such, many large US-based corporations have become so disheartened that they are now working with elite foreign universities in nations such as Russia, China, and France, which are more willing to offer favorable intellectual property terms.[181] In order to truly bring nanotechnology to the forefront of R&D, many venture capitalists and investors agree that Congress will have to move past Bayh-Dole and do even more to encourage and expedite federally funded academic research.

Reforms have resulted in some success. Private and public universities traditionally associated with major research programs have developed mutually satisfying relationships with businesses and industries. States have climbed onto the bandwagon. Some states have invested heavily in their own initiatives. Also, "over 20 states in the U.S. have realized that nanotech has economic potential and in 2002 made a commitment for nanotechnology that is more than half the NNI annual budget."[182]

Today, university research projects have been the most efficient in the field of nanotechnology and have yielded a great deal of publications and lab research. Glenn Fishbine attributes this to the fact that "universities work together much better than government agencies do."[183] Universities are able to pool research and intellectual and physical resources and, in recent grants, apply for funding from agencies like the NSF in loosely networked consortia. This enables a university to focus on what it does best, thereby maximizing results. Most, if not all, of the advancements in nanotechnology over the past decade have come as a result of research that at least began on the university level.

With nearly two-thirds of the overall funding for the NNI going toward federally funded university research, the remaining balance "is distributed among the national laboratories and grant programs administered by the [various Executive] agencies."[184]

The following sections examine the primary departments and agencies benefiting from the upward trend in federal support for nanotechnology research. The NSF and NIST were covered earlier in this chapter. The remaining departments and agencies are important players as well. The first two were there from the beginning, while the remaining three receive much less NNI support.

Department of Energy

The Department of Energy (DOE) and its Office of Science plays a major, often *the* major role in developing facilities and tools for nanoscale research. The Office of Science is responsible for planning, constructing, and operating the world's most advanced scientific user facilities, which attract more than eighteen thousand researchers annually.[185] Within the Office of Science, the Office of Basic Energy Sciences (BES) leads a broad program of fundamental research that applies these resources to the challenges of science on the nanoscale. It supports Nanoscale Science Research Centers that function as user facilities with resources for the synthesis, processing, fabrication, and analysis of materials on the nanoscale. These centers are located near one or more existing BES centers. The new buildings contain clean rooms, laboratories for nanofabrication, one-of-a-kind signature instruments, and other equipment. Each center began a limited-scope user program in FY 2003.

DOE's applied mission needs include R&D to produce lightweight materials, solid-state lighting, high-performance magnets, hardened cutting tools, surface material to reduce friction and increase resistance, smart materials, improved batteries and fuel cells, environment-friendly petroleum refining, and innovative systems for harvesting and storing energy.[186] The portfolio of research and user facilities includes applications in energy, national security, and the environment relating to the generation, conversion, transmission, and clean and efficient uses of energy. This portfolio is managed by the BES through DOE's national laboratories.

The Department of Energy Laboratory Act of 1993 helped pave the way for this system of national laboratories under the auspice of DOE, which would have eight expressed goals, including the improvement of science, mathematics, and engineering education. These laboratories have spearheaded research with academic institutions and have made some of the most important discoveries of the twentieth century in the fields of energy and fusion.[187]

According to Richard Russell,

> DOE is constructing five new centralized facilities associated with particular DOE laboratories. The DOE centers will offer peer-reviewed access to fabrication, testing and characterization facilities, and will do so at no charge to users doing non-proprietary research. These user facilities allow companies to experiment with high-risk, high-payoff nanotechnologies without the burden of sometimes significant capital investments and will foster industrial collaborations with academic and national laboratory researchers.[188]

The BES program also supports synchroton radiation light sources, three neutron-scattering facilities, and four electron-beam microcharacterization centers, which together host more than eight thousand researchers annually. With a budget of more than $3 billion, DOE's Office of Science leads the nation in supporting basic research in the physical sciences.

Long before the Clinton administration and the NNI, the energy-related potential of atomtech and nanotechnology had been recognized. Of all the areas of nanotechnology applications though, energy was and still remains the most underdiscussed, especially in light of its grand potential to solve many current problems that face the world. No one seems to deny that future energy needs will be linked to the ability to utilize solar energy and nanotechnology, and with its superior construction and refining capabilities, it is sure to open that frontier. Everyday, there is enough sunlight hitting the surface of the earth to more than provide for the energy needs of the world. "In the next 30 years, the energy supply will need to double just to meet the steadily increasing consumption in the industrial world and a moderate amount of catching up in poorer countries (from 14 TW [terrawatts] in 2000 to 28TW in 2050)." Smalley called this the "Terrawatt Challenge."[189]

The problem thus far has been the efficient conversion of light into electrical energy. If only a mere 50 percent of light were able to be converted directly into energy, solar cells suddenly become a very viable source of energy for the world in the twenty-first century.[190] In addition, nanotechnology will be able to create newer and more advanced materials that will replace silicon, which has made solar energy prohibitively expensive in the past. "By increasing the efficiency of solar energy collection while simultaneously lowering the power requirements of manufacturing, molecular nanotechnology may make solar power an unobtrusive and even sufficient source of energy for both home and industrial use, thus creating a viable, even desirable alternative to fossil fuels."[191]

One of the major reasons that DOE is rarely looked upon as a forerunner of nanotechnology research is that energy production in the United States during the 1990s had never been a major concern to most Americans. This mind-set didn't even change after summer 2002 when California experienced its energy crises, as rolling blackouts and power outages forced many Californians to spend weeks at a time without air conditioners and electricity during some of the hottest days of the year.

Then, Bush invaded Iraq. Gasoline prices soared. After the major fighting ended, over two thousand American men and women in uniform had been murdered in terrorist-related attacks in Iraq. These events made the United States highly sensitive to dependency on fossil fuel reserves in the Middle East. Then Hurricane Katrina hit the US Gulf Coast, and refinery output was interrupted, and gasoline and refined petrochemical products prices soared. As such, alternative energy options are being discussed again.

These events, along with strong lobbying by DOE and some industrialists, have prompted major changes in the amount of resources that the federal government is committing to the alternative-energy sector. This gateway has funneled and will funnel nanotechnology funding to DOE because of its promise to change the way America uses, as well as produces, fuel and energy.

In Bush's 2003 State of the Union address, he paved the way for the administration's new emphasis on quickly advancing R&D in the field of fuel-cell research in order to meet the government's goal of a ten-year plan to implement fuel-cell technology. This new doctrine helps explain the $55 million increase in DOE's FY 2003 scientific research budget, the bulk of which is being streamlined toward nanotechnology-related research on practical fuel cells.[192]

Fuel cells, which work by harnessing the chemical attraction between oxygen and hydrogen in order to create energy, would by contrast to the internal combustion engine, be fifty times as efficient and be able to produce large amounts of energy both cleanly and cheaply.[193] The major problem with fuel cells so far has been the safe storage of hydrogen, which needs extremely strong casing that can withstand high pressure for long periods of time. Nanotechnology comes into play by producing materials that are stronger than any currently available. Once nanotechnology enables hydrogen fuel-cell technology, the impacts will not only reach the traditional industries such as the automotive sector but also find new applications in businesses and homes. With a lurking energy crisis and fossil fuels becoming increasingly expensive as well as environmentally degrading, nanotechnology is being examined as fundamental to future energy needs.

Nanotechnology-related energy funding is very significant. In FY 2003 DOE received a 48 percent increase in its budget under the NNI for nanotechnology-associated research. Next to the Department of Defense (DOD) and the NSF, DOE consumed nearly a third of the entire NNI budget with a FY 2003 appropriation of nearly $139.3 million.[194] In 2003 BES funded research at more than one hundred sixty academic institutions in forty-seven states and at thirteen DOE laboratories in nine states. "Even if you look at the FY 2005 budget request, and it looks like DOE's Office of Science budget is pretty flat relative to FY 2004, we're still building the nano centers in each of the national laboratories that have these major facilities."[195]

DOE has nine research *targets* in energy-related technologies in which nanoscience is expected to have the major impacts:

- scalable methods to split water with sunlight for hydrogen production
- highly selective catalysts for clean and energy-efficient manufacturing
- harvesting of solar energy with 20 percent power efficiency and 100 times lower cost
- solid-state lighting at 50 percent of the present power consumption
- superstrong, lightweight materials to improve efficiency of cars, airplanes, and so on
- reversible hydrogen storage materials operating at ambient temperatures
- power transmission lines capable of one-gigawatt transmission
- low-cost fuel cells, batteries, thermoelectronics, and ultracapacitors built from nanostructured materials
- materials synthesis and energy harvesting based on the efficient and selective mechanisms of biology.[196]

Today, DOE is using a vast majority of the funds allocated from the NNI to work toward expanding and improving their national laboratories. In addition to these new funds from the government, DOE is also working to streamline and

DOE'S NANOSCALE
SCIENCE AND REASEARCH CENTERS

Center for Functional Nanomaterials	Brookhaven National Laboratory (NY)
Center for Integrated Nanotechnologies	Sandia National Laboratories and Los Alamos National Laboratory (NM)
Center for Nanophase	Oak Ridge National Laboratory (TN) Materials Sciences
Center for Nanoscale Materials	Argonne National Laboratory (IL)
Molecular Foundry	Lawrence Berkeley National Laboratory (CA)

combine older nuclear labs, which would allow the department to open newer facilities that can work on emerging technologies.

Some of the major 2004 efforts included "work on five DOE national laboratories in New York, Tennessee, Illinois, New Mexico, and California."[197] These centers have the expressed objective of "being the nation's critical focal points for the development of nanotechnologies that will revolutionize science and technology," according to a DOE budget document.[198] These laboratories will also provide for state-of-the-art nanofabrication equipment and allow for hundreds of visiting scholars, scientists, and researchers, and will attract some of the top graduate students and professors in the field.

Recent problems in the field of energy production combined with an increasing government priority for research of new energy technologies such as hydrogen fuel cells have certainly allowed for nanotechnology to dominate the government's research agenda at the beginning of the twenty-first century. Another major department associated with the NNI is defense.

Department of Defense and DARPA

Robert Trew helped direct the Department of Defense's nanotechnology effort. "Today's research offers such possibilities as materials that can be made defect-free in situ with predetermined properties, new sensors that can detect chemical and biological hazards and a range of new electronic instruments with advanced readouts and other features that might give a critical edge on the battlefield."[199]

Nanotechnology is a force multiplier. While military strategists have toler-

ated troop casualties, some lessons have been learned, especially after the first
Gulf War. High-tech warfare can save American lives, and Americans coming
home in body bags affect public support.

Anyone who questioned the $200 million annual allocation from the NNI given
to the DOD ceased doing so after 9/11. Chairman of the Defense Science Board
William Schneider said, "After the events of 9/11, there was an intensified interest
and availability of resources to address the military transformation for the twenty-
first century."[200] Threats today no longer come from conventional governments and
nation-states in distant lands; instead, our environment is now colored by terrorist
security threats. This diffused, decentralized problem is sure to alter substantially
how the military addresses technological advancements in its strategies.

So, while the benefits may be clear and ultimately lead us to improved battlefield dominance, the changes that this new technology will bring to the contemporary role of US defense may well provide for a new century that is far different
from the past. Even prior to the newly unstable geopolitical situation, the Government Accounting Office, in *The Industrial Base: Assessing the Risks of the DOD's Foreign
Dependence*, interviewed a group of experts, many of whom noted that "the U.S. may
have a long term national security interest in maintaining generic technologies
including nanotechnologies—the making of extremely small items."[201]

Despite large amounts of capital, however, the DOD's research prior to 1996
was "thought to be nearly all in top-down miniaturization" but has now increasingly begun to invest more in research that focuses on the development of
bottom-up, molecularly precise nanotechnology.[202] After the NSF and DOE, the
DOD is one of the largest supporters of nanotechnology research. According to
Josh Wolfe, "the DOD is committing $201 million for research in nanoelectronics,
magnetics, nanomaterials by design, and detection and protection against chemical, biological, radiological and explosive (CBRE) threats."[203]

It comes as no surprise that the military is on the cutting edge of nanotechnology, especially considering the government "has traditionally funded defense
over civilian research at a ratio of 58:42."[204] The 2003 nanobudget for the DOD
was $379 billion, which makes it a major player in nanotech research.

In order to facilitate research within the armed forces, Congress funds the
Defense Advanced Research Projects Agency (DARPA), which has very specific
research goals suited toward a military environment. In fact, military applications
of nanotechnology cover a wide variety of potential applications, almost all of
which currently revolve around battlefield dominance. "Improved electronic warfare capabilities, better weapons systems, and improved camouflage and intelligence systems" are all on the DOD's active research and development list.[205] In
the late 1990s DARPA began to gain traction in its quest toward battlefield dominance in the twenty-first century, especially with the new budget allocations
streamlined to the department from the NNI.

The army kicked off its research in 1990 by awarding a one-million-dollar grant to "develop the theoretical, modeling, and experimental tools to optimize manipulation of materials on a nanoscale."[206] Recent reports released from the military even suggest that the Army had some degree of success early on in their fledgling research programs. Later in the 1990s the Army dove further into the nanoworld by targeting particular technology areas for ground combat.

Some of its directives are very exotic, for example, "extending human performance by direct coupling of the human central nervous system to machines," and "developing soft kill products that disable machines or propulsion systems, change the soil or vegetation, or degrade materials."[207]

In March 2002 this program evolved into one of the Army's largest research programs in modern history. It involved the five-year, $50 million Institute for Soldier Nanotechnologies (ISN) housed at MIT. It is staffed by one hundred fifty researchers, including forty-four MIT faculty members from eight different departments (the ISN faculty members have started a total of thirty companies between them), more than one hundred students and postdoctoral researchers, and several industrial partners, including DuPont, Raytheon, Dow Corning, Carbon Nanotechnologies, Dendritic Nanotechnologies, Nomadics, and Triton Systems. As ISN's Ned Thomas put it, "We're a dating service."[208]

The ISN has three areas of focus: "protections (against bioweapons and gun shots), performance enhancement (exo-muscles helping to lift heavy objects) and injury prevention and cure."[209] As such, one of its research efforts involves "developing chameleon like uniforms and materials that could help protect soldiers against detection, threats, bullets, and chemical agents"[210] by "embedding nanosensors into ultra-strong and lightweight nanomaterials for the military uniform."[211] The Objective Force Warrior, a program at ISN, aims to produce "a lightweight, overwhelmingly lethal, fully integrated individual combat system."[212] The body armor would need to weigh less than the average 130 pounds of equipment today's Special Forces carry on a three-day mission. The two specific uniform systems under development are the Future Force Warrior System available in 2010 and the Vision 2020 Future Warrior System scheduled for ten years later. "The uniform from the waist down will have a robotic-powered system that is connected directly to the soldier. This system could use pistons to actually replicate the lower body, giving the soldier upwards of about 300 percent greater lifting and load-carriage capability." According to Jean Louis DeGay, a Soldier Systems Center representative, "We are looking at potentially mounting a weapon directly to the uniform system and now the soldier becomes a walking gun platform."[213]

Since 1999 the US Army has further invested in the concept of nanosoldiering outside of MIT in hopes of creating nanomaterials that would "simultaneously monitor a soldier's health, heat and cool the soldier as appropriate, and independently generate power so the soldier's *wearable* computer can wirelessly

DOD NNI CENTERS

Center name	Principal Investigator	Institution
Institute for Soldier Nanotechnologies	Thomas	MIT
Center for Nanoscience Innovation in Defense	Asschalom	University of California, Santa Barbara
Nanoscience Institute	Prinz	Princeton

remain in constant communication with headquarters."[214] Any one of these advancements would mean widespread changes in the way that wars are fought and how our soldiers act and react on the battlefield.

While much of DARPA's research has focused on battlefield combat, there has also been a great deal of resources generated toward the Micro-Air Vehicle (MAV) Project, which would create tiny unmanned reconnaissance aircraft that are hundreds of times smaller than the traditional drone surveillance aircraft that were used in Operation Iraqi Freedom:

> These miniature reconnaissance planes fly 20 to 60 minutes at speeds of 20 to 40 miles per hour. MAV's can serve as cheap airborne relays because, even though they have only a tiny half-ounce payload for a guidance system, video camera, and transmitter, they can communicate and transmit imagery over a considerable distance. Whatever is caught by the MAV's video camera appears on the screen of the soldier's laptop computer, providing him with valuable information from logistics movement as well as for his own battlefield supports.[215]

These vehicles can be further enhanced by nanosensors deployed in combat, "tiny mobile and cheap sensors that can detect troop movements."[216] The project utilizes intensive micromachine technology as well as nanotechnology utilized to make the airplanes components not only more lightweight than ever before, but also substantially more durable than traditional non-nanomaterials.[217] A wide array of prototypes exist today for the MAV program; however, full operational use of this technology may not be seen for several more years as scientists and researchers search for more lightweight material to construct the aircraft. This initiative is also high on the military's research agenda.

Since 2001 the Air Force has been using DARPA research funds to "create materials for its traditional jet fighters that are not only stronger and lighter, but also radar and sound absorbing as well as self repairing."[218] These properties will enable aircraft to travel at higher speeds and go on longer missions, with substantially less chance of detection.

On the maritime side of nanomaterials, the Navy is attempting to use synthetic yet molecularly perfect materials to create paints that will mimic the skin of sea animals such as sharks and dolphins, which will allow submarines to glide through the water more efficiently.[219]

Another major defense effort is nanoelectronics involving quantum computing. According to a press release on September 10, 2001, DARPA is funding a five-year project that will establish the Quantum Architecture Research Center between MIT and the University of California campuses at Davis and Berkeley. The project will examine possible methods to build a superfast computer that uses the properties of quantum physics.[220] "In theory, this would allow a computer to perform many different computations simultaneously at a much quicker rate than conventional computers...a supercomputer with a billion-fold increase in performance."[221] According to US Army Research's Henry Everett, "over the long-term, quantum computing will leave a vast footprint on national security and military conflict....For example, code-breaking: All conventional encryption would be vulnerable with quantum computing. We could protect against another 9/11 because we could break all codes and read all messages."[222]

But unsurprisingly, many of the new nanotechnology projects being spearheaded by the DOD have only recently produced preliminary results. With new advances in the military applications of nanotechnology, the government is starting to reassess its priorities. An NSF report on the status of nanotechnology and homeland security outlined a new vision for the defense applications of nanotechnology:[223]

- higher-performance platforms (aircraft, ships, subs, boats, and satellites) through stronger, lighter-weight structural materials, stealth materials, and low-maintenance and *smart* materials
- enhanced sensing through more sensitive and selective sensors of electromagnetic radiation, magnetic and electric fields, nuclear radiation, and chemical/biological agents
- information dominance through enhanced information technology. This is likely to take the form of smaller, lower-power memories, smaller and faster logic devices through improved processing, and enhanced secure communication systems with greater bandwidth
- safer operations involving hazardous materials or operations, through the use of remotely operated robots
- reduced manpower requirements through the greater use of automation in maintenance, management, and control of weapon platforms, systems, and hazardous functions
- lower life-cycle costs through the use of improved materials, coatings, and condition-based maintenance.[224]

Some particularly remarkable work is being done by small nano start-ups. For example, Nanosys announced a new product called Nanofur, a technology based on the tiny fibers on geckos' toes. According to Bob Dubrow, director of product development for Nanosys, "Military agencies have expressed interest in developing the technology into a product that would allow soldiers to use sticky gloves or boots to climb the walls of buildings."[225] Konarka "is supplying the Army with solar-powered battery chargers. The prototype devices used polymer photovoltaic plastics. Soldiers are more power hungry than ever and the Army believes flexible solar cells can provide the extra juice."[226] NanoLogix has been working on biodefense sensors called BioMEMS for rapid environmental microbial monitoring and biochips embedded in small transportable devices for rapid bacteria identification of disease-causing pathogens.[227] FAST-ACT, which is made of magnesium dioxide and oxygen, is marketed by Nanoscale Marterials, Inc., can break down deadly materials as VX nerve gas and sulfuric acid.[228] NanoProtect can be sprayed or rubbed on equipment, clothing, and people that have been exposed to biological agents neutralizing them, including anthrax and smallpox.[229] Researchers at the University of Louisville are developing nanowires needed by warhead guidance systems to find their targets.[230] And MIT researchers are developing a semiconducting organic polymer as a new technology "for explosives sensing which could help protect soldiers from improvised explosive devices, one of the greatest threats facing coalition forces in Iraq."[231]

In a post-9/11 world, however, "the line between national security, material science, health-care, and environmental applications becomes easily blurred."[232] For example, Army research and the ISN, if successful, would be certain to have wide-ranging ramifications on the healthcare and the telecommunications industry.[233] A recent article reported a DARPA project where I, Robot is building robotic swarms that may mimic the organized behavior of insects.[234] What we learn about this effort to streamline minesweeping and search and rescue may be useful in manned and unmanned space exploration as well. Any advance made by DARPA or other defense research labs would be certain to have a great deal of promise in the private sector, which looks toward both government and university research as potential starting points for major venture capital investment.

"Military needs often bankroll commercial technologies. Computers grew out of a need for more effective weaponry.... It will pick up the slack left as corporations retrench and clip R&D."[235] Other spin-offs in medicine and quantum computing would inevitably make its way into commercial sectors.

No longer are experts in the field questioning whether or not there will be applications of nanotechnology research on a military scale, but instead are disagreeing over when the applications of these technologies will reach the battlefield. Unfortunately, the war in Iraq caught the US military "in only the very early stages of the small tech revolution."[236] In fact, "most of the materials that would really make a difference are still in the research stage," says James Ellenbogen, a nanotechnology analyst for Mitre, a leading consulting company for the US Department of Defense.[237] Despite only limited development, some rudimentary nanoscale materials and components helped to topple the Iraqi regime, as they were utilized in communications systems and weaponry-guidance technology. Most other uses of small-scale materials during the war in Iraq involved the utilization of MEMS technology, which is considered by many to be a transitional advancement before full-scale implementation of nanotechnology. But overall, nanotechnology in Iraq was used to such a limited extent that it is likely this war will go down as the final battle fought without the benefits of this new technology.

Regardless of when nanotechnology is fully implemented and when this new global paradigm will occur, one thing is for certain: "nanotechnology will eventually alter warfare more than the invention of gunpowder... and will affect every aspect of weaponry," and forever change the role of America's armed forces.[238] In essence, our attempts to ensure that the United States is prepared to face the threats of an increasingly unstable twenty-first-century world may have been the largest single driving force bringing nanotechnology into the American economy. Now that America is concerned about homeland defense, many of these discoveries will have applications there as well.

Other Departments and Agencies

Similar to the burgeoning programs in the energy and defense sectors, the NNI has laid out new funding initiatives for other departments and agencies such as the Department of Agriculture, NASA, and Homeland Security, which previously seemed to have little interest in nanoscience.

In FY 2004 the Department of Agriculture found itself the big winner in the area of government funding, as the NNI granted it a 1,000 percent budget increase, to just over $10 million. While agriculture may not be the government's top priority at this point, the application of nanotechnology in this field "has only begun to be appreciated."[239] Roco seems to think "there are issues from conservation to the quality of food, to increasing the rate at which plants grow," areas in which nanotechnology can mean a great deal of changes for the Department of Agriculture. Opportunities for nanotechnology in agriculture and food and animal health-systems research would include food monitoring for pesticide residues, trace chemicals, pathogens, and the like; integrated rapid DNA sequencing to identify genetic variation and genetically modified organisms; and preserving the integrity of food during transportation and storage.[240]

On the NASA side, the recent breakup of the space shuttle *Columbia* has made the production of nanomaterials a major funding priority because all major space problems, including the cause of the *Columbia* crash, have been caused by the

NNI CENTERS (NASA)

Center Name	Principal Investigator	Institution
Institute for Cell Mimetic Space Exploration	Ho	UCLA
Institute for Intelligent Bio-Nanomaterials and Structures for Aerospace Vehicles	Junkins	Texas A&M
Bio-Inspection, Design, and Professing of Multifunctional Nanocomposites	Aksay	Princeton
Institute for Nanoelectronics and Computing	Datta	Purdue

weakening or failure of materials. In addition, Bush's new proposition to send an expedition to Mars demonstrates a renewed interest in space programs, especially those that will provide for safer and more effective missions. Nanotechnology can produce newer and cheaper nanomaterials that can make stronger and more resistant materials that could be used to avert disasters on future space missions.[241] The wildest work done by NASA is a project called ANTS for autonomous nanotechnology swarms. Presumably, NASA is "testing a robot that they hope to shrink to nanobots size and eventually form ANTS."[242] In the meantime, NASA has four NNI Centers of Excellence.

No government department is getting more attention than Homeland Security these days, and it has begun to nurse at the NNI teat. Small-tech approaches, many conceived in the 1990s and developed with federal support, promise to help protect a nation clamoring for solutions. Recent advancements in nanoscale research offer alternative technologies for detecting anthrax and other dangers.

"Nanotechnology provides an enormous opportunity to increase the sensitivity of sensors for detecting chemical, biological and nuclear threats," said Meyya Meyyappan, director of the Center for Nanotechnology at NASA Ames. "The bonus is that the product can come in ultra-small size, requiring only low power levels," Meyyappan added.[243]

"Homeland security will not be viable unless there is a method for (incorporating) microsystems," said Marion Scott, director of a microsystems program at Sandia National Laboratories in New Mexico. "Microsystems enable homeland security."[244]

CONCLUSION

Each time a new problem is faced by a department or agency, it seems nanotechnology is more and more becoming the answer to the problem and is being pushed to the top of the government's research agenda. Clearly, future economic changes from nanotechnology will likely come as a result of government-sponsored research and development programs. While executive departments and agencies allocate funding to specific programs, the NSF directs the bulk of the budget, and its spending is associated with large and creative efforts. The following chapter traces the NSF's role in promoting nanotechnology and reconstitutes a timeline of sorts to contextualize the NNI.

CHAPTER 4

GOVERNMENT INITIATIVES IN NANOTECHNOLOGY

Newt Gingrich, former House Speaker and a cochairman of the NanoBusiness Alliance, sees great promise in some national nanotechnology initiatives:

> Those countries that master the process of nanoscale manufacturing and engineering will have a huge job boom over the next 20 years, just like aviation and computing companies in the last 40 years, and just as railroad, steam engine, and textile companies were decisive in the 19th century. Nanoscale science will give us not dozens, not scores, not hundreds, but thousands of new capabilities in biology, physics, chemistry and computing.[1]

It is remarks like Gingrich's that help account for government-funded nanotechnology research "ballooning seven fold from under $500 million in 1997 to $3.5 billion in 2004."[2]

While Gingrich's hyperbole may be close to the truth, it is incredibly important to realize that "the critical breakthroughs needed to propel this technology into the mainstream are still illusive."[3] Because of the timelines involved and government interests in defense and economic development, there is a strong case for government support. Because industrial and commercial returns on basic research may be beyond the horizon for most commercial and industrial entities, there are initiatives. While states have spent heavily with some of their own projects (over $400 million), especially those in California, New York, and Texas, and huge nanocenters are opening up across the country, the following focuses on federal initiatives. Federal or national money is critically important in drawing states and universities into the investment cycle. States are investing mostly in building infrastructure, while the federal government invests in research, mostly at university nanocenters.

INITIATIVES AND SPIN

To take full advantage of the potentials in nanoscience and nanotechnology, three major national projects have been funded through the NSF in whole or in part. These initiatives and networks have had a profound effect on policy making because they function as institutional memory. Their strengths become codified in subsequent projects, and weaknesses guide calls and proposals down the line. The footprints they have left are substantial. The first was a policy initiative which in turn established two powerful networks.

National Nanofabrication Users Network (NNUN)

The first nanoprogram—on nanoparticle synthesis and processing—was initiated by the NSF in 1991. It focused on chemical processing, 1991 to 2001, $3 million to $4 million per year. The Nanoparticle Synthesis and Processing Initiative is hardly mentioned at all in government reviews of nanotechnology. While there are some reports on particle research,[4] there is very little written assessing this early initiative. Its primary function seems to be as a precursor to the user networks.

This nanoparticle initiative was followed by the National Nanofabrication User Network in 1994. Its mission was to make research easy and affordable so that the research could concentrate on the goals rather than the experimental techniques that allow the making of the structure needed to reach those goals.[5] Its Web page offers this mission:

> We accomplish our mission by providing the nation's researchers with effective and efficient access to advanced nanofabrication equipment and expertise. We enable research by providing state-of-the-art facilities, training, and project support. We help expand the application of nanotechnology by providing technical liaison personnel, education through workshops and short courses, and by acting as a bridge between disciplines to create research opportunities that might otherwise not be apparent to specialists in narrow disciplines.[6]

About six hundred fifty projects, with more than twenty-seven hundred faculty members and students, and twelve large centers were supported in FY 2000 with the NNUN as its centerpiece. During 2001 more than seventeen hundred users conducted a significant part of their research at NNUN facilities, including two hundred fifty small companies.[7] The 2002 report claimed a growth rate of approximately 20 percent, with a near doubling of the user population in the four years leading up to the report.[8]

The five NNUN sites were two hubs on the East and West Coasts at Cornell University and Stanford University, with three additional sites at the University of California–Santa Barbara, Pennsylvania State University, and Howard University.

As explained by Neal Lane,

> The NSF-supported NNUN was a good example of a distributed network or *virtual center* as some like to describe the arrangement. In the *virtual center* concept, high speed connections allow any researcher—regardless of where he or she may be located—to remotely use the capabilities and instruments of each of the five locations across the country that constitutes the users network.[9]

The sites did not host advanced technologies alone. "Its staffs had extensive experience in all phases of nanofabrication and fields ranging from nanophysics to biology to electronics with domain experts in micromechanics and biology to assist users in translating their ideas into experimental reality."[10] Part of its mission included the intellectual resources it provided to researchers soliciting to use the site.

National Nanotechnology Initiative (NNI)

The early story behind the NNI was reported in *Chemical and Engineering News* in 2000. It was forecasted in April 1998 when Neal Lane, then Clinton's science and technology advisor, tagged nanotechnology as the "most likely area of science and engineering to produce the breakthroughs of tomorrow" in his testimony before Congress. It was proposed in a presentation by Roco at an OSTP meeting in March 1999. It officially started in August 1999 when the NSTC (then the IWGN) released its first report, *Nanostucture Science and Technology*. A month later *Nanotechnology Research Directions* was released, and in February the National Nanotechnology Initiative report followed. (A review of these documents can be found in chapter 5.) "Together the three reports are a blueprint for the federal government to assess its strategic R&D investments in nanotechnology."[11] Thomas Kalil, a senior director for economic policy at the White House, issued a call for an R&D effort organized at the national level during a January 1999 workshop held in Arlington. In July 2000 the implantation plan was issued about the same time the IWGN was elevated to the NSTC. The NNI was established formally in FY 2001.

The goals of the NNI are to (1) conduct R&D to realize the full potential of this revolutionary technology; (2) develop the skilled workforce and supporting infrastructure needed to advance R&D; (3) better understand the social, ethical, health, and environmental implications of the technology; and (4) facilitate transfer of the new technologies into commercial products.[12]

Kalil said, "White House staff thought a nanotechnology initiative was a good idea for a number of reasons, including balancing the growing funding disparity between life sciences and physical sciences, training the next generation of U.S. scientists and taking an international lead in a transformation technology."[13]

Technology research in the private sector is driven by today's global economic realities. The pace of technological change is faster than ever before, and victory goes to the swift. These realities force companies to make narrower, shorter-term investments in R&D that maximize returns to the company quickly.[14] As such, for some technologies, especially those that may require decades of basic and applied research before producing a commercial application, either the government funds them or venture capitalists do. Sometimes, as well, the goals of government may not directly coincide with industry, for example, direct defense applications. At other times, the real nature of a start-up may not be adequate to attract venture capital funding. For these reasons and many more, a major initiative, like the NNI, seemed to be in order.

The NNI, through its broad range of managers (ten federal agencies with dedicated R&D budgets and four others as participants), planned to move nanoscience out of the labs and into the marketplace. "The time to take a concept developed through basic nanotechnology research to a commercial product is beyond five or even ten years—or, for truly fundamental research, may be altogether unknown. Therefore, private investors are not generally in a position to provide the necessary financial support."[15] According to Senator Ron Wyden, "Start-up investments are going to be key, and venture capital is hard to get right now in a slippery economy. Some modest investments in the start-up end can pay big dividends down the road."[16] Consequently, the NNI became a *true* initiative.

The NNI is managed within the framework of the NSTC. The NSTC is the primary means by which the president coordinates science and technology programs across the federal government. The NSTC's Subcommittee on Nanoscale Science, Engineering, and Technology (NSET) continues the plans, budgets, programs, and reviews for the NNI. The subcommittee is composed of representatives from each participating agency, the Office of Science and Technology, and the Office of Management and Budget.

The President's Council of Advisors on Science and Technology (PCAST) was originally established by President Bush in 1990 to enable the president to receive advice from the private sector and academic community on technology, scientific research priorities, and math and science education. Since its creation PCAST has been expanded and currently consists of twenty-three members plus the director of the Office of Science and Technology Policy who serves as the council's cochair. The council members, distinguished individuals appointed by the president, are drawn from industry, education, and research institutions and other nongovernmental organizations.[17]

In order to further strengthen the initiative, PCAST conducts an external review of the NNI, which includes a comprehensive assessment of current NNI programs and leads to recommendations on how to improve the management of the program.[18] PCAST

> will explore a wide variety of topics relating to nanotechnology and its potential benefits to the American public and the U.S. economy. Such topics may include the identification of metrics for measuring progress (and applying these metrics to continually assess program progress); social and ethical consideration of nano-technology; technology transfer issues and mechanisms; and comparisons of the U.S. program with international programs (in terms of both efforts and results).[19]

The NNI's funding strategy involves five modes: fundamental research, grand challenges (of which there are nine, such as chemical-biological-radiological-explosive detection and protection and nanoelectronics, nanophotonics, and nanomagnetics), centers of excellence that pursue broad interdisciplinary goals, infrastructure development (the NNUN, the Network for Computational Nanotechnology, NNIN, etc.), and research on societal implications and educational needs.

In FY 2001 the Clinton administration included six executive agencies within the NNI (Department of Defense, National Science Foundation, Department of Energy, National Institutes of Health, National Institutes of Technology, and NASA). The most extensive funding went to the Department of Defense and the National Science Foundation. By FY 2002, though, the new Bush administration expanded the NNI to include six additional agencies, including the Department of Justice and the Environmental Protection Agency.[20] Within each of the agencies, "about one-third of the funds are spent on fundamental research, one-third for grant challenges,[21] and one-third distributed among infrastructure facilities, and social issues development."[22] Funding from the NNI goes to numerous departments ranging from the Department of Agriculture to the National Institutes of Health, but nearly half of the overall funding for the NNI now goes toward the Department of Energy and the Department of Defense.[23]

In October 2000 the NSTC approved NSET's request to establish the National Nanotechnology Coordinating Office (NNCO). Besides being responsible for the day-to-day management of the NNI, the NNCO assists the NSET committee with identifying funding priorities, establishing budgets, and evaluating current NNI activities.[24] Also the NNCO develops and makes available printed and other communications materials concerning the NNI and maintains the initiative's Web site.

Beyond these grand challenges and infrastructure, an important component of the NNI is its centers and networks of excellence:

> To date, 15 centers of excellence have been established through the NNI. The primary objective of the centers is to enable research activities that cannot be

conducted through the traditional mode of single investigator, small groups, or with current research infrastructure. Each center is expected to establish part-nerships with industry, national laboratories, and other sectors, including state-supported nanoscience activities. The research activities of these centers are expected to enhance multidisciplinary research activities among government, universities and industry performers, which in turn, are expected to create a ver-tical integration arrangement that includes activities from basic research to the actual development of specific nanotechnology devices and applications.[25]

"The initiative has grown rapidly from an initial budget request of $464 million in FY 2001 to the $849 million required for FY 2004."[26] "The Adminstration's request for the 2005 fiscal year called for a 2 percent increase to $982 million,"[27] and Congress appropriated $1.23 billion. For FY 2006, the NNI is expected to get $1.05 billion in 2006, a 2.5 percent decrease from 2005, a relatively stable budget alloca-tion given the mounting costs of the war on terrorism, a sluggish recovery, and nat-ural disaster expenses.[28] "Between 2001 and 2003, NNI invested 65 to 70 percent of its funding in academic institutions, 25 to 30 percent in research laboratories, and about 5 percent in industry."[29] A recent review of the NNI by an advisory panel impaneled under the Twenty-First Century Nanotechnology R&D Act concluded the NNI was performing well, though "it encourages the NNI to broaden its impact by building bridges with more government bodies, economic development groups and industry." While giving the NNI high marks for it efforts, it warned the United States "is being challenged by other nations as they beef up their programs."[30]

Why, given the state of the US economy, does such an investment make sense? There are two answers to this question. First, "rescheduling of several projects at NIST and the reassignments of applied nanotechnology projects to the respective areas of relevance at the Department of Defense and NASA result in a decrease in funding at those agencies."[31] Second, spending in technology is a growth driver:

> The Congressional Budget Office estimation of the $1.3 trillion projected deficit that we're facing for fiscal years 2004–2013 would actually be $247 billion higher if it were not for improvements in productivity due to computers. If we succeed in our effort to harness the potential of nanotechnology, we will see productivity and revenue gains of a similar magnitude.[32]

Mark Modzelewski from Lux Capital credited the NNI for generating inter-national interest in nanotechnology: "The NNI worked everyone up. It's incred-ible that this once-obscure science is now the buzzword amongst the leaders of the free world."[33]

The NNI is not without its critics. Strangely, one is Drexler himself. He claims it was Feynman's vision and the rhetoric about it that motivated the NNI. "An NNI promotional brochure speaks of Feynman's vision of total nanoscale control calling it the original nanotechnology vision."[34] Drexler's complaint is that the NNI has compromised unrelated research, and the nanoscale technology funding coalition has obscured the Feynman vision. He claims the leaders of the funding coalition have attempted to narrow nanotechnology to exclude one area of nanoscale technology, the Feynman vision itself:

> One would expect that the NNI, funded through appeals to the Feynman vision, would focus on research supporting this strategic goal. The goal of atom-by-atom control would motivate studies of nanomachines able to guide molecular assembly. Leading scientists advising the NNI would examine assemblers and competing approaches to their design and implementation, generating road-maps and milestones. In the course of broad marshaling of resources, at least one NNI-sponsored meeting would have invited at least one talk on prospects for implementing the Feynman vision. No NNI-sponsored meeting has yet included a talk on implementing the Feynman vision, and the most prominent scientists advising the NNI [e.g., Smalley] have [sometimes] declared the Feynman thesis to be false.[35]

Drexler offers a few rationales: most technologists associated with the NNI were from unrelated fields and had no professional reason to understand the concept. Feynman's vision promised more than anyone could soon deliver, and public concern regarding its dangers might interfere with research funding. Drexler claims that the NNI, while having been sold as a technological revolution, has instead funded science, especially solution-phase chemistry, avoiding projects investigating molecular manufacturing. As such he claims "national policy is hampering dialogue, increasing security risks, and failing to deliver on revolutionary expectations."[36] This controversy was further detailed in chapter 2.

By and large, the NNI has been well received. This is due in no small part to the significant roles played by the last two presidents.

President William Clinton

On January 21, 2000, President William Clinton flew to Palo Alto, California, to announce his National Nanotechnology Initiative funded in FY 2001 at $497 million. He spread the NNI among six federal programs and departments: the National Science Foundation, NASA, Energy, Health and Human Services, Defense, and Commerce.

On the podium with Clinton were Energy secretary William Richardson; Rita Colwell, director of the NSF; Neal Lane, assistant to the president for science and technology; and Caltech president David Baltimore, and in the audience was the whole Caltech and Jet Propulsion Laboratory community. Clinton's primary justification involved "keeping America the world's leader in science and technology."[37] The speech was delivered during the heyday of the dot-com craze, and Clinton believed that federal funding would enhance America's chances of riding the growth wave. Unbeknownst to President Clinton was the eventual implosion of the dot-com enthusiasm. He admitted that "some of the research goals may take 20 or more years to achieve, but that is precisely why there is an important role for the federal government."[38]

Clinton also alluded to nanotechnology in his January 2000 State of the Union address when he spoke of materials ten times as strong as steel at a fraction of the weight and—this is unbelievable to me—molecular computers the size of a teardrop with the power of today's fastest supercomputers."[39]

President George W. Bush

Josh Wolfe wrote that when Clinton left office, the NNI fate was in limbo, but newly elected President George W. Bush rose to the occasion. "President Bush, in the first year of his administration, asked for another hundred million dollars for nanotech, and added another handful of agencies to the NNI. Bush's budget proposals for FY 2003 and FY 2004 further boosted the budget."[40] Bush's people even added post-9/11 rhetoric, claiming that "nanotech research can help build tools to detect weapons of mass destruction."[41] One of the agencies added to the initiative was the newly created Department of Homeland Security.

"Because nanotechnology is of such critical import to U.S. competitiveness, both economically and technologically, even at this early stage of development, it is a top priority within the Administration's R&D agenda."[42]

Bush's support seems genuine. He stated in his 2004 federal budget request, "The convergence of nanotechnology with information technology, biology, and social sciences will reinvigorate discoveries and innovation in many areas of the economy."[43] For Bush, technologies like nanotechnology may be the key to the United States' recovery from the market and employment doldrums in his first term of office. Wolfe predicts, "When the smoke settles in Iraq, look for Bush to start promoting the government's nanotech policy, similar to his State of the Union support for AIDS research, as the international race intensifies."[44]

National Nanotechnology Infrastructure Network (NNIN)

While there were at least three major bids to host the NNIN, it was awarded to a joint proposal spearheaded by Stanford and Cornell Universities. (They were the central hub of the NNUN as well.) They were joined by the University of Michigan, the University of Washington, Harvard, the University of California at Santa Barbara, Howard, the University of Minnesota, the University of New Mexico, and the Georgia Institute of Technology. The $70 million grant will award $14 million per year for five years to the participants in varying amounts. Each university will develop labs with a specific research focus and "will share their facilities to support research and education into nanoscale science, engineering, and technology."[45]

"The NNIN was supposed to be up and running by January 2004. But by early February, the NSF and Cornell University, the university leading the project, had yet to ink a contract because of delays in completing the necessary paperwork according to Lawrence Goldberg, a senior engineering adviser a the NSF."[46] Delays notwithstanding, "we are welcoming users regularly into our facilities. Some facilities have 5 to 10 new external users entering and getting trained a week," said NNIN director Sandip Tiwari, director of the Cornell Nanofabrication Facility.[47]

External users are provided on-site and remote access to top-down and bottom-up synthesis and self-assembly. The network allows many of its resources to be accessed remotely. It will provide a computation and Web-based infrastructure to organize and distribute the knowledge base. There will be a Web-based infrastructure that links NNIN initiatives in education, outreach, and societal and ethical studies. Network-based education and information tools and local hands-on activities will be developed to achieve a hyperlinked open textbook for advanced students, K–gray distance learning, a Web-based magazine for six- to ten-year-olds in science, modular teaching packages for undergraduates and teachers, and a specialized program for outreach to underrepresented communities. The NNIN will develop an infrastructure and research environment to support consideration of the societal and ethical consequences of nanotechnology. The topics of focus will include economic, education, environmental, health, legal, security, cultural implications, ethics, communication, workforce change, and industrial innovation. The various facets of this program will be highly linked, producing a cascade across the public and scientific communities.[48]

Goldberg says, "[I]t will cover more disciplines, such as environmental nanomaterials, and also will allow a greater range of users from academic to small businesses to international users...and reach out beyond its user base to provide educational information to elementary and high school students."[49] By September 2004 much of the educational initiative remained in the very early stages of development.

Twenty-First Century Nanotechnology Research and Development Act

"Politicians have been successfully sold the idea that nanotechnology is the industrial revolution of the twenty-first century."[50] The most recent manifestation of this trend was S.189, which worked its way through the House of Representatives and the Senate in 2003 and found itself incorporated in President Bush's budget for FY 2005.

According to Roco, "significant infrastructure has been established in over 60 universities with nanotechnology user capabilities."[51] The bill would add to this base. It began in the office of Senator Joe Lieberman but was shelved amid the chaos of responding to 9/11. In 2002 Senators Ron Wyden and George Allen and House Science Committee chairman Sherwood Boehlert (R-NY) donned the mantle. "The bill promised jobs and economic development, and in the short term delivered research projects to government labs and universities in potentially every Congressional district."[52]

On November 19, 2003, the United States Senate passed by unanimous consent a version of the Nanotechnology R&D that was negotiated with the House Science Committee. The bill enjoyed the backing of the NanoBusiness Alliance since its initial development in 2000. The House passed the bill and sent it on to the president, who signed it. Bush's support via the new law was "likened to President John F. Kennedy's push to land on the moon by the end of the 1960s."[53]

S.189 put the president's NNI into law and authorized $3.7 billion over the next four years for the program. The law also required the creation of research centers, education and training efforts, research into the societal and ethical consequences of nanotechnology, and efforts to transfer technology into the marketplace. Finally, the law included a series of coordination offices, advisory committees, and regular program reviews to ensure that taxpayer money is being spent wisely and efficiently.[54]

According to Congressman Mike Honda (D-CA), "The Act provides funding for basic research and development at universities and national laboratories so that we can study these most fundamental questions about the nanoscale world. The discoveries that are made by these researchers will be the foundation upon which the private sector can develop a new generation of products."[55] According the House Science Committee, "The hallmarks of [the law] are three-fold. It aims to increase interdisciplinary research, interagency coordination, and research on societal consequences."[56]

Specifically, the act authorized $809.8 million for FY 2005, $889.6 million for FY 2006, $955.4 million for FY 2007, and $1.0241 billion in FY 2008. Specific programs are designated at the NSF, DOE, NASA, NIST, and the EPA.

The justifications for such a large investment during slim economic times are at least twofold. First, "experts say this bill could help ensure US dominance in nanotechnology."[57] One can hardly ignore the arguments of its supporters, especially Senator Allen, who sees in it a way to provide employment domestically and to capture an important market in the global economy. Second, "nanotechnology needs nurturing. More than the Manhattan Project, nanotechnology is like the space program—without the government, it couldn't have taken off. Applications are 10, 15, 25 years down the line, and no corporation has interest in funding something that isn't going to bring back a profit ratio."[58] Only a handful of companies are investing in nanoresearch, and most of those are investing in nanoparticle applications. Venture capital investments in this field remain minimal.

The act represented an attempt to create a permanent bureaucracy, based in Washington, DC, to ensure that nanotechnology funding remains firmly in the budget for years to come. What we have learned about government bureaucracies is important here because bureaucracies are durable constructs. As such, the NSF and the Departments of Energy and Defense, among others, will be distributing the funds allocated under this act while the act itself will be under the direction of the executive department through the Office of the Science Advisor to the President.

Also, it

> contains a provision that would establish interdisciplinary Research Centers, funded in the range of $3–5 million per year, for the next 5 years.... The goal is to establish geographically diverse centers, including at least one center in a state participating in NSF's Experimental Program to Stimulate Competitive Research (EPSCoR). [The EPSCoR center is at Oklahoma State University in Stillwater.] Such centers could play a key role in developing a cadre of scientists and engineers trained in interdisciplinary studies to push the frontiers of nanoscience.[59]

In addition, the funding provisions of the act were to establish a formalized institute that had the duties and obligations to insure that societal and ethical implications of nanotechnology will be ongoing. The Center for Nanotechnology in Society (CNS) promised ongoing financial support for a five-year period during the duration of the concurrent NNIN. The SEIN component of the NNIN would be closely aligned with the act's center and may actually involve the same researchers and settings. "Such a center would ensure objectivity and substance. The truth is if we don't have this nano center a bunch of public relations firms are going to take the mantle of SEIN."[60]

During the debate in the House, Representative Sheila Jackson-Lee (D-TX) offered and then withdrew a related amendment that would have created the Center for Societal, Ethical, Educational, Environmental, Legal, and Workforce Issues Related to Nanotechnology. While it never made the final version per se, the CNS is being funded as a Nanoscale Science and Engineering Center in 2004. "We've seen a troubling development in the have-nots of our society finding them on the wrong end of the technological divide," Jackson-Lee said. "Change has not made its way into every area of our community. People are being left behind.... [I want to] ensure that nanotechnology works for all Americans."[61] As such, the act is incredibly sensitive to minority participation, and the CNS must ensure minority participation and outreach, especially to underrepresented populations.

More importantly, the act represents an attempt to address the funding concerns for nanotechnology. Start-ups are not attracting the venture capital they anticipated. Both a function of the current economy and risk-aversive investors, especially the near extinction of "angels" (friends and family willing to finance), funding for sustained research and development in nanotechnology demands creative and traditional approaches to basic research in science. Unfortunately, creative sources hardly abound.

Universities, especially public ones, facing tightening budgets have lost the flexibility to fund their own "Centers of Excellence." The act helped establish a network of these centers. Since 1991, for example, twenty-two new nanoscience research centers have been funded by the federal government. Absent a commitment from government and a handful of private foundations, the only other alternative would be business/university research relationships. However, these relationships have been seriously mitigated by university and state restrictions on licensing, how intellectual property is resolved by universities, and many other considerations. As such and in the immediate future, nanoscience- and nanotechnology-related research demands must be sustained by federal and state government funding. Whether this act comes close to this mandate, it did establish a substantial commitment in difficult financial times.

The act added an advisory panel to provide feedback to the president and senior officials as well. John Marburger announced that the National Nanotechnology Advisory Panel will be assumed by PCAST. While there are no well-known nanotechnology players on PCAST, cochairman E. Floyd Kvamme claims it has "folks with a lot of managerial experience managing technology projects, which will help them fill the largely oversight role established for the advisory panel."[62]

The act offers another important element. By formalizing the sustained support for a CNS, the commitment for societal and ethical implication of nanotechnology might serve as a template for science and technology decision making in terms of other newly emerging technologies. Whether it involves nanoscience and nanotechnology enhancement of products above the nanoscale or whether it involves the confluence of different fields, such as nanobiotechnology, novel challenges to the currently nonexistent public sphere in science and technology will arise. The act offered a rhetorical sensitivity to create and sustain a body of citizen-consumers who advise and consent on policy-related decisions (see chapters 10 and 11). The first opportunity for public participation came with the societal and ethical mandate associated with the Human Genome Initiative. Of course, when the initiative was completed, the mandate evaporated. Since nanotechnology is a general-purpose technology, its relationship to our lives is unlikely to end in five years. Much like the steam engine impacted industrialization for nearly half a century, nanotechnology, in one form or another, will remain intimately associated with us for many decades. The CNS might find problems and issues for many years to come and could have served as a model for other scientific and technological disputations. As this book goes to press, the CNS NSEC was just announced. It seems the CNS will be a distributed network like both the NNUN and NNIN rather than a stand-alone center per se (see chapter 10).

"Both sides of the aisle in the Senate should be commended for their foresight and hard work in getting this bill through," said Modzelewski. He added, "This nanotech law is a vital catalyst for the development and growth of what will become a $1 trillion piece of the global economy. This law will also allow the U.S. to continue to strive for global leadership in this highly competitive nanotechnology marketplace."[63]

Science Committee chairman Boehlert applauded the Senate's action as well. "The nanotechnology program will be a model of how government, universities and industry can work together to advance science and bolster our nation's economy."[64]

One of the consequences of this new act was its reading on Wall Street. For example, Nanogen's stock more than doubled on December 3, the day the act was signed by President Bush. Entrepreneurs say that the nanotechnology investment climate is warming up just in time to meet their growing capacity to put investors' money to work expanding research and bringing innovations to market. "We are really close to the take-off point where the industry could absorb a lot of money," said David Ludvigson, CFO of Nanogen.[65]

INTERNATIONAL ACTORS

It has become abundantly apparent that nanotechnology is being immersed into the US economy at an unprecedented pace, largely because of government efforts. However, the United States is not the only nation that has begun to place considerable importance on making nanotechnology a serious funding priority. According to John Marburger, Bush's science advisor and the man in the hot seat, "Nanotechnology and energy related technologies are fields where we're probably more vulnerable to being matched in innovations by some other countries and we have to pay attention to those, and they should be made a priority, and in whatever funding context we have, we should divert money to those things."[66]

The United States is concerned with Europe for much the same reason that it is concerned with Japan and China: the idea that another nation or region might be outspending us in an emerging technology that could lead to global dominance seems antithetical to America's competitive nature and its exceptionalist mind-set. Consider the following remark made by Craig Barrett of Intel: "U.S. leadership in technology is under assault. The challenge we face is global in nature and broader in scope than any we have faced in the past. The initial step in responding to this challenge is that America must decide to compete. If we don't compete and win, there will be very serious consequences for our standard of living and national security in the future."[67]

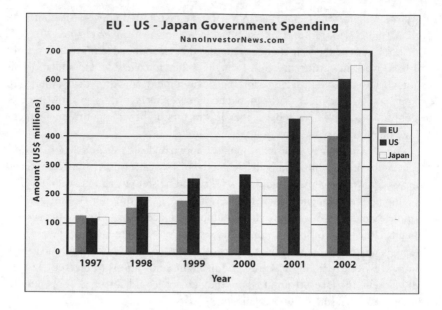

Nanotechnology investment expert Glenn Fishbine compares the quest for dominance in nanotechnology to the pirates of the sixteenth and seventeenth centuries (such as Sir Francis Drake) and their governments (in this case Queen Elizabeth) that provided the ships and finances necessary to make these acts of piracy possible.[68] In the twenty-first century, most governments are now hesitant to finance quests to plunder ships; nonetheless, the economics of piracy remain largely intact. Nations that are on the cutting edge of research look for world dominance by financing research and programs designed to discover and conquer burgeoning technologies. In today's world, it is nanotechnology that has the attention of major governments worldwide, and it is the new economic world that is ripe for exploration, discovery, and ultimately conquest.

It seems that despite a slow start in the exploration process in the late 1990s, the implementation of the NNI has allowed the United States to be first out of the blocks. Initially the United States was underspending its economic rivals such as Japan and China. In 1997 both the European Union and Japan provided far more funding for research than the United States.[69] For example, "in 1999 the NSTC reported that worldwide spending for nanotech was still at least twice that of the U.S. With the establishment of the NNI in 1999, the U.S. has significantly increased its funding for nano research" (see above).[70]

While 2003 and 2004 budget estimates place the United States in second place to the Japanese in overall spending,[71] it is difficult to determine true spending between nations because of the general lack of clear distinctions between nanotechnology and other small technologies.

After 1999 many nations initiated significant research and development with much of the same focus as the NNI in the United States, and while none of these programs has the breadth of the NNI, they are all designed to support economic sectors important to local economies.[72] For example, in 2005 India announced plans to put commercialization of nanotechnology on the fast track. President Kalam called on Indian companies to invest $300 million to research in nanotechnology. A set of R&D institutes are engaged in research and collaborative arrangements with government and industry that will focus on five areas: nanometal oxides, MEMS-based gas sensors and biosensors, nano solar cells, nano fuel cells, and nonporous silicon microcavity biosensors.[73] They announced a joint R&D memo of understanding to produce nano-related products with Israel.[74]

European Union nations as well as a number of Asian countries, including Japan and China, are trying to keep up with the pace of American spending on basic nanotechnology research, which consequently sets American investment in nanotechnology at only about 25 percent of the world's total. According to Lux Research, "although the U.S. government spends more in terms of raw numbers, it has fallen behind Asian competitors when their spending levels are corrected to reflect the difference in what a dollar buys there versus here. On that basis, while the United States invested $5.42 per capita in government nanotech spending in 2004, South Korea invested $5.62, Japan $6.30 and Taiwan $9.40."[75] While this metric might be correct, it seems to aggravate the issue of national leadership, which doesn't seem to have transformed itself into a problem; nonetheless, the nationalistic rhetoric continues to pervade discussions of national programs.

Even in university settings, the United States is finding itself in a vulnerable position. Major research labs in top foreign universities are beginning to take the edge off American research, and corporations such as Hewlett-Packard seem to be fleeing domestic universities, looking to commercialize foreign research. Clearly, the race is on. While programs have been initiated in Brazil, Israel, Pakistan, South Korea, Taiwan, the Netherlands, Thailand, Germany, and elsewhere, the following examines the United Kingdom and European Union, and Japan and China. "The U.S. government spent $1 billion on nanotech research in 2004, just ahead of China, Europe, and Japan, which each spent about $900 million."[76]

The United Kingdom and the European Union

Typically, global technology sectors in the twenty-first century seem to end up evolving into a bipolar battle between America and Asia for dominance in the marketplace. For some, Europe is written off as merely a speed bump on the road to real research and application. In nanotechnology, the situation is much different, as Europe has joined the race. Nonetheless, "Europe is perceived to be well behind the USA in both nanoscience and the transfer of nanotechnology to industry."[77] The United Kingdom has problems with its focus and its indigenous venture capital industry. France is simply not designed to integrate its labs into a larger project.[78]

As such, some government initiatives in Europe have used a network model to maximize efficiencies. For example, the Dutch announced the creation of a consortium through 2009 called NanoNed; its goal is to boost expertise with a budget of €235 million.[79] The Dutch consortia strategy mirrors the macro-initiative taken by the EU.

Despite shortages of skilled research personnel and interdisciplinary skill sets,[80] the EU as a body has made significant strides. Though entering the field late in the game, European nations, especially under the leadership umbrella of the EU, have become powerful players with large public investments in research and commercialization. While the EU is spearheading much of the current research, many countries have committed themselves independently to investing large amounts of capital into nanotechnology research long before the EU got into the game, as the table below illustrates.

Germany and the United Kingdom are clearly providing the bulk of spending for nanotechnology outside of the official EU program and using most of this research money for the traditional medley of studies in semiconductor fabrication and nanophotonics. Germany most notably has formed a program known as EXIST, which is a subsidiary funding program of its Ministry of Education and Research and is responsible for some of the most innovative start-up opportunities in the world. It offers among other things to pay the salaries of professionals and academics for five years while they begin start-up companies in nanotechnology.[82] But despite some individual efforts, the EU, through the Sixth Research Framework Programme, provided the bulk of leadership in dictating the direction of nanotechnology for the European community.

INDIVIDUAL SPENDING ON NANOTECHNOLOGY (IN MILLION EUROS)[81]

FY 1997–FY 2000

Country	1997	1998	1999	2000
Finland	2.5	4.1	3.7	4.6
France	10.0	12.0	18.0	19.0
Germany	47.0	49.0	58.0	63.0
Italy	1.7	2.6	4.4	6.3
Netherlands	4.3	4.7	6.2	6.9
United Kingdom	32.0	32.0	35.0	39.0
Total	97.5	103.4	125.3	148.8

The aim of the Sixth Framework Programme was to produce "breakthrough technologies that directly benefit the EU, whether economically or socially."[83] On a less grand scale, the program "proposed an integrated approach to strengthen Europe's R&D in nanosciences and help Europe to become the world leader in nanotechnology."[84] Much like the NNI, the Sixth Framework outlined broad themes and long-term goals for nanotechnology research, and it broke up funding into several categories as follows.

Europe's focus on nanotechnology is slightly different than its American counterparts. In the Sixth Framework, the focus is mostly on breakthroughs.[85] It still remains unclear as to whether or not this emphasis on application rather than science will pay off in the long-term for Europe. By focusing on breakthroughs, it is clear that the European community seems more interested in immediate applications than foundation-based research science, though that may be changing. In the short-term, this strategy seems successful, as the EU continues to run a close second to the United States in the number of patents accrued up to this point from 1991 to 2000, though that lead may have recently shifted to China.

THEMATIC PRIORITIES OF THE
SIXTH FRAMEWORK (IN MILLION EUROS)

FY 2002–FY 2006[86]

Thematic priorities budget	June 2002 final	Min. (%)	Total	Max. (%)	Total
Life sciences, genomics, and biotechnology for health	2,255	1.0	22.6	2.5	56.4
Information-society technologies	3,625	7.0	253.8	9.0	326.3
Nanotechnologies and nanosciences, knowledge-based multifunctional materials, and new production processes and devices	1,300	25.0	325	30.0	390.0
Aeronautics and space	1,075	0.2	2.2	0.2	2.2
Food quality and safety	685	0.2	1.4	0.2	1.4
Sustainable development, global change, and ecosystems	2,120	0.2	4.2	0.2	4.2
Citizens and governance in a knowledge-based society	225	0.2	0.5	0.2	0.5
Total	**11,285**		**609.7**		**781.0**

One of the intentional outcomes of this emphasis was the creation of the Innovation Relay Centers (IRCs), which serve the same broad function as the National Technology Transfer Center in the United States.[87] In Europe, there are nearly seventy regional IRCs that help to commercialize research and bring nanotechnology into the sphere of application.

The EU has made an increasing monetary commitment to nanotechnology.[88] The "program (2002–2006) devotes €1.3 billion to nanotechnology, and the EU aims to step up this effort in the broader context of the proposed doubling of the EU research budget in the 2007–2013 period."[89]

Nanotech leaders in the United States have asserted that European funding for nanotechnology may very well be greater than that of the United States.[90] European rhetoric at this point is feeding such a case. The European NanoBusiness Association (ENA) released a report titled *It's Ours to Lose: An Analysis of EU Funding and the Sixth Framework Programme.* It proclaimed that Europe is in fact measuring up to the United States in nanotechnology research and funding:

> A figure against which nanotechnology funding is often benchmarked is the budget of the US National Nanotechnology Initiative. At first glance, this appears to suggest that Europe's often quoted €1.3 billion budget over four years is tiny compared to the 2003 NNI budget of $710.2 million budget. Our analysis indicates that the top level figures do not reveal the whole story, that many of these headline figures are in fact misleading, and that European nanotechnology spending may in fact be significantly higher than that of the United States.[91]

From reading the ENA report it appears as though the main aspect of nanotechnology research currently underway in Europe is ensuring that they are in fact monetarily funding nanotechnology to a greater degree than its American counterparts. Of course, Europe has a broader definition of nanotechnology, hence its research includes non-nanoscale projects as well, so it becomes difficult to paint the entire picture based upon funding figures alone. This rhetoric about who's ahead and who's behind is simply another illustration of the function of hyperbole in promoting nanotechnology. Both the United States and the EU overtly or subtly tag national pride to their claims and counterclaims on who's ahead in the race to the bottom.

In general, it appears as though Europe is running a close second with Japan, nearly matching the United States at this point. However, unlike Japan, Europe's "R&D and commercialization infrastructure is similar to the infrastructure found in the United States."[92] This is sure to make Europe a far more friendly environment for venture capital and start-up companies than Asian nations.

The chart on page 141 helps illustrate the disproportional amount of nanotechnology funding that is coming from the government, as well as the problems it has created in the area of supporting new private-sector nanotechnology companies.[93]

Because of this disparity in private-sector investment growth, the United

FUNDING SOURCES FOR NANOTECHNOLOGY RESOURCESS IN EUROPE

FY 2003–2004

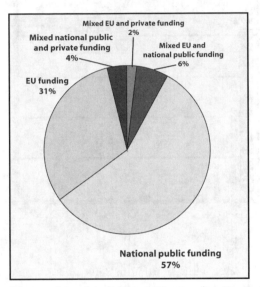

States has been able to enjoy a head start on Europe, but Europe is starting to get the message and making adjustments to encourage the flow of private capital.

Up to this point, "nanotechnology researchers in Europe know where to go when they need money to develop new projects: the government."[94] The role VC will play in Europe remains uncertain. Today, European nanoventures receive far more from the public sector than from the private, and while they may have the framework for broad-based venture capital in the future, it is a framework that thus far is not being utilized (see graph on page 142).

The EU has many projects about to take off. "The EU recently announced it would invest about $12 billion a year to finance 56 so-called Quick Start projects, which have the potential to produce economic results quickly."[95] Also, "Europe's small-tech community got its first look at the research funds earmarked for the EU's Sixth Framework Programme, the funding cycle that runs from 2002 to 2006."[96] They received nearly twelve thousand proposals from fifty countries in late 2003 for the $6 billion in research funds to be distributed in 2004. "The latest program for projects grouped under the heading nanotechnologies and nanosciences, knowledge-based multifunctional materials and new production processes and devices has a pot of €6 million. But a further €180 million is available for nanotechnology proposals that combine with information society technologies."[97] In April 2005, "the EU's Executive Commission announced that it wanted to almost triple spending on research to €10.4 billion a year from €3.5 bil-

VENTURE CAPITAL FUNDING BY COUNTRY
FY 2002–2003

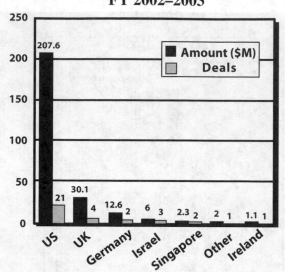

lion a year now for seven years, starting in 2007, and a bigger portion of that enlarged budget will go toward business research."[98]

Commercially established industries believe much more needs to be done. Recently, Nokia Corp. and STMicroelectronics signed a joint declaration with the EU "calling for $7.32 billion a year to be pumped into the development of nanotechnology to prevent Europe falling behind the U.S. and Asia."[99] This declaration is consistent with the recommendation of the EU paper *Towards a European Strategy for Nanotechnology*[100] for a threefold increase beyond the $1.6 billion the EU had earmarked for the Sixth Programme when the Seventh Programme kicks in. They argue, "substantial public funds will be needed to leverage the required level of private investment. This can involve European member states contributing directly through intergovernmental, national, and regional programs, and European funds."[101] This is quite a challenge given that "the Engineering and Physical Sciences Research Council, for example, wants to achieve 50 percent industrial participation by 2007."[102]

Here are two illustrations of the types of projects funded under the Sixth Framework. Nano2Life will be the first Network of Excellence in nanobiotech. It will develop joint research projects in four major technical areas: functionalization, handling, detection, and integration of devices. "One hundred and seventy scientists from 12 countries are integrated under this network to implement joint activities at the interface between the *nano* and the *bio* world. Nano2Life will set the basis for a virtual European Nanobiotech Institute with strong local establishments."[103]

COMPANIES ACTIVE IN THE NANOTECHNOLOGY SECTOR

FY 2003-2004

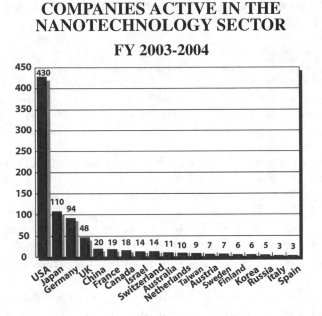

Another example is "the PHANTOMS Network, the Nanotechnology Network for Information Processing and Storage. It consists of an interdisciplinary platform in nanotechnology for the information processing and storage field, where its 228 members in 32 different countries are interconnected.... It is all about interaction with access to documents, simulation codes and reprints."[104] This platform provides both researchers and industry with access to the tools needed to create and maintain a multidisciplinary community in the forefront of the nanotechnology revolution and represents a single entry point for those seeking information about nanotechnology and, more specifically, bioelectronics.[105] Despite these efforts, the United States retains an astounding lead in commercialization (see graph on this page).[106] This trend suggests some serious structural problems in the European investment world.

Put simply, private investment in European industry has lagged. For example, the 2004 House of Commons Committee on Science and Technology's disparaging report on nanotechnology investment in the United Kingdom agreed: "The relatively low level of availability of venture capital in the UK has been identified as a significant barrier to innovation, which affects nanotechnology start-ups like those in other fields."[107] The report concluded, "We detected a strong suspicion that the relative lack of competition in the European venture market hampered efforts to secure funding on good terms in the UK. Compared with their US counterparts, European venture capital companies were held to be risk averse and myopic in outlook."[108]

There are some good reasons why this problem haunts British investment in newly emerging technologies:

> Americans had a more generous and long term approach to venture capital funding, chiefly because they managed much larger funds with the consequent greater leeway for taking risks. Because of the often long term nature of their R&D, nanotechnology companies will tend to stand at the riskier end of the venture capitalists' spectrum and thus have to work harder than most to attract funding.[109]

This does not mean the United Kingdom has relinquished leadership to the United States. For example, the innovative London Centre for Nanotechnology (LCN) can be viewed as an international hub. Andrew Chambers, technology director at Surface Technology Systems, portrays a recent investment in a new professional chair there, funded by the Japanese, as having the potential to place the United Kingdom among the world leaders in this influential field. The LCN brings together the University College of London and Imperial College London into eight departments in electrical engineering, materials, medicine, and so on.[110]

Globally, the United States seems to be the nation to catch in the race for nanotechnology. Unsurprisingly, however, the stakes are high for developing applications to a technology that most experts agree will change the face of economics, politics, and existing business models in the very near future. While many of the challenges associated with foreign spending on nanotechnology have been conflated by those advocating more research funding here at home, the United States has made nanotechnology a primary economic priority for the future. What Asia has accomplished in electronics and semiconductors might be repeated with nanotechnology.

Japan

John Marburger attests, "If you look at the amounts of money we budget explicitly for nanotechnology, it's not much different from what Japan is budgeting explicitly for nanotechnology. So you could say Japan is comparable to the U.S. in its investment in nanotechnology," though he does admit, "The U.S. is investing in a lot of other stuff that is relative to nanotechnology, but we don't count it in that budget."[111] Consequently, US spending may be higher than the nanotechnology-specific budgeting communicates. Japan is undeterred and remains very interested in nanotechnology. "The government regards successful nanotech development as a key to the restoration of the Japanese economy. Japanese government spending on nanotechnology reached $1 billion in 2003."[112]

As is true with most emerging technologies, Japan and China are investing such large amounts in research and development that they are at this point acting as the primary competitors for American industry. Because of Japan's dominance in the electronic-technologies market, it is not surprising that Japan has also emerged as the dominant power player thus far in Asia, with China currently finishing a very distant second.[113] However, that may be changing as well (see below).

Japan's investment in nanotechnology has a growth rate of 15 to 20 percent annually, putting Japan at only a slightly higher funding level than the United States in FY 2003.[114] Experts from the United States and Japan recognize that both nations are funding nanotechnology at exceptionally high levels, but the way in which nanotechnology research is funded and carried out is vastly different in each nation. The contrasting feature of the NNI, when compared with Japanese initiatives, is that the NNI is a special strategy that establishes the importance of nanotechnology to the nation's research and science strategy and includes various government agencies working together to draw up a long-term comprehensive plan.[115] In Japan, however, there is no equivalent to the NNI, and the predominate research and funding occurs within the Ministry of Economy, Trade, and Industry (METI), and the Ministry for Education, Science, and Technology (MEXT). At the same time, funding and research are spread across the nanotechnology field under the supervision of many different organizations, which makes it difficult for the government to formulate a set of long-term and comprehensive research goals in the field. Because of this, it is likely that Japan will formulate its own version of the NNI in the near future.

Another striking contrast between the NNI and various Japanese policy programs is the breadth at which the NNI attempts to develop nanotechnology. As the bar graph on page 146 demonstrates, Japan's commitment to molecular nanoscience is heavily focused in just a few key areas. Because of this, it seems plausible that Japan could one day find itself the global leader in just one or two key areas of nanotechnology, while generally lagging behind other industrialized nations in other areas.

Recognizing the United States' early lead in the field of nanotechnology, in 2001 Japan issued the Science and Technology Basic Plan, which "identified nanotechnology as a major field of concentration, along with the life sciences, information, and communication technologies, and the environment."[116] This initiative was later followed by Japan's Science and Technology Plan, which again listed nanotechnology as one of four major pillars needing to be explored. Under the vague auspices of these plans, most of Japan's research was to be carried out at university and government laboratories (not unlike in the United States).

In terms of specific fields of emphasis, Japan is focused on strong points within their research sphere, which includes nanomaterials and IT/electronics (see the bar graph on page 146 which reports sectors in billions of yen), both of which come out as clear winners in terms of financial backing.[117] And while government funding has by no means been interdisciplinary, there has been some

BREAKDOWN OF JAPAN'S NANOTECHNOLOGY BUDGET BY SECTOR

FY 2001–2002

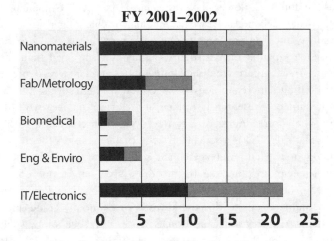

interest in a number of measures specifically aimed at creating a more broad-based approach to research and development, which increases the likelihood that public research can soon be adequately capitalized and exploited by the private sector. This new pseudo-interdisciplinary approach is most apparent in the National Institute of Advanced Industrial Science and Technology (AIST), which has the explicit responsibility of directing research programs of national interest in any number of its fifteen research institutions or twenty-three university laboratories. The most noteworthy of the government laboratories include the Institute of Physical and Chemical Research (RIKEN) and the National Advanced Interdisciplinary Research Laboratory (NAIR), both of which are located in Tsukuba Science City.

It is primarily the likelihood of losing the lead to Japan in nanotechnology that is motivating some of the nanoscience in the United States. But many overlook the fact that Japan is playing catch-up in the area of government-sponsored research. While the United States began to devote substantial government funding as early as 1996 with the NNI, Japan did not fully commit themselves to nanotechnology until late 2000.

Japan, as well as the United States, has made large strides in the field, both in funding and research. To say that the United States does not continue to enjoy an early lead in the field is simply untrue. The Japanese concept of nanotechnology is far broader than that of the United States. "It incorporates all the concepts of nanotechnology along with biotechnology. For example, technologies that deal with DNA at the molecular level fit cleanly with the Japanese definition of nanotechnology. A majority of Japanese activity focuses on either the semiconductor industry or the biotechnology industry."[118] In a world in which the United States took a similar definitional approach to nanotechnology funding, our current research and development in both the public and private sector would far exceed spending by other nations.[119]

In the United States, regardless of any potential fluctuations in government-subsidized research, the country can always rely on a very strong tradition of entrepreneurialism and venture capital, which will allow it to lead the pack regardless of any potential government neglect.[120] Up to this point, this has not been the case in Japan. Without some reforms it is doubtful that the commercialization of Japanese nanotechnology will ever progress along equivalent lines.[121] Unlike the United States, Japan has not offered much in the way of venture capital or other similar opportunities for small start-ups. There have been efforts to turn this around. For example, "Japanese giant, Mitsubishi, recently announced a $120 million venture fund to support nanotechnology. Mitsui announced plans to set up three units of nanotech research and has backed it with a commitment of $81.4 million."[122] But these efforts may be insufficient to have any major impact.

For Japan though, one of the major barriers to emerging technologies has not been scientific roadblocks but the basic structure of how technology was developed in general. In the past, Japanese research has often been "bogged down in bureaucracy and an organizational structure that dates back to the beginning of the century."[123] While still a problem, it has begun to slowly change. In 1991 the first major reorganization of Japanese government laboratories took place within the AIST. The plan included the reshuffling of several research institutes in Tsukuba City. While the changes were "intended to inject new life and greater flexibility into the government's research system,"[124] the program has had only limited success. The effort was also somewhat successful in "concentrating more efforts on basic research and trying to remove some of the internal divisions that typically existed within Japanese ministries."[125] Despite these efforts in the early 1990s, Japan's organizational difficulties may still greatly affect its research efforts. Even today, Japanese research laboratories are still very vertically structured, which leads to isolation and a lack of contact between groups that oftentimes even reside in the same building. Thus far Japan has shown little intermingling between fields, although they are attempting to curb this with the creation of laboratories such as RIKEN, which specifically focus on interdisciplinary research and development.

The main split that continues to divide Japanese researchers from their Western counterparts is the reduced contact between the organic and inorganic sciences. Since some of the most interesting breakthroughs in nanotechnology thus far have been occurring in organic chemistry and biotechnology sciences, this could be a potentially fatal flaw.

Another potential difficulty in nanotechnology research in Japan is the regulatory climate that currently exists in light of serious lapses over the past decade. After the repercussions from the 1995 Tokyo subway nerve gas attack and other

JAPAN—GOVERNMENT FUNDING FOR NANOTECHNOLOGY

NanoInvestorNews.com
1997–2002 (Millions US$)

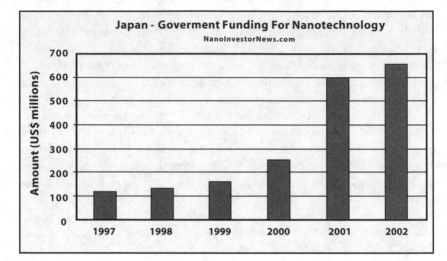

terrorist plots linked to fanatic groups such as the Aum Shinrikyo, public opinion is falling heavily on regulating new technologies far more closely than in the past. This type of regulatory environment will certainly not bode well for commercializing nanotechnology and may draw a great deal of foreign investment away from Japan and toward nations with a more favorable regulatory climate.

Many in the United States, however, overlook these limitations and continue to appeal to nationalism using Japan's increasing investments in nanotechnology to justify increasing expenditures (see the bar graph on this page). While these claims have certainly been effective in pushing nanotechnology onto America's scientific and economic agenda, they have also been vastly exaggerated and misinterpreted. Senator Allen in testimony before Congress asserted that he believes, "we're already behind [in nanotechnology research]. Japan, China, Korea, and the European Union are all ahead of us."[126] Using hyperbole to drive budget increases has been a ploy since the beginning of US involvement in nanotechnology and continues through the present. However, Allen's figurative language is moving a little closer to its literal meaning.

Truth be told, Japan has begun to make very positive strides and demonstrates a substantive interest in nanotechnology. Japan is matching the United States dollar for dollar. This comes as no surprise because "if Japan is to move up-market and survive…nanotechnology…and nanobiology are a must."[127] The two main government ministries responsible for about 90 percent of the country's nanotechnology research programs experienced budget increases. In general, the FY 2004 nanotechnology budget grew 3.1 percent to $875 million. METI's nanotechnology budget tops $101 million. Its overall budget is $1.15 billion, with nanotechnology designated one of four top priority areas. MEXT's budget will top $242 million with emphasis on fundamental materials research. Its Virtual Laboratories nanotechnology research project to develop strategic technologies was boosted to over $430 million.[128] In addition, MEXT established the Nanotechnology Process Foundry in 2002. It provides a platform access point for any industrial, academic, and national research institute group. Today, there are five nanofabrication groups in Japan supported by MEXT.[129] *Small Times* reports that in 2004

> a number of high profile consortiums set between industry, academia and government to develop nano- and bionanotechnologies surfaced. For example, Matsushita Electric Industrial Co., Tokyo Institute of Technology, Nara Institute of Science and Technology, and Osaka University teamed up to develop large capacity memory devices using protein circuitry.[130]

"In September 2003, Japan's two major government nanotechnology laboratories, the National Institute of Materials Science (NIMS) and AIST said they have founded a company, the Materials Design Technology Laboratory, to develop and supply new materials to Japan's automotive and electronics industry."[131] In late 2003, "the major trading house Itochu Corp. signed an agreement with one of Japan's top nanotech laboratories, AIST, to cooperate in helping small and midsize firms push research and development in nanotechnology and life sciences"[132] and recently

BREAKDOWN OF TOTAL JAPANESE PUBLIC BUDGET

FY 2002-2003

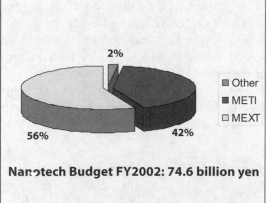

signed a memorandum of understanding to work with Sandia and Los Alamos
national laboratories and the University of New Mexico, New Mexico Tech, and
New Mexico State University. It plans to invest in companies and venture funds,
create licensing and distribution agreements and form collaborations with
research facilities to bring new technology to the global market. Itochu's primary
areas of interest are nanotechnology and biotechnology.[133]

Also in July 2004 NIMS launched an International Center for Young Scientists (ICYS) project designed to recruit non-Japanese English-speaking nanotechnology researchers in their twenties and thirties to Japan.[134]

Beyond government investment, there can be little doubt that Japan's corporate giants are not about to allow the United States to undercut leads in microelectronics and semiconductors. For example, Fullerene International Corp., a
joint venture between Mitsubishi, Materials and Electrochemical Research, and
Research Corporation Technologies (the last two both from Tucson) have combined their nanocarbon intellectual property and expect to produce three hundred
tons of nanotubes and fullerenes in 2005, aiming for full-scale commercial production of fifteen hundred tons by 2007.[135]

There are some indicators worth watching carefully. "Asian companies have
local suppliers of carbon nanotubes, so U.S. companies exporting large amounts of
nanomaterials is an indication that Asian companies are doing either more work
or a better job commercializing nanotech products than U.S. companies."[136] CNI
CEO Bob Gower claims that "companies in Asia, especially Japan, have migrated
to the front of the pack for carbon nanotubes. One-third of the material we are
sending out goes to Asia."[137] CNI teamed up with Somitomo to expand its sales of
single-walled nanotubes in Asia. Gower continues, "the first beneficiaries of
carbon nanotubes will be flat panel displays and 97 percent of those are produced
in Asia."[138] Also, Wolfe reported that Asian companies are attempting to safeguard
chip manufacturing dominance as well. As evidence, he cites Japan-based Takara
Bio as "the lead investor in a recent $10 million Series C for Chicago based Nanos-
phere."[139] The deal included exclusive distribution rights to certain Nanosphere
products in Asia. In addition, Japan is making inroads into the nanotextile industry.
Japan's Toray Industries says it has "developed a nano-scale processing technology
named *NanoMatrix* that puts an even coating of functional material onto each
monofilament of the treated fabric."[140]

Japan has become a major player. With some regulatory relief and other institutional reforms, it could well be on its way to becoming the major competitor
with the United States.

China

While Japan's dominance remains less questionable, its prominence in the East is not. China is making some major investments in nanotechnology, yet it still finds itself second to Japan in both private and public funding for nanotechnology, but it is catching up. China began a major public investment into nanotechnology shortly after the Ministry of Science and Technology conducted a survey in 2001 on the status of nanotechnology research and development as well as industry development in China as a result of President Clinton's NNI. The negative findings of the study led to a major initiative covering the next five years, which will lead to nearly $240 million in investment from the central government and $240 million to $360 million from local and provincial governments.[141] News from *China Nano 2005*: "China has put a total of 830 million yuan into the R&D of nanoscience and nanotechnology and started a major program on nanomaterial and microelectromechanical systems."[142]

So far, nanotechnology research in China covers the usual interdisciplinary areas such as nanomaterials and molecular assembly, but much like Japan, China is focusing particularly on its past strong points by using large portions of public funding for semiconductor fabrication technologies and the development of nano-based electronic components. In April 2005, it ranked third behind the United States and Japan in terms of patent application and ranks second in the number of papers published.[143]

There have been some significant high points, including China's universities. In 2003, "Hainan University in south China offered its first major in nanomaterials science and technology. Harbin Industrial University in Heilongijiany Province announced it has developed a robot capable of nanometer precision placement."[144] In 2005 researchers at the Northeast Normal University in Changchun reported making multiwalled carbon nanotubes by heating grass in the presence of oxygen.[145]

Beyond its universities, China is making some investments in its nanotechnology industry. China estimates there currently exist over eight hundred registered nanotechnology companies in mainland China and Hong Kong; among them, seventeen are currently listed on the stock exchange.[146] China "has made clear its nanotechnology research goals in its middle and long-term national science and technology plan (2006–2020)." According to Deng Nan, vice minister of science and technology, "we have made a decision to mobilize national research forces to reach some breakthrough."[147] China's High Technology Research and Development Program, known as the 863 Program, aims to enhance its overall research competitiveness. It has been especially useful in encouraging research at places like the Chinese Academy of Science, which received a multi-million-dollar investment from the government. That program has helped to fund the "hydrogen storage carbon nanotube program."[148] In late 2003 China started construction of the Nano Sci-Tech Industrial Park in Xi'an,[149] the capital of north-west China's Shaanxi province.

China is also involved in strategic partnering. "The Chinese Academy of Sciences said it had teamed up with Magma Design Automation, Inc. to create a nanoscale integrated circuit design lab in China."[150] According to government sources, "the nanotechnology market for Beijing alone will reach an estimated $60 million before 2005," though this cannot be verified.[151]

Most start-up companies that do exist are typically spun off major research institutions and generally lack truly professional leadership with substantial business expertise and global experience.

Purportedly, China had a weak research and industry infrastructure, but that may be changing. In addition, it offers abundant human resources, attractive market opportunities, and manufacturing competitiveness for nanotechnology businesses currently operating in other parts of the globe. For example, "the Chinese nanomaterials manufacturing companies are offering low cost nanomaterials globally with competitive quality. Moreover, many overseas Chinese are actively searching for opportunities in China to exploit their expertise and network overseas for developing a more successful career."[152]

Incentives, combined with China's increasingly liberalized economic policies, could one day make China one of the most competitive nanotechnology markets in the world. China's labor market and strong existing electronics infrastructure could make it difficult for the United States to compete in the area of foreign investment. In reality, it may very well be China and not Japan that ends up being the United States' largest competitor in nanotechnology in the new century.

While "China says its nanotechnology efforts have already spawned 50 universities, 20 institutes, and 100 companies,"[153] its venture capital industry is still in its infancy, and private investment for nanotechnology has been limited to just one major firm that exclusively sponsors nanotechnology-related start-ups.[154] To compensate in some way, China uses joint ventures as well. For example,

> the Chinese Academy of Sciences and Hong Kong-listed garment manufacturer Good Fellow Group formed a joint venture to develop nanomaterials. Good Fellow invested about $5.6 million in a venture called the Zhongke Nano Engineering Center, following on from investments from the Chinese government which budgeted $52 million since 2000 to develop the center.[155]

Even China is responding with hype. According to Li Minqian, senior researcher at the Shanghai Institute of Nuclear Research under the Chinese Academy of Sciences,

> Despite being a promising prospect, nanotechnology still has a long way to go in China before it can bring about any fundamental changes to people's lives.... Keeping in line with a government-initiated strategy, domestic enterprises and

other establishments involved in the nanotech sector should keep sober-minded to ensure that development of the technology is carried out in a cool, scientific and far-sighted way.[156]

The race is definitely on.

CONCLUSION

Not since the race to produce the fission bomb during World War II have so many countries invested as much rhetoric and resources to a project. In the new world order, the fields of war are being replaced by economic landscapes. Even al Qaeda knew well enough to target the World Trade Center towers in its attacks on the United States infrastructure. Their targets were the Pentagon, the Capitol, and the trade towers. It would seem the future resides with the strongest economy, and since nanotechnology is one of the United States' primary economic drivers, initiatives are bolstering this newly emerging technology. America's preoccupation with taking the lead, winning the race, and being exceptional has generated quite a bit of nanohype. The following chapter examines the rhetoric within the promotional reports on nanotechnology by governments and the business world.

PROMOTIONAL REPORTS ON NANOTECHNOLOGY

Without a doubt, the primary source of hyperbole on nanotechnology has come from governments, especially the United States'. The hype is hardly limited to overclaims about the impact nanotechnology may have on our lifestyle. Government agents have included a nationalist tactic building the argument primarily on warrants associated with economic growth and employment. They claim that without increasing federal appropriations, we would lose a competitive edge to foreign governments in some technological version of the Olympic Games.

They are not alone, and this chapter examines two additional categories of promotion. Business and industry have their own motives, often running parallel with governments' motives for promoting nanotechnology: profit. The evidence is found in the range of arguments they make in promotional documents like those developed for Credit Suisse Equity Research/First Boston and the *Nanotech Report* published monthly by Lux Capital's venture capitalist Josh Wolfe with *Forbes*. The final report covers the issue of risk: the Swiss RE Report is written from the perspective of the insurance industry. While many of the arguments are similar, they differ in degree and perspective, reflecting the sometimes unique interests of the authoring organization.

UNITED STATES

Government sources abound on nanotechnology, especially in the United States, where report after report has been released touting the importance of investing in nanoscience and nanotechnology. Supplementary findings on a broad range of issues have also found their way into government reports, especially in unique applications in Energy, Defense, and Homeland Security. Also, there have been a few reports on SEIN issues, especially nanotoxicology, a subject taken up in chapter 9.

In 1999 one of the earliest reports examined international efforts to invest in nanotechnology. Undertaken by the World Technology Division of Loyola College, it drew two major findings: "First, it was abundantly clear that we are able to nanostructure materials for novel performance. Second, there is a very wide range of disciplines contributing to the developments in nanostructure science and technology worldwide."[1] The findings were a function of the confluence of three important technological streams: control and manipulation of nanoscale building blocks, improved characterization of nanomaterials, and improved understanding of relationships and properties and how these are engineered.[2] The report concluded the United States was in the lead except for nanodevices and consolidated materials, in which cases the lead belonged to Japan. In most other areas, the United States had a slight lead or was on par with Europe. The researchers predicted we were "at the threshold of a revolution."[3] The executive summary concluded that "it is an absolute necessity to create a new breed of researchers who can work across traditional disciplines and think outside the box. Educating this new breed of researchers, who will either work across disciplines or know how to work with others in the interfaces between disciplines, is vital to the future of nanostructure science and technology."[4] This simple study framed much of the discussion in terms of producing a competent workforce to usher in the nanofuture.

In the same year, *Nanotechnology: Shaping the World Atom by Atom* gave a foundational description of nanotechnology with a heavy dose of hyperbole in four general areas. The four sectors were associated with both the economic health of the country and the health and defense of the individual.

- Electronics have driven entire regions of the country economically. It's also an area that involved the United States and Japan in stark competition. The hype: "By patterning recording media in nanoscale layers and dots, the information on a thousand CDs could be packed in the space of a wristwatch."
- Americans will do almost anything to improve human health. Fear of death is a great motivator. We've become a pharmaceutical society, with almost everyone popping two or more pills for miscellaneous purposes daily. Healthcare is the most inflationary section of our economy and supports a major potion of the insurance and trial lawyer economies. Except for stem cell research, Americans bar no holds on medical and biotechnological research. As such, these claims have legs. Unsurprisingly, some of the most noteworthy hype has been associated with human health:
    - "Nanotechnology will lead to new generations of prosthetic and medical implants whose surfaces are molecularly designed to interact with the body."

- ◆ "New nanostructured vaccines could eliminate hazards of conventional vaccine development and use, which rely on viruses and bacteria."
- ◆ "A slew of chip-sized home diagnostic devices with nanoscale detection and processing components could fundamentally alter doctor-patient relationships."
- The third area tags in to economic growth and the environmental lobby's agenda. While environmentalists have not been lulled, the argument that (in general) *green* manufacturing would overcome short-term pollution issues has been repeated regularly. "Bottom-up manufacturing should require less material and pollute less.... In effect, the constructed world itself could become sensitive to damaging conditions and automatically take corrective or evasive action like a hand recoiling from a flame."
- Finally, defense has become more pertinent with the post-9/11 war on terrorism and the invasion of Iraq. While the report predates these events, it was a concern in 1999 and became even more so in 2004. "Fighter aircraft designed with lighter and stronger nanostructured materials will be able to fly longer missions and carry more payload. Plastics that wear less because their molecular chains are trapped by ceramic nanoparticles will lead to materials that last a lifetime."[5]

Next, the IWGN submitted *National Nanotechnology Initiative: Leading to the Next Industrial Revolution* to Congress in February 2000. If the subtitle was not enough, Neal Lane opens the report hyperbolically expressing the intent of President Clinton and the administration:

> The President is making the NNI a top priority. Nanotechnology thrives from modern advances in chemistry, physics, biology, engineering, medical and materials research and contributes to cross-disciplinary training of the twenty-first century science and technology workforce. The Administration believes that nanotechnology will have a profound impact on our economy and our society in the early twenty-first century, perhaps comparable to that of information technology or of cellular, genetic, and molecular biology.[6]

It differs from the following report in tone. Much less technical, it uses metaphors and imagery to make the case. For example, referring to material and manufacturing and projections, it uses the tip-of-the iceberg metaphor to help the reader visualize the scale of possible developments. In energy, it notes green processing technologies.

In the appendix there is a joint letter by Neal Lane and cochair John Young. The letter includes three major claims. First, "in [their] view, the Federal government, together with academia and industry, plays a vital role in advancing nanotechnology." Second, "[they] believe that the science, technology, applications, products, and programs catalyzed by the NNI will inspire a new generation of young Americans with exciting new opportunities and draw them to careers in science and technology." And third, "the NNI is an excellent multi-agency framework to ensure US leadership in this emerging field that will be essential for economic and national security leadership in the first half of the next century."[7]

The third report, also to Congress, *National Nanotechnology Initiative: The Initiative and Its Implementation Plan*, was published in July 2000. It is a technical report. In it, Neil Lane, then assistant to the president for science and technology, introduced the request for a major financial commitment to nanotechnology and the NNI. Hyperbole appears in the executive summary, where claims are made such as: "Developments in these emerging fields are likely to change the way almost everything is designed and made." In another example, "the effects of nanotechnology could be at least as significant as the combined influences of microelectronics, medical imaging, computer-aided engineering, and man-made polymers."[8] Mostly this report outlines some of the things the NNI believes it will accomplish:

- transferring the entire contents of the Library of Congress to a device the size of a sugar cube
- making materials and products from the bottom up, that is, by building them up from atoms and molecules. Bottom-up manufacturing should require less material and create less pollution
- developing materials that are ten times as stronge as that steel, but a fraction of the weight for making all kinds of land, sea, air, and space vehicles lighter and more fuel efficient
- improving the computer speed and efficiency of miniscule transistors and memory chips by factors of millions, making today's Pentium IV seem slow
- detecting cancerous tumors that are only a few cells in size using nano-engineered contrast agents
- removing the finest contaminants from water and air, promoting a cleaner environment and potable water at an affordable cost
- doubling the energy efficiency of solar cells.

Endorsing the maxim *now is the time to act*, the research strategy needed to be balanced across five kinds of activities.

- long-term fundamental nanoscience and engineering research
- grand challenges
- centers and networks of excellence
- research infrastructure
- ethical, legal, and societal implications, as well as workforce education and training.

ALLOCATION OF BUDGET RESOURCES
AGAINST NNI ACTIVITY CATEGORIES
FY 2000–2001

	Fundamental Research	Grand Challenges	Center and Networks of Excellence	Research Infra-structure	Ethical, Legal, and Social Implications and Workforce Education and Training	Total
FY 2000	$87M	$71M	$47M	$50M	$15M	$270M
FY 2001	$170M	$140M	$77M	$80M	$28M	$495M

The document also introduced the National Nanotechnology Coordination Offices (NNCO), which would serve as the point of contact on federal nanotechnology activities, develop and make available printed and other material, and maintain the NNI Web site.[9]

The report stressed coordination and provided a list of key deliverables. For example, by 2005 the following should be first achieved (though we're still waiting on most of them):

- ensure that 50 percent of research-institution faculty and students have access to a full range of nanoscale research facilities
- enable access to nanoscience and engineering education for students in at least 25 percent of research universities
- catalyze creation of several new commercial markets that depend on three-dimensional nanostructures
- develop three-dimensional modeling of nanostructures with increased speed and accuracy that allows practical system and architecture design.[10]

The report emphasized international competition:

[T]he U.S. has a lead on synthesis, chemicals and biological aspects; it lags in research on nanodevices, production of nano-instruments, ultra-precision engineering, ceramics, and other structural materials. Japan has an advantage in nanodevices and consolidated nanostructures; Europe is strong in dispersions, coatings and new instrumentation. Japan, Germany, U.K., Sweden, Switzerland, and EU all are creating centers of excellence in specific areas of nanotechnology.[11]

Also, the report syllogistically claims that nanotechnology is expected to be pervasive in its applications across nearly all technologies—and technology is a major driving factor for economic growth—hence nanotechnology leads to growth.[12]

The report also makes the case for government involvement:

Private industry is unable in the usual 3–5 year industrial product time frame to effectively develop cost-competitive products based on current knowledge. Further, the necessary fundamental nanotechnology research and development is too broad, complex, expensive, long-term, and risky for industry to undertake. Thus, industry is not able to fund or is significantly under-funding critical areas of long-term fundamental research and development and is not building a balanced nanoscience infrastructure needed to realize nanotechnology's potential. By implication, the case is made that "government and university research can effectively fill the niche."[13]

It proposed a $225 million increase in federal expenditures for all participating departments and agencies.

BUDGET RECOMMENDATIONS

Agency	FY 1999 ($ M)	FY 2000 ($ M)	FY 2001 (compared to FY 2000) ($ M)	Increase (%)
DOC/NIST	16 (with ATP)	8	18 (+10)	125
DOD	70	70	110 (+40)	57
DOE	58	58	94 (+36)	62
NASA	5	5	20 (+15)	300
NIH	21	32	36 (+4)	13
NSF	85	97	217 (+120)	124
TOTAL	255	270	495 (+225)	83

A major part of the NNI has been interagency coordination. Nanoscience and nanotechnology are multidisciplinary, and the NNI stressed interagency collaborative activities as well.

It is worth noting the second category of activities—centers and networks of excellence. The vision was to fund ten nanoscience and technology centers and networks at about $3 million per year for approximately five years, with the opportunity for one renewal after the review. These centers played prominent roles in both the NNUN and NNIN and in a host of nanorelated research activities over the next decade and a half. In addition, the 2000 document introduced funding for societal implications, but it actually promoted education and training of nanotechnology professionals, especially in industrial careers.

EXAMPLES OF PROPOSED
NNI COLLABORATIVE ACTIVITIES

	DOC	DOD	DOE	NASA	NIH	NSF
Fundamental research		X	X	X	X	X
Nanostructured materials	X	X	X	X	X	X
Molecular electronics		X		X		X
Lab-on-a-chip (nanocomponents)	X	X	X	X	X	X
Quantum computing	X	X	X	X		X
Nanofabrication user facilities	X		X	X		X

Two years later, we get two additional reports. The first from June 2002, *National Nanotechnology Initiative and Its Implementation Plan*, mirrored the 2000 report. This report added a review of outcomes in year one, such as carbon nanotube field-effect inverters, fluorescent visualization with quantum dots, sequencing DNA with nanoporous membranes, and so on.[14] It added "the first year of NNI provided support to over 2,000 active university awards by the 15 participating departments and agencies," with "approximately 65 percent of the funds allocated to academic researchers, 30 percent to government laboratories, and 5 percent to industry."[15] Six major centers in nanoscale science and engineering were awarded $65 million over the first five years at Columbia, Cornell, Rensselaer Polytechnic Institute, Harvard, Northwestern, and Rice.

Three Grand Challenges were added: chemical, biological, radiological, and explosive protection and detection (CBRE); nanoscale instrumentation and metrology; and manufacturing at the nanoscale.[16]

The targeted outcomes extended through FY 2006 to 2007. For example, by FY 2006, outcomes potentially include the first terabit-per-square-inch memory chip tested in a laboratory, manufacturing at nanoscale for three technologies, and monitoring of contaminants in air, water, and soil with increased accuracy for improving environmental quality and reducing emissions.[17]

NNI CENTERS (NSF)

Center Title	Investigator	Institution
Nanoscale Systems in Information Technologies	Buhrman	Cornell
Center for Nanoscience in Biological and Environmental Engineering	Smalley	Rice
Nanoscale S&E Center for Integrated Nanopatterning and Detection	Mirkin	Northwestern
Center for Electronic Transport in Molecular Nanostructures	Yardley	Columbia
Nanoscale Systems and Their Device Applications	Westervelt	Harvard
Center for Directed Assembly of Nanostructures	Siegel	RPI

There is a brief but substantive discussion of efforts to involve small businesses under the Small Business Innovation Research (SBIR) and Small Business Technology Transfer (STTR) programs and a table of state- and local-funded nanotechnology programs.

Small Wonders, Endless Frontiers is from the National Research Council—it is an external review of the NNI. In its ten recommendations it calls for (1) an independently standing nanoscience and nanotechnology advisory board (NNAB) to advise NSET on policy, strategy, goals, and management; (2) NSET to develop a strategic plan with short-term to long-term goals and objectives, emphasizing long-term goals; (3) long-term funding in nanoscale science and technology; (4) NSET to increase investment at the intersection of nanoscale technology and biology; (5) NSET to create programs for the invention and development of new instruments; (6) the creation of a special fund for interagency collaborations; (7) NSET to support the development of an interdisciplinary culture for nanoscale science and technology; (8) industrial partnerships to be stimulated and nurtured, both domestically and internationally, and state initiatives to be leveraged; (9) NSET to develop a new funding strategy to ensure SEIN becomes integral and vital; and (10) NSET to develop performance metrics to assess effectiveness.[18]

Some of the highlights of the NRC's remarks: On the overall research port-folio, the NRC found that "more must be done to create the interdisciplinary and multidisciplinary culture," and "interdisciplinary interactions must be fostered and sustained over the long term."[19] The same group made the connection between nanotechnology and the genetically modified foods industry and com-plained that "NSET has not given sufficient consideration to the societal impact and developments in nanoscale science and technology."[20]

One of its more interesting claims had to do with international partnerships and our interest in collaborating with other countries. "The CIA estimated that approximately one in four new technologies is likely to threaten U.S. political, economic, and military interests by 2015.... U.S. participation in international col-laborations can lessen the risk that nanoscale technologies are turned against us."[21]

The 2003 NSTC report, *National Nanotechnology Initiative: Research and Development Supporting the Next Industrial Revolution*, began with a letter to Congress by John Mar-burger, director of the Office of Science and Technology Policy. In the letter, he con-cludes, "Investments in nanoscale science and technology R&D are essential to achieving the President's top three priorities: winning the war on terrorism, securing the homeland, and strengthening the economy. Programs such as the NNI will help ensure our global leadership in nanotechnology and the many areas it impacts."[22]

This report, much like *Shaping the World Atom by Atom*, is written to be more acces-sible and is much less technical. It includes recent achievements, such as nanotube fibers requiring three times the energy to break the strongest silk fibers and fifteen times that of Kevlar fiber, and nanocomposite energetic materials for propellants and explosives that have over twice the energy output of typical high explosives.[23] The report reviews departments and agencies and the budget request for 2004.

FY 2004 NNI BUDGET OVERVIEW (MILLIONS)

Agency	2002 (Actual)	2003 (Request)	2003 (Approp.)	2004 (Request)	Change from 2003 to 2004 ($)	Change 2003 to 2004 (%)
NSF	204	221	221	249	28	13
DOD	224	243	243	222	−21	−8
DOE	89	133	122	197	64	48
HHS (NIH)	59	65	65	70	5	8
DOC (NIST)	77	69	66	62	−4	−6
NASA	0	1	1	10	9	900
EPA	6	6	5	5	0	0
DHS (TSA)	2	2	2	2	0	0
DOJ	1	1	1	1	0	0
TOTAL	**697**	**774**	**770**	**849**	**79**	**10**

In addition, the report includes the Network for Computational Nanotechnology, an R&D user facility involving investigators from Purdue, Illinois, Stanford, Florida, Texas at El Paso, Northwestern, Morgan State, and the NIST Center for Neutron Research. The report also announced seed SEIN grants given to both the University of Virginia and the University of South Carolina, though only South Carolina's will be transitioned into a full-scale Nanotechnology Interdisciplinary Research Team (NIRT) grant the next year.

In the AAAS Report, Roco discussed the FY 2005 budget request. The FY 2005 request was for $982 million, a 15 percent increase over FY 2004.

NNI FY 2005 BUDGET REQUEST (MILLIONS)

Agency	FY 2003 (Actual)	FY 2004 (Approp.)	FY 2005* (Requested)
NSF	221	254	338
DOD	403	394	407
DOE	134	202	207
HHS (NIH)	78	108	144
DOC (NIST)	64	77	75
NASA	36	47	32
USDA	5	2	11
EPA	5	5	5
HHS (NIOSH)	0	0	3
DHS (TSA)	1	1	1
DOJ	1	2	2
TOTAL	**948**	**1,092**	**1,225**

*Before revisions

The request added the National Institute of Occupational Safety and Health (NIOSH) as a participating agency and nanoscale system, and its manufacturing, energy conversion, and agriculture and foods systems as areas of growth.[24] Roco included

> joint projects addressing nanoscale processes in the environment and implications of nanotechnology on the environment estimates to be approximately $50 million (about $25 million is primarily addressed to environmental issues; the remaining includes nanotechnology projects that have relevance to the environment and are part of the core R&D programs)[25]

to address the growing concerns over the health and environmental effects of nanoparticles, especially carbon nanotubes. A full page of the brief article reviews budgets and justifications for the Departments of Homeland Security and Justice and efforts to "improve the detection of explosives and chemical/biological weapons" and the development of "wearable low-cost devices to provide warning of exposure to unanticipated chemical and biological hazards."[26] In another 2004 article, Roco discussed targets for 2015.[27]

UNITED KINGDOM

There have been many reports coming out of Europe on nanotechnology. The ones that follow have been placed in chronological order by date of publication. All of them are available on the Internet. It is regretful that they are primarily drawn from the United Kingdom. However, there are a few reasons for this: they are representative of arguments associated with the EU's Sixth Framework Programme (see chapter 4), and the UK has taken the greatest interest in researching the nanotechnology phenomenon both from its role as an economic driver and how it plays among citizens and interest groups.

Economic and Social Research Council

The Economic and Social Research Council (ESRC) touts itself as "the UK's leading agency for funding research and training in social and economic issues."[28] It is "an independent organization established by Royal Charter," which is primarily funded by the Office of Science and Technology. Its 2004 budget exceeded £78 million, with which it planned to fund "over 2,500 researchers in academic institutions and policy research institutes throughout the UK and supporting over 2,000 postgraduate students" in an effort to produce world-class social scientists.[29]

The ESRC has a three-pronged mission: (1) to promote and support, *by any means*, high-quality basic, strategic, and applied research and related postgraduate training in the social sciences; (2) to advance knowledge and provide trained social scientists to meet the needs of users and beneficiaries, thereby contributing to the economic competitiveness of public services and policy and the quality of life; and (3) to provide and promote public understanding of the social sciences.[30]

Its four core strategic objectives are (1) to focus social-sciences research on scientific and national priorities; (2) to enhance the capacity for the highest quality in social-science research; (3) to increase the impact of ESRC's research on policy and practice; and (4) to deliver ESRC's activities effectively and efficiently.[31]

There are some striking similarities between the ESRC and the US NSF, which has directorates in the social sciences, especially in the area of nanotechnology research, such as the Social, Behavioral and Economic Sciences (BES) directorate. The rhetoric of economic competitiveness and leadership is found in the justifications for both the NSF BES and the ESRC.

In *The Social and Economic Challenges of Nanotechnology*, Stephen Wood and his colleagues made some interesting remarks. First, "a considerable degree of repackaging of existing research programmes [is occurring] in an attempt by academic scientists and commercial technologists to associate themselves and their area of work with a very fashionable *new new thing*."[32] They add that meanwhile, scientists working in nanotechnology are slightly bemused at the extent of the furor, the first example of a backlash prior to a technology's emergence. It is possible that scientists who have raised expectations about the potential of nanotechnology, in order to secure funding, share the responsibility for the emergence of this opposition. Meanwhile, other scientists are more cautious about what nanotechnology can achieve. For them the potential effects of nanotechnology, and even its nature, are less clear, or perhaps more mundane and incremental. Any discussion of its economic and social impacts cannot simply take the *new nanotechnology* as a given.[33]

The group refers to nanotechnology as evolutionary rather than revolutionary. The more revolutionary the claims, the greater the social effects and the dystopian constructions. They fear that "the dominance of radical perspectives has meant that the emerging debate surrounding nanotechnology has seemingly become polarized before it has been allowed to mature."[34]

They caution against inhalation and ingestion, though they admit many nanomaterials have been in use for years without ill effects. They warn against drawing conclusions about nanoparticles as a class because of the "wide variety of nanoscaled materials and the variety of potential exposure routes." They identify carbon nanotubes as similar to asbestos, complain about insufficient studies, and predict useful information to guide practices in the future including disposal.[35]

Their report projects a demanding agenda for social scientists. First, it

> must be broader than the public-science interface. This should be seen as part of the bigger issue of the governance of technological change. Second, the governance issue, while partly a question of enhancing the democratic processes, is also a question of social learning and of how we learn to evaluate risks and opportunities under uncertainty. Third, there is the perennial issue of equity and economic divides.[36]

In terms of specific research agenda, they direct social scientists to examine "issues unique to nanotechnology; the dependency for its development on interdisciplinary science and engineering; potential new risks; the human-machine-nature interface; specific ethical issues concerning artifacts which mix synthetic and living elements."[37] They warn against a single center to study the societal and ethical implications of nanotechnology since "the effects of nanotechnology are potentially wide-ranging while having context-specific dimensions, it is unlikely that concentrating resources in one center will provide the best value for money."[38]

In general, the report discusses enabling technologies, instruments, and likely commercial applications in the areas of tools, materials, electronic and information technology, medicine and health, cosmetics and food, and military, space, and security. It rejects Drexler's "radical" proposal of molecular manufacturing for many of the same reasons Whitesides,[39] Smalley,[40] and Ball[41] have offered. They conclude "the more radical the concept of nanotechnology and the more advanced its perceived possibilities, then the more revolutionary are its potential social outcomes."[42]

UK Royal Societies

"Wary of conservationists derailing a new technology in the wake of the genetically modified food debate, the British Government asked the Royal Society and the Royal Academy of Engineering to conduct an independent inquiry into the benefits and risks of nanotechnology."[43] In June 2003 the UK government organized a British nanotechnology working group involving the UK National Academy of Sciences and the Royal Academy of Engineering. The group's responsibilities included defining what is meant by *nanoscience* and *nanotechnology*, identifying specific applications, and considering where there might be health, safety, and environmental impacts of the technology.[44] Interestingly, the Brits are simultaneously studying public awareness through hosting public workshops. In addition, they instituted a month long Web consultation to allow everyone to engage with the project and inform the working group's thinking.[45]

The society set up a working group to carry out a study on the best practice of communicating the results of scientific research to the public. Chaired by Sir Patrick Bateson, biological secretary and vice president of the Royal Society, the working group's mission was to identify ways to increase confidence in research results and to identify alternative mechanisms for assessing the quality of scientific research before it is released to the public. The Royal Society solicited written evidence for the nanotechnology study on June 11, 2003, and held a workshop in London on September 30, 2003. Eighty-six individuals (20 percent were engaged in biological, medical, or pharmaceutical research; 60 percent in physics, chemistry, or engineering; 20 percent were from politics, law, regulation, the public, etc.) and organizations responded, and out of these, seventy-nine provided comments.[46] The respondents were European, Canadian, and American.

"The responses generally supported the terms of reference for the study. The confusion of science fiction with science fact was seen as a major issue. Several respondents, from industry, academia, and NGOs, highlighted the need to separate the hype from the hypothetical."[47] For example, "the scientists and engineers picked up on Drexler's ideas as having been a source for the current hype about the field; however, Gene Roddenberry and *Star Trek* were also identified as other sources."[48]

"Nearly every respondent that addressed health and safety issues highlighted nanoparticles as a potential area of concern."[49] John Carroll, an engineer from Cambridge University, agreed with many of the concerns expressed by the Center for Biological and Environmental Nanotechnology (CBEN) and others associated with nanotoxicology.[50] Another, Jason Wiggins, a program manager from Begbroke, expressed concern over the convergence of nanotechnology with biotechnology.[51] The applications seen as substantial drivers were computing, defense, fuel cells, and water purification. They noted that the United Kingdom lacked a strategic vision and networking on the commercial level.[52] Some touted capabilities. Still, others merely summarized definitions and applications found in the popular literature.

Others, like Vidhya Alakson from Forum for the Future, offered recommendations for citizen participation favoring partnerships with businesses actively developing nanotechnology applications in different sectors.[53] The National Consumer Council demanded more consumer-focused models of risk governance that were more equitable.[54] A group called Science, Technology and Governance in Europe (STAGE) broadened participation to NGOs and wider publics.[55] Many warned that nanotechnology might experience the same fate as genetically modified products.[56] Tim Harper of Cientifica added that "the earlier that the public and interested groups were involved in the debate, the lower the likelihood that false impressions of nanotechnology may become fixed in the public's mind."[57] Richard Balmer from Association of Liberal Democrat Engineers and Scientists (ALDES) concluded, much like legislators in Congress and American SEIN researchers, that "scientific research really needs to proceed on a twin track basis. Any move forward needs to be balanced by a resolve to identify hazards at the same time."[58]

Others offered some communications-related concerns. UCLA's James Gimzewski talked about the nanotechnology information issue in terms of a nanomeme with a layering of unlikely ideas over products fueled by a sensation-based media. "Mixed-up nano-memes have emerged, where the differences between science fiction novels, front cover stories and images of reputable journals are becoming differentiated by the proportion of fiction to fact rather than straight factual content."[59] Rachel Brazil of the Royal Society of Chemistry warned that "over-blown claims for nanotechnology will not help [in] communicating the reality of scientific progress and in differentiating between science and science fiction."[60] Peter Dobson, a director of the Begbroke Science Park at Oxford, agreed.[61] Others, like the Novartis Foundation, blamed irresponsible journalism for misunderstandings.[62] Stephen Wood and Richard Jones of the ESRC blamed "some scientists concerned to raise the profile of their field and increase funding."[63] Kristen Kulinowski of Rice and CBEN blamed companies trying to raise money from venture firms.[64] Others simply blamed greed.[65]

Some, like Kulinowski and S. B. Palmer of the Warwick Nanosystems Group, expressed concern over nomenclature.[66] Others, like Barry Evans of PVC Research and Enterprise at Surrey, believed no new legislation is needed to address these concerns.[67] While many advocated the application of the precautionary principle to risk assessment of nanotechnology,[68] Fabio Albertario, senior analyst at IAL Consultants, and Julian Snape from ExtroBritannia (a transhumanist organization) called for a more accommodating application of the precautionary principle.[69]

Some addressed more advanced applications of nanotechnology. Robin Hanson from George Mason University discussed the direct effects on nanofactories,[70] and Brian Wang from a software development and consulting company based in Silicon Valley warned of military uses and called for global power stability,[71] while Jürgen Altmann used the call to publish preliminary versions of a paper on military uses of nanotechnology.[72]

The CRN complained that "the report paid insufficient attention to a significant expected application of nanotechnology."[73] Simply put, the report does not give much credence to a Drexler form of molecular manufacturing. Unsurprisingly, Chris Phoenix claimed that recent work on limited molecular nanotechnology, especially in terms of the convergence of several current technologies, was evaluated unsatisfactorily. He continued that while a self-replicating nanobot may be "quite difficult to design, general purpose manufacturing in a limited domain based on carbon-lattice chemistry from special chemical feedstock may be significantly easier to achieve."[74] Phoenix also complained that while incremental discoveries seem to be the primary mode of getting to large-scale nanomanufacturing, "the development of a self-contained general-purpose nanoscale manufacturing capability would certainly constitute a dramatic breakthrough, and is not a likely result of other technologies."[75]

In January 2004, *Nanotechnology: Views of the General Public*, was prepared for the Royal Society and Royal Academy of Engineering Nanotechnology Working Group by the British Market Research Bureau (BMRB). The report is composed of "two elements of a qualitative strand, consisting of two evening workshops (mini-citizen juries), and a quantitative strand, for which questions were placed on BMRB's face-to-face omnibus survey (with a representative sample of 1005 adults aged 15 or over in Great Britain) from January 8–14, 2004."[76] Its results are reported in chapter 11. There is a breakdown by social grade that seemed to establish a "clear pattern" with "awareness of nanotechnology peaking at 42 percent for ABs (higher and intermediate managerial—administrative—professional) and falling to 16 percent for DEs (semi-skilled and unskilled manual workers on state benefit, unemployed, lowest-grade workers)."[77]

There was a recurrent comment from participants in the workshops: "Playing God" was a phrase that was used in a negative sense and one that respondents spontaneously reach for to disparage certain technology developments. However, they often found it difficult to be more specific about their use of the phrase.[78]

The report added that "spontaneous parallels with genetically modified organisms (GMOs) are being drawn by some.... The GMO parallel was one which was raised a number of times during the workshops" presumably because of the mention of manipulating matter at the molecular level to form entirely new materials.[79]

In addition, "parallels were drawn with the development of nuclear technology and plastics, which, it was felt, had both been hailed as *the future* in their time, but had proved to have serious long-term effects on individuals and the environment."[80]

UK House of Commons

In March 2004 the *Science and Technology—Fifth Report* of the House of Commons published findings highly critical of UK nanotechnology policy. It found "the Department of Trade and Industry [DTI] culpable of failing to build on early successful nanotechnology programs in the 1980s in order to maintain the UK's prominent position in the field." It added, "The levels of investment planned are insufficient to match those of other major international competitors."[81] According to Robert Key, a conservative member of Parliament and a member of the committee that issued the report, "The government is making up policy as it goes, day by day."[82]

This makes more sense if we back up a bit. There was the Taylor Report. John M. Taylor chaired the UK Advisory Group on Nanotechnology and issued a report, *New Dimensions for Manufacturing: A UK Strategy for Nanotechnology,* in 2002. In general, it recommended a "need to recast the scale and nature of nanotechnology activities" and "to raise awareness in industry of the enormous impact that nanotechnology could have [and] to ensure that investment and action by government, industry and researchers is fully aligned to maximize the benefits for the UK."[83] Taylor's group recognized the UK's efforts as "fragmented" and described efforts to facilitate the transfer of science from academia as *irregular,* which has "impeded the development of industrial awareness of, and support for, nanotechnology."[84] Specifically, they recommended a national strategy involving fabrication centers and other reforms. Its recommendations referenced the network of US facilities under the NNUN as a model:

> The most important obstacle to a more rapid application of nanotechnology in industry in the UK is the absence of facilities where researchers, companies and entrepreneurial thinkers can work together to assist established businesses in their adoption of nanotechnology, and to create and incubate new businesses triggered by advances in the science and technology.[85]

The reason the Taylor Report becomes important is that the 2004 findings of the Committee on Science and Technology impugn the "DTI's Micro and Nano Technology Manufacturing Initiative as an inadequate response to the Taylor Report."[86] In addition,

> the pressures for short term financial returns and for a wide regional distribution of funding will result in a disparate range of microtechnology facilities being supported instead of a few world class nanotechnology centers necessary to raise

the UK's nanotechnology profile. We do not have confidence in the RDA's [Regional Development Agency] commitment to supporting or delivering a strategy that is best for UK interests as a whole.[87]

The committee also reported that "universities and academics have been generally slow to respond to the challenge of supporting nanotechnology research.... [M]any of the barriers to successful innovation still exit. Venture capital is still a problem ... as is the apparent reluctance of major companies to sink large sums in R&D."[88]

In response, the DTI commissioned a survey of the UK industrial landscape in September to provide further evidence for a supporting initiative. The group reported in February 2003:

> The main conclusion of the survey was that few large UK companies were using, or had any real strategy for adopting nanotechnology. In the main, large companies regarded nanotechnology as somewhat peripheral to their core business, but expressed an increasing anxiety about their lack of information on its potential for product improvement and development. UK companies were found to be slow to identify or adopt *disruptive technologies* which require moving away from traditional approaches.[89]

The Committee on Science and Technology was not pleased with the knee-jerk reponse by DTI: "We question the need for the industrial survey commissioned by the DTI three months after the Taylor Report has been published. It did not add significantly to the body of the knowledge that was necessary to inform the framework for future funding. The DTI could have responded to the Taylor Report without this unnecessary delay."[90]

The UK government has put an end to the hope of creating a dedicated nanofabrication center in its response to a Science and Technology Committee report. While acknowledging that "the UK would have benefited from the establishment of one, if not two, nanofabrication facilities to give nanotechnology in the UK a distinctive focus, they did not think it right that the bulk of the available resources should be sunk in one center."[91] The government concluded that "government investment in nanotechnology is at present insufficient, poorly focused and by no means guaranteed to produce the overall levels of funding from the RDA's that are predicted."[92] As such, the United Kingdom "ranks significantly behind both France and Germany."[93] It added, "We have little doubt that levels of private sector investment in nanotechnology in the UK are nowhere near those of Japan."[94] "The sums of money currently committed by government and other agencies, spent in line with current strategy, will ensure that the UK continues to fall behind our major competitors,"[95] which underlies, to some extent, why the European Union initiatives have become more important than ever.

EUROPEAN UNION

A global report on nanotechnology released in March 2002 by CMP Cientifica, a Spanish research company, is "a snapshot of an explosion," according to the CEO of CMP. While current sales of nanocomponents are estimated at a modest $30 million a year or so, Barnaby Feder of the *New York Times* reported an estimate from Nanomat, a European nanotechnology consortium, that products using nanomaterials generated sales of $26.5 billion worldwide last year. These numbers are rising quickly.[96]

The Community Research and Development Information Service (CORDIS) 2002 meeting drew about seventy participants. Two of the more relevant issues discussed were workplace training and societal implications. Regarding workplace training:

> Efforts to eliminate barriers between traditional disciplines in academe, caused by differences in terminology and culture, are underway at many universities in the EC and U.S. The graduate students and postdoctoral researchers have effectively served as bridges to link faculty across disciplines. Partnerships of academe, industry and government around nanotechnology educational initiatives are also being developed. Re-training programs for current members of the technical workforce have been initiated.[97]

Equally compelling concerns were introduced about societal and ethical issues. It noted a positive impact on health may be counterbalanced by adverse effects. Nanotechnology may have systemic problems, and improved methodologies or models need to be developed. Studies on health and environmental implications were needed. Its impacts on privacy were noted and a polite nod to uncontrolled replication was made. It encouraged public participation venues such as "scientific cafes and consensus conferences."[98]

The conference made the following recommendations: (1) the exchange of personnel between the United States and the European Union; (2) the development of models for technical workforce recruitment and training; (3) an increase of joint exploration between industry and government in areas such as energy usage, protection of the environment, information technology, and biotechnology; (4) the joint development of instructional materials based on cutting-edge nanotechnology that could be used throughout the educational system and be facilitated through visits by graduate students, postdoctorals, and faculty members; (5) support for new initiatives that bring the technical community in contact with the media and other deliverers of informal science education such as museums and science centers should be promoted; and (6) the expansion of mechanisms that encourage a dialogue with community groups should be supported and evaluated to determine how this can be accomplished most effectively.[99]

In general, CORDIS seemed concerned about the absence of a general blue-print for nanotechnology research: "Scientific and technical roadmaps are needed to better focus research and development efforts targeting major social demands and avoid any possible barriers. However, it is difficult at present to define a single roadmap for nanotechnology. The possible applications in different areas can originate from developments that are unexpected today."[100]

The EU summarized some of the fundamental issues challenging development and commercialization and outlined the primary concerns confronting not only the United States but also the rest of the developed world:

> As for any technology, risks of inappropriate or even dangerous developments may exist. Adequate transparency in research is needed and a safe set of rules has to be set in place. The absence of control may result in irresponsible or even dangerous results, but unmotivated prohibitions would be counterproductive. Researchers might flee and constitute *technological paradises* in countries where less or no controls are enforced. Indeed, unlike nuclear energy, nanotechnology (as well as electronics, biotechnology, genetics, etc.) are developed by private companies, substantially outside direct public control. Public power, through politicians and public opinion, should have the cultural instruments and access to appropriate qualified expertise to assess, steer and—where appropriate—control developments.[101]

Another EU report was released in early 2004. The Community Health and Consumer Protection Directorate released *Nanotechnologies: A Preliminary Risk Analysis on the Basis of a Workshop Organized in Brussels on 1–2 March 2004*. The "Mapping Out Nano Risks" workshop involved seventeen experts who were asked to examine hazards derived from nanotechnologies within the next three to five years. It included four themes: risk determination, ecotoxicological implications, ethics and security concerns, and emerging patterns and methodologies to monitor risks. The report had two sections. The first offered a preliminary risk analysis, including a hazard-trigger algorithm as a potential prioritization for use by regulators. The second part was a collection of short essays by participants.

The report found "the adverse effects of nanoparticles cannot be predicted or derived from the known toxicity of bulk material." There was a "scarcity of dose-response and exposure data." It presented twelve recommendations, including:

- a call for a common nomenclature (this recommendation has garnered the full support of the International Union of Pure and Applied Chemistry[102] and the American Chemical Society[103])
- assigning a new Chemical Abstract Service Registry[104] number to engineered nanoparticles (since an engineered nanoparticle is like a new chemical and since a new chemical is assigned a unique CAS number, an engineered nanoparticle should be given a CAS number upon its creation)

- developing standardized risk-assessment methods
- promoting good practices
- establishing a dialogue with the public, industry, and many others.[105]

The report briefly reviewed mostly dated ultrafine particulate toxicology findings. When it examined ethics, the report concentrated on the importance of a two-way process of participation involving a full range of stakeholders, the need for monitoring research, and issues of equity and human/machine interface. Unlike many reports in this area, it articulated the difference between risk assessment and risk management, though it offered no model for communication. In terms of assessment and regulation, the experts advocated an incremental approach to "avoid preventable hazards and risks, set up a framework for participation, and monitor the development for further regulation."[106] They also recommended labeling of the products of nanotechnology and making them traceable in the market.

The major contribution is found in the concept note by C. Vyvyan Howard and Wim de Jong. They introduced a hazard-trigger algorithm to rank order-exposure assessment to triage products before introduction and marketing. Presumably, when a full assessment is not possible, this algorithm would allow regulators to be reasonably precautionary:

> The value of all the hazard triggers would have to be decided by an expert committee and would involve a review of the current literature and identification of data gaps. However, for many hazards including carcinogenicity, mutagenicity, and reprotoxicity there are quantitative classification systems in place. When there are gaps, the precautionary stance would be to assume that they are positive hazard trigger events until shown to the contrary.[107]

In the second part of the report there are a series of submissions. The highlights from a few of them are below.

Alexander Arnall from Sheffield and Douglass Parr of Greenpeace addressed societal impacts by tracking nanoscience and technology within three categories: disruptive, enabling, and interdisciplinary. They separated camps into nano-optimists and nanopessimists and forewarned "the public profiles of these opposing views are set to heighten in coming years."[108] They approached hype on multiple levels. "Nano-optimists have been charged with hyping figures to reckless and impossibly high expectations for economic benefits." As such, "there is a general understanding amongst industry that the level of hype surrounding nanoscience and technology has, to some extent, damaged investment potential."[109] They also rejected studies focusing on grey goo and ecophagy disaster and concluded: "There is a need, then, to move beyond current rhetoric: instead of debating the speculative, long-term technical possibilities and ramifications of nanoscience and technology, it is necessary to broaden the discussion for focusing on present-day developments and asking what are the real issues at stake here."[110]

Vicki Colvin from Rice University began her submission by stating "the pace of nanotechnology has outstripped the understanding of the risks." She admitted that "some industries have for years used colloidal pigments and nanostructured additives in products." Colvin believes that "society should adopt a watchful confidence to permit research in the area to proceed unimpeded so that an accurate picture of risk can emerge when products are being considered for distribution." She ended her submission with a call for standardization of nomenclature, the absence of which she feels "will severely hamper both risk assessment as well as communication efforts in this area."[111]

Annabelle Hett from Swiss RE offered an interesting review of the transfer of risk, the core business of the insurance industry. (She released a report for Swiss RE in the same year.) Hett called for a standardization of materials and applications that "would allow a comparison of scientific knowledge across industries and countries. It would also be a precondition for labeling requirements which may prove necessary for the insurance business in order to differentiate and insure certain exposures originating from different products or applications."[112]

"Nanotechnology, as an emerging risk, challenges the insurance industry because of the high level of uncertainty in terms of potential nanotoxicity or nanopollution, the ubiquitous present of nano-products in the near future and the possibility of long latent unforeseen claims."[113] Hett included insurers as a stakeholder in nanodebates and participation models.

Finally, Ortwin Renn from the University of Stuttgart noted "for more than 90 percent of the respondents in European as well as US surveys the term nanotechnology has no meaning and evokes educated guesses at best."[114] Expressing concern on how risk-communication experts may be able to craft messages given the lack of opinion, he suggested that "focus groups in which proponents and opponents of nanotechnology would be given the opportunity to develop their arguments in front of representatives of the general public or selected groups and then ask the respondents to share their impressions and evaluations."[115] His sense was that focus-group findings might enable risk-communication experts to "deduct potential interest violations, mobilization potentials and societal opportunities or constrains for political action."[116]

The Royal Society released *Nanoscience and Nanotechnologies: Opportunities and Uncertainties* in July 2004. One of the most compelling statements made involved the release of nanoparticles into the environment: "Perhaps the greatest potential source of concentrated environmental exposure in the near term comes from the application of nanoparticles to soil or water for remediation."[117] Overall, the report claims that "most of the social and ethical issues arising from applications of nanotechnologies will not be new or unique to nanotechnologies."[118]

There is a set of specific concerns related to the public acceptance issue. The report called for greater transparency as a variable for increasing trust. It warned that a close association with military applications might "significantly endanger the acceptance of a whole range of beneficial applications."[119] It challenged regulatory bodies "to consider whether the annual production thresholds that trigger testing and the testing methodologies relating in substance [on the nanoscale] be revised."[120]

In the end, they reject a moratorium "on the assumption that government will be minded to secure an appropriate regulatory regime as rapidly and effectively as possible."[121] The entire gray goo issue is relegated to appendix D along with a quote from Drexler, who commented, "it's very much the wrong issue to focus on for a variety of practical and sensible reasons."[122]

There is a remarkable distinction separating American and European reports. By and large, Europeans have a healthy regard for environmental considerations, while American reports are oriented toward moving nanoscience as efficiently as possible from the lab to the market.

BUSINESS STAKEHOLDERS

While public spending in 2003 totaled about $3 billion, "it is assumed industry in the same period invested at least the same amount."[123] Swiss RE predicts hundred of billions in sales by 2010, even a trillion in sales by 2015. As a result, many different business interests have entered into the fray. In 2002 In Realis published an investor's guide. The Canadian public-relations firm InfoComm was planning to as well but backed off in summer 2004 to reexamine the reports currently released. The following were selected as a representative subset of the many business and industry reports on nanotechnology.

Credit Suisse/First Boston

This report, one of the earliest released by industry, *Big Money in Thinking Small: Nanotechnology—What Investors Need to Know,* began with hyperbole: "Nano-led innovation will likely obliterate many of today's business models and lead to huge gains and losses in shareholders' wealth."[124] Calling nanotechnology a general-purpose technology (see introduction), it focused on sources of funding, start-ups, enablers, and major companies in nanotechnology. The report ended with a series of appendixes, a few of which are summarized below.

The report began with a repetition of hyped statistics from the NSF and elsewhere, such as "by 2010, half of all pharmaceuticals will be developed through nanotechnology, chips will have 100 times the storage density of current chips at a fraction of the cost, and nanotube-enhanced batteries may lead to the eventual destruction of the disposable battery industry and allow for electric cars."[125] In terms of industry, it noted that investment potential resides in three categories of business and industry. The report parallels these three categories in broad approaches to investing in nanotechnology (from least to most risk). "Major corporations are investing in nanotechnology." The report calls this capturing the nano-option, and it takes notice of "companies that make nanotechnology equipment" and "start-ups which are basically trying to capture as much of the intellectual property as possible, with an eye toward future commercialization."[126]

Before examining industries likely to be affected by nanotechnology, Mauboussin and Bartholdson observe:

> Creative destruction is the result of the discontinuity between corporate survival techniques and innovation.... Because of the advent of nanotechnology, we believe new companies will displace a high percentage of today's leading companies. The majority of the companies in today's Dow Jones Industrial Index are unlikely to be there 20 years from now.[127]

Citing the NbA and other sources, the report summarized a series of investment projections. NbA set annual sales in nanotechnology at about $45 billion. The NSF predicted sales of $1 trillion by 2015, and

> *In Realis* predicted $100 billion by 2005, $800 billion by 2010, and up to $2 trillion by 2015. *Evolution Capital* estimated that yearly nano sales are currently $20–50 billion, and growth will be $150 billion in 2005 and $1 trillion by 2010. The NNI estimated that 2001 sales were $30 million, and that 2015 sales will reach $1 trillion—a 110 percent compound annual growth rate.[128]

In appendix B, the report told investors to watch textiles, especially Nano-Tex, which partnered with Burlington Industries, which is behind the nanopants craze. A second area is nanotubes, which they characterize "as one of the most promising near-term nanotechnologies."[129] The report offers a few rationales. The cost of production fell twentyfold from 1996 to 2001; the Carbon Nanotech Research Institute (a subsidiary of Japan's Mitsui and Co.) has already opened the first nanotube fabrication plant, and researchers had nanotube-enhanced batteries on the market in 2004.[130] Other categories included plastics, sunscreen, and coatings.

This early report offered summaries and generalized projections. The next report has much more to offer. It is an online newsletter that provides the most comprehensive information on nanobusiness. Though the subscription price is fairly steep, there is nothing close to it today.

Forbes/Wolfe Nanotech Report

Josh Wolfe is introduced in chapter 7, but his impact on nanotechnology may be more affected by his monthly newsletter, the *Forbes/Wolfe Nanotech Report*, and the associated special reports published by Forbes than for any other reason. One of their announcements hypes: "Successful investing in nanotechnology—like any new emerging sector in its formative stage—will hinge upon your understanding of prevailing trends and ability to identify the market leaders' best position to catch the impending boom."[131] Presumably, the newsletter provides this service. It is edited by Wolfe, who is touted by Forbes as having "long-standing relationships at over 25 leading research-driven universities and his network of advisors includes leading technology professors, researchers and PhDs at preeminent engineering and computer science schools and research labs, as well as technology entrepreneurs and top executive and business professionals from leading Wall Street firms."[132] Forbes claims its "portfolio of nanotech stocks is up more than 256 percent since its inception in March 2002."

The role hype may play in investment was tackled in the December 2002 issue: "There are plenty of signs of nanotech hype. Hype is a dirty fuel that often feeds greed and fear, and can have unfortunate consequences on the market of stocks.... Consider what happens when excessive hype raises expectations too high. When they aren't met, there are serious repercussions."[133]

The hype in business is sometimes as exaggerated as that coming from nano-enthusiasts like CRN and early Foresight predictions to nanopromoters, like Roco and Bond. While the report tries to understate the hype, the advertisements for the report do not. They hype nanotechnology like everyone else: "The nanotech revolution will be far more world-changing than the microchip revolution or even the industrial revolution and will one day result in astonishing and primal changes to the world."[134] A page later in the same promotional document: "Bigger by far than the short-lived dot-com phenomena, and for real because it's founded on hard science that will yield real-world, practical benefits and cheaper/better, can't live without-'em consumer products. Think of nanotechnology in terms of a second—but greater—industrial revolution."[135] The flyer lists many incredible applications such as "nanoshells that protect transplanted pancreatic cells from attack by immune cells, enabling the reversal of diabetes" to "just around the corner is a safe, cheap, anti-virus, anti-bacterial, anti-fungal microbe-zapping hand cream that kills (even anthrax) on contact."[136] To top it all off, we get the usual examples of classic hyperbole, the extrapolations:

> In the very distant future, the coming of nanotechnology will mean that we could create out of the infinitely abundant atoms that wash around us... a plethora of medicines to treat and heal the sick, a regeneration of scarce natural resources

and an abundance of fuel for our cars and machinery.... Nanotechnology will soon—many say in the next 10 to 20 years—eliminate most of the more dreaded and common diseases.[137]

Wolfe offers antidotes to the hype on many levels as well. For example, stories in both the *Wall Street Journal* and *USA Today* warned that the nanotechnology bubble might be forming. Wolfe seems less concerned because he noted that "asset prices are not grossly overvalued... nanotechnology is coming to fore at a time when venture capitalists are bloodied and extremely wary of new start-ups... and there is no initial public offering (IPO) market to support overvalued assets."[138] Hence he warns investors to be wary but not to be alarmed over talk of a bubble.

Also, he discussed the hype curve: "Each time a so-called *breakthrough technology* comes along, its *hype curve* sends Wall Street towards a peak of inflated expectations."[139] The Forbes flyer adds, "When Wall Street analysts begin to understand the magnitude of what's at stake, we're likely to see hysteria for *all things nanotech* that could indiscriminately drive the trading prices of nano-stocks through the roof."[140] This hype is not without some mixed merit. When the Twenty-First Century Nanotechnology R&D Act was signed into law by President Bush, "it was a big day for many companies using nanotechnology mostly for marketing advantages."[141] Indeed, stocks in companies that had nothing to do with nanotechnology but their name experienced sharp increases in stock value. This was "the last time nano stocks popped in tandem.... So, if a single news release on a holiday week can generate a 20 percent spike... the nano space is a speculative space similar to the dot-coms," says Kenneth Reid, editor of *Spear's Security Industry.* "And it is the space that nanotechnology enters the year 2004."[142] For example, "*Nanogen's* stock jumped more than 10% during intraday trading on December 3 on 4.5 times its average daily volume. And *NanoPierce Technology,* another nano-in-name-only-company jumped as much as 24 percent on nanotech signing day."[143] Legitimate nanotechnology firms benefit as well.

Presumably to take advantage of this spike, the *Nanotech Report* will provide you with the information the serious investor needs. "However, investing in nanotechnology stocks is not for widows and orphans, and opportunistically timed press releases don't make for sound investment strategy."[144]

On a second level, Wolfe is vehement in his caution about risks associated with investing in nanotechnology: "Wall Street's day traders and message board manipulators are back. They are back to their old pump-and-dump tricks of the Internet bubble but this time it is the nanobubble.... Following the technology boom-bust cycle nearly by script, some nanotech stocks have begun their departure from reality."[145] He portends that "most of the movement to today's nanotech stocks is being driven by giddy enthusiasm and chartists. Watch out for the nanobubble."[146]

Wolfe also warns investors against companies that incorporate the prefix *nano* into their name in order to attract investment when the company has little, if any, value in nanotechnology: "Remember in the 80s PC boom, when a slew of companies changed their names to incorporate the word *tech*? And more recently, the addition of *dot-com* to brighten up boring corporate logos? Make way for the new *nano* companies."[147] In February 2004, Wolfe announced, "the prefix *nano* had become a surefire market cap turbocharger for many unscrupulous stock promoters."[148] He cited examples including NanoProprietary (NNPP.OB). While a nanopretender, it experienced a stock soar following the signing of the R&D Act. USGlobal Nanospace, a company that makes armor and other defense products, saw its stock, which started at $0.10 at the time of the announcement, hitting a high in January of $3.05. "Many of the nano-pretenders are not even remotely connected to nanotech advances."[149] Two years earlier, Wolfe outed a handful of firms. Nanometrics, Inc. (NANO), according to its own CFO, Paul Nolan, has nothing to do with nanotechnology. Nanogen (NGEN)—still, no nano. His call was correct as their real value won out, and these stocks dropped in value over time. "The best performing stock of the five was down 52.2 percent, while as a group, these stocks plummeted 66.1 percent."[150]

The *Nanotech Report* from its first issue in March 2002 begins with a feature story or two on some element of the industry; an interview of a corporate, government, or even academic leader; companies-to-watch pages; a follow-the-money page that summarizes which companies are getting grants or venture money, which universities are getting grants, and other types of investment; and the nanosphere, which lists traded and privately owned companies and their worth with a recommendation to buy or sell.

At the end of its first feature article, we find the mission for the publication: "This report, whose first issue you hold in your hand, is dedicated to seeing that you stay up to speed on changes that may beggar all those that had gone before."[151] With a salute to Gene Roddenberry and the opening remarks by Captain Kirk, the mission seems to have been well received. In 2003 the Newsletter and Electronic Publishers Association gave *Nanotech Report* an award for editorial excellence in newsletter journalism in the financial advisory newsletter category.[152]

In general, Wolfe believes nanotechnology "is no longer hype and hot air and he draws this conclusion for three reasons: political support, investor recognition, and the fact that start-up companies and incumbents alike have products on the market now."[153] And *Nanotech Report* is one the finest sources for information on new products and new companies in the nanotechnology business. Wolfe calls his shots clearly and bluntly. He debunks overclaims and uncovers companies overextended and overvalued, and he alerts investors and other interested parties in important developments and newly emerging concerns.

Wolfe's publication and e-mail announcements of newsworthy events are absolutely required reading by anyone in the field and investors interested in entering the market. The next and final report comes from an insurance company. In time, producers will need liability coverage, and insurers will need to determine risk. The Swiss Reinsurance Company began important dialogue on this issue.

Swiss Reinsurance Company

"Nanotechnology presents new challenges for insurance companies," said Rod Frail, spokesperson for the Insurance Council of Australia. "In general, nanotechnology was such a new development that we didn't even know if any members were insuring it. It hasn't even come up on our radar yet."[154]

The Swiss Reinsurance Company (Swiss RE) is the second largest insurer in the world. They rose to the challenge with a report prepared by Annabelle Hett, who works in Swiss RE's Risk Engineering Service. *Nanotechnology: Small Matter, Many Unknowns* was released in 2004. This report stakes out a claim for the insurance industry as one of the stakeholders in the debate over nanotechnology. Committed to a transparent risk dialogue with the various stakeholders, the report examines the general risk represented by nanotechnology as a whole.

The report began predictably enough with an observation about the confounding nomenclature. Observing that any particle reduced to the size of nanoparticles behaves very differently than it does in a larger state and the range of applications and drivers are broad and varied, it indicated concerns about the mobility of nanoparticles in the human body, especially in terms of inhalation and its effects on lung tissue and of passing through the blood-brain barrier and triggering neurodegenerative diseases.[155] It also expressed concern over the mobility of coated nanoparticles in the environment and the impact they might have once accumulated in the human body either through ingestion,[156] especially in terms of drug delivery or even water additives, and when ingested unintentionally as pollution.[157] The report also speculated whether another tissue barrier, the placenta, might be vulnerable as well.[158]

Like most other sources, the report is much less concerned with advanced nanotechnology. These scenarios "have been preying on people's imagination for some time, giving rise to horrific visions of intelligent, self-reproducing particles which might swarm all over the world and make it uninhabitable for humans.... [The report noted that this scenario] still belongs in the realm of science fiction." Nonetheless, the report notes a consequence of this misdirection: "The potential hazards relating to the manufacture of real and innovative materials or new applications, however, have attracted little attention."[159]

It warned that the analogy between naturally occurring nanoparticles and artificially and commercially manufactured nanoparticles is false. "Many of the nanoparticles occurring in nature, such as saline nanoparticles, are soluble in water. As soon as they are inhaled or come in contact with the tissue, they dissolve and lose their particle form. The nanoparticles from combustion processes—in motors, cigarettes or fireplaces—although not soluble in water, are very short lived." They aggolomerate into microparticles and become less reactive. However, "in order to prevent an agglomeration of the nanoparticles, commercially available nanoparticles are often specifically coated according to manufacturing specifications. Consequently, the particles in many commercial products—such as sprays and powders—remain reactive and highly mobile."[160]

Swiss RE dedicated a chapter to occupational hazards using the analogy of asbestos. Peter Binks, CEO of Nanotechnology Victoria, a government-funded consortium of three universities and the Commonwealth Scientific and Industrial Research Organization (CSIRO), defended the analogy: "The world's moved on since asbestos. I think the asbestos issue got out of hand because our industrial society at that stage didn't think about materials risks when it introduced materials."[161]

Swiss RE supplied an ominous warning:

The removal of nanoparticles from liquids such as drinking water is currently only conceivable by means of centrifugation and ultrafiltration. Neither method is suitable for processing large amounts, as both are cost-intensive. In order to filter nanoparticles from the air, new filtering techniques are required, because the air-purification filters currently in use in buildings and manufacturing plants generally have pores too large to "catch" nanoparticles.[162]

While some protective technologies and products are being researched, they are mostly untested.

Drawing commonalities, the report claimed only "a comprehensive plant air-filtering system could remove the particles, an extremely costly and, indeed, hardly realizable solution, given the current offer of air-purification systems for large buildings."[163]

In reviewing current regulation, the report reiterated the claim that current Material Safety Data Sheets cover larger particle forms. "Thus, the safety sheet for nano-titanium dioxide powder still recommends that a dust respirator be worn when handling the substance, although such masks are known to offer only limited protection."[164] It concluded that both the US FDA and the EC's Scientific Committee on Cosmetic Products and Non-Food Products Intended for Consumers have not established viable hazard guidelines.

The report also noted that "despite early warnings about the effects of asbestos on health, it took some 100 years to introduce internationally accepted asbestos standards."[165] They made the telling observation: "Nano-technologically manufactured products have been retailed for some time now without any particular labeling by regulators and have thus not been recognized by consumers for what they are."[166] Hence, the report made two broad-based recommendations. First, "ways and means must be found of permitting a proper assessment of the risk and ensuring that these new materials are handled safely." Second, "it is essential to have an internationally valid standardization of nanotechnological substances and materials as well as a uniform nomenclature.... Without standardization, even the labeling of products becomes an extremely difficult undertaking."[167]

Like so many of the reports, Swiss RE called for citizen participation and open and honest communication of risks:

> How people assess risks depends on a whole series of subjective perceptions. The so-called *fright factors* tell us whether an issue has the potential to create panic and hence is perceived as a threat.... Fear is also aroused by those risks that are forced upon the consumer, in which case, he cannot take an independent action. This underscores the importance of the product declaration, which enables the consumer either to accept a risk voluntarily or reject it.[168]

The report ended by observing that the public has not formed an opinion on nanotechnology because "little has been heard about the subject. Too few people are aware of developments," and uninformed people "find themselves hard put to distinguish between fact and fiction." The "reader and cinema-goer will probably associate nanotechnology with Artificial Intelligence and menacing scenarios."[169]

The last word on the issue comes in terms of the precautionary principle. "In view of the dangers to society that could arise out of the establishment of nanotechnology, and given the uncertainty prevailing in scientific circles, the *precautionary principle* should be applied whatever the difficulties."[170] To keep this closing remark in context, the report argued that "no reasonable expense should be spared in clarifying the current uncertainties associated with nanotechnological risks."[171] As such, Swiss RE called for an active rather than a strict application of the precautionary principle.

In early December 2004 SwissRE held a conference in Rüschlikon, Switzerland, not far from the lab where Gerd Bennig and Heinrich Rohrer developed the scanning tunneling microscope (STM), the workhorse of the industry. The presentations were published by Swiss RE in *Nanotechnology: Small Size—Large Impact* in 2005. Hett offers a summary and a conclusion.

CONCLUSION

Reports dominate our lives. Commissions establish committees that solicit testimony and collect data, which, in time, produce these reports. While the findings are not necessarily objective, what is reported reflects the ideology of the organizations that composed the report. Governments issue them regularly to decide what needs to be done, to involve stakeholders in the process, and even to avoid acting altogether. Industry and business reports are released to convince the community its interests are not at odds with public needs and values. Oftentimes, they are guides for the industry or even marketed to the public for a price. The next chapter takes a much closer look at business and industry in nanotechnology. While the world is always changing, these are some commonalities worth understanding as nanoscience creeps out of the labs and onto the commercial playing field.

CHAPTER 6

APPLICATIONS OF NANOSCIENCE

K ey indicators of private-sector economic interest in nanotechnology show that there is a great deal of newfound potential in the commercial development of nanotechnology. The NSF predicted annual sales of $340 billion for nanostructured materials, $600 billion for electronics and information-related equipment, and around $180 billion in annual sales from nanopharmaceuticals by just 2015.[1]

In March 2004 the nanotechnology industry held one of its gala events—the Nanotech 2004 convention in Boston.[2] As one might expect, the event was festooned by blue-sky predictions about the future of nanotechnology that would probably make most bystanders feel they were immersed in a world of science fiction rather than business. In many ways, this is one of the major problems in the business and investment field of nanotechnology. It's awfully difficult to find where the hype ends and where the hope begins. Many of these far-out predictions about the potential applications of nanotechnology have caused skeptics to note that "Nanotechnology will likely amount to nanoprofits...much like technologies ranging from artificial intelligence to virtual reality that looked cool in the lab but have floundered commercially."[3]

Actually, both the nanodreamers and the nanonaysayers have probably got it wrong. While "nanotechnology is still in its infancy, the science of simple atoms and molecules on one end, and the science of matter from microstructures on the other, are generally already well established."[4] This means that nanotechnology is certainly going to have substantial commercial applications, but perhaps not to the fullest extent imagined by those who are planning their retirement around it. "The biggest markets for nanoparticles remain in familiar products, from the black rubber filler in tires, a $4 billion industry, to the silver used in traditional photography." According to Lux Research, "only about $13 billion worth of manufactured goods will incorporate nanotechnology in 2005."[5] However, the trend will increase dramatically. "Toward the end of the decade, Lux predicts, nanotechnologies will have worked their way into a universe of products worth $292 billion."[6]

Sean Murdock, executive director of the Nanotech Business Alliance, argues commercial-strategies needs address consumer value. He tells a different story about genetically modified organisms (GMOs). He claims the reason they were resisted was "the manner in which GMOs were brought to market. It did little to create a strong consumer value proposition and instead engendered a negative response. The result would have been different if different strains had been brought to market that had consumer value.... The reaction may have been different." He adds, "There has been relatively little backlash against biopharmaceuticals, which in fact used many or most of the same techniques and technologies. The reason is that consumers can see a clear and direct benefit from the use of the technology, as almost all of us are consumers of healthcare."[7] Products need to come to market to establish a value chain. As Lux's Matthew Nordan and others admit, "there is no nanotechnology market, there is a nanotechnology value chain, going from nanomaterials (e.g., Southern Clay's Cloisite nanoparticles) to nano-intermediates (e.g., Basell's composite materials incorporating South Clay's nanoparticles) to nano-enabled products (e.g., GM Impala side body moldings made from Basell's nanocomposites)."[8]

In order to get a full sense of the strength behind the business sector of nanotechnology, it is important to understand the basic, and more importantly, realistic applications of nanotechnology. For the most part, the commercial applications of molecular nanoscience fall into six broad categories: manufacturing and materials, agriculture, healthcare, energy, electronics and computing, and luxury products.

INSTRUMENTS AND APPARATUSES

Before nanotechnology can affect everything we use, the science needs to be sufficiently advanced and the prototypes need to be designed, developed, and produced. The industry that gets us to that point is referred to as the "picks and shovels" industry. As such, major growth is expected in the nanotools industry in everything from advanced electronic microscopy to advances in conventional production technologies involving nanotechnology. "The US market for nanotech tools will increase nearly 30 percent a year through 2008 to $900 million. And by the year 2013 it will be worth $2.7 billion," according to the Freedonia Group.[9]

It begins with design and companies, like Nanostellar and Accelrys Software, that provide computational nanoscience software for designing nanomaterial and nanodevices.[10] Future designs of nanocatalysts, for example, could rely increasingly on computer simulations that would "reduce the costs of expensive experiments every time we need to test an idea or simply do calculations." Jens Norskov of the Technical University of Denmark continued, "Computer power and software have advanced far enough to calculate the complexities of how the hundreds of atoms on a nanocatalyst's surface, with their many electrons, might react to chemicals."[11]

"The companies at the forefront in supplying industries with the tools of atomic-force microscopes and advanced photolithography include Veeco Instruments, Keithley Instruments, FEI, Ultratech, and Nanoinstruments."[12] FEI is involved in a project to build the world's highest resolution scanning/transmission electron microscope with a resolution of half an angstrom.[13] IBM Almaden researchers are working on a magnetic-resonance force imaging microscope (MRFM), a hybrid of the magnetic-resonance imaging (MRI) and atomic force microscopy (AFM), "to provide a 3-D view of the nanoworld."[14]

There have been some interesting advances. A Northwestern team developed "an AFM probe tip with an integrated microfluidic system for capillary feeding of molecular ink." They anticipate "high-impact applications in the field of nanosensors, biotechnology and pharmaceuticals."[15] A Hawaiian company, Nanopoint, is "developing a system to conduct microscopy inside living cells at resolutions of 50 nm or less in the infrared, visible, and UV ranges."[16] A Chicago start-up is developing a holographic laser steering technology, the BioRyx Platform, which might make possible a nano-assembly line of sorts.[17] Another team from the Max Planck Institute for Solid State Research has succeeded in measuring vibrational modes of carbon nanotubes. "The vibrational motion reflects mechanical strength or softness of nanotubes and the unique electrical and mechanical properties are highly dependent on the present of defects on the atomic scale."[18]

Hysitron "supplies nanomechanical instruments to permit nanoscale measurements for advanced research and industrial applications."[19] Veeco Instruments and the Dow Chemical Company are developing a nanomechanical measurements instrument based on the AFM "to accurately and quickly map the mechanical properties of many materials on the nanoscale."[20] MIT and Virginia Commonwealth University researchers have developed a nanoprinting method that might enable the mass production of nanodevices and reduce the cost of each microarray.[21]

Zhang from Berkeley invented a silver superlens that can be used "to image structures with a resolution that is about one sixth the wavelength of light.... The lens could have many applications, such as imaging nano-scale objects with light."[22] A Purdue team produces an array of pairs of gold nanorods that they claim "could lead to optical superlenses that reflect no light and operate with sub-wavelength resolution."[23] Russian and Ukrainian scientists are using an extreme ultraviolet illumination source to create an optical microscope to image features as small as 100 nanometers.[24] In terms of production advances, a team from the University of Arkansas and the University of Nebraska has found a way to produce nanopores using standard STM techniques with *nano-EM*, a liquid medium, which is more efficient and less expensive than confidential business information.[25] The California Institute of Technology has licensed its dip-pen nanolithography of biological molecules on previously inaccessible or difficult-to-pattern surfaces to Arrowheads Research in March 2005.[26] Keithley Instruments collaborates with the Nano-Tech Center at Albany State University to research a variety of test equipment for high-tech production.[27] Another Berkeley team demonstrated it was "possible to transport individual pulses of laser light from nanowires to ribbon waveguides, a prerequisite if photonic devices are to be useful in communications or computing applications," including "a photonic Internet" and an "optical computer."[28]

MANUFACTURING AND MATERIALS

Nanotechnology may alter the way in which goods are mass produced in the United States. In fact, "the ability to synthesize nanoscale building blocks with precisely controlled size and composition and then to assemble them into larger structures" may inevitably revolutionize materials manufacturing.[29] "There's so much hype about nanotechnology, but certainly we don't have enough control yet to make many practical applications,"[30] said Kyeongjae Cho of Stanford's Multiscale Simulation Laboratory.

Most of the current market is in nanosized materials themselves: nanotubes, metal oxide nanoparticles, semiconducting nanocrystals, and fullerenes. "More than 200 companies worldwide sell nanomaterials." Matthew Nordan from Lux reported, "[I]t will take three years for competition to weed out ineffective suppliers and agreed-upon standards to take shape."[31]

However, in addition to changing the process of how goods are produced, the single largest application of nanotechnology is likely to be in the materials themselves. More than 20 percent of world industrial production is based on catalysts. In the world of catalysis, new studies have demonstrated that catalyst particles with a diameter of just ten nanometers are almost one hundred times as reactive as the same amount of catalyst particles that have a diameter of one micron.[32] This phenomenon is caused simply by the increased surface area of the matter that is on the nanoscale. Their reactions to various forms of doping can be quite different from the reactions of extended solids. For example, nanomaterials seem to be an excellent replacement for platinum. Three California companies (Catalytic Solutions, Nanostellar, and QuantumSphere) are developing nanomaterials for improved catalytic converters.[33]

Nanosized metals are used in many applications. As a diagnostic device, a Northwestern team attached gold nanoparticles to single-stranded DNA designed to home in on target DNA sequences. Upon bonding, the gold particles are pulled together enough to alter the wavelengths of light the particles scatter.[34] A team from NC State has developed a technique to induce nanodots of nickel to spontaneously assemble into three-dimensional arrays, and Kopin is applying a technique to make light-emitting diodes.[35] Applications are also evidence in defense. For example, "by adding materials known as superthermites that combine nanometals such as nanoaluminum with metal oxides, such as iron oxide, [the] power of weapons can be greatly increased."[36]

Nanomaterials of all sorts have entered the marketplace. QuantumSphere, Inc., is the leading manufacturer of metallic nanopowders used in batteries, fuel cells, air-breathing systems, and hydrogen generation cells.[37] Nanophase produces a broad array of nanopowders for a variety of uses in the chemical industry. With partners Altana Chemie and Rohm and Haas, it is making nanomaterials for use in chip polishing slurries.[38] In addition, they "make nanotech particles that go into everyday sunscreens like Oil of Olay's Complete line of UV moisturizers."[39]

Nanoparticles have been extracted from natural materials, such as clay. For example, Cyclics Corp. adds nanoscale clays to its registered resin for higher thermal stability, stiffness, dimensional stability, and barrier to solvent and gas penetration.[40] NaturalNano Inc. extracts nanotubes from halloysite clay for commercial additives "in polymers and plastics, electronic components, cosmetics, and absorbents,"[41] and they are "working on paints with a property that could be turned on or off to block cell phone signals."[42] Industrial Nanotech has Nansulate, a spray-on coating with remarkable insulating properties.[43] A Chinese group is planning to paint the exterior walls of World Expo pavilions with a "nanotech-based coating material which acts as a permanent air purifier. Exposed under sunlight, the substance can automatically decompose the major ingredients that cause air pollution."[44] There is even some speculation on smart paints that change color when exposed to changes in temperature or light. Recently, nanoceramic particles were attached by Victor Castano of the Universidad Nacional Autonoma de Mexico to the surface of Kevlar fiber, adding significant UV and chemical resistance without changing any other antiballistic properties.[45] Starfire Systems is developing nanostructured silicon-carbide ceramics as well.[46]

Fullerenes are 3-D molecules, usually of carbon, including C_{60}, C_{70}, C_{76}, C_{78}, and C_{84}. "Potential industrial applications include superconductive materials, electrochemical systems, polymer and meta nano-composites, electrolyte separation membranes for fuel cells, gas storage and separation membranes, longer cell-life lithium batteries, highly functionalized coatings and ultra-fine crystalline artificial diamonds for drilling and industrial polishing."[47] Luna nanoWorks uses carbon fullerenes discovered at Virginia Tech called *trimetaspheres*. According to Steve

Wilson, president of Luna nanoWorks, "They can hold 18 different metals alone or in combination giving the company up to 27 flavors and counting."[48] While the cost of producing fullerenes has been a barrier to commercialization, companies like Nano-C have produced second-generation technologies that reduce the expense.[49]

Carbon nanotubes have been associated with high-strength applications, and the race is on. Carbon Nanotech Research Institute in Japan has developed a mass-production method. Thomas Swan and Co. announced in 2004 it was prepared to make commercial quantities of single-wall carbon nanotubes. Carbon Nanotech-nologies, Inc., in Houston has been gearing up.[50] Carbon nanotubes are single-walled, double-walled, many-walled, and multiwalled. The tubes can be empty or filled. For example, a Drexel and TRI/Princeton team filled nanotubes with a variety of polar and nonpolar liquids and particulate fluid. This produced "magnetic nanostructures which could have application in memory devices, medicine and wearable electronics."[51] Carbon nanotubes, especially single-walled tubes, may "enable electro-mechanical systems such as micro-electric motors, nanoscale devices, and nanoconducting cable for wiring micro-electronic devices."[52] According the Cientifica, prices will decrease by a factor of ten to one hundred in the next five years, and the energy market will meet price barriers by 2008 to 2009. Production will shift from the United States and Japan to Korea and China by 2010, and the major supplier of all types of nanotubes will be Korea.[53]

Functionalizing the surfaces of carbon nanotubes offers some interesting applications. For example, researchers from the Weizmann Institute of Science in Israel attached DNA strands to carbon nanotubes and complementary strands to gold electrodes, constructing field-effect transistors.[54]

Currently, nanotubes are prohibitively expensive for most applications; hence there is an incentive to find less-expensive ways to produce them. For example, Kenji Hata from the Institute of Advanced Industrial Science and Technology has found a way to reduce production expenses by adding water vapor to the chemical vapor deposition process of making nanotubes, which reduces the buildup of amorphous carbon on the catalysts. "Water-stimulated enhanced catalytic activity results in massive growth of superdense and vertically aligned nanotubes forests ... with carbon purity about 99.8 percent."[55]

Research on carbon nanotubes pervades the literature. Nagoya researchers reported a one-dimension string of carbon atoms, carbon nanowires, "amid a welter of nanotube whiskers by shooting an electric arc between two carbon electrodes."[56] Nanoyarn was made from carbon nanotubes at UT Dallas. Researchers "adapted textile technologies used to spin wool and other fibers to produce yarns made solely of carbon nanotubes." They may "incorporate electronic sensors and actuators in the yarn that act as electrically driven muscles."[57] The McGowan Institute for Regenerative Medicine researchers have synthesized a molecule that curls up to form nanotubes, and bunches of these nanotubes assemble into a nanocarpet that responds to different substances by changing color and can be trained to kill bacteria such as *E. coli*.[58] Nanocables were invented by a group at UC Davis led by Pieter Stroeve. "Layers of semiconductors, such as tellurium, cadmium sulfide or zinc sulfide, are electrochemically deposited in a tube until a solid cable forms then the membrane that lined the tube is dissolved leaving finished cables behind."[59] Finally, researchers at UC San Diego have learned to control nanotube geometry and make carbon nanotubes bend in sharp predetermined angles.[60]

Nanoprocesses are already being used to clean up emissions from both cars and industry and have been crucial to reforming oil distillates to the appropriate chemical mix for use in airplanes and automobiles. "A study by Germany's Helmut Kaiser Consultancy has identified more than 65 nanotechnology applications likely to be introduced in cars over the next few years."[61] For example, "GE Advanced Materials and Dow Automotive have both developed nanocomposite technologies for online painted vertical body panels."[62] Mercedes Benz is using a clear-cost finish that includes nanoparticles engineered to cluster together where they form a shell resistant to abrasion.[63]

In terms of air filtration, a new product is marketed by Johns Manville and FibrMark Gessner. Called CombiFil Nano, it is "a composite of polyester spunbonds and eletrospun nano filaments which is said to exceed the established standards in industrial air filtration."[64] Water filtration has received much more attention. Potable water has been described hyperbolically as the oil of the twenty-first century, especially when more than one-third of the population of rural areas in Africa, Asia, and Latin American has no clean water, and two million children die each year from water-related diseases. "Investors poured $165 million into 44 companies focused on water purification and management between 2002 and 2004, according to Cleantech Venture Network."[65] Novation Environmental Technologies "holds a license which uses nanofiltration to help purify water."[66] The company eMembrane is "developing nanoscale polymer brushes coated with molecules to capture and remove poisonous metal proteins, and germs."[67] According to Modzelewski from Lux, KX Industries has developed antibacterial and antiviral water-filtering membranes that can turn raw sewage into clean water on the other end.[68] Aguavia is using nanopore membranes in a water-filtration system. "A six inch cube of membrane could purify 100,000 gallons of water a day."[69] Argonide's NanoCeram Superfilter uses nanofibers, which has multiple applications, including purifying water from biological agents and for industrial processing.[70] Israel, New Zealand, and Singapore are working on desalinization.

In addition, through these processes, newer materials might reshape entire industries. A team from NanoSonic has created "metallic rubber, which flexes and stretches like rubber but conducts electricity like a solid metal."[71] Vanderbilt researchers are using nanotech manufacturing to create diamonds. "The chemical vapor process makes a thin diamond film that can be built upon layer by layer."[72]

"Applications centering on tough paints, pigments, and functional nanofilms for scratch-resistant optical surfaces currently show the most potential for rapid development and are projected to have the highest revenue potential." Hrishikesh Bidwe from Frost and Sullivan expects revenues to reach about $6.5 billion in 2015.[73]

Applied Microstructures has a technique to deposit self-assembling organic nanofilms that has a broad range of applications, from moisture barriers in packages to anticorrosion glazes for machine parts, sensors, and displays.[74] Nanergy, Inc., is using PVNanofilm technology to illuminate safety house number signs that can recharge in low light.[75] Evident Technologies aims "to produce advanced quantum dots systems compatible with a range of inks, UV curable epoxies, and polymers used on currencies and their documents."[76] Green Millennium has a line of photo-catalytic coatings used to reduce infectious microbes and soiling of exterior surfaces, sun damage, and odors.[77] Ecology Coatings, NanoDynamics, and MetaMateria have partnered to develop liquid solids cured by exposure to UV light.[78] These polymer nanocomposites may "eliminate a lot of the expense involved in applying protective coatings to electronic gadgets or patio furniture."[79] Altair Nanotechnologies and Genesis Air are developing "surface-activated nanosized titanium dioxide for use in heating, ventilating and air conditional equipment... [for] environments with diverse atmospheric conditions such as casinos, tanning salons, and army barracks."[80] ApNano Materials, Inc., announced that its NanoLub lubricant "enhances the performance of moving parts, reduces the fuel consumption, and can replace current use additives to oils and greases which are very toxic."[81]

Second-generation nano-enabled products will differ by tapping many nanotechnology innovations instead of just one, employing active nanostructures, and requiring new manufacturing processes to exploit. For example, Purdue scientists have "created new wires out of DNA by attaching magnetic nanoparticles and cutting the structures into pieces, bringing the prospect of biological machines that are able to self-assemble."[82]

The impact of second-generation products may be substantial. Take the following illustration from a Lux Research report, *How Nanotechnology Adds Value to Products*:

> If six emerging nanotech innovations were all applied in one model year of a high-volume truck like the Ford F-Series, tier-one supplies to the auto manufacturer would win the most with $493 million in increment revenue for that model year. Consumers would rank next with $327 million in cost savings over five years of use, mostly from better fuel economy, as well as soft benefits in performance and safety. The truck manufacturer itself would follow with a net $248 million in cost savings and boosted resale value, plus points of differentiation against competitors. But incumbent supplies of materials like talc-filled composites and microparticulate platinum group catalysts would lose $297 million in combined sales displaced by nanoscale alternatives.[83]

AGRICULTURE AND FOOD PRODUCTION

While agriculture may not be the government's top priority at this point, the application of nanotechnology in this field "has only begun to be appreciated."[84] Some of these applications seem to be sufficiently provocative to produce a response from the ETC Group titled *Down on the Farm*.[85] In addition, "scientists could potentially construct much more complex nanoscale structures through the natural ability of cells to dock with different kinds of molecules." Agriculture could be used to grow nanoproducts. Robert Hamers for the University of Wisconsin–Madison added: "Such a potential would be superior to the painstaking manipulation of individual nanosized components, such as the microscopic wires and tubes that comprise the raw materials of nanotechnology."[86]

Advancements in this field could be broad. They could affect the production of foods and other grown products, such as timber, and products that make their way to our Publix and Home Depot. The report from the US Department of Agriculture identifies five areas of potential applications: "pathogen and contamination detection, preservation and tracking, smart treatment delivery system, smart systems integration for agriculture and food processing, and nanodevices for molecular and cell biology."[87] The USDA report, based on a 2002 workshop, is incredibly superficial, and much more can be gleaned from the popular reporting and from the ETC Group's report. The lumber industry road map[88] discusses the "development of intelligent wood- and paper-based products that could incorporate built-in nanosensors to measure forces, loads, moisture levels, temperatures, or pressures, or detect the presence of wood-decay fungi or termites."[89] In addition, it suggests "investigation [of] the ability of wood nanofibrils to be converted into carbon nanotubes, nanotubules, and nanowires"[90] and predicts that "directed self-assembly will allow us to use the building blocks available in the forest products industry to manufacture materials with radically different performance properties."[91]

There are a number of large food conglomerates investing in nanoscience, and their research extends from food safety through aesthetics. For example, Friesland Foods, a Dutch conglomerate, is researching ways to make cheese products more attractive to consumers. Their research "involves four areas: creating metastable micro-textured structures; controlling the component breakdown during eating; controlling the flavor release; and micro-encapsulating the nutrients contained in food."[92]

Among other advancements, nanotechnology would almost certainly provide for a whole new line of molecularly engineered chemicals that would nourish and protect plants and crops in ways that were never thought possible. Biological sensors could be capable of near-instantaneous detection of dangerous biological agents and microbial pathogens. Other senses could be used to monitor temperature changes, while other devices could be used to detect for pesticides and genetically modified crops (how ironic) within foodstuffs."[93] Scientists at the University of Mexico "are developing non-invasive bioanalytical nanosensors that could perhaps be placed in, say, a cow's saliva gland in order to detect single virus particles long before they have had a chance to multiply and long before disease symptoms are evident."[94]

In addition, nanofabricated detectors offer the potential to do thousands of plant experiments for simultaneous gene characterization and selection with only very small amounts of material. These more efficient methods of experimentation will "allow scientists to determine which genes are being activated or inhibited during the growing process or during disease."[95] Every species of plant could then undergo its very own version of a nonhuman genome project. Revealing which genes within plant species are responsible for certain processes, such as growth, will lead to substantial increases in production and agricultural efficiency.

Packaging has received a lot of attention. Helmut Kaiser claims there are over two hundred fifty nanopackaging applications on the market responsible for over $800 million in sales.[96] The Foundation for Scientific and Industrial Research at the Norwegian Institute of Technology (SINTEF) is currently "using nanotechnology to create small particles in [plastic film to] improve the transportation of some gases through the . . . film to pump out dirty air such as carbon dioxide. The concept could be used to block out harmful gases that shorten the shelf life of food such as oxygen and ethylene for deteriorating food."[97] Oxygen-scavenging materials could have significant implications as would the introduction of antimicrobial agents. For example, Robert Kwan of JR Nanotech attests: "Finely milled silver, with particles mere microns in diameter, can attack the RNA of microbes, preventing them from reproducing."[98] According to Helmut Kaiser, "by adding nanoparticles, bottles and packages with more light- and fire-resistances, stronger mechanical and thermal performance and less gas absorption can be produced."[99]

"The combination of DNA and nanotechnology research will allow companies to target parts of the human bodies and cells to which nutrition is to be delivered. Function foods will benefit firstly from the new technology, followed by standard food, nutraceuticals, and other types of nourishment," said Helmut Kaiser.[100] The creation of foods with embedded nanoparticles is also being researched. "Nanofoods are being embedded with *soft particles* those using common biological materials or with *hard particles* made up of non-organic substances. Edible nanoparticles can be made of silicon or ceramics, or materials that react with the body's heat or chemistry, such as polymers."[101] For example, "physicists are using nanotechnology to create tiny edible capsules that release their contents on demand. The capsules, called *colloidosomes*, assemble themselves into a hollow, sturdy, elastic shell with holes." They could be used to carry active ingredients such as fat blockers, extra nutrients, or even prescription medications. One of the researchers predicts that "they could hit supermarket shelves by the end of the decade." This market is expected to rise to $20.4 billion in 2010, and about two hundred companies are currently active in research and development.[102]

It may be making inroads in food preparation on many levels. For example, there is considerable literature "on dirt-repellent coatings at the nanoscale with applications for the safety of food production sites."[103] There is only one example in print on food preparation: "Sonny Oh has a product named OilFresh which is a mix of nano-size silver particles that allows restaurants to use deep-frying oil longer."[104] Oil Fresh is already used by Jeffrey's Hamburgers in San Mateo, and Oh hopes to "expand into large chains like McDonald's and Burger King."[105]

ELECTRONICS AND COMPUTING

Lux Research associate Will Arora said, "the 193-nanometer optical projection lithography chip is on its last stand...due to physical constraints. It will fail to keep up with Moore's Law."[106] Simply put, as chips become more powerful, the density of electrons becomes so high they hit against each other, causing copper atoms to move, producing a problem known as electromigration. The current approach known as complementary metal oxide semiconductor (CMOS) is virtually impossible to push below ten nanometers. Some companies, like Fujitsu, are "exploring the use of carbon nanotubes as one option for overcoming the problem in vertical copper wires used to connect layers of circuitry within chips." Yuji Awano at Fujitsu's Nanotechnology Research Center reported: "Carbon nanotubes can carry about 1,000 times the current density or the current per unit area, compared to copper. In addition, they transmit electrons about 10 times faster and dissipate heat much more readily."[107]

A study by FTM Consulting, Inc., reported future chips that use nanotechnology are forecasted to grow in sales from $12.3 billion in 2009 to $172 billion by 2014.[108] Lux reports that of the ten next-generation lithography platforms, seven are optically based, two replace light with electrons, and the last stamps features onto chips. Lux predicts that nanoimprint lithography is in the lead, and five companies are racing to commercialize nanoimprinting.[109]

For example, Harvard researchers "applied nanowires to glass substrates in solution and then used standard photolithography techniques to create circuits." According to lead researcher Michael McAlpine, "These advances could bring powerful electronics and computing to virtually all facets of life at low cost, and may open exciting doors for low-cost radio frequency tags or high refresh rate e-paper displays which are fully integrated on a single piece of plastic or glass."[110] According to NanoMarkets, "the market for nano-enabled electronics will reach $10.8 billion in 2007 and $82.5 billion in 2011."[111]

Electronics and computing represent the gem of most research and development units for major corporations such as IBM and Microsoft. Intel's road map includes at least four more silicon-based chip-manufacturing-process generations. It began working on carbon nanotube transistors in 1998. Hybrid silicon-base and carbon-nanotube transistors is one path among many under investigation.[112] Intel and Advanced Micro Devices already produce ninety-nanometer chips and are working on sixty-five-nanometer chips.[113] Intel's Strategic Research Project for carbon nanotubes is evaluating nanotubes as a method of building transistors.[114] Texas Instruments, Hewlett-Packard, Toshiba, and Samsung are reportedly working on sub-one-hundred-nanometer chips as well.[115] While "carbon nanotube interconnects won't generate much revenue until we reach the 45 nm node but for pilot plants that could be just a couple of years away."[116]

Integrated circuits made of organic materials are being investigated by Motorola, the Dow Chemical Company, and Xerox. "The devices will be thin-film transistors which are similar to light-emitting diodes."[117] Georgia Tech, the University of Manchester, and the Institute for Microelectronics Technology in Russia researchers have extracted individual ultrathin sheets of carbon atoms so thin they are only two-dimensional.[118] "This two-dimensional fullerene may make a great new transistor," getting us closer to "transistors made from a single molecule."[119]

Already, some of the challenges imposed on the industry as a result of working on the nanoscale are being tackled. For example, HP Lab's crossbar latch research project recently reported that it "discovered a method of ensuring silicon nanowires can continue to function if manufacturing defects partially sever the connections between the crossbar and the rest of the circuit."[120] In early June 2005, researchers at the University of Liverpool,[121] "the National Institute for Nanotechnology of the National Research Council and University of Alberta announced having designed and tested a new concept for a single molecule transistor."[122]

Specifically, "it is likely that molecular nanotechnology will play a vital role in the future of information and energy storage processes, as has been recently evidenced by the many developments in this sector."[123] IBM may be the first to commercialize nanoproducts across several different industries.[124] In August 2001, for example, "IBM researchers created a circuit capable of performing simple logic calculations via self-assembled carbon nanotubes (*Millipede*),"[125] a move that can be seen as the first real step toward nanocomputing. Millipede will be able to store forty times as much information as current hard drives. IBM pushed back commercialization until 2006 or 2007. Partnering with Infineon, IBM is working on magnetic RAM (MRAM), and this collaboration is moving faster than the Millipede.[126] MRAM preserves data after systems are switched off, is radiation hard, and is "inexpensive enough to serve as replacements for battery backed-up SRAM (static RAM) or to replace both an SRAM and Flash chip where both are used together."[127] MRAM promises to combine the speed of SRAM with the non-volatility of flash memory. Micromem is a small firm that believes it can make

MRAM cheap enough to install in radio frequency identification (RFID) tags.[128] "*NanoMarkets* predicts MRAM will rise quickly reaching $3.8 billion by 2008 and $12.9 billion by 2011."[129]

Companies like Nantero are banking on RAM based on carbon nanotubes, which could provide nonvolatile, high-speed, high-density memory that is both radiation resistant and low power. Called "universal memory," these bits are not encoded "by the direction of magnetic fields as in hard drive, but by the physical orientation of nanoscale structures."[130] NanoMarkets predicts an $8.8 billion market by 2011.[131]

"Cavendish Kinetics stores data using thousands of electro-mechanical switches that are toggled up or down to represent either a one or zero as a binary bit." Their devices "use 100 times less power and works up to 1000 times faster."[132] Iomega Corp. is designing subwavelength optical storage termed Articulated Optical-Digital Versatile Disc (AO-DVD), "encoding data on the surface of a DVD by using reflective nano-structures to encode data in a highly multi-level format." Another possibility termed Nano-Grating-DVD (NG-DVD) "uses nano-grating to encode multi-level information via reflectivity, polarization, phase, and reflective orientation multiplexing."[133] Zettacore is developing molecular memory that reads and writes data by adding and removing electrons off nanometer-sized molecules. "The molecules in their memory are porphyrins, a nano-sized cluster of which are used to store energy."[134] These molecular memory companies could experience $7.1 billion in revenues by 2011.[135]

Major corporations in the computing industry are looking to build upon these new discoveries and create a future with previously unimagined computer capabilities.[136] Nanosys has partnered with DuPont, Intel,[137] and Matsishita Electric Works[138] to work on thin-film technologies for electronics and memory systems. Lawrence Gassman of NanoMarkets thinks that "nano-storage chips will be worth $3.1 billion in revenues by 2006,"[139] and NVE has made inroads into flash memory chips,[140] as have Motorola and Silicon Storage Technology.[141] Currently, the most common nanostorage devices are based on ferroelectric random access memory, or FRAM. Data are stored using electric fields inside a capacitor.

There have been other noteworthy advances. Chemists in Glasgow engineered a "metal oxide-based nano-cluster with the potential to pack in 10,000 more storage units into a given area than is currently possible," with applications in "miniature mobile phones, camera memory cards, and nanoscale computers."[142] Another team from Konstanz and Hitachi created a hard-disk medium "putting down a layer of closely packed nanoscale latex spheres and sprinkling atoms of cobalt and palladium onto the spheres creating magnetic caps to store binary data as the polarity of a magnetic field."[143] A team from Case Western has approached production issues by "growing carbon nanotube bridges in [its] lab that automatically attach themselves to other components with the help of an applied electrical current." Lead researcher Massood Tabib-Azar discovered that "you can grow building blocks of ultra large scale integrated circuits by growing self-assembled and self-welded carbon nanotubes much the same way you'd a table."[144] UC San Diego researchers have "grown carbon nanotubes that change direction along their length." In addition to their use as tips for AFMs, they could serve as "electrical connectors in integrated circuits and as a replacement for the metal alloy solders that forms interconnects between microcircuit devices."[145] Even more exciting have been two recent events. First has been IBM's partnership with Stanford "to create the Spintronic Science and Applications Center aimed at establishing the company as a leader in spintronics—using the spin of electrons to store data—and could lead to a significant increase in the density of hard disk."[146] Second, researchers from the University of Science and Technology in China and the University of Heidelberg reportedly used "four photons to form a logic gate that can be used in quantum computers... though it will be 10 to 20 years before the logic gates could be used practically."[147]

Applied Nanotech licensed a single-layer photoconductive film from Texas at Austin to develop a ten-thousand-bit photoelectric memory chip.[148] Using electron-beam lithography to carve switches from wafers made of single-crystal layers of silicon and silicon oxide, researchers at Boston University have carved tiny switches out of silicon, fabricating mechanical switches that are thousands of times thinner than a human hair,[149] the results of which could mean:

- data storage capacity and processing speeds will increase dramatically and be cheaper and more energy efficient;
- biosensors and chips could become ubiquitous in daily life, monitoring every aspect of the economy and society;
- light-based electronics may be possible; and
- electronic paper will arrive in a few years.

A new age of computing and electronic power may not be too far off. Of all the applications of nanotechnology, it is likely that these will develop the most rapidly. In May 2002 IBM reported that it "had created carbon nanotube transistors that outperform models of even advanced silicon devices and even outperform previously designed nanotubes with an increased capacity for carrying an electric current."[150] The next several years will most likely usher in new and more sophisticated prototypes for nanotube computer chips that may alter the entire world of computing and electronics. NanoMarkets believes each of the display, sensor, and memory sections will include "more than $200 million in carbon nanotube based products by 2007."[151]

On a macro scale, for example, research in quantum dots may have an application "in large power switches found at utility plants where they can be used a wear indicators."[152] There are some excellent indications that carbon nanotubes might make highly efficient transmission lines. For example, a team from Fujitsu used bundled multiwalled carbon nanotubes and "produced a total resistance equivalent to that of tungsten, which has a resistivity around four times that of copper."[153]

Major growth is expected in display technology. For instance, "Korea's Samsung Group plans to produce TV displays featuring the most prominent building block of the Nano Age—carbon nanotubes—by 2006. If successful, these screens could be lighter, cheaper, brighter, and more energy-efficient than today's models."[154] Motorola has built a working prototype of a new color display using carbon nanotubes that grows them directly onto the display's glass substrate. They call the design a nano-emissive display.[155] A joint ATP venture in 2000 between Motorola, the Dow Chemical Company, and Xerox investigated flexible computer displays.[156] NanoChromics in Dublin, using proprietary nanostructured materials, has displays that "look like ink on paper and act with the intelligence of an electronic display."[157] Nanosys and Norel Optronics, Inc., claim to be producing a new generation of organic light-emitting diode (OLED) displays using fullerenes. "They are 50-100 percent more power efficient, have lower drive voltage and power consumption, improved lifetime and stability and better color tunability."[158] According to Adi Treasurywala of the University of Toronto Innovations Foundation, "Many consider this to be a breakthrough enabling technology and may very possibly be the first nanomaterial application to make it to mass market."[159]

"The tiny deformations or dents caused by collisions with various gas species can change the electronic properties of nanotubes. Kim Bolton of Göteborg claims 'carbon nanotubes can be used to detect gases that are very difficult to observe with current measurement techniques.'"[160] According to Eric Snow of the Naval Research Laboratory, these sensors will have implications in industry and defense, such as a nerve agent detector. After exposing nanotubes to DMMP (a chemical similar to the nerve agent Sarin), he reported the "detector was sensitive to one part of a billion of DMMP."[161] Even greater precision is anticipated. For example, the National Academies Keck Futures Initiative awarded researchers at UC Santa Cruz a grant to develop a new sensor technology that "would enable the optical detection of specific molecules with single molecule sensitivity using a compact chip-based device."[162]

Detectors and sensors that use carbon nanotubes can be used for environmental monitoring, respirator diagnostics, and biomolecule sensing are being developed by Nanomix.[163] NanoHorizons works in thin-film nanostructures "with applications in sensors and flexible micro-electronic applications in consumer, industrial, environmental, forensic and homeland security."[164]

Kopin manufactures CyberDisplay technology—ultrasmall screens.[165] Cambridge Display Technologies is rivaling liquid crystal display with its polymer organic light-emitting diodes technology. According to DisplaySearch, "the flat panel display market is expected to grow to $97 billion in 2008, representing a compounded annual growth rate since 2000 of approximately 19 percent."[166]

More revolutionary work is being done in superconductivity. "By incorporating nanowires as filament in beggar superconducting wires, more current could be carried without being destroyed by a magnetic field because nanoscale superconductors don't repel magnetic fields."[167] A team from Cambridge and Los Alamos reported a twofold to fivefold increase in the current densities of coated conductors in high magnetic fields as a result of a nanoscale effect.[168] Even more radical is the potential of Distributed Sensing Smart Dust as a Wireless Sensor Network. Developed from radio frequency identification platforms using nanocrystals, it offers a range of possible applications in computing. Vasiliy Suvorov of Luxoft Labs predicts that "Smart Dust is truly the vehicle that will deliver on the promise of pervasive computing. It's a distributed computing environment with a lot of artificial intelligence."[169]

HEALTHCARE

Living systems, whether they are humans, plants, or animals, are governed by molecular behavior at the nanoscale, where scientific concepts of chemistry, biology, computation, and physics all play an important role in converging to form basic life processes. The Freedonia Group predicts that "demand for nanotechnology healthcare products in the US will increase nearly 50 percent per year to $6.5 billion in 2009."[170] John Sterling of *Genetic Engineering News* predicts that "a number of drugs and innovative technologies will be among the first products to emerge from this revolutionary field."[171]

Both the National Institutes of Health and the National Cancer Institute have committed resources for nanotechnology research and development. "The NCI announced a $144.3 million, five-year initiative to develop and apply nanotechnology to cancer."[172] Undoubtedly, their initiatives are propelling research and development of nanotechnology in the health fields.

"There is growing activity in nanoscale approaches to target identification, target validation, high-throughput screening and lead optimization, dramatically reducing today's $800 million 10-15 year process for drug discovery."[173] As such, there have been significant investments to develop "arrays of nanosensors for detecting proteins, and utilized in conjunction with protein analysis techniques, providing research with a powerful analytical toolbox for drug discovery."[174] NanoMarkets claims "nano-enabled drug discovery solutions will generate revenues of $1.3 billion in 2009 and grow to $2.5 billion in 2012."[175]

For example, NanoTek is building a small, microfluidic machine to quickly and reliably synthesize drugs, medicines, diagnostic imaging agents and other compounds.[176]

The NIH has a "Roadmap for Medical Research" that includes nanotechnology. According to Michael Weiner from Biophan Technologies, "malaria, heart disease, and diabetes are serious diseases nanotechnology can help combat."[177] The NCI's goal is "to do away with pain and suffering caused by cancer by 2015."[178] Together they constitute a new Alliance for Nanotechnology in Cancer. One of its first steps is to create Centers of Cancer Nanotechnology Excellence. One collaborative research center funded is from Emory and Georgia Tech. In late 2004, the NCI awarded over $7 million to establish the Bioengineering Research Partnership in cancer nanotechnology. The NIH awarded nearly $10 million for researchers at Emory and Georgia Tech to develop a new class of nanoparticles for molecular and cellular imaging. "The goal of this exploratory program is to develop a new class of bioconjugated quantum dots that both image and target single-molecule processes in single living cells."[179]

Nanotechnology may enable a convergence of delivery and therapeutics. Right now, much of the research undertaken by major pharmaceutical companies involves new drug-delivery vehicles that promise to "enormously broaden the drug's therapeutic appeal."[180] Nanodevelopments seem to be ushering in the era of personalized medicine, where patients are given different treatments for the same disease. Starpharma is "developing dendrimers as pharmaceuticals that can be readily tailored to treat specific diseases, tissues or organs."[181]

The drug-delivery market is "growing 15 percent annually and is expected to be worth $28.8 billion in 2005—up from $14.4 billion in 2000."[182] Flamel Technologies has been testing its long-acting insulin, Basulin, which uses its Medusa nanoparticulate system.[183] A Cardiff team is "working on new nano-particle drug formulation for inhalers."[184] Nanovax has been working on a "micellar nanoparticle drug-delivery platform, a topical emulsion of oil, water, and lipids capable of being absorbed through the skin."[185] It is developing two hormone replacement therapies called Estrasorb and Androsorb.[186] Liquidia Technologies Inc. announced a process called Particle Replication In Nonwetting Templates (PRINT), which "enables fabrication of custom-sized, monodispersed and shape-specific particles of virtually any material and encapsulating nearly any active cargo.... This breakthrough allows for the production of nanoparticles that contain fragile organic matters such as genes and drug products," says Joseph De-Simone, cofounder.[187] Elan and King Phamaceuticals have had successes with NanoCrystal drug-delivery technology as well.[188] Johnson and Johnson will be using Elan's NanoCrystal technology in a Phase III trial for an injectable drug for schizophrenia.[189]

Dendritic NanoTechnologies and NanoCures' encapsulation approaches use dendrimers (branched polymers). Skypharma is working on a dendritic delivery system for a topical microbicide for the prevention of HIV, herpes, and other sexually transmitted viral diseases.[190] According to Wolfe, "If dendrimers make it to market for diseases like cancer, you could not only be saving treatment costs by piggybacking multiple drugs on one carrier, but you will likely be able to image, follow the progress and treat the cancer sooner and more effectively than ever before."[191]

"Sixty-one nanotech-based drugs and 91 devices or diagnostic tests have entered preclinical, clinical, or commercial development,"[192] according to *NanoBiotechNews*. Some nano-related drugs and products have received regulatory-authority approval. In January 2005, the first nanotechnology drug was approved. "American Pharmaceutical Partners, Inc. shares surged 50 percent on news that the Food and Drug Administration had approved the marketing of *Abraxane*, a nanoscale protein-based drug for the treatment of metastasic breast cancer."[193] It seems to be "positioned to capture a large share of this market, with forecasts rising to $500 million in 2006."[194]

While researchers at Texas Christian use silicon nanowires[195] and a Purdue team uses carbon nanotubes[196] to promote bone growth, Samuel Stupp at Northwestern uses "nanofibers to promote the growth of hydroxypatite crystals that form a primary component of bone."[197] And Purdue scientists used vascular stents with nanobumps on them to repair arteries. They learned "bone and cartilage cells in Petri-dishes attach better to materials that possess smaller surface bumps than are found in conventional materials used to make artificial joints."[198]

Early diagnosis has been targeted. "Nanotechnology promises the development of new analytical tools capable of probing the world of the nanometer, where it will be increasingly possible to characterize the chemical and mechanical properties of a cell,"[199] which is certain to aid in the treatment and perhaps eventual eradication of some of the world's deadliest diseases. "Patients can die waiting for test results. An Austin, Texas, start-up, LabNow, has the potential to speed AIDS treatment in much of the world—and let LabNow cash in on the $5 billion global markets for point-of-care testing. In October 2004, LabNow got $14 million in equity funding from a consortium led by George Soros. The company hopes to rolls out its systems in South Africa by the end of 2005."[200] A group at Arizona State is using monobase-conjugated nanocrystals to detect infection agents and to provide reliable forensic analysis.[201] Carnegie Mellon and Stanford researchers have experimented with nanosensors that measure the release of glutamate believed to contribute to conditions such as Alzheimer's and Parkinson's disease.[202] Another team at Northwestern is working on an assay to detect the presence of a protein (ADDL, amyloid plaque) normally associated with Alzheimer's. This "could also lead to a test to diagnose breast cancer…and form the basis for a new test for HIV and other diseases in blood screening."[203]

Miniaturization of chemical analysis and synthesis will improve throughput, performance, and accessibility and lead to significantly reduced costs. But experience has taught us that to maximize these benefits, it is not enough to work on the individual steps of a process (such as extraction, chromatography, or detection)—

we must scale down the entire system.[204] This initiative is often called lab-on-a-chip, and nanoscience is making significant headway on this project. Generally it involves two technologies: microfluidics and arrays of detectors. "Microfluidics involves microscale valves, pumps and channels that can separate and channel molecules in liquid.... Diagnostic chips incorporating microfluidics and microarrays are starting to make their way into the clinical setting."[205] "Molecular diagnostic testing is the fastest-growing segment of the in vitro diagnostic industry.... Sales are expected to exceed $5 billion by 2008."[206] For example, NanoLogix's FDA registered diagnostic test kits are designed to be used for the rapid identification of infectious human diseases and identify thirty-four disease-producing bacteria.[207] A Cornell and Harvard team has modified different nanowires within an array with receptors that are specific for different viruses, which allows multiple virus strains to be detected at the same time. "The Harvard researchers are developing arrays that can sense up to 100 different viruses.... The Cornell team also plans to create similar arrays."[208] A team from Purdue has produced a prototype of miniature devices to study synthetic cell membranes. The team hopes to produce "laboratories-on-a-chip less than a half inch square that might contain a million test chambers, or reactors, each capable of screening an individual drug."[209]

Acrongenomics Inc.'s Neo-EpCAM Cancer detection kit involves its Nano-JETA platform, which combines nanotechnology with molecular biology. They claim the "platform shows a potential in eliminating all known limitations of current technologies."[210]

Quantum Dot is developing clinical-grade quantum dots (semiconducting nanocrystals) with applications in ocular and cancer imaging.[211] In June 2005 an early-detection technology for the respiratory syncytial virus, which causes over seventeen thousand deaths annually, was reported by Vanderbilt University researchers using quantum dots.[212] Evident Technologies is working with non–heavy metal quantum dots.[213]

Eco-Tru is working on a surface disinfectant to reduce infections. In tests "it eliminated post-operative infections in 500 out of 500 patients."[214] Emergency Filtration Products is developing nano-enhanced filter media and "claims that one of the six types of nanoparticles it investigated had highly significant kill rates against all five pathogens (three types of bacteria and two species of fungi)."[215] Predeep Sharma from the University of Houston predicts "current quantum dots work will be to put to practice in the medical arenas in the next five to ten years."[216]

"Nucryst Pharmaceuticals engineers silver particles into infection-fighting bandages for burn victims."[217] EnvironSystems has been working with an Austin firm that grows human skin for burn reconstruction, and EnvironSystems says, "it eliminated the risk of organ rejection."[218]

A Penn State study reported ceramide administered through the bloodstream kills cancer cells. The researchers "encapsulated the ceramide in tiny bundles called liposomes." According to one of the lead researchers, "packaging ceramide in our nano liposome capsules allows them to travel through the bloodstream without causing toxicity and release the ceramide in the tumor."[219] Introgen Therapeutics has conducted animal trials of its nanoparticle-based therapy to suppress tumor growth and prolong survival of mice with lung cancer.[220] Stupp at Northwestern is studying the "injection of nanofibers and proteins" to prevent a blood vessels from forming, to starve a cancerous tumor.[221] USC and Children's Hospital of Los Angeles researchers have used nano-sized sugar polymers to target the growth-promoting gene for Ewing's sarcoma. They report the technology "effectively silences the gene and shuts down tumor cells' genetic machinery."[222] A group from Georgia Tech is using core/shell nanogels, which are functionalized to allow them to target cancer cells. "By applying a targeted heat source—like ultrasound—only to the tumor, doctors should be able to avoid killing healthy cells that happen to take in the nanoparticles."[223] Researchers from pSilvida of Australia announced in 2005 their BrachySil technology, an "experimental treatment for liver cancer that uses a biochip to release predetermined dose of radioactive molecules to kill cancer cells."[224] Remarkable anticancer therapies based on thermal ablation is being pursued by Nanospectra Biosciences and Triton Biosciences. Nanospectra's nanoshells are made up of an insulating core coated with an ultrathin metallic layer that "doctors can use to treat cancer patients by shining light that heats the nanoshell structure and burn tumors without affecting health tissues."[225] "Tumorous mice injected with nanoshells and then exposed to infrared light became cancer-free within 10 days."[226]

Nanotechnology also promises to abolish the current limitations of computer simulations. Increases in computational power, which is certain to be an inevitable result of nanotechnology, would create far more realistic environments and simulations that are essential for studying surgical and drug processes.[227] For example, Advanced Magnetics and Combidex use nanoparticles to illuminate magnetic resonance imaging (MRI) scans.[228] So does QuantumSphere, and it claims order-of-magnitude performance improvements.[229] A team from Kyoto anticipates the controlled product of endohedral fullerenes with applications in MRI and nuclear magnetic resonance analysis.[230] Researchers at the Lawrence Berkeley National Lab claim to "have developed a way to sneak nano-sized probes (crystalline semiconductors or quantum dots) inside cell nuclei where they can track life's fundamental processes, such as DNA repair."[231]

Cosmetics- and health-related products have seen their share of nanoproducts as well. Sales of facial treatments represented $7 billion of the overall $12 billion skin care market in 2004, and the cosmetic industry has targeted nanoscience. L'Oreal has been marketing nanocapsules since 1995 in its Plenitude line of cosmetics and in higher-end brands such as Lancome.[232] Oxonica has rolled out its Optisol UV absorber in 2003[233] and two years later added an improvement by "doping nanocrystalline titanium oxide with small amounts of manganese virtually eliminating free radical production."[234] Advanced Nanotechnology of Australia markets ZinClear in a product named Wet Dreams All-Natural Sunscreen.[235] Particle Sciences has developed Zatasizer Nano to investigate a delivery system for a new encapsulated form of retinol, vitamin A.[236] Male pattern baldness has also been approached. For example, Hosokawa Powder Technology Research has developed a hair-growth technology using two-hundred-nanometer nanoparticles to deliver active ingredients to hair roots and stimulate hair growth.[237]

Environmental cleanup is also on the nano agenda. Nanotechnology may usher in better sensing. According to NanoMarkets, "The total global market for environmental monitor/gas sensors is expected to generate $2.7 billion in 2008 and $17.2 billion in 2012."[238] Nanotechnology should lead to advanced water-filtering membranes. For example, Argonide makes alumina nanofiber filters whose positive charge attracts out negatively charged germs.[239] Nanoscale polymer brushes coated with molecules to capture and remove poisonous metals proteins and germs are being developed by eMembrane. Applied Nanotech is working with Japan to make nanocolumns of titanium oxide that are potent as photocatalysts.[240] Other innovations could produce remediation for some of our more carcinogenic toxic waste sites. For example, the Center of Environmental Kinetics Analysis (CEKA) at Penn State is studying how nanoparticles might be used to clean up environmental contaminants like nuclear waste.[241] A Temple University and Montana State team "uses iron nanoparticles plus visible light or sunlight to covert chromium-6 to insoluble chromium-3 which can then be filtered out."[242] Researchers at Rice and Georgia Tech identified nanoparticles of gold and palladium as a catalyst to remediate trichloroethene. Palladium converts TCE into nontoxic ethane.[243] Trane has field tested nanoparticles containing palladium on TCE in North Carolina.[244] Iron particles seem to have captured reportage on remediation. Researchers at Pacific Northwest National Lab and the University of Minnesota "found that iron particles can effectively destroy carbon tetrachloride."[245] Oregon's Health and Science University reports a "type of nano-sized iron may be useful in cleaning up carbon tetrachloride in groundwater,"[246] and Polyflon Co. announced the availability of the PolyMetallix line of nanoscale iron particles for use in environmental remediation projects in April 2005.[247] However, nanoized iron is not the magic bullet. For example, Rice researchers have warned "breaking down TCE with iron produces intermediate chemicals, like vinyl chloride, that are more toxic than TCE."[248]

Occasionally, an article surfaces like this one from the AARP, which quotes a Dartmouth-Hitchcock Medical Center doctor; the claims are pure hyperbole:

Nanobiotics would allow the creation of biological robots that not only permit observation of the human body at the most refined level possible, but that can serve as sentinels to identify and prevent disease before symptoms even appear. In time, sophisticated nanobots could be manufactured to enhance tissue or strengthen frail bones within the body, reversing more debilitating aspects of aging. Nanobots might even be programmed to maintain homeostasis.[249]

While the preceding may sound fanciful and fictitious, some of the recent news reports are not that far off the mark. For example, "Chinese scientists have tested a swimming device measuring 3mm × 2mm × 0.4mm." Tao Mei of the Chinese Academy of Sciences remarked, "We would like to make a 1mm one that could go inside the bloodstream. Maybe we can make it even smaller using nanotechnology."[250] Scientists from Nanomix and UCLA have discussed "an artificial eye—where the conversion of photons to electrons is achieved through a light-absorbing molecule or layer that is coupled to an electronic device attached to neurons." George Grüner of UCLA added, "It should also be possible to connect living cells directly to nanoelectronic devices. Such work opens up the avenue for what could be called *cellectronics*, the detection and modification of bio-processes at the cellular level."[251] Argonne already announced that its ultra-nanocrystalline diamond films "hermetically seal a silicon microchip implanted in the eye's corrosive environment.... It may help to restore sight to the millions of people with degenerative retinal disease." Argonne is "working with Second Sight Medical Products, Inc. and six other research organizations to refine a device to replace the eye's destroyed rods and cones as light receptors and optical signal converters."[252] Spire received a grant "to develop biocompatible surface structures for multi-channel neural electrodes, which could potentially help patients dealing with such debilitating neurological illnesses as multiple sclerosis, Alzheimer's disease, Parkinson's disease or spinal cord injury."[253] In 2005 Rensselaer Polytechnic researchers learned to grow nanotubes shaped like brushes. Prof. Anyuan Cao speculated they "might be used either to coat protective substances onto damaged surfaces in our bodies, for example veins, or to clean up unwanted deposits."[254] No less remarkable was an announcement out of Trieste, Ferraro, SISSA/ISAS (International School for Advanced Studies) and National Consortium of Materials, Science, and Technology (INSTM), where researchers reported to have grown nerve cells on substrates containing networks of carbon nanotubes. "In principle they could be used as assistive devices to functionally and structurally re-connect neurons."[255] With these new advancements (real or fictional), it is estimated that within the next decade, as much as half of all healthcare industry revenues may be in the field of nanomedicine, and this number will likely grow as nanotechnology and medicine become more intimately intertwined.

ENERGY

According to NanoInvestor News,

> Nano-engineering promises highly efficient new conductors and superconductors that could gradually replace current transmission facilities. Nano-enabled solutions, such as supercapacitors, will create entirely new opportunities for local electricity storage and may gradually lead to new distributed architectures for electricity grids. At the same time, nanoengineering is breathing new life into alternative energy sources—especially solar power and fuel cells.[256]

For example, researchers at UC Davis and Lytitek, Inc., have developed a process to make the multiwall nanotube supercapacitor film with a slew of applications in electric and hybrid vehicles, space, cell phones, lightweight electronic fuses, and so forth.[257]

With a growing concern over energy consumption in the United States, there are several new areas of interest. "Worldwide, 6 energy companies received $64 million in venture capital in 1999, 22 companies received $114 million in 2001, and 26 companies received $277 million in 2003."[258] The potential for nanotechnology to "significantly impact energy efficiency, storage, and production"[259] may be substantial, and the government has invested heavily toward meeting some basic energy goals, especially at its national labs. For example, Brookhaven National Lab reported that "double and triple carbon-carbon bonds promote strong electronic interactions and conduct an electric current with low electrical resistance making oligophenyleneethynylene nanowires good candidates for components in nanoelectronic circuits."[260] The University of Wisconsin and University of Massachusetts at Amherst researchers found species of bacteria that produce nanowirelike structures on one side of their cells.[261] "Microbial nanowires could be useful materials for development of extremely small electronic devices."[262] Nanowires may potentially double the efficiency of thermoelectric materials. "Nanostructured thermoelectric devices may be practical for applications such as recycling of waste heat in car engines, on-chip cooling of computer microprocessors and silent, more compact domestic refrigerators," says Heiner Lunke of the University of Oregon.[263]

Some of the anticipated developments may include:

- nanoscale photovoltaics for a hundred times as efficient solar power as thin layers or plastic and sprayable coatings for roofs;
- high-efficiency fuel cells, including lightweight storage tanks and the chemisorption of hydrogen;
- high-current cables (quantum conductors) to rewire the electrical transmission grid, enabling ballistic electricity transport;

- developing a nanoparticle-reinforced polymetric material that can replace structural metallic components in automobiles, which would lead to a reduction of 1.5 billion liters of gasoline consumption each year;
- semiconductors used in the preparation of light-emitting diodes can be sculpted on nanoscale dimensions; and
- conservation from lighter materials, improving car and airplane efficiency, and improved lighting sources like LEDs.[264]

The potential implications that nanotechnology will one day have on energy have been clear for quite some time, yet a renewed commitment from the government in looking for new and more efficient sources of energy has prompted a great deal of growth in the energy sector of nanotechnology. Indeed, the $1 billion FutureGen effort by DOE is attempting to develop a coal power plant having zero emissions. "Nanotechnology can supply more effective catalysts, advanced separation membranes and technologies suitable for sequestering carbon dioxide."[265] In addition, it might allow us to take advantage of ultradeep oil wells. "These wells will require new metals, coatings, drilling fluids and electronics able to withstand the high temperatures and highly corrosive environments encountered at these extreme depths."[266] It is likely that energy will be one of the first sectors to benefit from molecular nanoscience in the new century.

For example, half of all lithium batteries currently incorporate carbon nanotube fibers to increase energy efficiency. "By using nano-materials, the increased surface area of the lithium and widened bottleneck (lithium moving through electrolyte liquid from the negative electrode to the positive backs up on the surface of the liquid) allows the particles to pass through the liquid and allows the battery to recharge more quickly."[267] For 2010, according to Cientifica's *Nanotubes for the Energy Market* report, this figure will rise to 85 percent. In terms of fuel cells, 70 percent will use multiwalled carbon nanotubes in five years' time.[268]

Nanotechnology should play a significant role in the hydrogen economy. Efforts to catalyze hydrogen have been approached from many different directions. According to John Kennedy, director of the Center for Advanced Engineering Fibers and Films, "carbon filters with nano-sized pores can be used to achieve 30 percent of the DOE's hydrogen storage target at room temperature and moderate pressure."[269] Researchers at Rutgers use iridium to break down ammonia to extract hydrogen.[270] Other researchers at DOE's Northwest National Laboratory are taking a new approach to filling up a fuel-cell car with a nanoscale solid (a hydrogen storage material), which releases hydrogen from a solid compound almost one hundred times as fast as was previously possible. The team uses a nanoscale mesoporous silica material as scaffolding for ammonia borane.[271]

"One way around the hydrogen storage problem is to make hydrogen with nitrogen at a fuel processing plant to make ammonia, which is readily stored and transported. Cars could use ammonia as fuel and use the researchers' textured catalyst to extract hydrogen from ammonia with a fuel cell."[272]

Carbon Nanotechnologies, Inc.; Motorola; and Johnson Matthey FuelCells are working on *freestanding* carbon nanotube electrodes for micro–fuel cells.[273] Fuel-cell technology seems to be an important driver. Nanosys, better known for its failed IPO, has been working in solar-cell technology based on structures called nanotetrapods to be sprayed onto roofing tiles.[274] Integrated on the roof of a bus or truck, they could split water via electrolysis and generate hydrogen to run a fuel cell.[275] Nordan from Lux says, "nanomaterial solar energy may first be implemented on mobile devices like cell phones and laptop computers."[276] Nanosys is also developing fuel cells for portable consumer electronic devices.[277] Enerl, Inc., has a pilot facility to fabricate electrodes for high-discharge rate, lithium-ion batteries,[278] with applications in thin-film solar cells, catalytic and functional coatings, fuel cells, and thin-film transistors.[279] "Nanosolar has developed a material of metal oxide nanowires that can be sprayed as a liquid onto a plastic substrate where it self-assembles into a photovoltaic film."[280] NanoHorizons is working on nanoscale photovoltaic cell design with potential increases in the efficiencies in Organic LEDs and solar cells.[281] Altair Nanotechnologies signed an agreement with Advanced Battery in early 2005 to "create high-power, lithium polymer batteries for electric vehicles."[282] Front Edge Technologies is developing the world's thinnest battery, slimmer than a piece of paper. "The thinness of the batteries makes them ideal in the emerging areas such as radio frequency identification tags (RFID) and smart cards."[283] Further miniaturization is being undertaken by a team from the Naval Research Lab and the University of Maryland building the "first vesicle-based rechargeable battery...small enough to fit into nanoscopic devices."[284]

Quantum wires spun from carbon nanotubes could carry electricity farther and more efficiently, transforming the electrical power grid. Smalley's group at CNI has already produced one-hundred-meter-long fiber consisting of well-aligned nanotubes.[285]

Konarka and Evident Technologies are collaborating to increase the sensitivity of plastic solar cells to a wider range of the light spectrum. This power plastic "could be used for demanding energy, communication and military applications, such as battlefield or off grid power generation."[286] Incorporated into textiles, Konarka claims its dye-sensitized solar cells (DSC) "could be used for soldiers' uniforms, tents, field hospitals, covers for trucks and gun emplacements, and wearable electronics."[287] GreatCell of Switzerland with STI of Australia offers small-scale production of DSCs with nanoparticles of titanium oxide as the active components.[288] A team from Penn State also uses titanium dioxide nanotubes and claims significant efficiency improvements;[289] a Sandia team uses nanoscale deposits of platinum;[290] and a team from the University of Toronto is developing a sprayable infrared detector. "These flexible photovoltaics could allow up to 30 percent of the sun's radiant energy to be harnessed compared to 6 percent in today's best plastic solar cells."[291]

"PolyFuel has several fuel cell designs that utilize a nanostructured member delivering 10–15 percent more power than conventional membranes."[292] Nanotech is developing a hydrogen sensor for automotive fuel-cell applications in conjunction with KRI. Japan[293] and GM are in advanced discussions with Toyota "about building a joint factory to make hydrogen powered cars engines."[294] QuantumSphere predicts its nanonickel will replace platinum as the main catalyst in hydrogen fuel cells.[295] "This shift could lead to cheaper hydrogen fuel cells for home and cars in the growing alternative energy market."[296]

Transportation efficiencies are being maximized by nanocatalysts. "Headwaters NanoKinetix recently announced a nanocatalyst that can increase the gasoline octane number by five."[297] Oxonica is rolling out Envirox, a fuel-borne nanocatalyst (whose active ingredient is cerium oxide),[298] which was tested in Hong Kong several years ago and will be used by Cerulean International's entire fleet of seven thousand buses.[299] Acta has catalysts called HYPERMEC that are platinum-free and show good performance on a range of fuels, including methanol and hydrogen.[300] Green Plus, a liquid combustion catalyst that can be added to diesel, gasoline, marine, and other fuels, will be used by London's famous diesel black cabs. Green Plus Limited announced the cabbies will "save more than 10 percent of their fuel costs and their taxis pass the MOT emissions test with ease."[301]

LUXURY PRODUCTS

This category is controversial mainly because of concern that recent commercial products have hardly been earth shattering. While claims are made about significant improves in health and human welfare, here and abroad, especially among the poor and destitute, the developments have mostly been in the luxury-market field.

According to David Rejeski of the Woodrow Wilson Center, roughly five hundred products based on nanotechnology are now on the market.[302] The luxury-field entrants include Babolet's Nanotube Power and VS Drive tennis rackets; Inmat's polymer nanocomposites and accompanying tennis balls; Wilson's nCode badminton paddles, tennis rackets, and golf club driver heads; NanoDynamic's golf balls; AcuFlex's Evolution nanotech golf club shaft; Easton's Stealth CNT baseball bats;[303] Nanodesu's bowling balls; Apple's iPod Nano; Nanogate's clean bathroom tiles; step assists on 2002 GMC Safari SUVs; NanoTwin's NanoBreeze room air purifier; Nanergy's NanoSign illuminated house numbers; Toshiba's Senszoko odor-eating refrigerators; L'Oreal's Plenitude Revitalift anti-wrinkle creams; NuCelle's Sunsense sunscreen, and so on.[304] "Throughout 2005, companies large and small will be rushing more nano-based products from labs to the marketplace."[305]

And the reports keep piling up with a never-ending list of accomplishments associated with textile and clothing, such as Nano-Tex's spill- and stain-resistant khakis and NanoHorizons odor eaters. Presumably, "silver nanoparticles in socks may prevent foot odor by killing bacteria."[306] A Hong Kong group has used "a nano-thin layer of particles of titanium oxide as a substance that reacts with sunlight to break down dirt and other organic material to keep fabrics clean."[307] A Taiwanese undergarment manufacturer, Green-shield, claims its clothing "can eliminate up to 99.99 percent of bacteria, 90 percent of odor and 75 percent of sticky moisture within the cloth." This is accomplished by a process that "releases a constant stream of negative ions and far-infrared rays."[308]

CONCLUSION

It is abundantly clear that the applications of nanotechnology will likely be both numerous and far-reaching, but don't expect a nano-industry per se to develop. Nanotechnology enables products of another industry to be improved or enhanced. A Lux Report "predicts nano-enhanced product sales will grow from today's 0.1 percent of global manufacturing output to 15 percent in 2014. This amount equals the combined size of today's information technology and telecom industries, and it [would] exceed biotechnology revenues ten-fold. Within a decade, 11 percent of total manufacturing jobs worldwide will involve nano-enhanced products."[309]

While some nano-instruments and nanomaterials companies will rise and probably fall, and nanomiscellaneous companies, like Harris and Harris (a VC firm) will surface as well, there will be no nano-industry.[310] This may explain why Wolfe and others recommend investments in instrumentation, such as electron microscope providers; modeling, as in parallel computing and software; devices, especially automated self-assembly; materials (e.g., single-walled carbon nanotubes, etc.), and nanobio.[311]

The following chapter examines three categories of business and industry participating in the nanotechnology movement: major and established transnational firms, publicly traded nanotechnology companies, and start-ups that are dependent on venture capital funding.

CHAPTER 7

NANO-INDUSTRY AND NANO-ENTREPRENEURS

In 2004 a total of fifteen hundred companies announced nanotechnology R&D plans. "Overall the number of companies tracked participating in nanotechnology nearly doubled in the course between 2003 and 2004, with nanotechnology manufacturers and R&D companies up 75 percent."[1] Eighty percent, approximately twelve hundred, are start-ups, six hundred seventy of which are in the United States.[2] Daniel Colbert of NGen Partners warned: "A lot of companies are positioning themselves as nanotech companies, and credibility versus hype is a factor here."[3] However, fewer than twenty companies claim to focus solely on nanotech, but more than one hundred public companies have nanotech R&D initiatives. "Over 250 companies worldwide are engaged in commercialization of nanotechniques and tools, and over 10 percent of those have real products in the market now."[4] The NanoBusiness Alliance estimated that global nanotech revenue has already surpassed $45 billion. "By 2008, more than $100 billion in products will likely involve some type of nanotechnology."[5] Economists at the NSF estimate that nanotech could yield a trillion-dollar global market by 2015.[6] According to the ETC Group, "nanotech insiders predict the $1 trillion mark will arrive four years earlier—by 2011."[7] In 2004 Lux Capital predicted corporations and venture capitalists would spend over $5 billion on nanotechnology research and development.[8] The global carbon nanotube market alone may reach $500 million to $700 million in 2005.[9]

Are these projections meaningful? Given the penchant to include "all those technologies which use atomic scale precision to evolve products as nano," maybe they are not. As Thomas Abraham, vice president for Business Communications

Co., attests, "small tech or micro technology did not establish as a catchy title, although micro devices such as MEMS are increasingly becoming established markets. Since the boundary between micro and nano is narrow, some people also lump micro into nano markets."[10] The data seems to be very dependent on the motivations of the analyst or spokesperson making the claim. As such, these projections need to be evaluated critically. One of the cynics has been Alan Shalleck from NanoClarity. "The lack of nanoproduct commercialization in the U.S. where all the big money initially was spent overwhelms me." His diagnosis: "Nanotechnology companies in the U.S. are run today by technologists, not businessmen or entrepreneurs. Nanomanagement seems unenergized...too much available cash with no need to worry about the next payroll for the next two or three years."[11]

Or are the projections merely hype? Howard Lovy and others caution us about heightened expectations: "Nanotechnology could become the deserving victim of just another hype-and-deflate media cycle. We are already seeing a new skepticism in the way the popular media cover nanotechnology because there's a frustration with the disconnect that exists between promises and current reality."[12] He adds, "A nanotech industry built solely on selling nanotubes, nanowires, and the like, will end up with a few suppliers selling low margin products in huge volumes. That is not the picture that most business people and financiers have in mind when they invest in nanotechnology."[13]

Nearly everyone admits nano has produced little more than scratch-resistant paints and stain-resistant pants. Of course, only "by applying nanotech to existing markets can we establish the commercial viability of technology and *then* expand into new markets."[14] Investors aren't about to invest in the hyperbolic promises of the speculators. After the dot-com disaster and a series of corporate blunders by Enron and WorldCom and other public companies, the climate is risk aversive. As such, we are watching an industry growing by taking baby steps. It may be irrelevant that "buckeyballs will eventually occupy about the same space in a nano-economy that screws and nails do in today's economy" if the industry never gets off the ground.[15]

Claims associated with the business of nanotechnology are rife with exaggeration. "Experts predict it will drive progress in virtually every field, from computing to medicine, manufacturing, energy and the environment."[16] While false expectations can be problematic, this is not universally true; hype can whet appetites as well—hence hype is not all bad. "Hype helps this section to function and to be successful by attracting smart people and money," says Juan Sanchez, a nanotech analyst at Punk Ziegel & Co. Sanchez adds, "The risk with hype is the timing of it. Overhype at the wrong moment can bring dark clouds."[17] Unfortunately, those people responsible for the rhetorical flourishing may not be market savvy enough to time their remarks fortuitously, and there's the rub.

Merrill Lynch reported some positive indicators that seem to be driving nanobusiness. They included the financial success of Harris and Harris (see below), intense client interest, and heightened political and financial activity.[18]

The good news is that investment seems to be increasing even if cautiously so:

The dollars invested in nanotechnology increased in 2003, 42 percent over the year before, and nanotechnology accounts for an increasing proportion of small tech (the sector accounts for $982.5 million or 5.4 percent of the $18.2 billion in overall venture capital deployed in 2003). Investors participated in 34 nanotechnology funding events worth $301 million, or 30.6 percent of all small technology funding, although nanotech deal volume fell 17 percent, from 41 funding events in 2002 to 34 in 2003. This may be because investors also increasingly favored more mature nano firms.[19]

They are also favoring "larger, later-stage rounds as companies prepared for possible exit opportunities."[20] This amount represents over 5 percent of all venture capital (VC) funds distributed.[21] R&D funding doubled in 2004. "Most of the increase was driven by a big jump in corporate and private funding, which grew by 160 percent."[22]

"The revolution won't happen overnight, and even nanotechnology's biggest supporters acknowledge that the field could become the next craze—think dotcoms—in which hype outruns real application and business sense."[23] Everyone seems to be getting into the game. "For starters, the government, tech titans and ven-

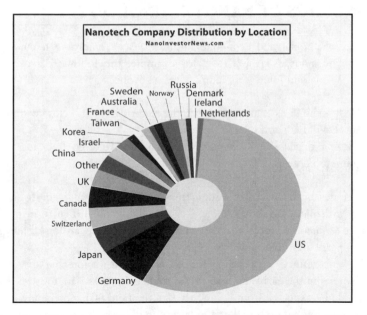

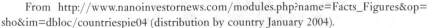

From http://www.nanoinvestornews.com/modules.php?name=Facts_Figures&op=sho&im=dbloc/countriespie04 (distribution by country January 2004).

ture capitalists are pouring money into the field, producing breakthroughs that have enabled several companies to make nanotechnology product announcements."[24]

The economic competitiveness associated with nanotechnology in the twenty-first century could not have come at a more opportune time for most Americans. Since September 11, "the patriotic meter is registering an American fervor, and in typical American style, it's [showing up] in the economy."[25] The post-9/11 surge in nationalism has clearly spread beyond barbeques and pep rallies and into a world where business leaders see America's economic dominance as a necessary prerequisite for ensuring our safety against terrorism and external threats in the new millennium. This new competitiveness and surging nationalism provide a ripe environment for nanotechnology investment, which many see as the gateway to prosperity in the New Economy. This is an environment that not even the most astute economists or technology experts could have predicted just ten years ago.

By any accounts though, nanotechnology is beginning to take hold in the private sector, placing the United States at a distinct advantage over all other nations. It is evident that the United States plays host to the vast majority of nanotechnology companies currently in existence, and enjoys a sizeable advantage at this point in the game (see the chart on page 215).[26]

ECONOMICS OF NANOTECHNOLOGY

Predicting the social and economic implications of emerging technology has proved to be a loser's game. Take, for example, the following excerpts from Erasmus Wilson and John Von Neuman, respectively:

- "When the Paris exhibition closes, the electric light will close with it and no more will be heard of it."[27]
- "A few decades hence, energy will most likely be free, just like the unmetered air is."[28]

Examples of historical misjudgments illustrate the single most significant problem with nanotechnology today—the fundamental difficulty in predicting the future of major technological and scientific advances, regardless of how meticulously a particular theory is researched.

The uncertainty surrounding the economics of nanotechnology led economic experts in the early and mid-1990s, such as Burgess Lair, to warn that very little should be expected from US industry in the field of nanotechnology.

"Nanotechnology, after all, is a long term high-risk undertaking that firms on their own will not readily undertake because of the cost, and because if they

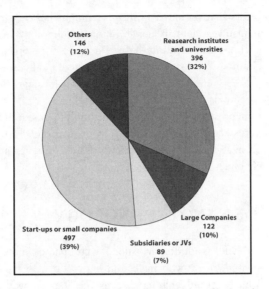

should realize any breakthroughs in it, those resulting discoveries will be greatly diffused. It's a classic public good, market failure."[29]

Even as early as 1995, it still appeared highly unlikely that corporate investment decision making would manage to stimulate any sizable interest in nanotechnology. Visionary Ralph Merkle complained,

> A hard nosed engineer, pretty much by definition, would have a planned horizon of under 5 years. In order to motivate such individuals it is necessary to either (1) describe something that they can do in the next five years that will provide them with some direct benefit (a product that makes money for example) or (2) pay them money to work on something that will not have a significant payback in five years.[30]

A variable often overlooked is one of obsolescent productive lines and the loss of a profitable aftermarket:

> Less well publicized is the challenge that established production companies will have in creating nanotech products which threaten their long term continuity. The issue is one of assessing the reality of and then managing the threat of disruptive technologies. The new problems arising are that we are not just dealing with a single technology shift but a diversity of approaches, each of which would require a different knowledge set to be successful, and a probable change in business model. Changing a company's business model and core competencies is a much bigger problem than evolving an existing product range.[31]

Government officials and agency research labs heeded the warnings of Merkle and others throughout the 1990s and worked to create joint research and development projects that included both government labs and private industry. Massive government investment in academic nanotechnology research since the passage of the NNI has also certainly helped to encourage private-sector investments in nanotechnology.[32] This private-sector growth has been evident in recent years, as business investment in the field now accounts for nearly 75 percent of all nanotechnology programs (see the chart on page 217).[33]

While government subsidization of nanotechnology research has done a great deal to spur private-sector interest, it has not been the exclusive driving force behind the popularization of nanotechnology. Entrepreneurs and industrialists have joined the charge.

BUSINESS OF NANOTECHNOLOGY

Lawrence Bock of Nanosys admits: "Most of the companies in the nano field right now are either *nano-pretenders* who have nothing to do with nanotechnology, or they're sort of what I would characterize as the nano-powders world, where they're making heterogenous sorts of powders which really address what I would call low value opprotunities."[34] Will nanotechnology transform the world business economy?

> Examination of current activity in nanotechnology suggests to us that the revenues from the nanotechnology industry are going to come from a relatively narrowly defined group of applications.... But the picture that is sometimes (perhaps) inadvertently painted of nanotechnology being an industry whose revenues are spread over an impossibly wide area of applications is misleading.[35]

Nevertheless, the hype proliferates and "VC and Wall Street strategists are among the biggest cheerleaders for nanotech."[36]

Business and industry have a vested interest in government initiatives, and this has been especially true for the NNI and its enabling legislation. They left their fingerprints all over the Twenty-First Century Nanotechnology R&D Act. According to journalist Howard Lovy, there is an unanswered central question underlying the NNI and the 2003 act:

> Why was this first-ever piece of nanotechnology legislation conceived, written, altered, and sold purely as a business proposition? For now, it is commerce that is driving the nanotech vision, redefining *real* nanotechnology to suit what is best for nano business. [Referring to Drexler's complaint that molecular manufacturing was slighted in the new act,] [B]usiness leaders and policy makers did this by carefully selecting which theories are the ones the general public is supposed to believe, then marginalizing the rest.[37]

While debate over the direction of the NNI was relegated to the previous chapter, Lovy's argument moves beyond this controversy. Who's driving the NNI? Is it science? Is it business? Or is it public policy? Many, including Lovy, grumble that decisions are made in terms of moving nanotechnology into the marketplace, maybe rightly or wrongly. For example, much of the rhetoric coming out of the offices of the NSF in Arlington relating to health and other societal and ethical implications are framed toward frustrating individuals and groups who might threaten the march to nanodriven economic prosperity—and we educate the workforce of the future.

There are major drivers behind this burgeoning industry. "The two major players are the U.S. government and the 300 plus companies developing instruments for nanoscale imaging, manipulation and manufacturing."[38] "An estimated five hundred nanotech companies are active throughout Europe, North America, and Asia including leading transnationals such as BASF, L'Oreal, Bayer, Exxon, IBM, and Hewlett Packard."[39] Josh Wolfe "speculates that IBM is devoting about 50 percent of its long-term research and development spending to nanotech."[40]

Major corporations are investing in nanotechnology to capture a nano-option, companies that make nanotechnology equipment, and nanotechnology start-ups.[41] For example, Intel reports $20 billion in revenues from products where nanotechnology played a key role in 2003.[42] Major corporations in nanotechnology include DuPont, which is developing nanocoatings; ExxonMobil working on fuel-cell technologies; General Electric in nanocomposites and nanostructures optoelectronics and biomimetics (reverse engineering biology); General Motors with nanocomposites; Hewlett-Packard in advanced computing; IBM investigating nanoscale integrated circuits; and Merck and Co. working on lab-on-a-chip technology, medical diagnostic sensors, and drug delivery. Leaders in nanomicroscopy include Hewlett-Packard, IBM, and Technical Instruments. Production tools are being developed by Intel, Molecular Imprints, and Nanometrics. Nanomaterials are being produced by Altair, Nanomat, and Nanophase Technologies.

Generally, the start-ups involve academics who are attempting to capture as much of the intellectual property as possible with an eye toward commercialization.[43] Unless bankrolled by a wealthy angel, they turn to the venture-capital world for funding. "Modzelewski said there are some 1,500 nanotech start-ups worldwide including 1,100 in the U.S."[44] In 2004, "over $400 million flowed into nano start-ups here in the U.S. and around the globe."[45]

But there's trouble in paradise. The first time this issue surfaced was in a tour of Zyvex a few years ago. During the tour, my party was informed that the company had developed a machine to test the tensile strength of carbon nanotubes. Upon questioning, we learned that quality control for carbon nanotubes had become a problem. "A 2004 Lux Research study found that many of the 200 global suppliers of basic nanomaterials failed to deliver what they promised," and according to Lux's Matthew Nordan, "as a group, they have a frighteningly poor track record." *Business Week*'s Baker and Aston extrapolated: "Until the industry puts a qualified supply chain in place, only innovators working with world-class labs can count on reliable material. This limits access to nanotechnology and hurts its growth."[46]

What these companies are making may give them value beyond their intellectual property (IP) and patent portfolios.

> For the potential investor, initially it is useful to consider nanotechnology companies as falling into one or more of the following sectors: tools, materials, sensors, biotech, and electronics. Although some companies are developing technologies or products that don't fit neatly into just one of these sectors, most of the nearer investment opportunities fall within these five groupings, and the categorizations help to define competitors, business issues, and opportunities.[47]

If you listen to the rhetoric of government enthusiasts and industrialists, in the twenty-first century, it is nanotechnology that may guide the new world economy, by creating thousands of new technologies, which may spur communication and integration in the global marketplace. In this new world the stakes are high, and for the United States to remain an economic hegemon it must create an economy with a strong foundation in nanotechnology.

As we have seen, other economically powerful nations such as Japan and China have invested heavily in nanotechnology, yet have done little to spur private-sector growth in the field. By capitalizing on these strategic shortcomings, the United States may have a unique opportunity in the twenty-first century to assert itself economically.

America has risen to the challenge. The private-sector investment in nanotechnology is coming from two main sources, each of which is integrally important in ensuring adequate business growth for nanotechnology.[48]

ESTABLISHED TRANSNATIONAL FIRMS

Lux conducted in-depth confidential interviews with executives from thirty-three corporations with more than $5 billion in annual revenue. "Global corporations are investing heavily in nanotechnology development, but most fail to tie these activities to an explicit strategy or coordinate their efforts across the company." According to Lux's Nordan, "uncoordinated nanotech efforts may be caught blindsided by better-organized, competitive, and fleet-footed start-ups." Surprisingly, "pharmaceutical companies are least likely to have an explicit nanotechnology strategy."[49]

The least risky of the major investment areas in nanotechnology and, as such, the most widespread at this point are the major corporations that are investing in in-house nanotechnology research in order to stay on the cutting edge of science and technology. "In recent years nanotechnology has become a more important part of research and development for many companies, and some are now devoting up to a third of their research budgets to nanotechnology."[50] The majority of Fortune 500 companies, including all the household names on the table below, are devoting substantial time and liquid capital into researching nanotechnology in hopes of gaining a competitive edge in their particular market.[51] For example, "19 of the 30 companies in the Dow Jones Industrial Index have launched nano initiatives."[52] According to the *Boston Globe*'s Michelle Rama,

> Large technology companies have been researching nanotechnology for years. Xerox and Kodak are studying replacing traditional toner with nanopigments. Fuji is looking into extremely thin coatings and magnetic particles for archival tapes. Intel, IBM and Texas Instruments are researching replacements for traditional transistors that are nearing their functional limits as they are shrunk. Hitachi, Seagate, and IBM are upgrading computer storage technologies with nanotech.[53]

SOME TRANSNATIONAL COMPANIES IN NANOTECHNOLOGY

COMPANY	DEVELOPMENT AREAS	FUNDING AND RESEARCH CENTERS
DuPont	Nanocoatings, color technologies, nanoelectronics and technologies	Partner in the Institute for Soldier Nanotechnologies
General Electric	Biomimetics, nanotubes, nanowires, nanocomposites, nanostructure optoelectronics	GE Global Nano Research
General Motors	Nanocomposites, hydrocarbon fuel cells	GM Nano Research Center
Hewlett-Packard	Molecular electronics, nanowires, nanodevices, nanocomputing architectures, biochips, AFM	Quantum Science Center
IBM	Chemical AFM, magnetic imaging, dynamic force microscopy, nanoscale integrated circuits, quantum computing, self-assembly	Millipede
Motorola	Biochips, molecular electronics nanotubes, AFM, self-assembly	Motorola Research Labs
Xerox	Nanoparticles, nanomagnets, nanoelectronics	Palo Alto Research System

Hewlett-Packard is researching nanotech to improve how ink interacts with paper and materials within the ink-jet head, among other things.[54]

Undoubtedly, there exists incredible potential in the business of nanobiology, especially nanomedicine. A recent market overview involving the intersection of nanotechnology and biology was done by Arnaud Paris discussing five markets:

- DNA microchip market—major high-added-value applications in the drug discovery and agrochemical fields
- lab-on-a-chip market—concept based on multiparametric analysis such as sensing to diagnosis
- nanocrystals market—more stable and sensitive than fluorophores and designed to label specific targets, they are expected to impact imaging technologies
- nanospheres market—block copolymers, dendrimers, and nano-emulsions have potential as a new drug-delivery system
- nanostructures market—pharmaceuticals and diagnostics to tissue engineering and organ reconstruction.[55]

In the past the largest difficulty in the field of nanotechnology was not application but investment. This seems to have turned itself around, and the firms involved are international. Nonetheless, the bulk of business in the nanorealm is in the United States.

START-UPS AND VENTURE CAPITAL

The second source of private nanotechnology investment involves the venture capitalists. In general, government funding flows through university systems and "gets converted into kinetic energy as exciting new start-ups surge on the scene."[56] There are "over 95 investment companies involved in nanotech,"[57] and they have "pumped $711 million into the sector worldwide in 2002 and 2003."[58]

Not everyone agrees that venture capital is so important to national leadership in the world of nanotechnology. According to Chevron-Texaco Technology Ventures' Molecular Diamond Technologies vice president Waqar Qureshi, "Investment from venture capital is only a small percentage of the total investment going into nanotechnology. From the perspective of advancing the science and technology, it doesn't matter that there is less investment from venture capital" in some year compared with others.[59] While Qureshi's opinion is overwhelmed by claims to the contrary, the hype is contradicted by the data.

According to Jurvetson, "venture capital investors look for one-of-a-kind companies that will offer 100 fold or greater returns."[60] According to the National

Venture Capital Association, "acquisitions and mergers are the most common way for venture investors to liquidate their investments. In 2004, nearly 60 percent of the money returned to investors was from acquisitions."[61] Put simply, start-ups are funded and acquired by larger and more secure firms.

In late 2004 nano-experts predicted a sharp rise in acquisitions, failures, and mergers among the roughly fifteen hundred companies worldwide involved in nanotechnology R&D.[62] Lux's Nordan monitors the nanotech landscape and noticed "signs of getting close to a *tipping point* with nanotechnology—18 to 36 months away."[63] As Nordan put it, "A lot of start-ups are underfunded, on a wing and a prayer, maybe part-time, and these guys are not going to be sustained. There's going to be these five to ten massive consolidation rounds people don't see, business failures and companies picked up for a song for a patent or some process technology."[64]

The range of approaches is significant as well. "Some start-ups are addressing multiple markets with a core, versatile technology. Some are orienting themselves vertically toward particular product markets."[65] However, of the twelve hundred nanotech start-ups, "only about 10 percent have ever attracted venture capital, and just 10 percent of those have received more than one round of funding," according to Lux Research.[66]

Coming off the dot-com collapse of the late 1990s, many venture capitalists are somewhat averse to the idea of investing in a *new* technology economy. Luckily, the flourishing biotechnology industry helped to restore some of the confidence in high-tech investment that was eroded during the dot-com collapse. In fact, US venture capital investment seems to have grown from just $100 million annually in 1999 to over $780 million in 2001.

Indicators of VC interest at this point show that nanotechnology is advancing at a sustained pace. Being hyped as *nano* has benefits and drawbacks. Bruce Kisliuk, a director at the USPTO, believes "people looking for venture capital money will call anything *small* nanotechnology."[67] This comes as good news for start-ups given some reluctance by venture capital investors. On the other hand, "Colbert and other venture capital investors say the *nano* label actually increases their skepticism and generates no premiums as it might for momentum investors."[68] Nevertheless, while government spending is vital at this point to provide a basic foundation, in the long term, if nanotechnology is to become an integral part of the American economy, it is going to be the venture capitalists who prove to be vital variables in the economic calculus.

"As most nanotechnology companies are still in phase one, private investment is a significantly smaller source of funding for nano start-ups and research think tanks than government spending. For example, in 2002, venture capitalists only committed $282 million to nano deals,"[69] but the trend is growing. "In 2001, for example, 76 percent of nanotech deals were classified as either start-up/seed or early stage, with 24 percent expansion and no later stage deals. By 2003, a mere 9 percent of nanotech deals fell into the first category, while expansion funding increased to 47 percent and later stage had climbed to 18 percent."[70] In 2004, "some 45 venture capital firms invested nearly $200 million in 12 nanotech start-ups. That's hardly an investment frenzy. Venture capitalists call it a *measured strategy* to invest in the best,"[71] but it is good news for start-ups.

Because nanotechnology is a general-purpose technology with applications ranging across a variety of sectors,[72] nanotech start-up companies at this point cover a largely interdisciplinary field. So far, VC funding has been dedicated mostly to devices and instruments, which are building the foundation for future research.

"2004 was a year in which many investors first tuned into the nano theme. Demand for information grew. So-called nanotech experts appeared on the scene, explaining how nanotechnology would radically change our lives. Stock recommendations focused on illiquid stocks—that is, very small stocks with few shares available. As a result, such stock rapidly became the playthings of retail investors with short-term mindsets."[73] Despite better-than-expected investment, nanotechnology continues to emerge onto the VC scene from the fringe, with experts in the field proclaiming the "next industrial revolution," while many in the investment arena continue to take a wait-and-see approach before placing significant faith in the technology. In this kind of environment, it is a daunting task for venture capitalists to separate the hope from the hype and find legitimate sources of applicable technology. On the other hand, even bubbles achieve momentum, investing suspends analysis, and then it is driven more by the fear of missing a rise in the stock. Venture capitalists "definitely can benefit from environments where the public markets are more speculative."[74] Hence, they may have a motive to perpetuate the hype.

However, despite blips on the stock markets associated with new government policy or breakthroughs, hype does not seem to actually be driving the investment funds. According to Deloitte and Touche's Ed Moran, if hype buying "is the improvident and widespread investment of VC money in early-stage companies, ... nanotechnology has not received a disproportionate amount of attention from the investment community.... This has not occurred in the nanotechnology sector—investors are being cautious in their allocation of resources."[75] Waqar Qureshi agrees: "Herd mentality is not yet an issue in nanotechnology.... There is some hype, of course, but that hasn't affected most people's perception that although there is a huge potential, this is a long-term play."[76] The VC sector in nanotechnology, though, has boiled down to a struggle between two competing, but not necessarily equal, factions—the visionaries and the pragmatists. Insofar as VC and investment are concerned, this is the dominant economic as well as ideological struggle that currently drives the economic debate concerning molecular nanotechnology, and pragmatists win out hands down.

It is in this "enabler" sector of nanotechnology where you see major investment firms coming onto the scene. "Over 60 U.S. venture capital firms, in addition to numerous corporate venturing operations, have invested in nanotech-related companies. Venture One tracked almost $500 million in nanotech funding to start-ups in 2002."[77] Companies like Merrill Lynch, Smith Barney, and Credit Suisse have all funded nanotechnology companies whose primary focus is on producing microscopes and materials that are necessary to develop and commercialize nano-technology. "Over the past few years, J. P. Morgan Partners has provided VC to five nanotechnology companies, in diverse market applications like textiles, drug delivery, electronics, and flat panel displays. JPMP led a $30 million C round of funding in Optiva, which was one of the largest nanotechnology funding rounds for all VCs in 2002."[78]

More importantly, there are several VC companies that have begun to dedicate large portions, if not all, of their investments toward nanotechnology-specific companies. There are currently more than ninety-five investment companies involved in nanotechnology. Though the $282 million invested in 2002 is only "a tiny fraction of the total $17 billion in venture money spent last year,"[79] this ratio is meaningless. According to David Forman of *Small Times*, "VC interest in small tech remained steady in 2004 despite an overall cooling in the VC market; that money is shifting to the expansion and latter stage companies and that nanotech-nology is getting a bigger share of the bucks."[80]

What may be keeping the lid on nanostocks is "lingering skepticism" drawn from the bursting of the dot-com bubble and may help account for rewriting predictions pushing the industry's tipping point further in the future."UBS Warburg's chief tech-nology strategist told Barron's that nanotech's tipping point would not be realized until after 2020."[81] On the other hand, there are some clear differences between the dot-com fiasco and nanotechnology according to Neil Aronson of Mintz, Levin, Cohn, Ferris, Glovsky and Popeo PC. First, concept companies with flawed business plans are being examined with much greater scrutiny. Second, barriers to competi-tion in nanotech are real. Decent patent protection can create considerable value. Third, markets are huge and nanotech has a place in almost every industry.[82] Aronson is not alone in his point of delinking nanotech from the dot-coms. This may help explain the investment growth curve; "the year 2003 saw some $304 million in ven-ture capital funding for nanotechnology, a 42 percent increase over 2002."[83]

Steve Jurvetson of Draper, Fisher and Jurvetson (see below) also sees distinct dis-similarities between the dot-com bubble and what is happening in nanotechnology:

> So from the entrepreneurial side of the equation, the formation of new ventures tends to correlate with university spin-offs. Second to that are government labs, but in both cases they're usually federally funded research for a number of years and [have spent] a substantial amount of money before they ever enter the start-up phase. That's very different from the Internet.[84]

A different rebuttal was made in a recent Lux Capital report. "There is no bubble in nanotech venture capital funding. VC firms invested $79 million in nanotechnology companies in the first half of 2004, down from annual totals of $325 million in 2003 and $386 million in 2002. VC investment will total approximately $200 million in 2004."[85] There may not be actual contraction going on at all. By the end of the third quarter in 2004, only $122.1 million had been invested, which puts funding on track to be well below the $200 million predicted by Lux. However, while the dollar amount comes down significantly, there has actually been a dramatic uptick in deal volume. David Forman reported, "By the end of the third quarter, investors had already invested in 30 nano companies, on pace to significantly outstrip the total of 34 funded during 2003.... While the amount of money invested fell a precipitous 35 percent in 2004, the number of companies receiving funding increased 32 percent."[86]

Small Times's Forman continued: "The more mature of the companies founded in 2001 and 2002 raised fat rounds in late 2003 in anticipation of a possible exit window in 2004 or 2005. Consequently, they had no need to raise money. ... A new generation of nanotechnology companies turned to venture backers in 2004. That new crop sought earlier stage funding, which account for more deals, but fewer dollars invested."[87] Early-stage funding accounted for 36.4 percent of all nanotech funding, up from 26 percent in 2003. Expansion-stage funding was responsible for 45.5 percent and later-stage for 13.6 percent, compared with 47 percent and 18 percent, respectively in 2003.[88] Overall, Nordan observed "the companies that have been successful in raising money this year have all been late stage fund-raising where typically they are not showing ideas or prototypes, but a large-scale manufacturing concept."[89]

Furthermore, Peter Hebert from Lux adds, "most of it has to do with the overall market for technology stocks."[90] Colleague Nordan agrees: "This is in line with a sustained decrease overall in venture capital spending. The percentage of total venture capital devoted to nanotechnology has held more or less constant between 1.5 and 1.7 percent" since 2002.[91]

Of course, there are a few investors who still see Drexler's vision as something other than science fiction. Because of the exciting nature of these visions, the financial endeavors of visionaries generally occupy more than their fair share of the mainstream media. Almost all of the "visionary" research in nanotechnology is being funded through government grants, and only a very limited amount of private funding is being given to those companies who predict long-term returns. In fact, many top investors in the scientific field agree that the nanotechnology hype that is being associated with Drexler's vision of molecular assembly has the potential to doom nanotechnology as a burgeoning industry.

Most nanotechnology proponents have had a difficulty finding even a handful of scientists who support the claim that a molecular assembler can be constructed, regardless of the time frame. Considering that most VC investment in any technology relies upon the general standard, that investors will begin to see some type of financial solvency approximately five years down the road, it comes as no surprise that better than 95 percent of the VC associated with nanotechnology is coming from the pragmatists who are focusing on what most would deem as *the enablers* of nanotechnology (see below).

NANOTECHNOLOGY ENABLERS[92]

Microscopy	Nanomaterials	Production Tools
FEI	Altair International	Intel
Hewlett Packard	DOW Capital	Molecular Imprints
IBM	Materials Research Corp.	Veeco
Technical Instruments	Nanomat	Zygo
Molecular Imaging	Nanophase Technologies	Nanometrics

In fact, this market now represents millions of dollars in investments, practically none of which include the "stealthy" nanotech applications that you might read about in some scientific literature.[93] The bottom line is that venture capitalists simply don't want to fund scientific vision—they want to fund sound research.

An additional caveat comes from Vinod Khosia of Kleiner, Perkins, Caufield & Byers and one of the founders of Sun Microsystems, who claimed "an inordinate amount of capital had been poured into the industry causing inefficient and sometimes unwise deployment of funds."[94] As such, many of these start-ups are not profitable at all and may never be.

Today's venture capitalists are concerned with profits, not visions. It is for this reason that the lion's share of nanotechnology companies are focusing on small-scale and oftentimes rudimentary nanotechnology that will provide the tools used by universities and research groups to study some of the more sophisticated nanotech. But it is misleading to conclude they are the sole beneficiaries of positive rounds of funding. "As 2005 began a core of nano start-ups showed they were progressing solidly across the chasm. Nantero, Nanomix, NanoOpto, Nano-Tex and other nano companies announced impressively fat funding rounds in the first quarter [approximately $78 million]."[95] In addition, there are some indications that flow of venture capital will increase in 2005. "Many investment groups are sitting on large amounts of cash raised during the Internet era, and they want an outlet. Money is more likely to flow when they see an outlet for their investments."[96]

An interesting sidebar has to do with Japanese development, VC, and American start-ups, especially start-ups with IP value. Wolfe broke the news: "Asian investors are at it again, but this time it's American nanotechnology intellectual property."[97] Asia is breaking into the American start-up market. First, Wolfe has "seen a flurry of activity involving Asian giants and U.S. nanotech start-ups," and he references Matsushita's deal with Nanosys to develop flexible photovoltaics. Second, he says, "federal officials told him of their concern over the growing trend of exporting know-how rather than products."[98]

He crafts a wonderful argument beginning with the observation that "U.S. researchers discover and develop leading edge technologies in the lab, but due to a lack of available funds, Asian corporations are the ones to commercialize the product. It's a pattern that pre-dates nanotech."[99] It seems while we remain ahead in terms of research, we may lag slightly behind in thinking of commercial applications, mainly because "Asian conglomerates possess the necessary long-term investment horizon for funding early stage nanotech developments."[100]

Modzelewski believes there is nothing to cry over what's happening here. "The fact that Asian companies are reliant on basis research in American universities is not necessarily a bad thing. Developing relationships in Asia is extremely important for nanotech start-ups because Asia gives us an exploding market—particularly in China—as well as an inexpensive manufacturing base."[101]

Before we get to some of the leaders in the VC world, three events are worth discussing briefly: the arrival of the Punk Ziegel, Merrill Lynch, and Lux Capital's Nanotech Indexes; the Nanosys initial public offering (IPO); and expected trends set by nanotechnology companies in the market.

Punk Ziegel and Company Index

"Punk Ziegel & Company may very well be the leading investment bank in the industry."[102] In 2002 it was the first firm on Wall Street to initiate nanotechnology industry coverage, and they employ two full-time nanotechnology research analysts. In March 2004 Punk Ziegel launched a stock index posted on the firm's Web site that initially tracked fifteen publicly traded nanotech-related companies. In August it added four more. The companies making the cut, according to JoAnne Feeney, an analyst with the firm, have "future prospects we believe are going to be largely driven by their nanotech operations." She reported that very large companies, despite having leading research endeavors, are not included because nanotech doesn't currently drive businesses. About 40 percent of the index is in life sciences. Another 38 percent are in instrumentation and equipment; 11 percent are in materials; 7 percent are in devices, and 4 percent are investment firms, including Harris and Harris.[103] The index is down 20 percent since the start of 2005, but Feeney claims "there are signs of growth, which is always good for the technology areas although there are still mixed signals coming out of the semiconductor industry."[104] Much like the disclaimers given by Merrill Lynch, Feeney says, "the aim is to give the investor a way to examine the current performance of nanotechnology as it's appearing across the economy and clarify the opportunities that are emerging from nanotech innovations."[105]

Merrill Lynch's Nanotech Index

In 2004 Merrill Lynch introduced the Nanotech Index (Amex: NNZ). It trades on the American stock exchange and has already gone through a series of transformations. Merrill Lynch "states that its index consists of companies where nanotechnology represents a significant component of their future business strategy."[106] Merrill Lynch comments "nanotechnology could be the next growth innovation following on the heels of information technology," and while it "may be overhyped in the short term, high barriers to entry and low barriers to adoption should make for a profitable sector long term."[107] Thiemo Lang, a German money manager, reported, "as a direct consequence of the launch, retail investors jumped on the most illiquid stocks in the index, betting that hopefully soon-to-be-released new nano funds might just track the index and pump up the most illiquid stocks even more."[108] In response, Merrill's Steven Milunovich responded that "the firm covers only three of the stocks in the index and compiles the list not to recommend investment vehicles but as a way to track the young industry."[109]

How useful is the index? Steve Jurvetson merely noted it was inevitable and reserved comment: "Nanotech is becoming the nexus of the sciences, so it's appropriate that the technology strategy team at Merrill Lynch is taking the lead in researching this multidisciplinary field."[110] Less generously, Josh Wolfe seemed unimpressed and offered this characterization: "When it comes to nanotechnology, there's more marketing baked into NNZ than market in the financial sector sense. Merrill may have jumped the gun in an effort to be [one of the] first. Many of the most important nanotechnology names are still private."[111] Of course, this index does compete on some level with his own *Nanotech Report* and the Lux Index. However, Wolfe is not alone in criticizing Merrill Lynch's effort. For example, Sun Microsystems' Khosia "points the finger at Merrill Lynch, one of the IPO's primary underwriters, for taking advantage of an emerging nanotech bubble."[112] Regardless of how one feels about NNZ, it is noteworthy that it actually surfaced as an investment service. That alone attests to some economic importance to the commercial state of nanotechnology.

Lux Nanotech Index

Launched in March 2005, the Lux Nanotech Index includes twenty-six publicly traded companies, and some are large. According to CEO Peter Hebert, its function "is to serve as a benchmark for the value that markets ascribe to emerging nanotechnology."[113] Grounds for inclusion include listing on NYSX or AMEX or quoted on NASDAQ or Small Cap Market Systems, having a minimum $75 million market valuation, and having a minimum average trading volume over the preceding three months of fifty thousand shares. According to Lux Research vice president Nordan, it differs from Merrill Lynch and Punk Ziegel: "It represents only applications of emerging nanotechnology...it contains a balanced selection of companies across the value chain (such as makers of tools, manufacturers of raw materials, intermediate products with nanoscale features, and nano-enabled end products)."[114]

The index frenzy seems to have subsided. Nonetheless, concerns remain that

> stock indexes or funds, which are normally created for definable industrial sectors, will mislead unsophisticated investors into believing there's a nanotech industry in which to invest. The fallout might be more hype, a herd mentality toward momentum investing, and ultimately, disappointment when reality sets in. All this could spell a backlash and setbacks for nanotechnology development.[115]

"2005 now looks promising. Investors are more familiar with nanotech stocks and have gradually abandoned their *nano* fixation, instead adopting a more informed evaluation of the companies."[116] Concerns over momentum investing by investment professionals do seem to have subsided. Even unsophisticated investors are able to access much more information about companies carrying either the *nano* prefix or claims to be heavily involved in nanotechnology. According to Thiemo Lang, "The transition from a nano fixation to a more sober analysis based on profitable and sustainable business models will continue through 2006 and beyond, successively retrenching any of the hype that has lingered on within company valuations. Additional support will come from a steadily improving quality of news flow from the industry and research labs."[117]

The 2004 Nanosys IPO story is a case study on deciding to go public when value is based exclusively on an IP suite rather than product revenue. In general, the consensus of experts supports a conclusion that the Nanosys story is more about "market cycles and timing" than a foreboding conspiracy theory on nanotechnology, but the debate is inherently instructive.[118]

THE NANOSYS IPO STORY

The final area worth mentioning is start-ups that go public. "Bankers and venture capitalists are pushing for initial public offerings of nanotech start-ups."[119] There are only a few, but one has garnered most of the press attention: Nanosys.

Nanophase (NANX) went public in 1997. It owns or licenses twenty-five US patents and patent applications and forty-one foreign patents and patent applications; it has maintained an aggressive schedule for new nanomaterial development for targeted applications and new markets.[120] Lumera, Inc., and Immunicon Corp. both went public in 2004. Lumera's value lies in proprietary nanotech and related IP to make wireless antennas and systems, biotech, and electro-optic devices.

Immunicon developed blood diagnostic products that use magnetic nanoparticles.[121] Cambridge Display has joined the list. "The average stock in the *nanosphere* has been trimmed 20 percent since 2004's first day of trading."[122]

Nanosys has a broad technology platform, with over two hundred fifty patents and patent applications. It collaborates with industry leaders, such as Dupont and Intel, to produce products in computing, optoelectronics, communications, energy, defense, and life and physical sciences.[123]

In April 2004 Nanosys, a research lab and patent pool, filed for an initial IPO worth an estimated $115 million.[124] In July the prediction was that its shares would price between $15 and $17 when it went public. If true, "the company could have had a market cap of around $350 million to $370 million after the offering. That's more than 100 times 2003 revenues of $3.1 million."[125] Herein lies its difficulties. According to Mark Brandt and Drew Harris, Nanosys may have misjudged the reach of its "*talking loud*" strategy. While they admit that "attracting publicity for your research advances may make it easier to get financing, whether your company is seeking early stage seed money or about to go public," they recommend stealth. "Actually getting a patent or developing commercial applications may still be years down the line, or might not be successful at all, making it foolish to talk about your chickens before they hatch."[126]

Nanosys' actual profit record was fairly dismal. "Since its founding in 2001, the company had burned through $25.8 million. As of June 2004, it had $35 million in cash. Nanosys reported a loss of $8.78 million on revenue of $2.5 million for the six months ending June 30."[127] Its IPO disclosure indicates incurred net losses from July 2001 through December 2003 of $17 million.

Experts predicted that "the Nanosys IPO would be a litmus test for the sector and set the tone for those that follow." Others "argue the company is not the best yardstick given its early stage and plan to license patents rather than build products."[128] Nonetheless, its failed IPO garnered a lot of attention and ink in the business reportage.

While it would be fortunate if the second group were correct, we will have to wait and see.

Nanosys pulled its IPO on August 4, 2004, leading Juan Sanchez to comment, "This is going to delay nanotechnology for a while. A black cloud will be around for a while."[129] "The risk with nanotech is that early movers might sour the market for companies that are waiting in the wings for an offering in a couple of years."[130] Donn Tice, CEO of NanoTex, added, "if it doesn't go well with Nanosys, then there's a stain they've created for the others to follow."[131] Small Times offered an interpretation on the Gartner hype cycle: "Depending on the Nanosys IPO, nanotech could climb to new heights on the hype curve or possibly lose some of its luster."[132] This becomes especially daunting when Nano-Tex, ZettaCore, and Molecular Imprints are on deck.

Vinod Khosia criticized the IPO from the beginning. He claimed that "it was too soon for Nanosys to make an IPO. Nanosys doesn't have profits, doesn't have revenues to speak of (in 2003 it brought in $3 million in grant and contracts but posted a net loss of $9 million), and doesn't really have anything to sell."[133] Khosia used the standard that a firm doesn't go public until it has established a solid revenue stream and posted several quarters of profits. Nanosys hadn't.

Answering these criticisms, the business media offered some reasons why Nanosys' decision to go public was not without merit. First, it has big names on its board: Charles Lieber from Harvard and Paul Alivisatos, who directs Berkeley's Molecular Foundry; big-name companies have also hired Nanosys for research projects, accounting for its revenues rising from $283,000 in 2002 to $3.1 million in 2003. Second, "the company owns or controls rights to 200 patents"[134] by "commercializing IP coming out of labs at Harvard, MIT, the Hebrew University of Jerusalem and other institutions."[135] Third, "the commercial team has working partnerships with a blue-chip roster of companies, including Intel, Matsushita, DuPont, and In-Q-Tel, the CIA venture capital company."[136] For that matter, a few weeks after the withdrawal, on August 19, 2004, Nanosys announced a $14 million award from DARPA to develop flexible low-cost solar cells.[137]

So, what happened? There are three reasons for the withdrawal. First, Nanosys said it was withdrawing the offering "due to volatility in the public capital markets."[138] Wolfe wrote that the "decision says less about nanotechnology than it does about the stock market in general. Of 28 offerings to date in July, 13 are trading below their IPO price and others slashed their offering prices to squeeze out and get published by the markets."[139] "A closing IPO market window, terrorism jitters, and the poor performance of comparable biotech IPOs in 2004," says Lux Capital's Robert Paull, "resulted in a *perfect storm* that doomed the Nanosys launch."[140] By mid-August, "nine IPOs had already been withdrawn, and seven more were withdrawn in July, up from an average of 1.8 per month for the first six months in 2004."[141] In truth, NASDAQ did drop 6.8 percent in July 2004, its worst monthly decline since December 2002.

Second, Kita Bindra of Burns, Doane, Swecker & Mathis believed that apprehension from the dot-com IPO busts of the '90s might be the cause: "Because Nanosys was planning to go public so close to what happened a few years ago, there was a negative connotation to the offering."[142] Nanosys' IPO sought "to raise about $155 million. That's probably not enough to get Nanosys to the point where it's self-sustaining. IPO investors would have faced the further risk that additional financial rounds will dilute their holdings."[143] Maybe Kohsia and others were correct in that Nanosys' IPO was premature. Thiemo Lang, a manager at Activest said, "Wall Street more likely favors companies at a more advanced stage of development with existing products."[144]

Perhaps Wolfe might want to retract his comment of June 2004: "I now believe that 2004 will be compared to 1995, the year Netscape's IPO set off a boom in web stocks."[145]

One of the benefits of the Nanosys fiasco may have been the deflation of the potential nanobubble. "The aborted Nanosys IPO revealed the difficulties the financial markets have in evaluating a pre-product company years away from profitability and a business model based on a technology platform strategy. Although many hope a Nanosys IPO would serve as a catalyst for others to follow, the withdrawal had the opposite effect of slowing things down."[146] Nanosys is not alone.

And as Forman reported, "Having a fistful of VC money doesn't guarantee passage through the valley of death. Nanotech saw an early and promising start-up meet its demise earlier this year—Optiva, Inc.... Over the past three years Optiva raised more than $41 million in venture funding from some of small tech's top backers and gained some notoriety in early 2003 for being the first nanotechnology investment led by JPMorgan Partners after an 18-month-long search among start-ups."[147] Douglas Jamieson from Harris and Harris said, "its initial produce provided a lower margin than desired, and more profitable products needed additional time and money for development. In the end, Optiva couldn't find investors with the additional capital willing to bet on its doing so."[148]

For that matter, Forman reported that despite fat first-quarter funding rounds in 2005, "no one's claiming their intent to go public, mind you—especially not, well, publicly. In fact, executives are actually going out of their way to deny any plans for stock offerings.... But the ramp-up in coverage of the nano-tech start-up community by investment banks is a sure sign they anticipate new opportunities to float some shares."[149]

The model for IPOs will probably be along the lines of a commercial-product start-up rather than the IP licensing model used by Nanosys. Of course, nothing is universally true, and Nantero and a number of small companies continue to pursue an IP licensing model, while others, like NanoOpto and Nanomix, are currently very focused on particular product markets.[150]

"It is highly probable that as nanotech firms begin producing more commercial products in the latter part of the decade, that a nanotech frenzy will engulf the stock market. And in the absence of a statute preventing profitless firms from going public, the door will inevitably open for future [Nanosyses] to go public."[151] However, Business Week's Baker and Aston concluded the Nanosys IPO withdrawal has had a dampening effect. "Now the IPO market appears stalled, even for stars such as Nanofilm, a manufacturer of optical films, and Konarka Technologies, Inc., a leader in solar panels."[152] Nevertheless, "analysts and investors have a short list of nanotech-related firms that might soon be able to go public. On the list are Nanofilm, NanoDynamics, Nantero, Konarka, Molecular Imprints, ZettaCore, Zyvex, and eventually, Nanosys."[153]

The front runner seems to be Nano-Tex despite its repeated claims that it has no IPO plans.[154] It has "licensed more than 80 textile mills worldwide to use its treatments in products sold by more than 100 leading apparel and interior finishing brands."[155] Actually, "Nano-Tex is one of the few nanotechnology companies with real products, real customers and real revenue today,"[156] says Jeff Crowe of Northwest Venture Partners.[157] Marcus Mainord of Stephens, Inc., predicts if Nano-Tex goes public, there might be "a flurry of three or more."[158]

As mentioned earlier, there remains a quartet of publicly traded nanotechnology companies led by NanoPhase Technologies, which went public in 1997. Recent IPOs have included Immunicon, which is working on diagnostics using meta nanoparticles and recently received FDA approval for CellTracks Analyzer II. It has been joined by Cambridge Display Technology,[159] which is working on polymer light-emitting diode displays, and Lumera, which is a platform technology company. How these companies fare will be telling. For example, "a poor performance by CDT would support the contention that the platform technology model is out of synch with the current stock market." According to Forman, "It would suggest companies promoting the promise and potential of nanotech have less appeal than companies with new and better products tailored to a specific existing market, with scalable processes, solid relationships, and sustainable and growing revenues." "By moving forward with a specific product in a proven market," adds Immunicon's Eric Erickson, "you prove your technology platform.... Companies should put the horse before the platform cart, not the other way around."[160]

According to Mainord, "in the past 12 months, there's been a sell-off, with nanotech related stocks down about 25 to 35 percent. A large portion of the hype is already priced out of the stocks."[161]

INDIVIDUALS

Venture capitalists represent the pragmatists of the industry. Their focus is not on the Drexlerian science fiction vision of nanotechnology but more on the short-term, profit-driven, enabler component. Three of the most prominent minds in VC today are Josh Wolfe, Steve Jurvetson, and Charlie Harris.

Josh Wolfe—Lux Capital

Josh Wolfe, founder and managing partner at New York–based Lux Capital, has been crowned by many as the "pretender to the throne" and "wannabe scientist."[162] Wolfe, born in California and raised in Brooklyn, where his mother encouraged him toward public service and science, was a Westinghouse scholar, which he used to get a bench slot in Dominick Auci's research program at the SUNY Health Science Center of Brooklyn. Wolfe published two peer-reviewed papers in medical journals on the role proteins play in HIV infection. Dr. Auci introduced him to futures markets. Wolfe said, "I was hooked by the complexity of it. All of a sudden, I wanted to make money."[163] So he has.

Beginning as a premed major at Cornell, he switched to finance and worked on Wall Street during summer breaks. His gigs included Salomon Smith Barney, the Prudential Insurance Company of America, and Merrill Lynch, where he worked municipal bonds, derivatives, and even real-estate investment banking on deals ranging from billion-dollar hotels to trailer parks. His big break came in 1999 when he hooked up with Lux Capital.[164]

Red Herring reported, "He has been called *hype artist* and *adventure capitalist* by observers of the nanotech industry. Wolfe is just the kind of guy people love to loathe. He's young, short, well-coifed, confident about something that befuddles scientists and investors twice his age."[165] Many in the scientific community love to hate Josh Wolfe as well, for, while being a proponent of nanotechnology research, he is quick to voice his skepticism about much of the hype surrounding it. He is quick to point out that nanotech has its share of "sci-fi whackos."[166]

As a member of the Smalley camp who sees limited but nonetheless life-altering innovations stemming from nanotechnology, Wolfe burst onto the scene in August 2001 when Lux Capital published the *Nanotech Report*, which took an "in depth look at the technology as well as the people and companies spurring this development."[167]

Thus far, Lux Capital has made a reputation for itself in the VC industry as the first major operation whose sole focus is on commercializing federally funded academic research. Wolfe claims that his interest right now relies on "the picks and shovels of nanotechnology."[168] Like many of his colleagues, the picks and shovels section—the enablers, such as Veeco instruments—are the best way to ensure some type of return in the next three to five years.[169]

Most recently, Wolfe and his associates at Lux Capital released their economic expertise in nanotechnology in the much anticipated *Nanotech Report 2003*. Available to the investment community with a hefty price tag of $4,750 a copy, the report examines near-term prospects in the field. The report has not only received high acclaim but also sold a great number of copies, yet Wolfe insists that there is not an immediate profit motive behind the report. Instead, the nine-month project was simply meant to raise awareness of Lux Capital, and demonstrate its understanding of nanotechnology, to "establish Lux as a thought leader and an early mover in nanotechnology."[170] So far, Lux has achieved just that goal. After spending nearly two years searching for its first start-up investment, it is finally beginning to take hold in the nanotech community.

Most investors have high hopes and expectations for a number of existing start-up companies, yet investors such as Wolfe warn that investors must look beyond the hype: "Most small tech companies in business today won't be around in five years, and will be victims of the creative destruction of economic forces."[171] Wolfe's view on hype is fairly straightforward. He divides nanotech mavens into two categories: "There is a scale that moves from left to right. There are the *cosa nostra* or the *mental enthusiasts*—like Kurzweil—the extropians and the transhumanists, and then there are the down-to-earth folks, trying to *de-nanobot* the field."[172] He admits receiving business proposals based on self-replicating miniature robots: "It's utter nonsense—thoughts that you can change the economy because you can manufacture things instantaneously at your desk by just hitting a button."[173] He adds that "any technology based on the Drexlerian version of nanotech—that is, the self-replicating assembler—should be put in its place. These far-out ideas should promote ethical debates and get people involved, but investors should not be looking at that type of thing."[174] Wolfe's publication, the *Forbes/Wolfe Nanotech Report*, is featured in chapter 5.

Steve Jurvetson—Draper, Fisher, and Jurvetson

Wolfe and Lux Capital, "with a modest $50 million at its disposal, is certainly not alone in recognizing the opportunities in VC."[175] The large and prestigious Silicon Valley VC capital firm of Draper, Fisher, and Jurvetson, with an estimated $1.2 billion in capital, is a major player looking for deals in nanotechnology. Managing partner Steve Jurvetson not only is an investment guru in emerging technology but also speaks and writes regularly about the investment and economic opportunities that are arising in nanotech.

Jurvetson has always been a tinkerer. "The first thing I ever bought with my own allowance was a book about chemistry."[176] His first experience with nanotechnology came in his high school debating club. He earned a Bachelor of Science in electrical engineering in two-and-a-half years as the top student in his class. His master's research included the study of new semiconductor materials. After graduating in 1989, he landed a job in Hewlett-Packard's computer-design division, and seven of his communications chip designs were fabricated at HP. After a brief period with Bain & Co., he returned to Stanford for an MBA. Presumably, he beat out two hundred fifty applicants for his position at Draper and Fisher. Six months later he was a partner. Jurvetson "was the wunderkind who in 1996 invested $300,000 in former hardware engineer Sabeer Bhatia's idea for an e-mail service after more than 20 other VC firms passed on it."[177] The story is legend. "Hotmail was acquired by Microsoft for more than $400 million."[178]

In 1999 he attended a conference hosted by the Foresight Institute and reportedly took Drexler's nanotech class at Stanford, and his interest from that point on hasn't seemed to wane. He believes nanotechnology is the next big thing in the economic world: "Since the dot-com bubble first began to deflate, Jurvetson was among the first venture capitalists; his firm, Draper, Fisher, and Jurvetson, has funneled more than $73 million into more than a dozen companies pursuing nano or MEMS products."[179] In late March 2001 his company invested in Arryx, Inc., a Chicago firm that uses holographic Optic Trap technology, which "provides the means to intelligently and independently manipulate thousands of microscopic objects with tiny beams of laser light."[180] "FlexICs Inc. fabricates semi-conductors on plastic for thin displays that can be rolled up, while Konarka Technologies, Inc. cranks out solar panels on flexible plastic."[181]

"All of DFJ's nano investments are spinouts from universities and other research laboratories. And not one of them is in Silicon Valley."[182] Most of the companies in which Jurvetson and his partners invest are enabling companies utilizing basic nanotechnology principles.

Draper, Fisher, and Jurvetson is not dedicated exclusively to nanotechnology; nonetheless, it is a pioneer in funding nanotechnology-related start-ups. Jurvetson is treating nanotechnology in much the same way that he treated the Internet in the 1990s—with limited enthusiasm. Jurvetson and his associates see a great deal of promise in nanotechnology, but much like Wolfe and Lux Capital, they are careful to separate the hope from the hype. It was this strategy that allowed Jurvetson to stay relatively prosperous even during the dot-com collapse of the late 1990s.

"Jurvetson has become one of the nanotech industry's biggest evangelists" and "has managed not to alienate himself from either camp in the debate between Drexler and Smalley."[183] Nonetheless, he admits the earliest fruits from nanotechnology are pretty mundane, though he testifies that "the US government absolutely believes that this is the future technology wave," and he thinks "it'll be more important than the industrial revolution itself."[184]

As such, DFJ has taken a lead in VC investment, and it "has become one of nanotech's leading advocates and investors. "About 20 companies, or 17 percent of its total portfolio, are in the nanotechnology, microelectromechanical, and novel materials area."[185] In 2003 its portfolio companies included Arryx, Zettacore, Nantero, Imago Scientific, Konarka Technologies, NanoOpto, NanoCoolers, Solicore, and D-Wave. Nearly "a third of the firm's dollars and deals in 2003 involved nano or MEMS companies."[186]

His firm makes near-term early-stage investments. Jurvetson is careful to point out that "if it can't make revenues over the next three to five years, it's more likely to be a government research program than a venture backed start-up."[187] This credo is certainly nothing new to government leaders, who, recognizing the potential problems in attracting private-sector investment, have allowed most of the more "visionary" research to be conducted under the watch of universities and government laboratories.

Charlie Harris—Harris and Harris Group

Prior to 1984, Charlie Harris had an eighteen-year career in the investment industry, including serving as chairman of Wood, Struthers and Winthrop Management Corp., the investment advisory subsidiary of Donaldson, Lufkin and Jenrette. He is currently a trustee of two not-for-profit entities, Cold Spring Harbor Laboratory, a research and education institution in molecular biology and genetics, and the Nidus Center, a life sciences business incubator. He is a life-sustaining fellow of MIT and a shareholder of its Entrepreneurship Center. Harris was a member of the advisory panel for the Congressional Office of Technology Assessment. He graduated from Princeton University and the Columbia University Graduate School of Business. He has been chairman of the board and chief executive officer of the Harris and Harris Group since 1984.

In 1994 the company invested in its first small-tech company, Nanophase Technologies Corp., a spin-off from Argonne National Laboratory, which subsequently completed an initial public offering in 1997. At the close of 2001, "the firm [had] $13 million in cash and an annual cash burn rate of $2 million."[188] The company sold its interest in Nanophase in 2001 and invested part of the proceeds of the sale, approximately $500,000, in August 2001 in privately held Nantero, Inc.,[189] a Harvard University spin-off developing advanced memory devices using carbon nanotubes.

The Harris and Harris Group changed its name to Small Technology Venture Capital and its ticker symbol to TINY in early 2002. It started the year at $1.98 per share and by midyear found its values hovering just below $4. "Discipline has helped Harris turn $37.4 million in invested capital since its inception into $105.7 million.... Of the 35 deals Harris exited, 46 percent made money."[190]

In February 2002 the company invested $700,000 for a 15 percent stake in privately held Nanopharma Corp., a research-based Massachusetts General Hospital spin-off founded to develop advanced drug-delivery systems. On March 13, 2002, the company invested $635,000 into NanoOpto, a privately held producer of novel-materials compositions to enable a unique optical integration platform. In 2003 it had ten "tiny tech" holdings in its portfolio, including some of the best capitalized and buzzed about start-ups: Nanosys, Optiva, and Nanotero.[191] In 2003 the Nanotech Report predicted "Harris and Harris Group's Net Asset Value will rise sharply in the coming years as more of its portfolio companies raise additional capital at higher prices while some of them are acquired in trade sales."[192]

In 2003 TINY experienced a 600 percent meteoric rise in its stock price. According to Wolfe's *Nanotech Power Elite for 2004*, experts say Harris is "one of the true investment sages of the nanotechnology business" known for "opening up nanotech VC to the masses."[193] According to Wolfe, he "is setting himself up to be the Berkshire Hathaway of nanotechnology."[194] Harris and Harris Group has

almost fourteen million common shares outstanding and exercising the overallotment option, recently sold over three million shares to raise over $36 million
which it plans to invest in tiny technology.[195]

"Harris and Harris Group built franchise value in nanotech and believes that
TINY's new investment strategy will pay off in spades."[196] Its strategy is built on
a strong foundation. "Its assets were just over $79 million at the end of 2004."[197]

"Harris and Harris invests only in *tiny tech* and has a current portfolio of 20
companies.... Their investments fit in the classic venture capital time horizon of
five to seven years."[198] "The company hopes to raise [even more captial], which it
says it will use to make new investments in tiny technology, including nanotechnology."[199] The firm "continues to expand to keep up about 150 potential deals
every six months."[200]

Nevertheless, Wolfe advises against investing in TINY until its shares fall
much closer to their net asset value (NAV). While it has had a stellar investment
record that started in 1983, "by nearly every valuation method that exists TINY is
wildly overvalued...priced beyond even perfection."[201] Wolfe complains that
"thanks to nanotech's hype, TINY now trades at more than 3.3 times its NAV of
$2.26."[202]

Wolfe, Jurvetson, and Harris are truly pioneers of the nanotechnology community. Their status as hopeful skeptics in the nanotech industry have gained them
much acclaim, as their investment views represent the vast majority of venture capitalists at this time. Despite their current "pioneer" status in the nanotechnology
community, they are not alone, as nearly one hundred venture capital firms laced
from Silicon Valley to New York City are expressing a keen interest in nanotechnology. None of the investment leaders is focusing on start-ups that subscribe to a
more visionary approach to the technology. Instead, investors are taking the lead of
Wolfe, Jurvetson, and Harris, establishing themselves as pragmatists and taking a
measured approach, especially in the wake of our uncertain economic times.

NANOTECHNOLOGY BUSINESS ALLIANCE

While there is certainly a great deal occurring that *pulls* nanotechnology into the
pop-culture world of the twenty-first century, there is also a *push* by venture capitalists and investors. One of the major pushes to place nanotechnology into the
mainstream of business investment came in October 2001 with the formation of
the NanoBusiness Alliance (NbA) by Mark Modzelewski, Nathan Tinker, and
Josh Wolfe.

The NbA "is the first industry association founded to advance the emerging
trillion-dollar business of nanotechnology."[203] By creating a collective voice of
the nanotechnology business community, the NbA has aligned itself as the

industry leader in research, education, and awareness concerning the emerging innovations of nanotech. Today the NbA's advisory board is headed by former House Speaker Newt Gingrich; he is joined by Herb Goronkin of Motorola and leading venture capitalist Steve Jurvetson. It has over two hundred fifty members and has offices in Washington, DC; New York City; and Denver. Its *NanoBusiness News* has nearly forty-five thousand subscribers.

In 2002 the NbA expanded its breadth through the creation of the Nano-Business Development Group (NBDG), which is the for-profit, consulting arm of the organization. Combining the expertise of renowned leaders in accounting, law, and consulting, the NBDG works to help start-up companies, government agencies, and Fortune 500 companies prepare for investment and research in nano-technology. It also has over two hundred fifty members, including General Electric Co. and J. P. Morgan Chase & Co. Beyond this, not much is reported about it.

Mark Modzelewski was its executive director through 2004. He also served as director of Niehaus, Ryan, Wong NYC (NRW), heading strategic communications, investor relations, and issues-management efforts for a host of leading technology companies, including Yahoo, Covisint, PanAmSat, SmartRay, eSpeed, Technanogy, Organic, and the Motley Fool. He grew the new office to over $8.5 million in annual billings. In addition to leading the start-up and development of the New York office, Modzelewski helped create the Business Intelligence Unit and the Interactive Bureau. Before joining NRW, he was account leader for issues management, communications, and investor-relations strategies at Golin/Harris International for the DaimlerChrysler merger, MasterCard, Sprint, and a series of clients undertaking business reorganizations, M&A efforts,[204] and other complex financial and legal undertakings. During this time, he also advised numerous technology start-ups on business development, marketing, and investment strategies.[205]

Modzelewski was a Clinton policy guru. He served as special assistant to Secretary Henry Cisneros at HUD and Secretary Dan Glickman of the USDA, developing policy, legal, and communication strategy efforts on issues ranging from Y2K preparedness, US mortgage and banking systems, the online transformation and integration of HUD and FHA programs, digital divide efforts, biotech and organic food standards, and regional economic-development programs. While with the administration, Modzelewski also led program management, policy, and communication efforts for the president and CFO of the Resolution Trust Corporation (RTC), the quasi-governmental corporation that tackled the multibillion dollar savings and loan bailout. At closing, the RTC disposed of its entire asset portfolio of over $450 billion. In addition, Modzelewski claims a friendship with Senator Hillary Rodham Clinton (D-NY).

According to an unidentified venture-backed nanotech CEO, "Most of the industry participants use him for advice and as a conduit for messages and information."[206]

Modzelewski seems to have little patience with promoters and naysayers who exaggerate claims and counterclaims about nanotechnology. Witness the role he played in the Drexler debate on nanotechnology from chapter 2.

> Mark's tireless efforts on three fronts: disseminating nanotech information; reducing business hurtles for nanotech companies; and influencing governmental policy through lobbying, made him one of the most important power brokers. No one understands the worlds of government, public policy, and their marriage with science better. If someone's got an agenda, they've got to pass through his non-profit industry clearinghouse before it will be seen by government officials and business leaders. This is why Wolfe referred to him "as the central hub of the nanotech industry."[207]

As executive director of the alliance, he oversaw "all strategy and development, and organized the Alliance's Angel Network and Hubs Initiative for regional development of nanotech."[208]

His rhetoric usually involves claims that Americans should not fall behind. For example, when asked who the United States' closest competitor may be, he answered, "Europe. They discover, manufacture and outspend us nearly 2 to 1. The Asians are reliant on our basic research, so they can't ever quite maneuver the U.S. from that strategic standpoint."[209] In addition to concerns encapsulated in overblown nationalistic remarks, he seems to have very little patience for opponents: "Many in the environmental movement have become Luddites. They think because nanotech will change the world that it is inherently bad."[210] This rather oversimplified categorization of all environmentalists was overshadowed by remarks he was alleged to have made about Glenn Reynolds and other individuals who expressed concern with the direction of the NNI (see chapter 2).

However, the NbA has suffered a few setbacks. For example, "efforts to launch regional Hubs have so far failed to gain the kind of traction the Alliance has hoped for."[211] And, its HEITF has accomplished little beyond organizing a committee of advisors (see chapter 10).

In April 2004 Modzelewski left as executive director to become a managing director for Lux Research to "focus on public policy and corporate consulting efforts."[212] Lux Research, Inc., will be backed by the Lux Capital Group, LLC. Peter Hebert, chief executive of Lux Research described its relevance:

> The company's business model is to provide corporations, investors and governments with ongoing and actionable business advice and information services of all facets of nanotechnology, to offer large companies and organizations nanotech's big picture. The firm would anticipate the macroecnomic ripples and complex reverberations nanotech is likely to cause in the business world providing a steady intelligence feed.[213]

According to its own release, "Lux Research is the world's premier research and advisory firm focusing on the business and economic impact of nanotechnology and emerging technologies. It provides continuous advisory services, customized studies, and consulting to large corporations, start-ups, the government and financial institutions."[214] Upon his exit, Modzelewski was uncharacteristically gracious toward Drexler and the Foresight Institute. At the 2004 meeting of the NbA he said, "Foresight has created some frameworks and guidelines for going forward that people should be looking at."[215]

While NbA cofounder Nathan Tinker will continue to help lead, Modzelewski's successor is Sean Murdock. Murdock is a founding member of AtomWorks, an Illinois coalition to foster nanotechnology, and chairman of the Chicago Microtechnology and Nanotechnology Community (CMNC). As a former management consultant for McKinsey and Co. in its Chicago office, Murdock has an MBA and an MA in engineering management from Northwestern. Murdock "sees the Alliance as embarking on a new phase with a new set of challenges—among them, engaging a broader range of businesses and inviting them to the table. He plans to create a richer dialogue with the general public on environmental and other issues of general concern."[216] As he put it at the 2004 meeting, "It is good business to understand and avoid risk even while nanotechnology takes on some of human civilization's thorniest challenges, such as renewable energy and better health care."[217] Tinker added, "The NbA will encourage more trade, exchange and manufacturing opportunities with nanotech firms from around the world."[218] He expects that "Murdock's leadership to focus on building membership as well as developing more services such as group insurance for start-ups."[219] In addition, the NbA has recently "entered into a strategic partnership with working-in-nanotechnology.com to jointly develop a career portal. The two organizations will work together to develop surveys and workforce studies, symposia, training materials, seminars and webinars, and education initiatives."[220] In addition, in 2005, NbA under Murdock's leadership completed its annual Public Policy Tour with key legislators and administration officials. Key policy issues on its agenda included "a pro-innovation regulatory regime, incentives for commercialization of nanotech breakthrough, and high priority applications of nanotech in national defense, homeland security, electronics and information technology, life sciences and environmental protection."[221] Murdock has deep shoes to fill.

CONCLUSION

Where are we likely to see the first applications? Credit Suisse/First Boston highlights some areas. In materials, it directs attention to textiles from stain-retardant materials to bulletproof military uniforms with embedded sensors that can detect biological and chemical threats. Nanotubes seem promising for their strength, and the fuel-cell and battery industry will use nanotechnology to improve storage. They project nanocomposite plastics and coatings that are stain and scratch resistant. In medicine, nanotechnology may have it first major impact in nanoceuticals and delivery. Semiconducting nanocrystals to quantum dots may have tagging applications. Nanowires will reform semiconductors, and the nanochip will improve speed and storage in computers. Even nanophotonics, light-based electronics, may be in the offering.[222]

The projections are astounding:

> Nanotechnology is approaching a phase change that will see it spread exponentially across manufactured goods in the next 10 years. In 2004, $13 billion worth of products incorporated emerging nanotechnology, less than one-tenth of global manufacturing output. In 2014, Lux Capital projects this figure will rise to $2.6 trillion exceeding biotechnology by ten times and having an economic impact on par with information technology and telecom.[223]

With nanotechnology coming to the forefront of mainstream America, a massive private investment in nanotechnology is simply inevitable:

> Nanomaterials will find new homes both as replacement materials and as enablers and new products. Nano-intermediate products in electronics and IT will be joined by burgeoning applications in other product categories from structural metal to orthopedic implants. In 2014, manufacturing output incorporating emerging nanotechnology will total $2.6 billion—4 percent of manufacturing output, 50 percent in electronics and IT and 16 percent in healthcare and life sciences.[224]

America's free-enterprise system and its conducive atmosphere for risky VC has placed us at a major advantage over economic rivals. Currently, the quest for economic dominance in nanotechnology is being led by the pioneers—the likes of Josh Wolfe, Steve Jurvetson, and Charlie Harris. While the three have differing visions about where the hope ends and the hype begins, they all agree that nanotechnology is likely to change the world. The only question is when.

The next chapter studies organizations that have entered the debate over nanotechnology. Four were selected. Two support a revolutionary point of view, while two others are more conservative and vigilant.

CHAPTER 8

NONGOVERNMENTAL ORGANIZATIONS AND NANO

W hy do individuals or groups of individuals decide to function as watch-dogs? Determining what motivates people is always a challenging task. However, as individuals join organizations, motives are more easily discernible because they are shared, usually as a condition of membership.

An organization might be motivated by a personal vendetta; one group leader may dislike or mistrust another. There is a palpable tension between Drexler and Whitesides/Smalley. When they speak in private and publicly about the other the tone and nonverbal markers are stark. The animosity has translated into clique building, spawning groups of people divided into camps.

On another level, an organization may be personally motivated by actual or perceived self-interest. For example, Zac Goldsmith, an editor for the *Ecologist*, adviser to Prince Charles, and devotee of the ETC Group, is heavily involved in the organic food industry through his wife and might have been especially vigilant against GMO foods for that reason. Goldsmith is associated with groups like the Soil Association and the Picnic. In addition, Goldsmith has a strong dislike for Tony Blair and vice versa and almost anything the Blair government does like supporting investments in nanotechnology.[1]

The organization may be misdirected whereby the data upon which some response is predicated may simply be incorrect. During April 2003, article after article about Prince Charles's fear of grey goo referenced academic reports and studies indicating some propensity for the uncontrolled replication of nanobots while failing to identify the sources per se. It seems the primary source was the ETC Group's *The Big Down* report, which Goldsmith and some British ecologists lauded as "the definitive work on nanotechnology."[2] However, vetting many of the environmental concerns expressed in the report demonstrates their argument is weak, and the editorializing made some of their conclusions highly suspect. For example, the authors associate carbon nanotubes with asbestos,[3] citing an inhalation study from Warsaw. ETC cited one line from the study and followed with a parenthetical statement related to tobacco-smoke toxicity, attempting to associate well-known concerns about cigarette smoking to carbon nanotubes. The Huczko study reference used to establish the similarity with asbestos is totally incorrect.[4] Silvana Fiorito's immunology study about the piggybacking or hitchhiking phenomenon was very preliminary research with ETC making an obscure reference to *uninvited* nanoparticles in an attempt to produce as much fear as possible. Fiorito actually studied graphite particles of about one micrometer in diameter and discovered an immune response could be triggered by their presence. When carbon nanotubes were added to a cell she discovered that they accepted the nanotubes without developing inflammation. She reported that she "doesn't know why some forms of carbon are more readily tolerated by cells."[5] Nonetheless, her work is packaged by ETC as the basis for a hazard claim.

The organization may simply want publicity associated with protest to further another, often totally unrelated, agenda. This claim was made about the ETC Group when it got into the nanogame, especially when it released its 2004 *Communiqué*.[6] Media coverage can be used to leverage donations and other resources to enhance the bureaucratic structure, increase salaries, and provide funds for parallel protest initiatives.

The organization may be acting altruistically and be legitimately concerned with the subject of the protest. Greenpeace has a long and distinguished history uncovering environmental problems, protecting forest lands in Argentina, and frustrating Japanese whale hunting. There has been growing concern over the environmental and health consequences of nanotechnology, especially nanoparticles, though the evidence is not sufficiently authoritative to claim a causal relationship with certainty. There are sufficient findings to justify wariness, and not unexpectedly, Greenpeace shows up.

Add to these observations the fact that these organizations are self-selected representatives of the public good—their claims of selflessness should constitute grounds for suspicion. This in no way means they are evil or even that their protestations are without value. It means only that what they say must be vetted carefully and critically.

While many organizations out there are associated with nanotechnology, four were selected as fairly representative of the community of watchdogs. While the Foresight Institute (FI) and the Center for Responsible Nanotechnology are cautious in their criticism and advocate a vibrant nanotechnology policy under international control, the ETC Group is not. Joined by colleagues such as the *Ecologist*'s editor, Zac Goldsmith, the ETC Group has called for a moratorium. The final group is moderate though critical; recently, Greenpeace has called for a regimen of regulation and case-by-case evaluation of nanotechnology.

PROPONENTS

Who proposes nanotechnology? While it is valid that an industry associated with nanotools and nanomaterials directly benefits from vibrant and growing government and private spending, a lobby hasn't actually surfaced (though a case can be made that the NanoBusiness Alliance come close—see chapter 7). As such, proposing nanotechnology is more about belief and values than about profitability at this time. Both the Foresight Institute and the Center for Responsible Nanotechnology, currently seen as outliers, are trying to keep molecular manufacturing as a variable in the public debate. Both organizations are linked to Drexler and his advocacy, and for Drexler there is a strong ego protective function to advocacy in this setting. They are both proponents of their own brand of nanotechnology.

The Foresight Institute

The Foresight Institute played and continues to play an important though lesser role in the promotion of nanotechnology. K. Eric Drexler is the founder and chairman emeritus, and Christine Peterson, Drexler's ex-wife, was the founder and president of FI. Though it perceives a role for itself in the foresight movement, it denied being a metagroup per se.[7] FI is neither a department of nanotechnology nor an office of nanotechnology. It is also not a corporation or a lobby group. If anything at all, Drexler and FI envision themselves as a simplified organizing structure, a foundation.

They "will foster the emergence of a broad family of groups loosely linked in a flexible organizational framework."[8] As an "organizational development project [they] will refine plans...through discussions and the circulation of draft proposals; [they] will also help organize initial groups within the metagroup and help develop needed metagroup institutions."[9] One of FI's early board-of-directors member Jim Bennett described the vision of FI. In an interview, he reported that "the Board[10] is concerned with high-level strategy and oversight of the Institute's activities,"[11] and he saw "the future of the FI as an organization dedicated to educating people about nanotechnology and stimulating discussion about nanotechnology and its consequences."[12]

The confluence of these three factors lead the FI membership, especially Drexler and Christine Peterson,[13] to send multiple annual mailings and to dedicate some space in *Foresight Update* (FU) to more and more efforts at solicitation in the early years. As if the novelty had faded and interest waned, FI moved from a group of interested scientists to a not-for-profit organization employing aggressive fund-raising using a three-pronged rhetorical strategy.

Early FI rhetorical strategies were traditional. First, they used claims of inclusion. Peterson unabashedly wrote, "But *we* know that without progress [in addressing] 'thorny issues' [and] through the education efforts of the FI and the policy development work of the Center for Constitutional Issues in Technology developing nanotechnology will not solve the problems *we* want to solve."[14] Efforts toward inclusion actually included some giveaways donors could share. Peterson distinguished between a giver and a nongiver as a thinker and a nonthinker.[15]

Second, they used some peer pressure: "But Foresight activities and success would be small if we had to sustain our efforts based on the minimal donations alone."[16] Furthermore, Peterson wanted the readers of FU to know that a "good percentage of our members are at the higher level [rather] than the mere minimum."[17] She prodded the donor to reconsider that paltry check of $25, and she inferred a direct relationship between large contributions and FI successes. Much like the televangelist who tells his viewers that a large donation is related to increased chances of redemption, she wrote, "When you give at the $50, $100, $250 or $500 membership level, you give Foresight a much greater ability to make things happen."[18] Peterson, Drexler, and FI hawked FI as the answer to some of the world's most perplexing, earthshaking problems.

Third, they used modified fear appeals. FI's argument is consistent with its *leading force* thesis: whoever leads wins the race. If Japan's ahead, we are led to believe they will win. Peterson and Ralph Merkle suggested that the Japanese are much less myopic in their commitment to nanotechnology. The deep text of the trend toward hailing Japanese achievements and despairing US efforts smacks of ethnocentrism and national exceptionalism, an observation also made, if not more so, with contemporary government boosters. This phenomenon transferred to the nanotechnology movement today with bureaucrats using rhetoric linking nanotechnology to national economic growth and security to issues of leadership, often exaggerating who's ahead to justify programs and funding decisions.

In 1993 an appeal for money and in-kind assistance included this prediction from Peterson: "[T]he farther in advance an action is taken, the greater effect it can have by a given date. Those of us who influence nanotechnology development and policy now will have a tremendous effect compared to those who start years later."[19] Categorizing the research in basic science and engineering related to nanotechnology as a race suggests a finish line, FI awarded the winner the laurels of the *leading force*. Nevertheless, pitting forces against one another, FI makes a threatening claim to contribute now or be led by another leading force.

This fund-raising behavior typifies foundations of all sorts. They tend to view themselves as uniquely essential for some prize. Somewhat provocative was a change in their solicitation strategy proven effective in telemarketing. Whereas traditional efforts by FI involved mailings, which were easy to trash, Peterson revealed another approach at least having been contemplated: "So when we mail you an invitation or call you on the phone to talk about the idea, I hope you'll respond strongly and join now."[20] Since it doesn't seem the calls were made, there was nothing to really avoid. However, someone might have assumed a check today might preclude an unwanted phone call later.

Though in the mid-1990s its solicitation was toned down considerably, FI's early efforts smacked of hucksterism to many in the field. Today, FI has found some major contributors and its much more sedate in FI's pitches. Some of the corporate contributors to FI's conferences are Ford, Sun Microsystems, Apple Computer, Park Scientific Instruments, AMP, Foley and Lardner, JEOL, and others. At one point, FI had a deal with Amazon Books to receive a donation for purchases made through FI's Nanotechnology Bookstore.[21] FI seems to be doing well with membership dues, conference fees, and gifts as well.

Besides hierarchies of gold, silver, and bronze sponsorship associated with club-instigated special events, FI has added a level of membership for senior associates, including corporate senior associates. They are defined as "a community working to guide powerful emerging technologies to improve the human condition and the environment. [They] try to maximize and spread the benefits of coming technologies while minimizing their downsides." Actually, they are a diverse group of individuals who gather annually or semiannually in Silicon Valley and elsewhere to discuss policy issues relating to molecular manufacturing. In 1999 the annual gathering of the associate members was called a Group Genius Weekend, though this term seemed to fall away soon thereafter.

Bennett articulated the missions of the FI in both short and long terms. The short-term goals of FI were, first, to increase awareness of nanotechnology, and, second, to develop the reputation of the institute as the focal point, as the center of expertise to make people aware, as they become aware of nanotechnology, that Foresight is a place that they can turn to for information and answers. In the long term, the goal of Foresight is to create an environment for discussion in which the consequences of this technology and the precautions that one could take in anticipation of those consequences can be discussed in an intelligent and calm fashion.[22] This noteworthy set of goals has become part and parcel of current government initiatives that generally include clearinghouse and education functions as a requisite for receiving funding.

Further clarification of the message of FI was made by its former executive director, Jamie Dinkelacker.[23] Foresight has refined its identifying message to preparing for nanotechnology. This is a more focused statement of its current mission than the previous preparing for future technologies.[24]

More recently, FI became an ardent advocate. The *Foresight Perspective and Policy* statement reads:

Foresight Institute's mission and fundamental goal is betterment of the human
condition especially as it is related to molecular nanotechnology.... It is our
policy to prepare the future for nanotechnology and to pursue this mission by:
  • Promoting understanding of nanotechnology and its effects;
  • Informing the public and decision makers;
  • Developing an organization base for addressing these issues and commu-
    nicating openly about them; and
  • Actively pursuing beneficial outcomes.[25]

FI organized a fairly advanced information-distribution system. For example,
FU listed upcoming and recent events (seminars and conferences), publications of
note, a Web watch, and a media watch. FI added *Foresight Online* whereby users can
reach FI by e-mail. FI reported an over-twenty-four-thousand-reader electronic
discussion group on nanotechnology on the World Wide Web and Internet in the
mid-'90s, and FI maintains a substantial Web presence. A Web enhancement project was discussed in 1997. Presumably, annotator code developed by Wayne Gram-
lich would be used in developing a hypertext publishing system to handle real
debates on complex issues. Terry Stanley was working on the annotator project for
Foresight, but its completion was never reported, though there was some experimentation with CritLink.[26] In 1998 it hosted the Openness and Privacy Strategic
Discussion.[27] In June 2000 FI announced the Engines of Creation 2000 project
scheduled to be completed in 2001. The project would update *Engines* and develop
software for the design of hypertext systems. While this effort may have mostly
floundered, Nano-dot (http://nanodot.org/) is a great success and offers news and
discussion of emerging technologies and is hosted by FI. It links inquiring minds to
recent articles and testimony related to molecular manufacturing. The most recent
FU was published in August 2004.

    In an effort to both promote itself and to encourage research, today's FI
resorts to recognition and awards. For example, at a 1992 conference, the idea of a
prize "for research work leading to nanotechnology" surfaced, and at the 1993
conference, Charles Musgrave, a PhD candidate in chemistry at Caltech, received
the first Feynman Prize for his work on a hydrogen abstraction tool used in nanotechnology.[28] In 1995 the Feynman Prize, including a $10,000 cash award, went to
NYU professor of chemistry Nadrian C. Seeman. "Seeman received the award for
developing ways to construct three-dimensional structures, including cubes and
more complex polyhedra, from synthesized DNA molecules. Seeman said he is
hoping to use architectural properties of DNA to direct the assembly of other
molecules."[29] Some other recipients were Ralph Merkle, Stephen Walch, Bill
Goddard, Uzi Landman, and James Heath.

    FI announced a Feynman Grand Prize of $250,000 "to the first individual or
group to achieve specific major advances in molecular nanotechnology. Entrants
must design and construct a functional nanometer-scale robotic arm with speci-

fied performance characteristics, and also must design and construct a functional nanometer-scale computer device capable of adding two 8-bit binary numbers."[30]

The prizes proliferated. The Feynman prize was expanded into a set for experimental advances and another prize for theoretical advances in nanotechnology. In 1997 a $1,000 award was established for a distinguished graduate student. In 2000 an award in communication was created.

Foresight has both a board of advisors and a board of directors. Its advisory board has included Stewart Brand, Doug Englebart, Ray Kurzweil, Amory Lovins, Marvin Minsky, and others. Its current directors are James Bennett, Glenn Reynolds, and Brad Templeton.

In 2003 Tim Kyger, with experience as a staffer on both the House and Senate sides who had worked directly with the relevant science committees, served as Foresight's "man in Washington." FI maintains an international presence with executive director, Foresight Europe, Philippe van Nedervelde. According to his Web page,

> He does so primarily by means of multimedia presentations throughout Europe to audiences of all sizes and compositions (academic, business, governmental, civilian); by interfacing with the media; by representation at European Union level events such as EU parliamentary hearings and through assistance and support to U.S. Foresight Executives when they visit Europe for presentations, etc.[31]

FI did grow up. Peterson explained in 1997:

> When Foresight started in 1986, we hoped to be able to jump right into discussions of what nanotechnology and other powerful coming technologies will mean for individuals, organizations, and nations. But the time was not ripe for these kinds of discussions. So, Foresight backed off from [its] original strategy, focusing instead on educating the technical community on nanotechnology. It looks like this paradigm shift is now well advanced.[32]

FI has contributed much to the world of nanotechnology (specifically molecular manufacturing). Its most noteworthy achievement may have been its published guidelines for the development of nanotechnology. FI spawned two subgroups; they are examined below to evidence the breadth of FI.

FI: *Institute for Molecular Manufacturing*

The Institute for Molecular Manufacturing (IMM) was described by Drexler as "the only non-profit organization dedicated to funding nanotechnology research directly. It has both the long-term focus and the near-term contacts needed to make the right research happen."[33] A 1991 FU claimed the "IMM will: (1) fund

seed grants to scientists for research and development; (2) act as an interdisciplinary clearinghouse for U.S. and international developments in technologies and
processes leading to molecular manufacturing; and (3) disseminate educational
materials and sponsor seminars on molecular manufacturing."[34] E. Lynne Morrill
Steigler, the original executive director,[35] reported that they were "asking individuals, corporations, and foundations for tax-deductible contributions to further the
state of the art in this cutting-edge technology."[36]

IMM is noteworthy for its aggressive fund-raising efforts as well. Its senior
associates program provides opportunities for membership at different dues levels.
To sweeten the membership, at one point, IMM senior associates could receive
autographed copies of Drexler's book and even a framed color graphic of a nano-
device signed by designers Drexler and IMM advisor Merkle.[37]

IMM was just beginning. "Neil Jacobstein, a director of the Institute for Molecular Manufacturing, called for a peacetime Manhattan Project to spend billions of
federal dollars to develop nanotech. He conceded, 'the idea is not on anyone's agenda
except my own.'"[38] Nevertheless, IMM has tried to do its small part by fostering projects by awarding grants. One of the first grants made by IMM was to Drexler. His
book *Nanosystems*, which he self-touts as "the first technical book on nanotechnology,"
was reportedly partially supported by a grant from IMM.[39] Gary Stix claims, "most
of the Institute's grant money has gone to pay Drexler to work on projects such as
computer simulations of molecular gears, bearings, and other parts."[40] A second
award was given to Markus Krummenacher of IMM, and others followed.

The first IMM board of directors included Marc Stiegler, vice president of
technology at AutoDesk; Ray Laden, former president of United Telecommunications; and James C. Bennett, cofounder of the American Rocket Company.
Advisers to the board of IMM included Ted Kaehler with Apple Computer;
Michael Kelly, consulting professor at Stanford's Department of Material Science
and Engineering; Ed Niehaus of Niehaus, Ryan and Haller Public Relations, in
charge of FI's press relations; and Harden Tibbs of Global Business Network.[41]
The current advisers are Neil Jacobstein, David Forrest, Paul Melnyk, Jim Bennett, Linda Vetter, Ed Niehaus, and Steve Vetter.

IMM deserves much greater study. Frustrated by bureaucratic largesse and
corporate shortsightedness, FI chose IMM to encourage research into nanotechnology. IMM is FI's not-for-profit foundation. This approach is interesting on
many levels. Formal organizations (government, industry, etc.) did not seriously
invest in nanotechnology in the early '90s. On the other hand, academics and scientists seemed willing to redirect personal and business revenue to nanotechnology research, seeing something that other organizations did not.

Bennett helped enumerate the likely causes of government reluctance that
IMM sought to avoid or circumvent. These antecedents still haunt programs like
the NNI today.

First, there was and is categorical bias. Government and industry have become accustomed to certain categories. If something like nanotechnology does not fit a category, or even more problematically, crosses between categories, it fails to have sufficient power to attract and sustain interest and attention.

Second, there was and is customary bias. Anything that requires rejection of prevailing and customary norms and standards of evaluation fights a heavy presumption. What we know might not be ideal, but what is unknown carries substantial risk. A bad decision is much easier to make than correct. Decisions tend to empower an idea, philosophy, group, or organization, which nearly immediately perceives a self or vested interest in the decision. The longer the decision remains in force, the greater the momentum or inertia. Therefore, presumption mostly favors customary practices.

Third, there was and is a time bias. Merkle, summarizing the concept of time horizons, aptly suggested that it is this concept that might define an important distinction between research and development in the United States and Japan:

> [I]n Japan there is a concept of *basic technology* as opposed to the kind of *basic research* concept which we have in the U.S. In basic technology, you are pursuing technological goals and technological objectives, but with long time horizons, you are no longer tied to a specific short range project. Nonetheless, what you are pursuing is research where there is an expectation that there will be practical benefits in the long run.[42]

All organizations have a functional time limit beyond which decisions have no perceptible meaning. Government organizations seem to have no greater foresight than the term of office associated with its appointed directors. Venture capitalists and other investors want to see quick profits. The time frames associated with breakthroughs to production far exceed these time barriers. This argument frequently appears to justify an increase in government spending within the NNI. Nanotechnology, like other advanced technologies, has a very long lead time before any profitable result ensues. Unless parties like foundations or governments take up the slack, the technology may never move beyond its conceptual stage. According to James Bennett,

> The remaining option is funding from individuals and foundations. Fortunately, these sources can make a major difference to nanotechnology research at the early stage. When the IMM was founded to do nanotechnology research, it was set up to receive contributions from individuals... we knew that informed individuals would make the biggest funding impact in the early days of the field. We will tap additional funding sources for IMM as it grows, but for now, individual donors are making the biggest difference to IMM's research.[43]

The IMM's research direction seems to involve four project areas: the Molecular Nanotechnology Theoretical Feasibility Project, the Molecular Nanotechnology Demonstration Project, the Assembler Safeguards Project, and the Nanomedicine Project. It also offers recognition and prizes.

The IMM prizes in computational nanotechnology are awarded at the annual Foresight conferences. Some recent recipients include Santiago Solares, Mario Blanco, and William A. Goddard III at Caltech; Carlo D. Montemagno at UCLA; and Lawrence Fields and Jillian Rose from Phlesch Bubble Productions. Its current Web page links and cross-links to other Foresight locations and maintains an archive of the conferences as well as news and events associated with the IMM.

The IMM has published a collection of reprints and unpublished papers on its Web site. The *IMM Report* appeared in FU. The contributions have been dominated by Robert Freitas, J. Storrs Hall, and Jim Lewis, with a slant toward discussions by Freitas on nanomedicine. Actually, the writing of *Nanomedicine*, volume 1, was partially funded through support from Foresight and IMM. The IMM announced the Freitas Research Award to support work on the subsequent volumes.

The IMM predates the NNI, though there remains an important distinction worth rementioning. The NNI, especially with the Twenty-First Century Nanotechnology R&D Act, shied away from molecular manufacturing involving Drexlerian mechanosynthesis, opting for more conventional approaches to nanotechnology, a distinction discussed in chapter 2.

FI: Center for Constitutional Issues in Technology

A second spin-off FI group was founded in 1991—the Center for Constitutional Issues in Technology (CCIT). It was hailed by Drexler as the organization that "will address the public policy implications of [technical and scientific projects in pursuit of nanotechnology], both in the US and other nations, concentrating on the people who make decisions."[44] The reason it is discussed here has more to do with the recent preoccupation with SEIN by the US federal government than as a vital organizational component of FI. Indeed, there is very little evidence it has accomplished much at all.

Its "purpose is to support the study of policy issues relating to emerging technologies. The organization will sponsor position papers, and will hold conferences, workshops and other educational events in pursuit of its goals."[45] Initially, it planned to raise and distribute money "to support studies of past episodes of the emergence of new technologies…to assess the successes and failures of the public policy responses to these technologies, in hopes of better guiding the evolution of policy for future technologies."[46] There was a policy conference involving senior associates that took place annually, and other FI meetings often include a policy component, but the role played by the CCIT seems minimal.

James C. Bennett was CCIT's president. Bennett defined the focus of this group in his 1992 paper to the first general conference: "Currently, CCIT has prepared an initiative regarding near-term issue standards in emerging nanotechnology. This program would examine some of the questions regarding product standards in emerging nanotechnology and molecular manufacturing, comparing governmental versus industry roles in emerging standards."[47] Bennett reported that the CCIT "was formed as a member of the Foresight family in order to provide a specific focus for research, thinking, discussion, and ultimately advocacy in the area of public policy."[48]

Its founding board of directors consisted of Ted Kaehler, Ed Niehaus, and Charles Vollum of Cogent Research, Inc. Back in 1992 Bennett reported "CCIT is still going through its start-up phase. We are beginning to define a number of research projects which we feel will provide the first open discussion of policy questions in regard to upcoming issues spurred by the emergence of molecular manufacturing as a precursor to developments."[49] The CCIT was to "foster discussion of what kinds of research are done, what kinds of institutions are funded, and by what means funding is allocated."[50]

The CCIT had a long-term perspective and aspiration to serve as an advisory group to a coordinating agency once nanotechnology took off. When government and industry turned to academia and the research world for guidance, they'd be there. "When the *gold rush* starts, we can have a considerable impact by evaluating the various options for regulation (or approaches to regulation that are presented by the situation at hand) and trying to work with the agencies we think are most appropriate."[51] That didn't occur.

Both these organizations were established in mid-1991. Of both the IMM and the CCIT, much less has been reported and is known about the CCIT. Nonetheless, it is impossible not to notice many features of the NNI that have parallel equivalents with FI and its spinoffs. We may never learn how influential Drexler was to the birth of the NNI. However, when we examine FI, we have an easier time seeing analogues between FI as an organization and the NNI as an initiative.

FI: The New Foresight Nanotech Institute

Before October 2004, "Foresight had promulgated the concept of quadrillions of molecular assemblers creating macroscale objects. Yet Drexler and Foresight abandoned the concept in order to emphasize desktop nanofactories, and the desktop nanofactory paradigm has become Foresight's new vision."[52] At the November 2004 Foresight meeting, Drexler was soliciting support for an animation of just such a desktop system. Its raw version debuted at the same Conference of Advanced Nanotechnology. Mark Sims of Nanorex and the Foresight Institute put up matching funds to ensure it completion. They subcontracted Lizard Fire

Studio Productions (Burch's company) to do the production, and it was produced by John Burch and Drexler. It is called "Productive Nanosystems: From Molecules to Superproducts," and its finished version was released on July 12, 2005.[53]

On May 23, 2005, the Foresight Institute changed its name to the Foresight Nanotech Institute. Along with the name change came a change in the organization's mission and purpose. It may or may not be cosmetic, but a press release posted on the Foresight Web site explains the change of direction. Unabashedly proclaiming that Foresight had "largely succeeded in its original mission of educating policy makers, professionals and the general public about the potential of nanotechnology," its new president Scott Mize unveiled the new mission as "ensuring the beneficial implementation of nanotechnology."[54]

In redefining its new mission, Foresight laid out six "challenges" for which it believes nanotechnology will provide solutions: "(1) Meeting global energy needs with clean solutions, (2) Providing abundant clean water globally, (3) Increasing the health and longevity of human life, (4) Maximizing the productivity of agriculture, (5) Making powerful information technology available everywhere, and (6) Enabling the development of space."[55] Mize claims that Foresight will continue its work on the "long-term vision and potential of nanotechnology [read: the Drexlerian vision]," but will "concentrate more on near- and medium-term applications."[56]

Typically cynical, Tim Harper's blog railed on Foresight after the change. "While this new direction shows some considerable ingenuity, it still begs the question over whether Foresight has abandoned the Drexlerian creed, or whether the nanobot ate their homework."[57]

The question becomes, does the 2005 version of Foresight really have much in common with the original Drexlerian mystique of molecular manufacturing, or has the organization been hijacked by more thoughtful and politically savvy people? Maybe the latter since it launched the *Technology Roadmap for Productive Nanosystems* and announced a grant of $250,000 from the Waitt Family Foundation started by Gateway founder Ted Waitt.[58]

Center for Responsible Nanotechnology (CRN)

This group, which got started in December 2002, is associated with two individuals, Michael Treder and Chris Phoenix. Treder came to the CRN with a background in communications from the New York Transhumanist Association and began the CRN with Phoenix after communicating over e-mail about nanotechnology.[59] Though they were separated by over twenty-five hundred miles, their virtual organization remained a strong voice supporting the Drexlerian version of nanotechnology. The CRN is affiliated with World Care, a 501c(3) organization.

Treder is the executive director and was a member of the executive advisory team for the Extropy Institute and served on the boards of directors of the Human

Futures Institute and the World Transhumanist Association. Phoenix is the director of research and holds an MA in computer science from Stanford and is a self-touted inventor, entrepreneur, and published author in nanotechnology. CRN's board of advisors include Jose Luis Cordeiro, author of *The Great Taboo*; K. Eric Drexler; Jerome Glenn, executive director of the American Council for the United Nations University; Lisa Hopper, president of World Care; Douglas Mulhall, author of *Our Molecular Future*; and Rosa Wang, founder of Geographic Engine.com.

The CRN calls itself an information-ethics organization and claims to act to raise awareness of the issues associated with nanotechnology. "CRN studies, clarifies, and researches the issues involved—political, economic, military, humanitarian, and technological."[60] The aspirations of the CRN are conservative and involve producing information and publishing it widely, mostly coauthoring and publishing papers usually first on their Web site (crn.org) and eventually in a book.[61] When they publish others' works, they publish it freely and do not compensate authors unless a prior agreement has been reached. They are forthright in their intentions to stay on course and accept grants only for things they want to do anyway.[62] Anyone can join their mailing list. They accept memberships and are pleased to accept contributions.

Foremost, the CRN supports the Drexlerian version of nanotechnology and the prominence of the assembler (limited or otherwise) as its centerpiece: "You need to start with a working assembler, a nanoscale device than can combine individual molecules into useful shapes.... A human-scale nanofactory will consist of trillions of fabricators, and could only be built by another nanofactory. But an assembler could build a very small nanofactory, with just a few fabricators."[63] They refer to building blocks of any physical product as a nanoblock with a surface covered with mechanical fasteners, hence pushing two blocks together would make them stick. Each nanoblock might be a cube two hundred nanometers on a side, large enough to contain a CPU, a microwatt of motors or generators, or a fabricator system flexible enough to duplicate itself if given the right commands. Almost cartoonishly, the nanoblock would be mostly solid but would unfold like a pop-up book or inflate like an air mattress.[64] They perceive nanotechnology manufacturing as being less than a half century away.

At a 2005 National Academy of Science meeting Phoenix argued that the NNI needed to research the plausibility of molecular manufacturing, an argument that received little traction from the committee. Both Treder and Phoenix have criticized the NSF for the direction of its research agenda. Treder claimed: "Molecular manufacturing needs to be addressed. The NSF presents themselves as asking the right question, but the answers are more than wrong: they are simply off topic." Phoenix concurs: "Molecular manufacturing is an evitable consequence of advanced nanotechnology. We need to prepare for revolutionary changes, not just incremental improvements like new nanoparticles."[65]

They defend the metaphor of the nanofactory: "a hierarchy of machines wherein machines would operate at successively larger scales. They would be assembled into a desktop-sized unit that would have software input for controlling the factory and special containers of basic materials used for manufacturing. Nano-factories could be put to work building more nano-factories, so that technology would spread quickly." How quickly? Phoenix claims "with perhaps only a year passing from the first nano-factory to worldwide deployment."[66]

The CRN took a stab at nano-ethics.[67] It is incredibly simplistic and borrows heavily from Jane Jacob's *Systems of Survival.* The CRN compares and contrasts guardian, commercial, and information ethics. Phoenix and Treder add remarks unsupported by their observations, such as "applying a consistent system of ethics to the wrong situation can be as bad as mixing and matching ethics for convenience," and "to force an organization to adopt alien ethics (or to solve alien problems) is to force it to act unethically." While these statements might be warranted, those warrants are missing in their document. Also by focusing on mature molecular manufacturing, they sidestep many of the difficult ethical questions that need to be addressed in a world of applied nanoscience and nanoparticles. Their ethics paper ends with much of the same tired claims and projections of nanofactory manufacturing and an imperative: "Now is the time to begin designing the procedures, organizations, and technologies that will be required to make this transformative infrastructure a success." The transformative infrastructure they refer to is a "worldwide network of restricted nanofactories." While interesting reading for some, once again its long-term focus seems to undercut the immediate demands of nanoscience and nanotechnology.

The CRN rejects any patchwork of regulation and responded to *Minding the Gap*, a report by the University of Toronto Joint Centre for Bioethics, which warned "that a backlash against nanotechnology development is gathering momentum and needs to be addressed."[68] Phoenix argued that "calls for complete relinquishment of the technology are no less danger-provoking, and irresponsible, than is the cry for entirely unfettered development." Treder added, "We have much to gain, and much to lose. The need for sober and responsible public discussion of the implications of this new technology is acute and urgent."[69]

The CRN likes to communicate "a powerful sense of urgency" in its publications,[70] suggesting "the time from the first assembler to a flood of powerful and complex products may be less than a year."[71] Elsewhere they write that "a crash program started today could complete the first working nanofactory within a decade at a cost of between five and ten billion dollars."[72] This time frame is not referenced to another source and seems mostly speculation at best.

They advocate a major crash program fearing "the rapidly falling cost will allow several players—corporations and/or nations—to pursue independent development projects. Environmental or social concerns, or simple Luddism, could delay the research. Spending large amounts of money requires either political will or corporate boldness, which could be lacking at the crucial time."[73] When the window of opportunity will close is not resolved by the CRN, and why current US efforts beyond marginalizing molecular nanotechnology are not sufficient to keep the window from closing is obscure. They argue that an opaque Manhattan-style program might move the leader so far ahead of the pack that researching nanotechnology defensively would be foreclosed. On a second level, they argue for mature defensive nanotechnology as a solution to rogue military attacks. Trying to balance the deterrence value of transparency against the opaqueness of the massive investment program is left unresolved. Rather naively, they write "as long as all molecular nanotechnology capability is administered by an international body, with product designs being reported and tracked, it will be possible to verify the offensive and defensive capabilities of each."[74]

They want the program to be "closely guarded, with a high level of security, and it should be international in nature; an international program can absorb national or corporate programs, reducing the total number of programs."[75] They have developed an anally retentive outline for the international administration of molecular nanotechnology. Of course, this is premised on a mutual-inspections regime to verify the status of any national program, and they actually argue the international body must be trustworthy which would seem to be a given. In addition, they expect this international administration to "work simultaneously to increase national security, increase economic security, and fulfill many other positive goals, while at the same time applying the necessary restrictions and policies to avoid instability in several different domains."[76] This is quite an agenda!

The international nature of this project is presumed to be sufficient to preclude corporate monopolies by some sort of regulation.[77] They advocated Embedded Security Management (ESM)[78] as a platform for protecting international property rights and called for "automatic scanning for designs that violate trademark, copyright, or patent to prevent illegal development of protected designs."[79] Presumably, "each new design would have to be approved before it could be manufactured. Designs would be divided into classes, each with their own approval scheme."[80] They discussed flagging undesirable designs, nanoblock patent infringement, responses to efforts at cracking security systems, self-destructing nanofactories, and tracing break-ins using a GPS-like system.

They assume this protection will maintain incentives to innovation and that harmful products can be carefully controlled. Furthermore, they suggest that these securities would allow "government to track the design and manufacture of products within their jurisdiction. Illegal product designs could be vetoed from further manufacture, and their designers arrested or blacklisted."[81] They also suggested that "a hardware key could be required, so the holder of a certain key could build certain products."[82]

This is based on their observation that product design would be "on more or less a specification of combinations of nanoblocks, and automated design analysis systems can be an integral part of the patent system for these designs," and to encourage innovation they advocated "a 'patent holiday' for a few months" to reduce the patenting of "the most obvious and useful designs [so they] could not be owned by opportunists."[83] To reduce harmful products, they advocated licensing and checks. They continued, "In more dangerous classes, designer and builders, and the products themselves, they may need to be licensed as *probably safe*."[84]

They assume that distribution of benefits will eliminate incentives to develop independent programs. They advocate free use of products for humanitarian purposes. Nodding to Big Brother, they advocate an interesting mix of international closeness and faith in government. They ignore ignominious incentives from agents like organized crime, profiteers of all sorts, terrorists, and governments without altruistic motives. Simply put, they advocate peace.

Six statements summarize their positions: (1) effective administration may distribute benefits; (2) restrictions may control unwise use of nanotechnology; (3) nanofactories can be constructed with built-in restrictions; (4) careful study leads to better planning; (5) institutional reform must precede development; and (6) planning must begin now.[85] The CRN themes its work into areas of interest: nanoweapons, human rights (basic needs and surveillance and control), socioeconomic impacts, environmental issues (from reclamation and recycling to gray goo), and educational outreach. Unsurprisingly, the last two themes have figured substantially in recent NSF calls.

They defend the precautionary principle as a model for regulating nanotechnology. This approach is premised on the assumption that "costs of delay (opportunity cost) are significant and even outweigh the risk of development."[86] In turn, this is derived from a secondary assumption that "molecular manufacturing could save millions of lives per year."[87] The CRN defends a more *active* form of the principle which they define as "choosing less risky alternatives when they are available, and for taking responsibility for potential risks."[88] Furthermore, "if damage is likely but not certain, the lack of certainty is no excuse for failing to mitigate the damage."[89]

When it comes to nanotechnology, Phoenix and Treder agree that a permanent and global prohibition of molecular nanotechnology would be infeasible.[90] They have repeatedly criticized moratoria: "Bill Joy and others have proposed halting nanotechnology research entirely. This would not actually work; instead, it would relocate the research to less responsible venues. The risks might be delayed by a few years, but would be far worse when they appeared because the technology would be even less controllable."[91] As such, they advocate the "creation of one—*and only one*—molecular nanotechnology program, and the widespread but restricted use of the resulting manufacturing capability."[92]

They admit a program of this sort would demand "different kinds of administration. Some possibilities include built-in technical restrictions in portable nanofactories; intellectual property reform; and international cooperation or monitoring of various kinds."[93] They have concluded, "An early, closely guarded, international development program is probably the approach that retains the most control in the long run."[94] Elsewhere, they reference former Foresight head Christine Peterson's testimony to the US House Committee on Science and assert, "The safest course appears to be a single, rapid, worldwide development program by an organization that recognizes the necessity of wise administration."[95] The CRN assumes that global regulation may "avoid the internal corruption that naturally accompanies so much power."[96]

Elsewhere they write, "any regulation that imposes severe penalties will increase curiosity and rebellion, while providing at best a partial deterrent. Secrecy will likewise increase curiosity, and spark independent research. Any regulation that increases cost will increase the economic motive."[97] These remarks seem to be at odds with other claims, though they do write there are risks from both too little and too much regulation.[98]

They seem preoccupied with bureaucratic centralization. "A single project provides a single point to monitor and control."[99] For example, "the number of nanotech nations in the world could be much higher than the number of nuclear nations, increasing the chance of a regional conflict blowing up ... because nanotech weapons could be developed much more rapidly due to faster, cheaper prototyping ... raising the possibility of horrifically effective weapons."[100] Treder recently added: "The tenuous balance of mutual assured destruction (MAD) and the worldwide network of commercial trade are both threatened by the rise of advanced nanotechnology."

As they put it, "the idea is to have the only game in town—so that everyone feels the need to support it."[101] Not unlike the leading force position advanced by Drexler in *Engines*, by powering into the lead, a benign power could use the technology, much like blue goo,[102] as a check on itself. "The safest course appears to be a single security infrastructure, designed and implemented with a maximum of scrutiny from military, commercial, and private experts, applied to all nanotechnology that could be used to create unrestricted molecular manufacturing systems."[103] For the CRN, "this requires a body with global jurisdiction, perhaps analogous to the International Atomic Energy Agency."[104] To reduce the incentive to develop other national or corporate programs, transparency is accorded by sharing the fruits of the technology without sharing the lines of code. Unfortunately, this strategy sounds a lot like Microsoft's marketing.

They acknowledged the Greenpeace report on nanotechnology (see below) and called it a step forward. Included in their praise was Greenpeace's support of a citizens jury to determine scientific priorities on nanotechnology and its referencing of the precautionary principle, though they disagree with Greenpeace's statement that there is no need for new mechanisms of public involvement. They promised a supplementary commentary,[105] and their "Technical Commentary" came in September 2003.[106] They defend limited molecular nanotechnology or LMNT (serial nonreplicating nanofactories), which they claimed reduced the emphasis on nanobots or did not require nanobots at all. "Some products of LMNT may be small robots, but product robots require no onboard manufacturing capability." LMNT still uses an assembly line, but the single nanofactory is fastened down. The nanofactory used *simple* feedstocks, and "the main barriers to development of LMNT are matters of policy." In the last section of this report, there is an on-point discussion of the Greenpeace report. They rejected the timeline for molecular nanotechnology and return to their defense of LMNT. They attempted to discredit the issue of barriers to development by rebutting Smalley's fingers issue. They warned about intentional abuse of manipulators as weapons and found solace in the open-source software movement as a resolution to the nanodivide and called for international management. In general, their response was that Greenpeace's "analysis is based on an early understanding of molecular nanotechnology." Simply put, it was not their nanotechnology.

Recently, they seem to want to enter the societal and ethical implications realm of nanoresearch and at the same time return molecular manufacturing into the limelight after it was excised as a research priority in the new Twenty-First Century Nanotechnology R&D Act. Treder posted an announcement in late June 2004 about a program of research projects in collaboration with the CRN: "Under supervision of an instructor, students will research some aspect of molecular manufacturing and write papers suitable for publication. CRN will provide advice and review, especially on the scientific aspects. Participating students will have direct access to Phoenix."[107] The CRN has already prepared a list of thirty essential studies that must be performed, and the first project involved a team from Arizona State University Law School's Center for the Study of Law, Science, and Technology titled *Transnational Regulatory Models for Molecular Nanotechnology*.[108] The project is directed by law professor Gary Merchant,[109] whose current research interests include the use of genetic information in environmental regulation and toxic torts, the precautionary principle, and legal issues relating to genetically modified foods.

Oddly enough, I have met most of the individuals discussed above, and they are eloquent and dedicated and very creative and clever. In the cases of both FI and the CRN, their unwillingness most of the time to step back from molecular manufacturing (read as *mechanosynthesis*) and cartoons of *working nanofactories* have led to unbridled ridicule from the mainstream scientific and policy-making critics. Focusing past the immediate and near-term instances of applied nanoscience has resulted in their overall marginalization. However, of both these proponents, FI seems to be reinventing itself in terms of its public meetings and conference agenda as well as the increasingly sophisticated nanoscientific developments and national- and international-initiatives reports. Currently, the same cannot be said of the CRN, and that's unfortunate since Treder and Phoenix are two of the most passionate and committed proponents in the game.

OPPONENTS

Opposing nanotechnology is a bizarre idea since it raises images of begrudging stockingers breaking spinning and finishing machines in the early nineteenth century. None of the opponents have advocated an Earth First! response to nanolabs. Protests have been limited most recently to the groundbreaking ceremony of the Berkeley Nanofoundry and some near-naked young people exposing their thongs at an NbA conference. By and large, the opposition has taken the form of documents posted at Web sites or distributed more conventionally. The ETC Group and the Greenpeace Environmental Trust were chosen because they are earnest in their rhetoric, and much of what they claim has some merit.

ETC Group

The group most insistent on stopping research in applied nanoscience is the ETC Group. They make the most absolute statements and draw the most closed conclusions. Pat Mooney, the executive director, made the claim, "In the absence of any kind of laboratory protocol for handling nanotech materials, the only sensible course of action is to call for a total moratorium."[110] They are motivated by a desire to extend their ideology into the burgeoning field of nanotechnology for reasons unknown, though after a less-than-successful attempt to put an end to globalization and corporate concentration, they may see in nanotechnology an opportunity to gain influence and prestige. Their publications are not carefully vetted and subject to overclaiming research findings that agree with their position while underreporting contrary points of view beyond discounting them. "In March 2002, ETC acted on fears calling for a worldwide ban on nanotechnology research,"[111] which was later categorized as "a moratorium on the use of synthetic nanoparticles in the lab and in any new commercial products until the governments adopt best practices for research,"[112] and "their ideas have reached many thinkers in the EU."[113]

The ETC Group was known as RAFI—the Rural Advancement Foundation International. Its full name is the Action Group on Erosion, Technology and Concentration, hence the acronym. Erosion umbrellas issues on human and cultural rights. The technology agenda covers new technologies, including human genomics and neurosciences, nanotechnology, and biological warfare. And concentration explores "New Enclosures," corporate strategies beyond intellectual control to strengthen monopolization. The group plans to retain its more traditional work in agricultural biodiversity, biotechnology, biopiracy, intellectual property, and issues of international governance.

The first time that the staff of RAFI and the staff of the Dag Hammarskjöld Foundation sat together was over lunch in the temporary Parliament building in Stockholm in 1981. The Hammarskjöld Foundation was shaping a third system, the citizen's perspective on society in contrast to those of the state and the business community. RAFI and Hammarskjöld organized together the Civil Society Organizations (CSO) consultation on biotechnology at Bogève, France, and in 1988, *Development Dialogue* published the results.

According to executive director Pat Mooney, the new name was the result of a Web site contest in 2000, though the majority of respondents suggested no name change was necessary. Research director Hope Shand admits that the change was done to secure nonprofit status in the United States. Headquartered in Winnipeg, it is incorporated in Canada and Europe. Its agenda can be found in one of its more ambitious publications, *The ETC Century*, the treatise of the ETC. In 1993 RAFI became the first CSO to document the collection of indigenous genetic material and the patenting of human cell lines around the world. As such, RAFI remained committed to discussing the intersection of new technologies, corporate strategies, and the implications for humanity.[114]

Its new president is Tim Brodhead, former secretary-treasurer of RAFI and past head of the Canadian Council for International Cooperation. Brodhead replaced Sven Hamrell (Hamrell led the Dag Hammarskjöld Foundation for thirty years), who headed the group since its incorporation in 1985. Its board includes Olle Nordberg, the current director of the Hammarskfjöld Foundation.[115] ETC's advocated global moratorium on nanotech research stirred Prince Charles to intervene.[116] Jasper Gerard recaps:

> So what's going to destroy the world this week, according to Mystic Charlie Wales: GMO crops, bans of overly ripe brie, any architecture that isn't neo-Georgian? Or has he grown up sufficiently to worry about the rise of Islamic fundamentalism in Iraq, or perhaps SARS? Nope. He thinks the end is near due to the threat from nanotechnology. He has read a report that has left him jibbering…. The Royal Society is sufficiently worried to hold a conference on the subject to stave off GMO-style technophobia, but now Chazza has intervened, fat chance.[117]

The ETC entered the nanotechnology debate with the publication of *The Big Down* in January 2003.[118] Its full title is *From Genomes to Atoms, Atomtech: Technologies Converging at the Nano-scale.* "We've had a file on its since 1988," said Mooney, "but it was on the back burner until we did a patent search in 2000. We were shocked at the number of patents, how fast it was accelerating and the range of big companies involved."[119] The ETC has been very active in the position on nanotechnology. It immediately began building a network of interested parties. "ETC began telling its network of advocacy groups that reputable researchers had deep concerns about the environmental risks."[120]

Reviews are mixed, though Philip Shropshire's remark that "the name *The Big Down* would be a great name for a band" introduced a series of divergent criticisms and comments. "Sure, it's *Just Say No* prescription for nano is unreal and won't happen. But the report is an excellent primer on a number of cutting-edge movers driving nanotechnology forward. You could go even as far as to say that this is one of the best summaries of nanotechnology ever—its history, directions, scientific foundations and more."[121] It is unclear how ironically the previous statement should be read, though it seems more literal than not. While describing *The Big Down* as one of the best summaries ever written seems more a comment on Shropshire's literacy on nanoscience than not, it is a good read.

One of their strongest forewarnings in *The Big Down* involves the merging of biotechnology and nanotechnology called *convergence*, which would produce dramatic and personal changes. Mooney and the ETC predict a "longer-term industry scenario, 2010–2020 will witness the commercialization of nanotechnology and its convergence with biotechnology."[122] Bionic-Nano-Atomtech might create both living and nonliving hybrids. It could blend people with robots.[123] "How many human chromosomes can we put in a harp seal before Greenpeace sides with the cod? If you put three human chromosomes in a rat, can it run for office?"[124] Their Web site warns,

> Through the manipulation of biological materials, it is now possible (or scientists believe it some day will be possible) to: Craft synthetic DNA; Use the synthetic DNA to create unique living organisms; Construct new artificial amino acids that can be built into unique proteins; Add a fifth letter to DNA;[125] Write DNA code; Use DNA to build nano machines capable of exponential self-assembly; and design exponentially self-assembling nanomachines.[126]

It is difficult to decide what level of nanotechnology the ETC is addressing in its fear-mongering over *convergence*. If it's assemblers and nanobots, we may need more than a decade or two to get there. If not about bots, then which specific tools or particles are being questioned?

Overall, its prescription is not very persuasive. As Shropshire put it, "But just saying, *No*, and asking for some vaguely stated goal for more community dialogue doesn't cut it. Quite simply, *The Big Down* has all the wrong targets. What we need to do is bring advanced technology back into the public domain. And this requires improving our democracies."[127] The treatise doesn't add much to the debate over the risks associated with the development of nanotechnology beyond its commitment to a precautionary principle construction of policy to guide nanoscience and technology. ETC's reading of the principle "says that governments have a responsibility to take preventive action to avoid harm to human health or the environment, even before scientific certainty of the harm has been established."[128] This strong reading of the principle is at odds with the balanced reading given to it by the CRN.

While it does emphasize convergence of technology, the report is weak on details and strong on rhetorical flourishes, such as "When GMOs meet Atomically Modified Matter, life and living will be never be the same."[129] Its criticism is over-inclusive as well:

> The problem is that ETC dilutes the concept of nanotechnology. Anything from cosmetics with nanoparticles to speculative molecular machines is viewed as nanotechnology, and supposedly risky.... By promoting the most dramatic promises and threats of nanotechnology while at the same time extending the term to cover even standard processes ETC makes it look like potentially devastating risks are being ignored.[130]

However, this group should not be underestimated. They are media wizards. "What Mooney lacks in scientific evidence, he gets the popular press to fill in with what-ifs."[131] When speculation is unavailable, there are always conspiracies. Mooney says, "Rick Smalley and other leading nanotech researchers are trying to sneak nanotech into science, industry, and society before it's fully vetted."[132] In one of their publications, *Genotypes*, they claim efforts "to marginalize St. James' Palace by arguing that the Prince's concerns are either non-existent, centuries distant, or exist only in pulp fiction ... may only be the latest of a series of technical and tactical mistakes made by nanotech's over-eager proponents."[133] Of course, the poor prince has been lampooned mostly by the media. His understanding of science and technology has seldom become the grist for serious scientific rebuttal simply because the science world won't take him seriously beyond the occasional barb. Most provocative seems to be Charles's membership in the British Royal Family and his outlier status.

The ETC does warn about "the privatization of science and a staggering concentration of power in the hands of giant multinational enterprises."[134] They fear that these nanopolies are ready to patent the elements of nature. Patents have been granted on human genes and single nucleotide polymorphisms (SNPs). With patent litigation costs so high, intellectual property has become a nontariff barrier to market entry for smaller innovators. Since 1995 the number of intellectual property lawsuits reaching federal courts has risen ten times as fast as other legal actions.[135]

However, Mooney, conceding some regulatory success over controlling the proliferation of patents on life, predicts that

> new patents on nanotechnology—*atomic patents*—suggest that we could win the battle over life patenting but still surrender monopoly control of agriculture and health to the nanotech industry. The industry is looking for sweeping patent

claims that will dominate this technology. In some cases, the claims will not involve life. In many cases, the claims will relate to bionic matter. There is an urgent need to re-think the framework of the intellectual property debate in order to challenge the new technologies now.[136]

Mooney and others warn about drift-net patenting. While patent issues are important to any discussion of nanotechnology, there are many steps in the extended argument before it reaches bionic matter. The first overclaiming patents have not been vetted yet in court. The assumption that only a few transnational corporations will run the show remains premature, but not wholly unlikely.

Learning from the global concentration of industry, they caution that "the biotech boutiques of the '70s and '80s are now being replicated with the nano-nichers of the *Dot-Aught* decade."[137] The ETC categorizes risks as horrendous and warns that industry will monopolize all animate and inanimate matter, dumping failures in the third world. In addition, the successful application of nanotechnology will radically transform local economies in developing countries. The ETC complains that atomtech has evaded regulatory scrutiny, and impacts are unknown based on insufficient data and study.

A second major level of criticism is associated with military use of advance technologies, especially biotechnology. Referencing high-sci corporate gurus from companies such as Nanotronics, Inc., Mooney tables companies against contracts and grants from 1998 and 1999 cataloguing their activities, and he warns that technologically enhanced biowarfare will be cheap, easy, and opaque; will involve limited stockpiling; and is inevitable. He warns of the weaponization of crop and livestock pathogens and ethnically targeted viruses and asserts that a dozen countries are researching their use.[138] Of course, these claims are not inherently dependent on nanotechnology but are the product of biotechnology, which is conveniently conflated by the ETC.

Mooney's criticism becomes more topical as he approaches the postbiotech era in *The ETC Century*.[139] First, he asserts that nanotechnology will move past carbon as its basic building block to encompass the entire table of elements, a claim made nowhere else by anyone else. Second, he shares the mechanosynthesis views of nanotechnology using the metaphor of Lego blocks:

> Nonliving matter can also be Lego-built, atom by atom and molecule by molecule. Depending on how the Lego is put together, the end product could be a diamond, a daffodil or a dinner for two. In theory, nanotech can pull its atomic raw materials from garbage dumps or thin air to manufacture houses and hair-dryers that are stronger and more durable than any products available in the marketplace today.[140]

Third, he linked the commercialization of nanotechnology to self-replicating nanobots. Beyond admonitions of uncontrolled replication, he found the risks mind-boggling. "Governments in our privatized world will act to secure monopolies for the enterprises undertaking these ventures. So-called democratic societies will surrender much of their freedom in return for the safe use of nanotech of colossal projects, such as re-upholstering the ozone layer, desalinating the oceans, etc."[141] Much like the misdirection associated with proponents above, the case the ETC makes is far removed from applied nanoscience, nanoparticles, and nanopowders.

Mooney wrote, "Just as with biotech, we are *not* suggesting that this field of research be abandoned. But now—before the commercial hype and corporate pressures are too great—society should establish the benchmarks and ground rules for its investigation. Extreme care should be taken that, unlike with biotech, society does not lose control of this technology."[142] While appearing as a shift in argument, Mooney and Shand want an immediate moratorium on applications, not on research. Mooney argued nanotechnology could adversely affect four industry sectors: agriculture, transportation, macromaterial (including mining, construction, and heavy industry), and energy. The extent of the adversity is premised on the speed of its introduction. For example, "there is no need for Wal-Mart if there is no need for walls."[143] The ETC visualized nanotechnology as another, if not the quintessential, technology that would provide business and industry with the power to stranglehold the working class, even the government.

As such, *The ETC Century* ended with this recommendation: "Governments should impose a moratorium on the development of self-replicating nano-machinery unless and until intergovernmental agreements can be adopted that set standards and guarantee the safety of nanotechnologies."[144] It argued, "one mistake has already been committed—the mishandling of regulation and safety considerations of nanoparticles.... Scientists failed to establish a common laboratory protocol to ensure the safety of workers exposed to particles.... A second mistake may prove unforgivable."[145] Once again, the action devolves to uncontrolled replication when the issue is corporate concentration, the antecedent to their claims.

In June 2004 the ETC published what they call a *communiqué* titled *Nanotech News in Living Colour: An Update on White Papers, Red Flags, Green Goo, Grey Goo (and Red Herrings)*. It begins with framing the most pertinent issue for the communiqué: "nanotech products are already commercially available and laboratory workers and consumers are already being exposed to nanoparticles that could pose serious risks to people and the environment."[146] Hyperbole is not a stranger to the ETC Group, and on page 1 they overclaim the conclusions of the Swiss RE Report (see chapter 5), for Annabelle Hett never concluded that "unknown and unpredictable risks associated with nanotoxicity or nanpollution could make nanotechnology un-insurable."[147]

In general, this communiqué adds very little to their arguments, though they add quite a bit of data and warrants to their claims. For example, they make the claim "in recent months, governments in the U.S. and Europe have reluctantly concluded that current safety and health regulations may not be adequate to address the special exigencies of nano-scale materials."[148] Marburger has already gone on record denying this claim in terms of US regulatory agencies and without a direct reference or any citations; it is unclear who is speaking for Europe.

To their credit, however, they do note one of the most troubling realities about nanoparticles: "At present there are no regulations explicitly targeting the products of nanotechnology anywhere in the world, even though hundreds of products have reached the market and hundreds more are in the pipeline."[149] However, "hundreds" may be an overestimate at this time, though the ETC has an unattributed and unofficial EPA document that lists well over one hundred products.[150] Also, regulatory entities are not ignoring this problem and are mustering efforts to address the subject.

The ETC is correct in claiming that "the usual method of controlling toxic substances—based on thresholds calculated by weight or percentage/weight—is not relevant since the toxicity of materials appears to depend on properties other than mass."[151] This conclusion seems to be held by all scientists researching nanotoxicity issues. It has not led to hand-wringing but to studies of regulatory options and research projects to address this phenomena. For example, only a few pages later they cite John Howard, the director of the US NIOSH who announced the preparation of a best-practices document for working with nanomaterials.

Once again, don't underestimate the ETC; they have supporters. Unsurprisingly, their supporters hold views consistent with their philosophy. For example, *Resurgence*, published in Devon, UK, questions trade without responsibility and money without morality. *Resurgence* continues to publish articles that are on the cutting edge of current thinking, promoting creativity, ecology, spirituality, and frugality. The online magazine asked its readers to "demand an immediate comprehensive nanotechnology moratorium of their governments." They added corroboration by claiming this view was held by a growing number of experts. "The call for a moratorium on nanotechnology research—at the very least until a transparent public debate is held about the momentous moral and societal issues involved—is being proposed by a growing number of environmentalists, ethicists and philosophers, disarmament experts, skeptical scientists and others."[152] It is aping the ETC.

Although Mooney in *The ETC Century* criticized biotechnology, especially Lazarus-link seed (sterilizing the plant seed at harvest), he clearly favors organic farming and new technologies associated with it, which encourage democracy and decentralization.[153] More than any other issue, the organic farming link may best help explain the relationship between the ETC and Goldsmith, but it also explains what led the ETC to oppose nanotechnology. For the ETC, nanotechnology may be associated with industrial concentration, and concentrated industrial power is associated with many of, if not all, the ills of development. Nanotechnology takes the worn neo-Marxist agenda and provides it with some new currency. Recently, ETC has published a report on nanotechnology and agriculture, *Down on the Farm*,[154] and another on nanotechnology and patents, *Nanotech's Second Nature Patents: Implications for the Global South*.[155] Both these recent works use worst case scenarios to make their arguments. The good news, as Ole Peter Galaasen of the *Plausible Futures Newsletter* put it, "the fear mongering of the ETC group [is] finally being replaced by serious scientific inquiry."[156] Some of us are less certain that this is true.

Greenpeace Environmental Trust

Given the increasing number of toxicology studies published about nanoparticles, especially carbon nanotubes, anticipating a mainstream environmental-interest group might find a critique of nanosciences (read: nanotechnology) intriguing and reasonable. Enter Greenpeace and a temperate, reasoned argument.

One of the best features of the report was observed by Glenn Reynolds:

The report's not-entirely-unreasonable worries about the dangers of nanomaterials are distinguishable from more science-fictional concerns of the Michael Crichton *Prey* variety. And that means that it will be harder for Greenpeace to conflate the two kinds of concerns, itself, as has been done in the struggle against genetically-modified foods where opponents have often mixed minor-but-proven threats with major-but-bogus ones in a rather promiscuous fashion.[157]

Reynold's reaction epitomizes concerns expressed by SEIN researchers. Some of the voiced objections tag low-incidence, high-impact scenarios to exaggerate risks of high-incidence, low/medium–impact scenarios. For example, using the gray goo story as an impact to the claim against the increasing use of carbon nanotubes in products marketed today. The warrants for the claim are overwhelmed by the impact story. Conflation of applied nanoscience across multiple time frames has tended to produce a misdirected debate on environmental implications. Greenpeace avoided this.

As such, this work is a much more usable guide indeed. Reynolds added, "Take out the code words and phrases that are tailored to Greenpeace's audience and you'll find some sound advice in there for the nanotech industry."[158] He goes further to claim that the report "is likely to do more good than harm at blocking Luddite efforts to turn nanotechnology into the next GMO food."[159]

Others, like Howard Lovy, are much less charitable in their reviews:

Greenpeace's just-released report on nanotechnology is vintage advocacy-group treatment of scientific research: Grab the available facts, then make them conform to your predetermined conclusion. That, after all, is what advocacy groups do. And most intelligent readers are able to keep that in mind when they come across any *study* that comes out of an organization that filters information through its preset worldview. It's true for Greenpeace, the National Rifle Association or the Save the Bog Turtle Foundation.[160]

According to Modzelewski, Greenpeace "switched their battle from genetically modified foods, and are now targeting other emerging technologies, including nanotech. As much as nanotech is the tech buzz word du jour, we're also the new target du jour."[161] He characterized their report as "industrial terrorism," which is not only incredibly defensive but also misleading. According to Modzelewski:

> It's a great way to raise new funds and pretend they care about something. The reason these groups care about nanotechnology is because they view it as the next industrial revolution. And, to them, slowing it down, creating fear and upsetting people is their means of creating a choke point on the development of industry and technology. They saw how it worked on genetically modified foods, and so this is a great way for them to do the exact same thing.[162]

Ascribing this motivation is simply unjustified. Greenpeace's argument is more precautionary than prescriptive. In general, "the report seemed to take a hard line. It didn't dismiss out of hand efforts by some green groups to place a global moratorium on nanotech. It warns industry may find a moratorium *virtually* self-imposed if they do not take public acceptance seriously. It warns of public backlash if the nanoindustry does not persuade the public of the technology's safety."[163]

The Greenpeace report is called *Future Technologies, Today's Choices.* Its secondary title is more enlightening: *Nanotechnology, Artificial Intelligence, and Robotics: A Technical, Political, and Institutional Map of Emerging Technologies.* It is primarily the work of Alexander Huw Arnall from Imperial College. It was released in July 2003. The part of the report associated with nanotechnology is limited to the first forty pages.

Doug Parr, Greenpeace's chief scientist, claims that concerns over the environmental implications of emerging technologies led to the commissioning of the report, which he calls a comprehensive review, touting the reputation of Imperial College London for its technological reputation. Parr made two observations worth dissection. First, there are signs that nanotechnology is receiving more governmental support stimulating industrial interests without concurrent public debate. Second, nanotechnology offers a huge opportunity to involve the public in setting scientific priorities. As such, he advocates a citizens' jury approach to invigorate a public sphere in science and technology decision making.[164] He added that research groups such as the Biotechnology and Biological Sciences Research Council and the Engineering and Physical Sciences Research Council "could commit to considering results and utilizing the insights from the findings of such a jury."[165]

He added that since "the development of nanotechnology will go through various different stages, and thus societal debate will need to be an ongoing process rather than a single outcome, there will need to be continual incorporation of the insights from such a debate into policy and product development as the prospects become more tangible."[166] It is not difficult to notice that Parr sees in nanotechnology an opportunity to create a standing infrastructure that may have implications beyond nanotechnology. For years, Greenpeace has articulated a vision of humanity's interface with nature inclusive of some public sphere of checks and balances.

Greenpeace has a list of reservations about nanotech. The new materials may constitute a class of nonbiodegradables, and environmental and toxicological data on nanoparticles are insufficient. This observation leads to the remark *no data, no market*. The materials should be considered hazardous until shown otherwise, beckoning the precautionary principle as the methodology. Parr appended two additional trepidations: nanotechnology driven by military agenda and the ownership of nanotechnology techniques and products, adding fuel to the criticism from the ETC Group.

At the end of his foreword, in an attempt to pull together the themes of nanotechnology, artificial intelligence, and robotics, he warns of the convergence of nanotechnology, biotechnology, and self-replication. Greenpeace's goo is not *grey*, it's *green*. Their concerns are amplified as nanotechnology and biotechnology merge or converge.

Greenpeace made a few interesting cases in their report. They "reject the idea of a moratorium on nanotechnology ... the question then becomes one of how, not whether, to develop nanotechnology."[167] They accurately examine the hype surrounding claims and counterclaims of nanotechnology by both enthusiasts and detractors and summarize environmental challenges adding to some of the ETC Group's concerns.

By and large, the report "concludes that the prospect of self-replicating nanobots remains way off and it is acting as a distraction from far more immediate concerns about nanotechnology, such as the potential effects on health and the environment from nanosized particles."[168] While not wholly discounted, its low probability relegates it to the back of the shelf of Greenpeace's concerns. This perspective amplifies the rational voices of toxicologists and environmental scientists who believe gray goo should be the grist of school debating clubs rather than dominating the debate over the health and environmental applications of instances of applied nanoscience today.

At the end of the section on nanotechnology, Arnall briefly discussed self-replication. He offered one paragraph on sociopolitical concerns, tagging the term *paradigm shift* without contextualizing it and hyperbolizing the impacts nanotechnology may have. He added another paragraph on genetic discrimination, failing to make a case unique to nanotechnology. He also discussed the nanodivide and weapons applications. In his discussion of the nanodivide, he failed to make a case beyond industrial concentration and spent little ink on discussing the patent and intellectual property issues. He briefly mentioned nanotechnology refinements of existing nuclear-weapons designs without offering illustrations or scenarios. The report comes down to three pages (39–41) where regulations and public participation are discussed. In 2005 Doug Parr commented on how the Royal Society report was received by Parliament. While the report recommended new funding to investigate the effect of nanotechnology on people and the environment, which was not forthcoming, "the longer-term issues of the social, environmental and ethical impacts of nanotechnology are barely recognized in the response."[169] Also in May 2005, Greenspace helped launch a public debate on nanotechnology known as *UK NanoJury*. Generally, Greenpeace has mostly removed itself from the nano debate in 2005, rejecting invitations to participate in oversight organizations and international conferences on nanotechnology.

CONCLUSION

These four groups or organizations attempt to make objective claims but draw their warrants from the ideologies that define them. Each of them is motivated to make arguments that cohere to the viewpoints that have become concretized in not only their goals and mission statements but also their publication history. Many of the observations they make have merit. Indeed, in the next chapter we discover some of the same considerations expressed by researchers in scientific disciplines. Nonetheless, each of these four groups is biased toward forwarding arguments that are consistent with their motivations and their claims. As such, their arguments need to be subjected to the same strict and critical scrutiny afforded to government hyberole on the subject.

CHAPTER 9

NANOHAZARDS AND NANOTOXICOLOGY

One of the trepidations expressed by government, industry, and health and environmental groups about nanotechnology is that "it may backfire, threatening human health and unleashing new forms of pollution."[1] It's happened before: thalidomide, the pesticide DDT, asbestos proofing, ozone-destroying chemicals in spray cans and refrigeration, and genetically engineered seeds and modified foods were all considered benign until their harmful effects came to light. While some people may see no downside per se to issues like genetically modified foods, others vehemently oppose them.

It may be becoming less important how scientifically valid the criticisms may be that scare the powers that are intimately associated with this technology. Many individuals and groups have clutched onto negative speculation and hearsay about nanotechnology for a multitude of reasons, and both the advocates and opponents, scientists and laypersons, have been speaking directly to the people. Others simply misreport findings of studies, exaggerating conclusions or ignoring them altogether: Advocacy groups that could impede progress in nanotechnology research have already begun to form. According to Vicki Colvin, they "often specialize in misinformation. They take information about risks and blow it out of proportion."[2]

The challenge confronting risk communicators and even perception-management specialists is daunting. Denying a claim is much more difficult than making one, and that is true irrespective of the integrity of the data and the logical rigor of the warrant. In a debate where one side is restricted to evidence that is both reliable and valid, while the other side is not, can produce not only a lopsided set of arguments but also lopsided debates. Only an audience with a sophisticated understanding of the subject and the argumentative subterfuge can be persuaded by the denials based in reality against the fictitious assumptions of the opposition.

If that fear were not enough, consider the possibility that nanotechnology will have wholly unintended consequences because it will be used for purposes originally unintended. As engineer Lester Lave of Carnegie Mellon University warned at the 2001 NSF SEIN conference, "When you introduce a new technology, it's almost impossible to foresee what the consequences will be." One everyday example of unintended consequences, Lave says, may be sitting in your driveway: "The notion of driving vehicles designed to be off-terrain [yet] have 90 percent of those vehicles never leave the road shows us people are using these things for a purpose they were never engineered for. That leads to the question: how are people going to use nano-technologies?"[3] There are incredible anecdotes usually told by hospital emergency room personnel. People put things where they don't belong. People do things that are incredibly stupid. The entire addiction and substance-abuse problem of huffing attests to that phenomenon. As such, nanotechnologies must be foolproof as well.

"History is packed with examples of skepticism that turned out to be unfounded, and sanguinity that was misplaced." IEEE's Society on Social Implications of Technology past president Clinton Andrews recounted: "Typists at first resisted carbon paper, thinking it would threaten their jobs. The first electronic computers stirred fears the companies were building *electric brains*."[4] Strangely enough, the sectoral dislocation implications of nanotechnology have received little discussion, except for the ETC Group, especially Pat Mooney. He complained, for example, "in Bolivia, an economy dependent upon its mines for its survival, the demands for copper or for tin, or for zinc, are going to drop because nanotechnology will not require the same quantities of those commodities."[5] Notwithstanding, most of the focus has been on Armageddon scenarios, like grey goo, and toxicology-related issues.

TIME FRAMES AND ETHICAL CALCULI AMONG THE FRAMES

There is a shell game being played about nanotechnology. For example, when a claim is made that carbon nanotubes may be an occupational hazard for workers manufacturing them, one response claims that generally the health advantages from mature nanotechnology might eradicate many debilitating and life-terminating diseases. The primary claim is immediate, while the subsequent claim is both speculative and a function of nonexistent technological developments. In another debate, a respondent redirects gray goo by arguing it is impossible given current knowledge. In both these instances, phenomena from two very different time frames are used, and the result are incoherent.

One of the overwhelming challenges confronting anyone trying to understand what is probably true or false in the debate over nanotechnology extends far beyond what constitutes nanotechnology. Today, nanotechnology is applied nanoscience. What one thinks about nanotechnology in terms of its benefits and drawbacks is wholly dependent on which version of nanotechnology and which stage of development are being discussed.

As such, many theorists approach the discussion of nanotechnology in terms of waves or time frames in order to break down the phenomena into more reasonable pieces. Much like a puzzle enthusiast who separates pieces into shapes and colors, theorists have discovered that discussions about societal and ethical implications make more sense when the topic is more manageable.

For example, take Berkeley's J. Bradford DeLong's approach: "Nanotechnology is coming in three waves: first—materials-wave over the next two technological generations, a second—biologicals-wave between one and four technological generations from now, and a final—Drexlerian-wave that may or may not ever come to be."[6] DeLong separates the events into materials, nanobiotechnological convergence, and assembler nanotechnology. We are well into the materials revolution, and we are beginning to see some implications already. While the convergence of biotechnology and nanotechnology has begun to receive some notice, it is still decades away, and no one seems to be able to set a date when Drexlerian self-assembling nanobots will get here.

Another example is the simple breakdown with clear demarcations provided by Credit Suisse/First Boston. In its report on the state of investment in companies claiming to be working in the field, the authors decided to offer three simple ranges: in the next five years, five to ten years, and ten years and later. One could easily have called this range the near term, the intermediate term, and the far term. Over the next five years more rudimentary technology will emerge as tools, materials, nanoparticles, composites, sensors, and fuel cells. Five to ten years from now, we may expect precision building, two-dimensional nanoelectronics, hybrid bionano-functionality (such as implantable devices), and sensor proliferation. Ten years and beyond, we may be dealing with self-replicating machinery, three-dimensional nanoelectronics (computers with three dimensional displays), and tissue and organ regeneration.[7] Credit Suisse/First Boston avoids the ambiguity of the wave model and the Drexlerian conception of self-assembling nanomachines altogether.

Regardless which is chosen, the importance of working with a time frame in our discussions about nanotechnology is that each carries with it a different set of ethical dilemmas, hence the SEIN are different within different time frames.

While some of the same questions asked in the next five years may still be asked ten years or more from now, by focusing on the societal issues of a technology that might be mature in a century distracts us from the immediate and locks us into studying an advanced technology with tools that have not had the benefit of one hundred years of development. A false assumption in risk studies has been that future consequences should be examined against present capabilities to ameliorate the degree of hazard. As such, while nanotechnology in 2250 may include replicating nanobots, it will also include advanced means to mitigate the impacts they could produce. Without much imagination, it is easy to conceive of a series of preventative measures, though none has yet been developed. For example, we could make self-replicating nanobots that are dependent on an exotic vitamin that must be present for replication. While we haven't developed this vitamin, we also haven't developed the self-replicating nanobots. Simply put, grey goo might be easily avoided given advances in nanofabrication and software development. When they become viable, controls might develop right along with them.

On another very important level, the frames provide a guide for determining what set of principles might be relevant in testing the societal and ethical implications of existing nanotechnology. In the immediate future, we must apply what we have learned and know about environmental ethics to the nanoproducts moving into the market, including how they are made, how they are used, and how they are disposed of.

In five or ten years, nanotechnology will be not only associated with products from cosmetics, stain-resistant pants, and brake linings to toothpastes that repair enamel and self-cleaning shoes, but also found in the production lines of many current products. The list is fairly extensive:

> stain resistant fabrics for clothing and bedding, cosmetics and sunscreens, tennis balls and racquets, bowling balls, odor-eating socks, time-released perfumed fabrics, paints, capsules carrying hemoglobin, sensors to test water impurities, ski wax, Humvee turrets, long-lasting paper, nanotubes for flat display screens, artificial silicon retinas, several drug delivery systems, flash memory devices, [and] diagnostic agents for use in MRI scans.[8]

With its increased efficiency and durability, we may find that significant social dislocations will be the primary problem confronting policy makers. Sectoral unemployment from obsolescent products and services will demand an additional ethical calculus. So while retaining environmental vigilance, another series of sociopolitical ethical questions will arise.

In the long term, we will need to remain observant of the effects of nanotechnology on the environment and sociopolitical dislocations and distortions, but we may need to consider the question of what it means to be human given advances that bring humankind to a posthuman threshold. As we become integrated into machines and vice versa, what we are, what we think, and how we feel about the world may change. Much like the quandary associated with future generations, we are wholly unprepared to debate a philosophy when we do not understand the needs and demands of future variables in the calculations. "The potential social implications of assemblers and replicators are fascinating areas. The intellectual discussion about assemblers and replicators in the futuristic nanoworld may offer insights in many areas including ethics of near-term nanotechnology."[9] However, Wei Zhou warns us that

> focusing on the ethics discussion on the long-term potentials is not enough. Ethics discussion about near-term applications is clearly needed to guide further development of nanotechnology applications. Ethics researchers and legal scholars should be encouraged to study the many ethics issues arising from the near-term nanotechnology research and commercialization.[10]

The preliminary inference from these observations is that while it might be an interesting academic exercise to ponder the ethics of problems that exist decades from the present, it represents a serious devaluation of problems in the present. To deny life-saving nanopharmaceuticals or another nano-enhanced sustaining technology to a desperate person because there might be an ethical consideration at some time in the future is the worst form of academic arrogance. Uncertainty is a feature of newly emerging technologies, especially the predictions we make about them. In our attempt to live without uncertainty, we may lock ourselves into a paradigm that condemns millions to suffering and death. As such, examining SEIN must progress beyond a parlor game or a philosophical debate at a coffee shop. While these may have value, they are not germane to the task at hand. As such, the following examines the most pressing immediate SEIN issues: the health and environmental implication of the commercialization of nanoscience in terms of the use of nanoparticles in current production lines, from manufacturing the particles to their disposal.

FEAR AND TREPIDATIONS

Ken Donaldson from Edinburgh, an expert in occupational and environmental medicine, called for a new discipline—nanotoxicology. "Efforts to untangle science and science fiction regarding the risks from nanotechnology are needed."[11] In addition, we are finding more nanoparticles in the environment every day. The Institute of Medicine estimated that British workers are exposed to nanoparticles in the academic labs and in emerging nanoparticle-producing companies. This may be up to two thousand individuals.[12]

A recent Lux Research report claims that while large companies spent over $3 billion on nanotech research and development in 2004, "corporate EHS (environmental health and science) officers are mostly unaware of this activity." In addition, "dozens of start-ups are mostly dodging the topic in public presentations for fear that they may held legally liable in the future for any admissions of risk made now."[13]

We would assume nanomaterials are mostly kept enclosed and workers wear gloves and masks handling them. At least, this may be true in countries that set a high priority on worker safety. We also notice that some materials are shipped in a liquid form to reduce exposure to air-borne nanoparticles.[14] Unfortunately, some may escape. Other nanoparticles are simply rubbed into the skin as cosmetics. Some of these and others associated with drug delivery or to reinforce artificial hips, for example, may relocate or slough off from the applications to enter across barriers, blood-brain or placental, within the human body. Actually, "nanoparticles have been demonstrated to cross the blood-brain barrier with little difficulty hence the interests in using them for anesthesia and other therapeutics."[15]

INVESTMENTS IN NANOTOXICOLOGY

According to the ETC Group in *The Big Down,*

> The U.S. Environmental Protection Agency currently allots no more than 10 per-
> cent of its nanotech research grants for the environmental benefits and potential
> harmful effects of nanotechnology at the societal level. Given the potentially huge
> implication for the environment and human health, the societal implications of a
> completely new economy of manufacturing needs to be understood—now.[16]

The situation is worse in Europe, where only "5 percent of the European Community's funding of nanotechnology was engaged in examining environ-mental, social and ethical dimensions of these technologies."[17]

While the ETC feels this is insufficient, the US federal government seems to disagree:

> Health and environmental research makes up about 11 percent of total federal
> nanotech funding this year.... For example, the EPA in 2004 has launched a $4
> million research project that will study the toxicology of manufactured nanoma-
> terials. It will also look at the fate, transport and transformation of manufactured
> nanomaterials and the impact of human exposure to them. Also, the National
> Institute of Environmental Health Sciences' National Toxicology Program initi-
> ated a $3-million project to examine through inhalation exposure the toxic and
> carcinogenic potentials of quantum dots, nanotubes, and other materials in labo-
> ratory animals.[18]

To give the ETC their day in the sun, they do a fine job explaining the shell game:

Given that *applications* and *implications* are lumped together in the U.S. government's 11 percent estimate, it is difficult to tell how much funding is devoted to determining risk and toxicity and how much is devoted to developing products to be used in the environment or in medicine. The tandem goals of developing commercial products and determining impacts on human health only go hand-in-hand in the most responsible regulatory environment. Given the absence of regulatory oversight presently, the 11 percent figure is not reassuring.[19]

Steve Crosby, managing editor of *Small Times*, agrees. "None of the nanotech funding legislation introduced into the US Congress sufficiently addresses that shortcoming.... The situation will not change soon given the nano government funding devoted to environmental research."[20] Crosby and others, including the ETC Group, are probably correct—there needs to be more resources invested in funding nanotoxicological and other environmental studies into nanoparticles.

According to Tom Kalil, former technology advisor in the Clinton administration, an implication to the production of studies like those mentioned below is "to heighten the interest in funding initial research in nanotoxicology to find out what synthetic nanostructures are safe and what structures raise environmental and human health issues."[21] As more studies are released, more studies will also be funded. Time will tell whether the research pumps have been sufficiently primed.

A PRIMER ON NANOTOXICOLOGY RESEARCH

We begin with a set of reservations. First, it is important to realize that every couple of months more research on the environmental and toxicological implications of nanoparticles is being published, and we may never know enough about them to be certain; hence, the debate over what proof levels need to be met to provoke some regulatory response becomes a substantive part of this debate. It's also important to realize that Rice's CBEN is in the forefront of this research, and its primary spokesperson has become Vicki Colvin, who is not without detractors. For example, she was described as having a "rather pro-technology bias."[22]

A second assumption that should be challenged involves asking whether or not a particle or compound or an element or an approved drug is the same as a macroparticle of the same substance. "Nanoscale materials," observes the National Institute of Environmental Health Sciences' deputy director of environmental toxicology, John Bucher, "don't act like particles and they don't act like chemicals. They take on properties that are either intermediate or they are unique, and we're just beginning to sort through this."[23] The nanoparticle and the macroparticle may be called by the same name, and they may be composed of the same stuff, but they won't look or act the same.[24] Simply put, "the way molecules of various shapes and surface features organize into patterns on nanoscales determines important material properties, including electrical conductivity, optical properties, and mechanical strength. So, by controlling how that nanoscale patterning unfolds, researchers are learning to design new materials with new sets of properties."[25]

There are some astounding characteristics found at this scale. For example, "electrons...no longer flow through electrical conductors like rivers."[26] Also, "as objects become smaller, the proportion of their constituent atoms at or near the surface rises. Collections of very small particles, therefore, have high surface area compared to their volume;...catalysts with labyrinthine interiors with nanoscale features means there's more surface area for thousands of chemical transformations."[27] This led A. Nemmar and his colleagues to claim that since "ultrafine particles have also a much larger surface area [they] hence [have] a more toxic potential."[28]

Consider two specific illustrations. In 1994 Gunter Oberdörster studied rodents and "demonstrated that ultrafine particles administered to the lung cause a greater inflammatory response than do larger particles per given mass."[29] Or consider this simpler anecdote: "If you take aluminum, like we have in aluminum cans, and bring it down to about 20- to 30-nanometer grain size, the material properties dramatically change. It explodes spontaneously in the air. It becomes a rocket fuel catalyst."[30]

On a different level, there is a dichotomy at play when studies are done on incidental rather than manufactured nanoparticles. "Engineered nanomaterials differ in significant ways from the heterogeneous and polydisperse particulates and in significant ways from ultrafine particles."[31] As Rice's Kristen Kulinowski told me, "Much of the research done on particles of this size has been on incidental particles, which are not associated with nanotechnology in the way that engineered nanoparticles are. There may be some important limitations in our ability to extrapolate from the incidentals research to draw conclusions about nanoparticles."[32] The literature supports a difference. "The primary difference between ambient and engineered nanoparticles is that the former have widely varying shapes, sizes, and compositions, whereas the latter sire single uniform compounds."[33] Whether this is sufficient to delegitimize all research associated with ambient nanoparticles is a different case yet to be affirmed.

Nonetheless, it is important to observe that "few indications of problems have turned up yet among the limited results so far. The Rice Center reports, 'We think that most nanoparticles will probably be relatively inert' in the biological realm."[34] However, research on environmental issues relating to nanotechnology are strikingly absent from the literature. "In a field with more than 12,000 citations a year, CBEN was stunned to discover no prior research in developing nanomaterials risks assessment models and no toxicology studies devoted to synthetic nanomaterials."[35] As of late 2004, "only 50 peer-reviewed research papers on environmental and health effects of engineered nanoparticles [had] been published."[36]

Range of Concerns

Though Prince Charles and other critics of the research fear that further down the line nanotechnology could have unexpected—and potentially disastrous—consequences, very few are suggesting that currently available products are dangerous. Nonetheless, some environmental concerns have been expressed by both the ETC Group and Greenpeace International. The ETC's call for a moratorium is primarily premised on the uncertainty associated with the safety of nanoparticles, whether in the near or distant future. Their argument is allied with references to the precautionary principle, which is more commonly employed in European environmental debates, though in the United States it can be found to some degree in the occupational safety and health regulatory process used by the FDA and in some international agreements, many of which the United States has failed to ratify.

Do the unknown risks of engineered nanoparticles, in particular, their environmental impact, outweigh their benefits? In another way, are the benefits of foregoing engineered particles worth the opportunity costs?

"The U.S. government estimates that 2 million Americans are already exposed to nano-scale materials regularly."[37] Nanomaterials include carbon nanotubes, fullerenes, nano-oxides, nanocrystals, nanopowders, and so on. "The market for nanomaterials is forecasted to grow by 30.6 percent per annum for the next five years, but the real driver of growth will be carbon nanotubes and fullerenes with an annual growth rate of 173 percent."[38]

We know "exposure is growing as the nanotechnology industry increases demand for small, size-controlled particles."[39] It's not just carbon nanotubes. "Today, an estimated 140 companies are producing nanoparticles in powders, sprays, and coatings.... The world market for nanoparticles is projected to exceed $900 million in 2005,"[40] and "at least 44 elements in the Periodic Table are commercially available in nanoscale form."[41] For example, "a US company, Nanophase, has been selling nanoparticles of various metal oxides since the mid-80's."[42] "Burn dressings treated with silver nanoparticles are used in over 100 of the 120 major burn centers in North America.... The fuel lines of many US and European cars, some Toyota auto bodies, and Renault's plastic side panels all incorporate nanoscale particles."[43] In 2004, "the estimated total amount of nanoparticles in commerce is a few thousand tons."[44]

Over half—and perhaps as much as 80 percent—of nanotechnology is going to be based around carbon.[45] Actually, "Japan's Mitsubishi Corp. is already making hundreds of tons of the material for industrial purposes, and a dozen or more companies around the world are also ramping up production efforts."[46]

While much remains unknown, the following examines the environmental implications associated with nanoparticles currently being used or soon to be found in products, reviews the issues of disposal and toxicity, and focuses on carbon nanotubes (CNTs) since they are having the greatest impact commercially.

"The global market for carbon nanotubes alone was estimated at $12 million in 2002, but is expected to grow to $430 million in 2004."[47] The price for carbon nanotubes has been dropping precipitously. "Three years ago, the price of bucky-balls was about $600 a gram; the cost has come down to $30 a gram—20 fold decrease and it's still dropping."[48]

GLOBAL MARKETS FOR RESEARCH-GRADE SINGLE-WALL AND MULTIWALL NANOTUBES FOR SHORT-TERM APPLICATIONS, 2001 THROUGH 2006[49] ($ MILLIONS)

2001	2002	2006	Adjusted Annual Growth Rate % 2001–2007
7.0	11.9	231.5	> 100

The number of applications for CNTs continues to grow as well. "A mixture of 90 percent Zylon and 10 percent nanotubes will produce a 60 percent increase in strength and a doubling of energy absorption."[50] Zylon is twice as strong as Kevlar, and it is used in bulletproof vests. "Fuel cells with 10 percent composition of nanotubes yields a 60 percent increase in the amount of power available."[51]

American companies are producing CNTs as well:

Carbon Nanotechnologies Inc. built its first plant three years ago, it could make 1 pound of nanotubes a year. Six pilot plants later, the company is in final testing of a unit capable of making 20 pounds a day. Later this year, the company plans to complete a unit capable of continuous operation, making 100 pounds a day. If all goes well, a full-scale plant, likely to be built in the Ship Channel Area of Houston, could begin manufacturing 1,000 or even 10,000 pounds a day by late 2005.[52]

According to Charles Choi,

Nanotubes are pursued globally. Producers in the US include Bucky USA, Car-bolex and Carbon Nanotechnologies, while France has Nanoledge, Russia has Nanocarblab, Cyprus has Rossette, Japan as Mitsui and Showa Denko, Korea has Iljin, and China has Guangzhou Yorkpoint and Sun Nanotech. Production volume runs a little more than 2.5 metric tons per day and is expected to grow to roughly 1,750 metric tons per year as early as 2005.[53]

We begin with the fundamental observation that "new forms of carbon—called nanotubes and fullerenes—are being manufactured for the first time and their impact on the environment is unknown."[54] While fullerenes resemble soot, they might have significant health implications given their architecture, concentrations, and functionality. Colvin seems to agree: "The exposure to nanotechnology workers is a near-term concern. . . . Longer term, there is an opportunity for a much wider exposure to the entire ecosystem to engineered nanomaterials through water and soil."[55]

On top of all this was an observation made by the Royal Society. As they reviewed the cosmetic industry selling us nanoparticles in sunscreen, they also suggested that while the industry "insists that these products are safe, it has refused to release any test data into the public domain."[56] While it remains unclear whether this means that there has been no safety research or that research that has determined there may be some problems has been squelched remains unknown.

Are Nanoparticles Biodegradable?

First, there are concerns about disposal and whether carbon nanotubes are biodegradable. Barnaby Feder from the *New York Times* offered this adage: "Don't ask, don't tell is the operating mode for much of the nanotechnology industry these days when it comes to where discarded products end up."[57] As a rule, Colvin suggests,

> increasing concentrations of nanomaterials in groundwater and soil may present the most significant exposure avenues for assessing environmental risk.... Materials present in water can become degraded, transformed and accumulated in a variety of ways.... Even at low aqueous concentration, polyaromatic hydrocarbons can have pronounced environmental effects, and fullerenes may have similar properties.[58]

Conclusions are problematical given that "the ways in which nanoparticles behave in groundwater environments or water treatment plants are as varied as the diverse molecules or atoms used to assemble them."[59]

In the process of producing and in terms of disposing of carbon nanotube products, waste will be produced. The potential problems include

> whether the nanomaterials would pass through filters in a water treatment plant, whether the substances bind to various contaminants such as pesticides or PCBs, and whether the new particle could enable already present chemicals to become more mobile and possibly cause more harm, and whether bacteria take up the nanomaterials, and whether this might open a route for the particles to move up the food chain, and whether... [when a protein is] attached to the surface of a nanomaterial,... [a protein's] shape and function can change.[60]

Another Rice researcher is much more concerned:

> In a small-scale project, led by environmental chemist Mason Tomson of Rice University, researchers investigated how buckyballs travel through soil. The team suspended them in water and then poured them through a soil-like material. When the balls were clumped together to form particles a few micrometers big,

they were absorbed into the soil like any organic compound. But when they were dispersed, the water formed a protective sheath around each buckyball, allowing them to travel through the soil without being absorbed. "It's completely unexpected," says Tomson.[61]

Rice's Helene Lecoanet and Mark Wiesner studied "the transport of three varieties of fullerenes in porous media and the effects of flow on nanoparticles deposition"[62] and reported that "C_{60} aggregates are capable of migrating through a well-defined porous medium analogous to a sandy groundwater aquifer."[63] They do conclude: "Deposit structure and the integrity of the fullerene aggregates upon deposition will likely play an important role in determining bioavailability and persistence in natural systems while these same factors might be controlled in engineered systems to tailor nanoparticle aggregates for separation or fractionation."[64]

Greenpeace's Alexander Arnall has postulated that "nanomaterials provide a large and active surface for absorbing smaller contaminants, such as cadmium and organics. Thus, likely naturally occurring colloids, they could provide an avenue for rapid and long-range transport of waste in underground water."[65] Finally, an obscure reference to Chiu-Wing Lam's research appeared in an issue of *Yomiuri Shimbun*: "Nanomaterials are more absorbent than other molecules. Consequently, if they were to spread to the environment, there is a danger that they might absorb pollutants and disperse them widely."[66] The impacts on potable water supplies might be serious.

At a March 2002 meeting with some fifteen EPA-interested officials, Wiesner described a phenomenon involving hitchhiking bacteria: "We know nanomaterials have been taken up by cells. That sets off alarms. If bacteria can take them up, then we have an entry point for nanomaterials into the food chain."[67] Beyond that, he challenged researchers "to examine whether nanoparticles absorbed into bacteria in any way enhance the ability of other materials, including toxic ones, to piggyback their way into the bacteria and cause damage."[68] "The concern that nanomaterials could bind to certain common but harmful substances in the environment, such as pesticides or PCBs, leads to the short-term worry of such material infiltrating humans."[69]

In March 2004 Wiesner "presented preliminary findings indicating that different kinds of nanoparticles do not flow in uniform ways in water [suggesting] it will be difficult to predict how nanoparticles behave in groundwater environments or water treatment plants and that generalizations are not tenable—nanomaterials will have to be studied case-by-case, not as a class of materials."[70] According to Wiesner, the implications should not be substantial. "Even the nanoparticles best suited for travel appeared capable of moving underground no more than about 10 yards before sticking to larger grains of sand."[71]

Rice's Christine Sayes and her colleagues studied water-soluble fullerenes, seeing them as "essential for many emerging biomedical technologies which exploit the unique chemical properties and physical structure of C_{60}.... Water-soluble fullerene species are also important for understanding the eventual fate and environmental implication of fullerenes used in consumer products... such as coatings and fuel cells." Her team concluded: "Fullerene materials can in some cases generate superoxide anions that could be the agent responsible for membrane oxidation and cytotoxicity." However, "hydroxylation of the C_{60} cage could be used as a remediation for the possible unintentional biological effects of the pristine fullerenes."[72]

In a recent issue of *Environmental Science and Technology*, Rice's J. D. Fortner and his colleagues found a wrinkle in the research protocols. They reported "upon contact with water, under a variety of conditions, C_{60} forms a stable aggregate," which they termed nano-C_{60}. The team concluded that guidelines for disposal "need to be revisited" because "nano-C_{60} behaves neither as an individual molecule nor as a bulk solid."[73] They wrote that "the environmental fate, distribution, and biological risk associated with C_{60} will require a model that addressed not only the properties of bulk C_{60} but also that of the aggregate form generated in aqueous media." They reported that "C_{60} upon contact with water can form negatively charged colloids which are stable over time." They concluded: "This work illustrated the limitations of the current guidelines for the handling and disposal of C_{60}, which are based entirely on the properties of bulk carbon black."[74]

On the other hand, Colvin reported some mitigating preliminary results of her own: "Typically high crystalline inorganic solids, fullerene nanomaterials may be susceptible to biodegradation, a factor that controls their long-term persistence in the environment."[75]

Prior to complete biodegradation, there is another route of concern. A *Nature* article by Geoff Brumfiel added these ominous comments from Colvin: "Unpublished studies... show that the nanoparticles could easily be absorbed by earthworms, possibly allowing them to move up the food chain and reach humans."[76] Greenpeace reiterated this concern but added, "it is becoming increasingly clear that work by CBEN and other organizations, such as NASA and the EPA, alone are insufficient for the scope of the issues."[77] Hence, more research is called for.

However, there is a flip side to this debate: "The EPA is already funding 16 universities and 11 private companies to research nanoscale answers to pollution and waste."[78] There is a strong case being made that atomic-scale precision would mean less waste because many nanotechnologists consider their work green chemistry.[79]

So instead of concluding that nanoparticles, especially carbon nanotubes, are dangerous to health and welfare, we are led to compare the nanosolution to more contemporary alternatives. Indeed, comparative assessments seem to be in order.

Are Nanoparticles Toxic?

The following examines three categories of exposure: workplace (including labs), direct consumer exposure, and indirect consumer exposure.

The area that seems to be generating the most concern is workplace exposure to workers producing, packaging, and unpackaging nanoparticles. Exposure can occur in three ways: ingestion, inhalation of agitated particles, and contact with exposed skin.

This brings us to the debate on ultrafine particulates (UFP) and whether research on UFPs contributes anything to the environmental health and safety (EHS) debate over engineered nanoparticles.

Research indicates a plethora of problems associated with inhalation of ultrafine and nanosized particles, including fibrosis or scarring, the abnormal thickening of brachioles, the presence of neutrofils (inflammatory cells), dead macrophages,[80] and some chemical hitchhiking (metals[81] and hydrocarbons[82]). Research has suggested links between ultrafine particulates in air pollution and lung disease[83] and brain damage.[84]

There is a large body of research on ultrafine particles (less than 2.5 micrometers in diameter). Particles in the under-one-hundred-nanometer range would be a subset of this class of particles, but there is some reservation involving extrapolations from the health and safety research on ultrafine particles to nanoparticles, as mentioned earlier. Not the least of which are (1) engineered particles are designed so that they do not agglomerate or clump into larger particles, and (2) engineered particles often have their surfaces functionalized to carry another material, such as a protein.

Many argue as well that some extrapolation from ultrafine research may be justified given engineered nanosized particles would be unlikely to dispel concerns and reservations associated with size since it remains a relevant (though not the sole) variable in damage and hazard discussions of nanoparticles.

The primary area of concern is associated with the impact nanoparticles will have on humans and other species, especially to workers producing carbon nanotubes. Vyvyan Howard, a toxicologist from the University of Liverpool, and others have observed that nanoparticles can pass into the body by three means: inhalation, ingestion, and transdermally. While ingestion may not be an immediate concern per se, it may be noticed on the toxicology radar soon: "Two hundred companies are already working on inserting nanotechnology into food. The report, by Helmut Kaiser, a German consultancy, concludes that, "with nanotechnology, industry is set to design food with much more precision, and lower costs and sustainability. The technology is already used to preserve foods, and boost flavor and nutritional values."[85]

Most toxicologists research inhalation and transdermal absorption. Colvin warned that "[t]hey're not inert.... One thing they're definitely going to do is absorb material and they can insinuate themselves into cells."[86] While Wiesner seems concerned about piggybacking bacteria, no research has considered what the effects might be when viruses or prions are the hitchhikers. Rice's Jennifer West added, "Researchers need to screen more materials for toxicity, screen more materials for their ability to piggyback on different synthetic molecules, and investigate their effects not just in cell culture but also in living animals."[87]

Many researchers are studying the effects of the instillation of carbon nanotubes using animal subjects, having the sample injected into the trachea and lungs rather than inhaled. Inhalation experiments are expensive and difficult to control, and "no one has expressed a willingness to fund them."[88]

For example, Andrzej Huczko and his colleagues compared carbon nanotubes to asbestos fibers. This similarity can only be sustained in their aspect ratio: "SWNTs [single-walled carbon nanotubes] are much smaller and more flexible meaning there is no evidence that they exist in the instillation as long rigid rods, but rather more like a string of spaghetti all folded up on itself. There are significant weaknesses with this analogy."[89] Beyond the alert sounded by such a comparison, there are some surface similarities: they are long and needlelike. As such, "the potential health risks of inhaled nanofibers cannot be dissociated from the well-known adverse effects of asbestos fibers."[90] In addition, in an effort to reduce clumpage, research into ways to solubilize nanotubes to declump them is under way, which also allows the fiber to remain unagglomerated.

In terms of inhalation, there is a growing body of literature that speculates about the similarity between carbon nanotubes and other fibrous minerals. Whether linked to images of black lung, asbestosis, or meliothelioma, writers use rhetorical flourishes to hearken cases of debilitating diseases to amplify their claims. It is fortuitous some of the first toxicology experiments should involve CNTs, for not only are these nanoparticles being produced at a greater rate than other nanoparticles, but also there is some research on improving their commercial safety.

The first major CNT issue seems to be inhalation during production, an occupational health issue. "Bob Gower, head of Nanotechnologies Inc., says engineers from the National Institute for Occupational Safety and Health visited his company and found that because the tubes clumped together, they weren't likely to be inhaled by workers."[91] Of course, once we learn of ways to reduce the clumping phenomenon, we may increase the risks associated with inhalation.

Instillation studies are used because inhalation is more expensive and incredibly messy since researchers would need to pump a suspension into an enclosed vessel in which the lab animals were located. This approach would more accurately resemble airborne concentrations in the workplace. For example, David B. Warheit and his colleagues evaluated their own findings with this caveat: "The physiological relevance of these findings should ultimately be determined by conducting an inhalation toxicity study."[92] Simply put, "Inserting carbon nanotubes in rats' lungs provides data on how toxic they might be but does nothing to answer the question of what level of exposure would be necessary for a rat to breathe in damaging amounts of such particles."[93]

Many of the instillation studies reference earlier experimental work associated with extrapulmonary effects of air pollution. A. Nemmar and his colleagues, using individualized particles in the five-to-ten-nanometer range, discovered "the smallest particles translocate from the lungs into the circulation and thus influence cardiovascular endpoints more directly…[and] may pass into the circulation and thus exert direct effects on the heart and vessels."[94] The researchers concluded their "findings provided plausible evidence for particle translocation from the lung into the blood and then its distribution to the organs."[95] In the same year, Nemmar and his colleagues reported a second study involving intratracheal administration of ultrafine particles in hamsters. They found "homeostasis may be affected by the presence of ultrafine particles in the circulation and that this phenomenon is dependent on the surface properties of the particles." They found, for example, "positively charged amine-polystyrene particles led to an increased prothrombotic tendency [clotting]."[96] Nemmar with another group of researchers observed "inhaled particulates may affect the autonomous nervous system directly, by eliciting a sympathetic stress response or affect the heart and blood vessels directly, through inflammatory cytokines produced in the lungs and released into the circulation."[97]

Gunter Oberdörster and colleagues observed in their rat studies that "translocation pathways include direct input into the blood compartment from ultrafine carbon particles deposited through the respiratory tract." Their results demonstrated that "effective translocation of ultrafine elemental carbon particles to the liver after inhalation exposure."[98]

Human mortality studies "in fourteen different locations have shown overall daily mortality increases as the concentration of small particles in the air increases."[99] A. Seaton and his colleagues suggested "ultrafine particles are able to provoke alveolar inflammation, with release of mediators capable, in susceptible individuals, of causing exacerbations of lung disease and of increasing blood coagulability, thus also explaining the observation in cardiovascular deaths associated with urban air pollution episodes."[100]

According to the ETC Group's reading of Huckzo and, in turn, Zhang,[101] "due to their similar aspect ratios, carbon nanotubes physically resemble asbestos fibers, which, by reacting with cellular components can produce dangerous byproducts, and are hazardous to humans."[102] Unfortunately, the reference to Zhang in Huckzo cannot be found. The *Journal of Immunology* article does not make the claim Huckzo cites, which, in turn, fuels the ETC reference in *The Big Down*. Huczko and his colleagues draw the following conclusion: "Intratracheal instillation of fibrous carbon nanostructures does not change pulmonary function and does not induce any measurable inflammation in bronchoalveolar space in carbon nanotube-exposed guinea pigs. Thus, working with soot containing carbon nanotubes is unlikely to be associated with any health risks."[103] Also, "autopsies didn't reveal significant differences in the animal's inflammatory reactions."[104] Nonetheless, the ETC adds the spin: "Setting tobacco smoke to the same standards, smoking would probably be declared a perfectly safe activity!"[105] An interesting study by S. Takenaka and his colleagues found that whether inhaled or instilled, ultrafine particles tended to "promote particle coagulation [which] results in larger agglomerated particles."[106] The study notes that particles are overall eliminated via three routes (all of which raise some caution flags): ingestion into the gastrointestinal tract and excretion, translocation into lymph nodes, and transfer of the material into the blood. If there is some good news, it is the speculation "that because of their very small size, the ultrafine particles were not efficiently phagocytised[107] by macrophages[108] and instead were cleared rapidly through the circulatory system."[109]

Chiu-Wing Lam of Wylie Laboratories in Houston, a senior toxicologist at NASA's Johnson Space Center in Houston, and his team have been studying the health effects of carbon nanotubes.[110] They found, "Single-walled carbon nanotubes (SWNTs) caused a dose-dependent reaction consisting of inflammation and tissue death."[111] They studied SWNTs and tested toxicity of three products made by different methods and containing different types or amounts of residual catalytic metals. "All the SWNT products induced a dose-dependent formation of epithelioid granulomas [abnormalities that interfere with oxygen absorption and can progress to fatal lung disease] in the centrilobular alveolar septa. The granulomas consisted of aggregates of macrophages laden with black SWNT particles.... [T]he researchers found that mice inhaling micrometer-sized clumps of tangled carbon nanotubes had the same reaction they would have to ordinary dust. But when they were exposed to individual carbon nanotubes, the mice developed lesions in their lungs and intestines. 'Carbon nanotubes are not innocuous,' says Lam."[112] Lam worked with three types of SWNTs: "All three types caused lung granulomas. And although each mouse got just one exposure, the lesions got worse over time, with some progressing to tissue death."[113] The researchers attempted to limit overreaching conclusions by noting that the acute toxicity might have been caused by "the nickel and yttrium, surrounded or coated by NTs, freed by ultrasonication."[114] A few pages later, they admit that conclusions drawn from instillation studies in terms of the dangers of workplace inhalation are not the same but close enough, citing an earlier study by Lam's team.[115]

At the 2003 ACS meeting they reported, "if SWNTs reach the lung, they can be more toxic than quartz. If airborne SWNT dusts are present, respiratory protection should be used to minimize inhalation exposure."[116] This led the researchers to conclude: "Exposure protection strategies to minimize human exposures should be implemented."[117]

Lam's findings have been exaggerated even before they were published. For example, an article in the *Yomiuri Shimbun* claims that Lam "showed that nanotubes, when inhaled in large quantities can cause inflammation of the lungs." This was an instillation, not an inhalation, study. In addition, the newspaper article places this citation immediately after another paragraph comparing nanotubes to asbestos.

Another study questions Lam and colleagues' findings when they noticed fewer adverse effects. Researchers from the DuPont Haskell Laboratory near Newark, Delaware, led by David Warheit, tested the hypothesis whether significant lung toxicity and the hazard potential increases with decreasing particle size.[118] The results of their rat SWNT-instillation study, after discounting 15 percent mortality associated with simple mechanical blockage, demonstrated that SWNT-related toxicity was transient and intermediate between quartz and carbonyl iron particles. Their interim results indicated that SWNT exposure did not produce any sustained lung inflammatory effects, unlike quartz particles.[119] While "all the surviving rats developed granulomas," they did "without the inflammatory responses that usually accompany those lesions."[120] They also reported "all the surviving rats seemed completely normal within 24 hours"[121] and that "the inflammatory response only lasted one week."[122] They also noted that "the nanotubes' tendency to clump rapidly led to suffocation for some rats exposed to high doses, but it also kept most tubes from reaching deep regions of the lung where they could not be expelled by coughing and could cause long-term damage."[123]

The uncertainty is reported not to suggest there are no consequences from inhaling carbon nanotubes but to note that the amount of evidence that exists is minimal and conclusions are difficult to reach at this point. Warheit and his colleagues wrote in their conclusion, "due to their electrostatic nature and tendency to agglomerate into nanorope structures, exposures at the workplace to respirable-sized carbon nanotubes are extremely low."[124]

While most of the studies address workplace implications, others go beyond this setting in their conclusions. For example, the University of Rochester's professor of environmental medicine Gunter Oberdörster "has shown that rats exposed to a mist of nanometer-sized polytetrafluoroethylene, or 'Teflon' particles experience respiratory irritation."[125] While he noted that micrometer-sized clumps of nanoparticles were not reactive, individual particles can pass from lungs into the bloodstream and were reactive. He reported recently that the translocation of ten-to-fifty-nanometer-sized particles from the respiratory tract to interstitial sites and to other organs, including the central nervous system, has been demonstrated, yet many unanswered questions remain regarding their potency to cause health effects at these sites.[126] He warned that anthropogenic ultrafine particles at the workplace and in the ambient air will illustrate human exposure scenarios.

More recently, he argued that "nano-sized particles can translocate across epithelial and along axons and dendrites of neurons . . . [and] it appears that mitochondria are preferred subcellular structures where nano-sized particles localize."[127] He concluded that this had mixed blessings since "the findings give us ideas about the intriguing possibilities that nanoparticles offer for potential use as diagnostic tools or as therapeutic delivery systems."[128]

In discussing inhalation issues, we can hypothesize that there are at least two ways they may be inhaled: during the production process or as a result of sloughing. For example, Jennifer West from Rice reserves concern for carbon nanotubes because carbon is such a fundamental element in the body. Nonetheless, she and Wiesner have expressed apprehension that companies are now looking into using carbon nanotubes in radial tires, which might produce airborne particulates as they wear down. Whether at the level of concentration they are conjecturing it might constitute a legitimate health concern simply is not known.

Workers can be protected via a myriad safety protocols, and the *Economist* was unconcerned about most other concerns. "Since they are, in essence, a form of soot, this is not surprising. But as most applications embed nanotubes in other materials, they pose little risk in reality."[129] However, researchers remain guarded in their assessments of safety. "The actual risk will depend not only on the intrinsic hazard of the agent but also on the likely exposure. However, one should not conclude too rapidly that exposure will be negligible, certainly not if the material proves to be highly toxic."[130]

The next major issue is associated with transdermal absorption and the situation of nanoparticles where they might provoke undesirable reactions. Hyped by Vyvyan Howard of Liverpool as a possible route that may allow nanoparticles to enter the body,[131] toxicologists, especially Colvin, find it implausible given the number of layers of skin tissue the human body has. Nevertheless, some studies suggest caution.

In 1997 R. Dunford found that chemically modified titanium oxide nanoparticles from over-the-counter sunscreen products induced the formation of hydroxyl radicals and oxidative DNA damage.[132] On the other hand, Huckzo and Lange found no significant health hazard related to skin irritation and allergic risks from fullerene soot with a high content of CNTs.[133] Most studies suggest caution. A 2003 study looked at the effect of SWNTs exposed to human epidermis. Exposure was associated with "oxidative stress, which was confirmed by the formation of free radicals species...and a decrease in total antioxidant reserves in the cells." There was also some iron loading, which is associated with infection and inflammatory disorders. Consistent with findings that graphite workers tended to have an increased number of incidences of skin diseases, "exposure to SWNT produced oxidative stress and cellular toxicity."[134] Nancy Monteiro-Riviere and her colleagues found that "a detectable fraction of chemically unmodified MWCNTs are capable of intracellular localization as well as causing irritation in keratinocytes."[135]

For workers, some element of care should be able to reduce exposure. "The use of gloves and other personal protective equipment will minimize dermal exposure while handling the material; the propensity for large clumps to become airborne and remain so for long periods may lead to dermal exposure in less protected regions."[136] However, "based on studies of naturally occurring nanoscale particles such as ultrafine particle aerosols and surgical wear debris from implants, we can speculate that nanoscale inorganic matter is not generally biologically inert."[137] Indeed, "wear processes cause implanted biomaterials to produce particles, ranging in size from tenths to several microns.... These particles under some circumstances are cytotoxic and can produce inflammatory response and bone loss."[138]

On a different level, research has demonstrated that "nanoparticles deposited in the nose can make their way directly to the brain. They can also change shape as they move from liquid solutions to the air, making it hard to draw general conclusions about their potential impact on living things."[139] This becomes significant as nanoparticles seem important in pharmaceuticals, and CNTs could be used to strengthen implants, among other things.

There are a few interesting remarks made about toxicity issues associated with medical implants. For example, Jessica Gorman reports, "there is the possibility of nanotubes leaking out into body tissues and causing problems."[140] Silvana Fiorito, an immunologist in Patricke Bernier's lab at the University of Montpellier, has added some evidence in her study of implants: "When blended with polymers, the nanotubes could make implants stronger. However, there's the possibility of nanotubes leaking out into body tissues and organs." Fiorito studied graphite particles of about one micrometer in diameter and discovered that an immune response could be triggered by their presence. Though her work is preliminary, when carbon nanotubes were added to a cell she discovered that they accepted the nanotubes without developing inflammation.[141] She warns the reader that "she doesn't know why some forms of carbon are more readily tolerate by cells,"[142] but that reservation is sufficiently ominous to justify replication and further study. The ETC, more ominously, spins her work this way: "The ability to slip past the immune system may be desirable for drug delivery, but what happens when uninvited nanoparticles come calling? In other words, once nanotechnologists have figured out how to distract the bouncer guarding the door, how can you be sure you're still keeping out the riff-raff?"[143]

There is significant concern that nanoparticles making their way into the human body may settle where they might be harmful. "Tests revealing accumulation in the livers of laboratory animals demonstrate that nanoparticles will accumulate within organisms."[144] Within the bloodstream, there are other concerns. West explains, "If nanoparticles are present in the bloodstream, proteins in the blood will attach to the surface of the nanoparticles. When the proteins envelop the nanoparticles, the protein's shape as well as their function may change. Parcels of nanoparticles in the bloodstream may be useful for some medical applications, but the changes in the proteins could trigger other unintended and dangerous effects, such as blood clotting."[145]

Rice's Mark Jenkins warned that the large surface of nanoparticles offers an ideal place where oxygen reactions can occur in the airway and lungs, resulting in the formation of free radicals, with subsequent cell damage or cell death, followed by inflammation.[146] Still another consideration was raised by Bice Fubini in his study of particulates: "Adsorption of xenobiotics prior to inhalation may transform the particle into a carrier of carcinogens in the lung which may act synergistically with the particle."[147] As an example, he referred to the adsorption of polyaromatic hydrocarbons at the surface of asbestos fibers. An impact of this could be DNA damage caused by free radicals generated at the particle surface itself.

A 2004 study sheds doubts on the inhalation findings. "While laboratory studies have indicated that with sufficient agitation SWNT material can release fine particles into the air, the aerosol concentrations generated when handling unrefined material in the field at the work loads and rates observed were very low."[148] According to Colvin, they "found no respirable levels of small nanoparticles in a variety of workplaces that processed the materials. Rapid and irreversible aggregation of engineered nanoparticles in air may increase their mean size significantly and thus limit the inhalation or exposure of organisms to isolated nanoparticles."[149] Contact with exposed skin can be reduced with care and personal protective equipment, such as gloves. The researchers cautioned that "although the use of gloves and other protective equipment will minimize dermal exposure while handling this material, the propensity for large clumps to become airborne and remain so for longer periods may lead to dermal exposure in less well protected regions."[150] Simply put, with sufficient care the ambient concentrations of fine particulates may not be large enough to cause egregious harm, especially when care is taken in handling the carbon nanotubes.

Regarding direct intentional contact with skin, such as in cases of nanoized sunscreens, mild caution seems to be in order. J. Landemann and his colleagues studied "coated titanium dioxide microparticles commonly used as UV filter substances in commercial sunscreen products" and concluded that "penetration of microparticles into viable skin tissue could not be detected."[151] M. Tan and his colleagues also studied "the percutaneous absorption of microfine titanium dioxide from sunscreens.... The results from this pilot study showed that levels of titanium in the epidermis and dermis of subjects who applied microfine titanium dioxide to their skin were higher than the levels of titanium found in controls."[152] While they made no claims regarding whether the accumulation was statistically significant, S. Tinkle's group's findings suggest caution as well. Their studies warned: "The cutaneous immune response to chemical sensitizers is initiated in the skin, matures in the local lymph node, and releases hapten-specific T cells into the peripheral blood."[153] While they studied chronic beryllium disease, their investigations of cutaneous exposure to particulates as a potential route for sensitization to beryllium warrants comparable studies with other ultrafine and nanosized particles that also come in contact with stressed skin.

According to Vyvyan Howard, "There appears to be a natural passageway for nanoparticles to get into and then subsequently around the body. This is through the caveolar openings in the natural membranes which separate body compartments.... There is considerable evidence to show that inhaled ultra-fine particles can gain access to the blood stream and are then distributed to other organs in the body."[154] Howard cites studies from traffic pollution[155] and research from pharmaceutical companies that have designed ultrafine particles to deliver apoliproteins to the brain.[156]

"A second possibility is related to the ability of nanoparticles to slip past the human immune system unnoticed, a property desirable for drug delivery, but worrying if potentially harmful substances can attach to otherwise harmless nanomaterials and reside in the body in a similar manner."[157] Colvin discussed these problems when she examined water soluble fullerenes, which she categorizes as having "many promising medical applications based on their unique free-radical chemistry and antioxidant behavior."[158] Citing two renal studies by H. H. C. Chen,[159] she adds, "the LD_{50} for one type of water-soluble fullerene was found to be 600 mg/kg body weight from direct intraperitoneal injection and an unusual form of kidney damage was observed in lower doses."[160]

At the 2003 ACS meeting, on the other hand, David Allen and colleagues from Texas Tech and Kentucky's Schools of Pharmacy reported no adverse effects of nanoparticles at the blood-brain barrier.[161] Nevertheless, there remains some concern about nanoparticles crossing the barrier: "New research hints that particles can become more permanently lodged in the brain. This could permit minute particles to accumulate with potentially toxic consequences."[162] Presumably, "other scientists have wondered at recent meetings whether nanoparticles can cross the placenta and get into the developing fetus."[163]

"In January 2004, Ken Donaldson, a professor of respiratory toxicology at the University of Edinburgh, told the Royal Institution seminar in London that, once inhaled, ultrafine carbon particles can move to the brain and blood."[164] Presumably, "in the olfactory bulb and other parts of the brain, levels of particulate matter although lower than in the lungs in rat experiments remains relatively stable over six or seven days."[165] There are additional studies addressing nanosized particles selectively transported to the brain via the olfactory neuron.[166] For example, G. Oberdörster concurred with Donaldson and showed that "nanoparticles can make their way from a rat's throat into its brain apparently via the nasal cavities and olfactory bulb."[167] In a recent follow-up, he reported,

> approximately 20 percent of ultrafine particulates (UFP) deposited on the olfactory mucosa of the rat can be translocated to the olfactory bulb. Such neuronal translocation constitutes an additional not generally recognized clearance pathway for inhaled solid UFP, whose significance for humans, however, still needs to be established. It could provide a portal of entry into the CNS for solid UFP, circumventing the tight blood-brain barrier.[168]

Major support for this thesis appeared two months later at the March 28, 2004, meeting of the American Chemical Society in Anaheim, California. Eva Oberdörster (Gunter's daughter), a lecturer in the biology department at Southern Methodist University who led a recent study in collaboration with Colvin and Christie Sayes from Rice, presented her prepublication results. She reported that "modest concentration of buckyballs in water caused significant harm to two aquatic animals. Water fleas were killed by the addition of the tiny carbon balls and fish showed up to a 17-fold increase in brain damage compared with unexposed animals."[169]

Daphnia, or water fleas, are a common test organism for aquatic toxicity, and eight hundred parts per billion is slightly above concentrations encountered with more common pollutants. "Half of the fleas were killed at a concentration of 800 parts per billion in a three week test."[170] The decimation of water fleas may not appear important unless we note that these "tiny animals fill an ecologically crucial niche near the bottom of the aquatic food chain."[171] This might be important as a disposal issue. However, it is noteworthy that the *Daphnia* data was not included in her published bass study, though the data may be published in an upcoming review article on nanotoxicology to be published by *Environmental Health Perspectives* (EHP).[172]

"In a second test using 9 juvenile largemouth bass, a concentration of 500 parts per billion led to a 17-fold increase in a form of cellular damage in samples of brain tissue. This damage, known as lipid peroxidation, can impair the normal functioning of cell membranes and has been linked to illnesses such as Alzheimer's disease in humans."[173] The peroxidation occurred "after 48 hours though there were no outward behavioral changes in the damaged fish."[174] Oberdörster did admit: "We don't know if the fullerenes are directly causing lipid peroxidation in the brain tissue or whether it is a secondary effect caused by inflammation,"[175] and some of the damage "may be inflicted by immune system cells responding to the exposure."[176] She also found "higher levels of other detoxifying enzymes known as P450s" in the gills and liver of the fish, indicating the presence of some kind of xenobiotic agent.[177]

In the published version, she also makes a brief reference to minor turbidity in water from the presence of bacteria, and then two pages later she suggests "uncoated fullerenes may be bactericidal to beneficial bacteria found in aquaria."[178] Her findings are not unlike those of Yoko Yamakoshi and colleagues, who reported on the bactericidal properties of fullerenes.[179] Oberdörster pre-emptively suggests fullerene coatings may break down, and she references a study on quantum dots whose coating degraded when exposed to air and ultraviolet light (see below).[180] Nonetheless, she argued environmental contamination from consumer products and spillage should be studied further, especially since they are being produced by the ton.

Hers was the first study to show that nanoparticles may cause brain damage, though these were not human studies. Oberdörster wants more funding "from the EPA to conduct additional studies to look at such issues as how buckyballs actually get into the fishes' bodies and cause damage."[181] Oberdörster also reports that she "found a variety of genes that were turned on or turned off in the livers of exposed fish, indicating a whole-body response to fullerene exposure," adding that "these studies represent the first steps in a longer process of studying changes in gene expression."[182] Even she notes, however, that "the amount of buckyballs used in the study 'are actually pretty high concentrations to find out in the environment' and are unrealistic for long-term exposure unless a spill occurs."[183] In addition, "the buckyballs used in Oberdörster's experiment were uncoated."[184]

There have been some reservations voiced about Oberdörster's findings. According to Colvin, who distanced herself from its findings, "the surface characteristics of the lab's buckyballs, which are not a form that is commercially available, needed further study. She said that they had not been coated, a process that is commonly used to limit the toxicity of such materials in applications like drug delivery."[185] Oberdörster herself noted that her "study does not in any way support that a sci-fi horror thing is going on. Don't stop using nanotech, because there are so many important things it can offer."[186] In a telephone interview she described her study as justifying "a yellow light, not a red one."[187] She ends by noting, "in vivo studies with fullerenes and other manufactured nanomaterials will have to be carried out to determine whole-body tissue attributions, and potential corresponding effects, to determine safety."[188]

John Marburger, director of the White House Office of Science and Technology Policy, "did not believe [E. Oberdörster's] study would affect funding for nanotech research or harm nanotech's image among lawmakers, who are in the process of crafting annual appropriation bills that will provide funding for nanotech research."[189] Even so, when her findings received extensive news coverage and water cooler chatter during a major Infocast NNI Conference on March 31 to April 2, 2004, in Washington, DC, speaker after speaker expressed interest and concern. Much of the coffee break discussion centered on her fish study and its effects on commercialization.

More studies are needed, and some seem to be underway. For example, Colvin reported that "when her team exposed human fibroblast cells to nanocrystalline C_{60}, they found that the cell membranes were degraded and that the cells jacked up their production of glutathione, a small protein antioxidant that snuffs out free radicals."[190] This indicates the cells were heavily challenged.

As mentioned earlier, Feder of the *New York Times* reported "some companies making nanoparticles have conducted toxicology studies… [but] the extent of those studies is not known, and some results have not been disclosed, either for competitive reasons or because the costs of preparing the data for publication in scientific journals."[191] Feder offered the example of C Sixty Inc., a Houston company working on drugs and drug-delivery systems using buckyballs, whose rodent studies by Laura Dugan of Washington University indicated that "buckyballs collected in the kidneys and livers are excreted like other wastes," and "unreported data on its coated buckyballs in zebra fish embryos and adult rodents showed toxicity levels comparable or lower than many existing medicines."[192]

Toxicity research is not limited to carbon nanotubes. An additional consideration is directly associated with some of the professed diagnostic and drug-delivery implications of nanotechnology. Recently, researchers from the University of California, San Diego, found quantum dots with a core of cadmium selenide to be extremely toxic. Quantum dots are biological labels for long-term tracking of cells in vitro. According to Sangeeta Bhatia, "In this application, the toxicity of zinc sulphide–capped CdSe quantum dots is negligible—we showed that quantum dots remain intracellular and fluorescent even after a week in culture with sensitive primary cells, with no effects on viability or function.... *In vivo* application, long-term ultraviolet exposure resulted in high levels of cadmium-ion formation and cytotoxicity in hepatocytes [liver cells]."[193] While the researchers learned to reduce the cytotoxicity following exposure to air by adding two monolayers of zinc sulphide as a surface coating, UV light still induced toxicity.[194]

Overall, concern has been expressed with regard to populations having contact with nanoparticles. For example, "while scientists in South Africa handle nanoparticles as if they were dealing with the AIDS virus, other researchers, including some in Europe, wear only a 'Japanese subway mask' as protection. 'This is like wearing a volleyball net to keep out mosquitoes,' says Pat Mooney" of the ETC Group.[195] However, that is not true overall. In the United States, according to Cabot vice president of research and development David Bonner, "When we work with carbon nanotubes, we're handling them very conservatively using the same procedures when working with asbestos."[196] What level of protection this actually affords the workers remains uncertain.

All this research has led to a flurry of regulatory regimens. While the CRN advocates cautious vibrant development, Greenpeace, Caroline Lucas (UK member to the European Parliament), and others call for caution, even a modified moratorium, specifically on products applied to the skin.[197] They claim, for example, nanoparticles of titanium dioxide are an important ingredient in transparent sunscreens.[198] However, "tests show that they are highly reactive, generating chemically *hot* free radicals that can literally burn up bacteria."[199] We know that "knocking out microbes can both create serious environmental pollution and also impoverish the soil for many decades."[200] Nanoparticles of zinc oxide are used as well. And Nanophase shipped out over 450 tons of zinc oxide nanoparticles in 2002.[201] While "existing literature rules out direct absorption of micronized titania through the skin, these materials are active photocatalysts and can damage biological molecules,"[202] although sun exposure is probably worse. Nonetheless, Colvin and CBEN agree that "information relevant for particle sizes below 100 nanometers is not available."[203]

An estimated one hundred forty companies worldwide are producing nanoparticles.[204] Nanoparticles are being used in some new paints, certain lines of stain- and water-repellent clothing, a few kinds of antireflective and antifogging glass, and some tennis equipment.

The ETC Group, especially Hope Shand, demands a total moratorium based on concerns about the potential dangers of new forms of carbon discussed at the World Summit on Sustainable Development in Johannesburg (August 26–September 4, 2002): "We feel at this point there needs to be a mandatory moratorium on synthetic nanomaterials until we can see laboratory protocols to protect workers and protect the environment."[205]

A recent UK report made this recommendation: "The scarcity of published research into how nanoparticulates behave in air, water, soil and other environmental media make an assessment of environmental exposure pathways difficult. Nanoparticles and nanotubes that persist in the environment or bioaccumulate will present an increased risk and should be investigated. The release of nanoparticles to the environment should be minimalized until these uncertainties are reduced."[206]

RISK ANALYSIS

Vicki Colvin wrote, "It is far too premature to complete a formal risk assessment for engineered nanomaterials—in fact, it may never be possible with such a broad class of substances."[207] Her colleague Mark Wiesner agrees: "What's becoming obvious is while it's trendy to talk about nanomaterials in broad-brush terms, we can't do that. We need to discuss them on a case by base basis."[208] Belgian researcher P. H. M. Hoet agrees as well: "Engineered nanomaterials must not be considered as a uniform group of substances. Differences in size, shape, surface area, chemical composition and biopersistence require that the possible environmental and health impact be assessed for each type of nanomaterials in its own right."[209] These reservations, if validated, make a case for much more vibrant research programs on health and environmental implications.

Risk analysis must factor benefits as a counterbalance to hazards especially when recent studies suggest that coatings and hydroxylation may be able to mitigate hazard values. According to Rice's Kulinowski, "if we can control surface properties of nanoparticles, we may be able to tune out their toxicity."[210] On the other hand, "ultraviolet light from the sun might break off the hydroxyl groups, rendering the spheres toxic again, yet in the body, the coated buckyballs might remain intact and safely serve as drug-delivery vehicles."[211]

There is a significant body of speculative claims that have been recently validated by research in the remediative potential of some nanoproducts.[212] "Because of their size, they can move through microscopic flow channels in soil and rock, reaching and destroying groundwater pollutants that larger particles cannot."[213] For example, researchers at Oregon State have suggested one type of nanoized iron may be useful in clearing up carbon tetrachloride in groundwater.[214] "Government scientists at Argonne National Laboratory in Chicago documented how they had engineered nanoparticles capable of cleaning buildings exposed to radioactive materials."[215] A recent article went as far as to conclude: "If these technologies prove successful, they could reduce cleanup costs at selected Superfund sites by 75 percent."[216]

In addition, nanoproducts may be able to improve our capabilities to detect contamination and prevent contamination as a substitute for current manufacturing processes and products.[217]

Another concern is the speculation involved in hazard estimates:

> Traditionally, risk assessment begins when the source of a contaminant and its exposure pathways are well known. From this starting point a multitude of possible outcomes and their risks can be calculated. Clearly for nanotechnology this process must expand to include a wide range of *what-if* scenarios for possible products, nanomaterials, and exposure routes. All these factors will lead to more general risk assessments with less accurate risk projections.[218]

This entire issue becomes incredibly problematic given that so many of the concerns expressed are done so as maximin or mini-max fallacies. These are low-incidence/high-consequence calculi whereby incredibly small probability against hazards computes to catastrophe. This reasoning has maintained much of the rhetoric on the precautionary principle.

A second major challenge involves the incorporation of net benefits in a risk calculation when the hazards and benefits do not accrue to the same populations. The term *environmental racism* has been used to describe the location of waste sites in areas populated by certain racial categories. In contemporary risks studies we have begun to examine *environmental class-ism*. For example, while blue-collar workers may come in contact with raw nanoparticles, white-collar families may benefit the most from the products. It is important we do not lose sight of this consideration as more and more nanoproducts enter the marketplace.

A third major challenge is to decide on burdens of proof and the assignment of presumption in the debate on environmental health and safety of nanoparticles. The Royal Society and the Royal Academy of Engineering "recently concluded that factories and research laboratories should treat manufactured nanoparticles and nanotubes as if they were hazardous."[219] Nicholas Stine and Peter Lurie from the Public Citizen's Health Research Group concurred. This raises concerns over precaution and caution and the precautionary principle, which might have fore-closed the discovery of fire had it been followed millennia ago.

While this may not be the best setting to explore a final reservation, keep in mind that laypersons determine risks differently than experts. For example, intu-itive toxicology refers to the assignment of risk that involves biases that may exclude both probabilities and assessments of hazards quantified by empirical research. While Murdock of the NanoBusiness Alliance and others claim "with the right processes in place, we're going to be able to deal with all the risks, we're going to mitigate those risks, and we're going to realize the upside of the poten-tial,"[220] many of us remain unconvinced because we understand that the citizen, as layperson, reasons differently from the scientist when it comes to risk.

Oberdörster and colleagues have put together a fairly comprehensive review of studies in the field of nanotoxicology. In particular, I recommend page 34 of their report, which summarizes the research in a single page of readable text. They conclude: "We should strive for a sound balance between further develop-ment of nanotechnology and the necessary research to identify potential hazards in order to develop a scientifically defensible database for the purpose of risk assessment."[221] No one who is seriously studying societal and ethical implications of nanotechnology (unless they are shilling as marketing and public relations technicians) disagrees.

The toxicological studies that have been reported in recent years are a signal that the biological response to these materials needs to be considered. "That doesn't mean that we put a halt on nanotechnology," Rice's Joseph Hughes says. "Quite the opposite." While my colleagues and I are wary of the nationalistic rhetoric that has been articulated in recent business and government reports on nanotechnology, they do have a point. If nation-states with some record of envi-ronmental stewardship, as imperfect as that may be, elect to cease work on applied nanoscience, other nation-states that are less concerned with environmental health and safety will fill the void. The lesson learned from the stem-cell funding ban must not go unheeded. If the federal government persists in banning federal funding of new embryonic stem-cell lines, the intellectual nucleus for this tech-nology will be built in Seoul. A similar thing could happen with nanotechnology.

By using a coevolution model, research in EHS must be undertaken along research in applied nanoscience, or the technology will be built offshore. Further-more, as technological research proceeds, it will pull EHS research along and vice versa. Nanoscientific research will serve as a sample for study, and EHS lessons learned along the way may serve as a template for approaching technological study elsewhere.

"As information becomes available, we have to be ready to modify these regu-lations and best practices for safety," Hughes continues. "If we're doing comple-mentary studies that help to support this line of new materials and integrate those into human safety regulation, then the industry is going to be better off and the environment is going to be better off."[222] Regulation and liability regimes will develop alongside nanotechnology. However, at this time, regulation per se may be premature. As the Oberdörsters put it, "[G]overnmental regulation is not possible given the lack of needed information on which to base such regulations."[223]

ENVIRONMENTAL CONCERNS AND THEIR ETHICS

The absence of a regulatory framework and a woeful lack of technical data enable "both nanotechnology proponents and skeptics alike to make contradictory and sweeping conclusions about the safety of engineered nanoparticles."[224] These concerns have led to a call by government, industry, and academia to develop a societal ethics that will provide insight into SEIN. It remains incredibly unclear what this mandate would entail.

A universal ethical system that would apply to all technology may be wholly inapplicable to one as ubiquitous as nanotechnology. The ethical use of an automobile is much easier to discern; an automobile is easier to recognize than nanotechnology per se, hence individuals who are affected by its use would recognize its presence and make informed decisions on acceptable risks. That is much more difficult to do with nanotechnology, which can best be described as an enabler or enhancer technology. Simply put, applied nanoscience is joined to current product lines. While developments in areas such as computation and medicine will become possible only with nanotechnological developments, those applications are still years away. Nanobots building anything we could envision may be decades away, if ever arriving at all.

Even if an ethical algorithm for nanotechnology is impossible to design, attempting to build one has profound benefits. An important "aim of discussing ethical issues is to foster sensitivity among executives, research scientists, and other decision-makers to ethical issues."[225] Hence, social scientists, especially ethicists and communication academics, are trying to develop a way to work with government and industry to produce an informed consumer, a socially responsible business and commercial setting, and a model for regulation that would allow opportunities made by nanotechnology to benefit the most people while reducing unacceptable risk to a point where a decision can be made whether to produce or forego a nanoproduct.

CONCLUSION

SEIN research is powering a dialogue. Hopefully, findings will be vetted in open venues, and citizen-consumers will feel sufficiently empowered to participate. This challenge is massive since it involves a newly emerging technology, and as such there is no scientific and technological history to study except by analogy, and that is always troublesome. To make matters worse, ethicists and communicators are confronting negative straw persons created by watchdog groups. The dominant vision of nanotechnology has been constructed hyperbolically by individuals and groups who often base their proclamations on insufficient and inadequate data, producing fear rather than attempting to stimulate a critical dialogue.

In addition, it's not only the quack to worry about. The proponents and boosters have used hyperbole as well, overclaiming the near-term impact of nanotechnology, in terms of health, wealth, and security. To make matters even more complicated, when rebutting the exaggerations of detractors, proponents make the argument that nanoscience is really nothing new at all. It's just chemistry! The debate is weighed down by contradictory rhetoric that frames nanotechnology as both revolutionary and evolutionary.

SEIN researchers and scholars and toxicologists and ethicists will need to correct for the misinformation and release objectively verifiable information. They will need to redirect the debate to the present and near term and away from the speculative science fiction of the long term. Finally, they will need to learn that facts are not enough—citizen-consumers are influenced by arguments peripheral to the subject at hand—persuasion has always been part art and part science. Are they willing or able to that? The next chapter examines the commitments made by the federal government, business, and universities in SEIN studies.

CHAPTER 10

SOCIETAL AND ETHICAL IMPLICATIONS OF NANOTECHNOLOGY RESEARCH

"The bloom has been coming off the rose since *Silent Spring*," said Dr. John H. Gibbons, one of President Bill Clinton's science advisors. Until then, "people thought of science as a cornucopia of goodies. Now they have to choose between good and bad. The urgency," he said, "is to re-establish the fundamental position that science plays in helping devise uses of knowledge to resolve social ills."[1] A recent Harris poll found that "the percentage of Americans who believe scientists have *very great prestige* had declined 9 percentage points in the last quarter century, down to 57 percent from 66 percent."[2] "Some experts warn that if support for science falters and if the American public loses interest in it, such apathy may foster an age in which scientific elites ignore the public weal and global imperatives for their own narrow interests, producing something like a dictatorship of the lab coats."[3]

"Despite serious questions about the safety of nanotechnology when the science gets into the wrong hands, some policy makers are sold already. The Clinton administration was first to push for nanotechnology, and the Bush administration has asked Congress for a big boost in research spending."[4] According to one of the most recent evaluations of US government policy toward nanotechnology, there is an assumed relationship between marshalling resources, intellectual and others, in addressing negative impacts to technology and the level of benefits that may be reaped. "With any new and disruptive technology, and particularly one that has significant potential for extremely broad impact, there will be societal and ethical implications. Understanding these implications and ensuring that their consideration is integrated with the development of the technology is vital to achieving maximum societal benefit."[5]

While contextually footnoting recent concerns about nanoscience and nanoparticles of all sorts, the National Science and Technology Council (NSTC) universalized their claims to distance any particular reservations they may have about nanotechnology. George Whitesides, outspoken professor from Harvard, while rejecting Drexler's molecular machinery model and ridiculing fears over free-thinking nanobots, claims that "legitimate social issues do exist including privacy concerns, health risks from nanoparticles in the environment and alienation from technology by those who distrust it."[6] Unfortunately, "the barrier between scientific reality and science fiction is only as high as the imagination of a talented cartoonist. Pictures abound on the Internet of nanobots and other imaginary things."[7] JP Morgan's Alan Marty posted a similar concern on the focus of SEIN research: "We support the 21st Century Nanotechnology R&D Act's efforts to further address the social and environmental impacts of the science, but we would caution that this effort be focused on real science, not well-read science fiction."[8] SEIN researchers are equally concerned. Debating about low-incidence/high-consequence scenarios should be the subject of disaster B movies and not SEIN research.

The National Academy of Science (NAS), in its 2002 review of the NNI, recommended that the research on the social implications of nanotechnology be integrated into nanotechnology research and development programs in general. The Academy noted that rapid technological development will affect how we educate new scientists and engineers, how we prepare our workforce, and how we plan and manage research. Moreover, "accelerated nanotechnology developments could have broader social and economic consequences that may afford an opportunity to develop a greater understanding of how technical and social systems affect one another."[9] That's quite a mandate. While it does not necessarily follow that what we learn about nanotechnology might prepare us for subsequent confrontations with other forms of emerging technology, it may help as we prepare a workforce and establish best standards for research and commercialization.

The NAS panel noted,

> The increasing rate of innovation associated with nanotechnology developments has the potential to compress the time from discovery to full deployment, thereby shortening the time society has to adjust to these changes. The panel recommended that research on the potential societal and ethical concerns associated with nanotechnology, and research directed toward improving the understanding of how technical and social system affect each other, should be an integral part of any federal nanotechnology R&D program.[10]

As the length of time between discovery and commercialization shrinks, oversight as a methodology for checks and balances becomes less useful. Hindsight teaches us to avoid making the same mistakes, and if the mistakes are not fatal, then that's fine. However, there are so many serious SEIN concerns that the only alternative is foresight, and that demands speculation and a host of uncertainties that irritate scientists to no end.

If there is a window of opportunity for SEIN research, it won't stay open for very long, hence the recommendation for concurrent research.

Chairman Sherwood Boehlert, who held hearings on societal and ethical implications, agreed with the report: "Nanotechnology will be neither the unallied boom predicted by technophiles nor the unmitigated disaster portrayed by technophobes. The truth will be in between.... Our record, when it comes to technology, is not very good, and how can we expect it to be? The social consequences of technology, the most subtle and far-reaching impacts, are the most difficult to predict and even more difficult to forestall. But that is not a reason to do nothing."[11] The process itself is meritorious.

The undertext to SEIN research, at least as conceptualized by the US government, is based on the fear that misinformation or the absence of any information on SEIN will threaten the momentum of nanotechnology, as an industry, a national initiative, and a movement. "The NNI implementation plan recognizes that the societal implications of nanotechnology must be taken into account so as to ensure that technical advances will be adopted."[12] SEIN researchers are viewed as enablers. "This effort will go a long way in limiting the effectiveness of groups that seek to unfairly portray nanotechnology R&D as too dangerous to press forward with. These organizations attempt to create fear and paranoia by blurring the lines between legitimate societal risks and imaginary science fiction."[13]

This warning by the NSTC has been teamed with a major commitment to societal and ethical implications concerns associated with nanotechnology. The NNI mentions it.[14] The NSF has awards and grants to study societal and ethical impacts. Other agencies, including the EPA, the DOE, and the DOD, have done so as well. The nodes in the NNIN include a similar requirement. The Twenty-First Century Nanotechnology R&D Act[15] calls for a Center for Nanotechnology in Society to "evaluate workforce and ethical issues."[16] Presumably, an NSEC will be established purportedly to coordinate national SEIN research.

These warnings are not restricted to the United States. The Office of Technology Assessment at the German Parliament (TAB) came to the same conclusions:

> The state of research into potential environmental and health impacts of the production and use of nanotechnological processes and products is unsatisfactory. Considerably greater research efforts are urgently needed here, as the lack of knowledge of the environmental and health consequences could create barriers to the market launch of nanotechnologies. Research into societal and ethical aspects of the development and widespread use of nanotechnology should be initiated now.... Comprehensive information for the general public is a prerequisite of a rational societal debate on nanotechnology. Something to aim for would be a central information point for the general public on the topic of nano-

technology. Access could be had here to the information provided by the individual competence centres and other information portals at national, European and extra-European level.[17]

This new interest in ethics could be due to increasing concerns expressed by individuals and groups, often mediated from both scientific and unscientific circles, amplified by a *new* transnational generation of protestors who have targeted globalization, genetically engineered and modified foods, and the US attack on Iraq. Some policy makers fear nanotechnology might be the next locus for their angst. These miscreants see nanotechnology as another extension of dehumanizing technopoly into their lives, which disregards nature and simpler ways of living and is aesthetically problematic.

Also, it could be due to the mostly favorable publicity received by the biotechnology industry when they incorporated similar considerations (Ethical, Legal, and Social Issues in Science [ELSI])—a "fruitful STS inquiry in terms of the many social/ethical issues as a result of the Human Genome Project" and an analogous model would seem prudent for nanotechnology.[18] Paul Thompson, an ethicist from Michigan State, believes, "Nanotech researchers should build on the example of the ELSI project." He suggests we "take a hard look at potential ethical and cultural issues, but follow through much more carefully and get ahead of the public."[19]

Julia Moore was a scholar with the Woodrow Wilson Center and returned there after a brief stint at the NSF. She has studied biotechnology policy making for many years. Her summary follows:

> Whether nanotechnology research results in the ultimate doomsday machine or in mankind's salvation is up to us. In the mid-80s, there was considerable opposition to the now-celebrated Human Genome Project. The increased availability of genetic information raised difficult questions about how the information is used by insurance companies, law enforcement agencies, schools and employers. The founders of the Human Genome Project acknowledged that they did not have answers to these significant societal questions. So they set aside 5 percent of the project's annual budget for a program to define and deal with the ethical, legal and social implications raised by this brave new world of genetics—creating one of the largest such efforts ever.[20]

Vicki Colvin of Rice agreed with Thompson and Moore in her 2003 testimony to Congress: "Mapping of the human genome carries with it many of the same potential concerns as do other fields of genetic research. The founders of the Human Genome Project did not try to bury these legitimate concerns by limiting public discourse to the benefits of new knowledge. Instead, they wisely welcomed and actively encouraged the debate from the outset."[21] Maybe we can learn something from this object lesson.

There are parallel and novel concerns about applied nanoscience, especially in terms of both occupational and environmental exposure to manufactured nanoparticles such as single-walled carbon nanotubes. The NAS seemed to have agreed when it recommended SEIN research. "Some technologists, such as those in the nuclear power and genetically modified food industries, have ignored these kinds of challenges and suffered the consequences. Others, most notably those in the molecular biology community, have attempted to address the issues and to use their understanding to stimulate an informed and objective dialogue about the choices that can be made and the directions taken."[22]

Industry is on board for many reasons. "The US VC firm Draper Fisher Jurvetson has said it would not invest in a nanotech business unless the products had already been proven safe."[23] Beyond VC support, "if the industry fails to persuade the US public that nanotechnology is safe," Daniel Ritter of Preston Gates Ellis & Rouvelas Meeds LLP warned, "the US could risk losing its leadership position."[24] L. Val Giddings from the Biotechnology Industry Organization reminded an audience gathered at an annual NbA meeting that "the moral high ground is presumed to be in the hands of the critics."[25] If that is true, then there is little time available to strategize how to respond to critics who are misguided on how to accommodate the interests of those who are not.

On a much more cynical note, SEIN may simply be perception management by legislators, regulators, and civil servants to distance themselves from culpability should anything disastrous ensue. This cynical, even skeptical, view in our minds is a healthy one. What reservations we express along the way might release critics from apologizing later. The distance between advocacy and critical studies is not sufficiently great to preclude co-option. The scenario of government using experts to serve as public relations spokespersons is not that unlikely. The following traces the interests and speculates on motives suggesting cynicism as the modus operandi for researchers in this new field.

Nonetheless, a special panel of the National Academy of Sciences, the National Academy of Engineering, and the Institute of Medicine chaired by Samuel Stupp of Northwestern University wrote *Small Wonders, Endless Frontiers*,[26] in which they reviewed the status of a set of recommendations of the NNI in 2002. Their ninth recommendation to the National Science Technology Council's Subcommittee on Nanoscale Science, Engineering and Technology (NSET) was to require the inclusion of SEIN research in funding proposals under the NNI.

The panel recommended support specifically tailored toward SEIN research and a SEIN component within other grants, much like the requirements for minority participation and educational outreach. The rationale seems to be predicated on the apprehension that commercialization of nanoscience might provoke a response similar in some ways to that experienced by American biotechnology companies when they marketed genetically modified seeds and food products abroad.

DIAGNOSIS

The diagnosis is public participation. Uniformly, the European reaction seems predicated on two elements. First, genetically modified organisms (GMOs) use biotechnology, and the root of the word is technology—*technology* is the problem. "None of the supposed problems raised in the biotech debate were either substantial or specific to biotechnology. Rather, they expressed more a disenchantment with modern life. Instead of recognizing and dealing with these wider concerns, too many supporters of innovation took the criticisms seriously as *ethical problems*."[27] Much like the ETC Group's criticism, the cause was corporatism, not necessarily biotechnology. "As a consequence, scientists were distracted into researching and debating mythical risks, and there is an elevation of the absurd concerns of green campaigners who present themselves as speaking on behalf of the people. Instead of phony involvement in the fears about new technologies, we need a serious debate about how best to take science forward."[28] These stakeholders are unlikely to be persuaded to change their attitudes. The second reason was that the public was closed out of the decision-making process, so the risks were involuntary. Especially in the United Kingdom, observers see a strong parallel between "governments taking up the commercial bonanza that nanotechnology promises while ignoring the discussion of the risks or importance of public participation. That's almost exactly what they did with the GMO debate."[29]

Others offer more poignant assessments: "The real lesson to be learned from the GMO debate is that the demand for public involvement is the most dangerous element."[30] For that to happen, the debate will need to be focused and on point. Flooding citizen-consumers with academic treatises on Kant and Von Neumann and bizarre extrapolations that address science fiction plotlines will be totally unproductive.

Others have been much more pessimistic: "The biotech industry learned the hard way that ignoring the ethical considerations of genetically modified foods, stem cells, and gene therapy cut profits, hampered investment, and slowed innovation. Odds are the same [that this] will happen to nanotech, despite some efforts to head off the backlash."[31] However, the window of opportunity remains open. But these SEIN researchers must act quickly; otherwise their work will be mostly historical criticism.

DEFINING SEIN

One of the goals of ELSI was to spur a debate about genomics outside of the laboratory and courtroom. ELSI researchers did not tell scientists how to do science, and as long as humanities and social science scholars worked to understand the science, their contributions would be supported.[32] If one were searching for a "father of SEIN," it might be Professor James Watson, of double-helix fame, who helped champion the inclusion of ELSI funding in genomics budgets.

SEIN encapsulates research by toxicologists, ethicists, and futurists into a range of issues from environmental impacts, regulatory regimes, workplace and economic dislocations, bionanotechnology convergence, and transhumanism and posthumanism. It involves experts from philosophy, communication studies, law, and political science, as well as fiction studies and art. It also includes the less expert as well as the self-proclaimed technophile, the social critic, and the crank.

It involves an overabundance of issues, for nanotechnology can "threaten social structure, economic stability, and spiritual beliefs and values."[33] Others, like Representative Eddie Johnson (D-TX), go even further. Nanotechnology "holds the potential to change *everything*."[34] On the other hand, it may be prudent to temper this drama. Ray Kurzweil provides a useful example about what can happen with informed planning:

> As a test case, let me bring up an example. We can take a small measure of comfort from how we have dealt with one recent technological challenge. There exists today a new form of fully non-biological, self-replicating entity that didn't exist just a few decades ago, the computer or software virus. When this form of destructive intruder first appeared, strong concerns were voiced that as they became more sophisticated, software pathogens had the potential to destroy the computer network medium they live in yet the immune system that has evolved in response to this challenge has been largely effective. No one would suggest we do away with the Internet because of software viruses. Our response has been effective and successful, although there remain, and always will remain a concern, the danger remains at a nuisance level. Keep in mind, this success is in an industry in which there is no regulation, no certification for practitioners.[35]

Kurzweil offers a splendid illustration. The knee-jerk response of a moratorium on research not only would be as effective as stopping the flow of water with a sieve but also does not address the greater concern that as citizens in a technological world, it may be much too late to return to an earlier state of development when technology wasn't so important. We need to determine how to live with our technologies the best that we are able.

Mihail Roco and others advocate the inclusion of many different voices and stakeholders:

> The inclusion of social scientists and humanistic scholars, such as philosophers of ethics, in the social process of setting visions for nanotechnology is an important step for the NNI. They are professionally trained representatives of the public interest and capable of functioning as communicators between nanotech-

nologists and the public or government officials. Their input may help maximize the societal benefits of the technology while reducing the possibility of debilitating public controversies.[36]

Roco seems to perceive SEIN as a college of academics who can mollify the negative consequences of outrage. Outrage is an important variable in algorithms built to communicate risk to the public. Since the process of communication becomes more problematic and expensive as outrage increases, dampening the outrage might be a very wise strategy. Unsurprisingly, America's primary proponent, Roco, finds in SEIN research an opportunity to control the intensity of the outrage US nanotechnology policy might encounter: "Research on societal implications will boost the NNI's success and help us to take advantage of the new technology sooner, better, and with greater confidence."[37]

In an April 2003 NSTC publication, we uncover a second purpose for SEIN research: "Research on implications for human health, society, and the environment is increasingly being emphasized as tangible new nanostructures and nanomaterials are discovered and new nanotechnology products are developed. The results of such research are being taken into consideration by those Federal agencies whose work is directed at regulatory issues."[38] As such, SEIN research may play some role in the regulatory process. Considering the ominous challenges confronting regulatory agencies like the EPA, FDA, and OSHA under their current budgetary restrictions and the lack of a clear mandate appropriate to most nanoproducts, it would seem SEIN research might actually help. This would allay some fears expressed by government officials that "anti-globalization advocates may cause fear in consumers or get ill-informed legislators to pass senseless regulation."[39] Who would lose out if an anti-nanotechnology movement of sorts arose? Roco would claim consumers would lose out if nanoproducts failed to materialize because nanotechnology was set back. He might also regret the negative impact on the US economy and its global leadership.

Josh Wolfe addresses another group of potential losers: "That means nano-materials producers—like Nanophase and Altair or carbon nanotubes CNI—lose. And lose big. Even larger targets like General Electric, IBM, and DuPont are in environmental crosshairs."[40]

On the other hand, all this SEIN research might serve as perception management to debunk negative claims rather than to engage regulators in a dialogue that may actually make the introduction of some products more challenging. This masking function of SEIN research perturbs the social-science community and threatens to discourage participation by objective researchers while attracting perception-management professionals, issue wranglers, and spin PhDs.

Academics, accustomed to scholarly freedom, are watching SEIN research with a lot of skepticism. Too many times in the past, they have found themselves intimate with the commercial and industrial communities, in the process losing their impartiality and credibility. Before we commit time and interest, the ethics of handing out money for ethical-studies programs needs to be vetted.

SEIN researchers are very concerned. They do not want to function as the public relations division for commercial nanotechnology. They do not want to be apologists for government agencies. And they do not want to be window dressing. They want to be taken seriously and have input into science and technology decisions and policy making.

There is sufficient pessimism that SEIN research will do little but pave the way for a nanotechnological future. According to Langdon Winner, "I fear that the manner in which the work is done will reproduce the backwards logic that has shaped far too much of American technological development in recent decades. It is a logic that justifies the creation of a wide range of flashy new gadgets but cannot be bothered to examine the most urgent facts about the human condition in our time."[41]

The rhetoric of SEIN must be associated with an equally compelling effort to provide access to decision making. This is the first rub because there is no public sphere for science and technology policy making (see chapter 11). As such, the deduction to be made is simply this—if the public hears a more balanced debate on nanotechnology, they are less likely to withdraw support for it, whether governmental or commercial. If this is true, then SEIN becomes an important component of the NNI.

Consider the call by Prince Charles and the ETC Group for a comprehensive moratorium on applied nanoscience, especially the fabrication and application of single-walled carbon nanotubes and some other nanoparticles. P. Singer and his colleagues noted, for example, the call for a moratorium coupled with increased expenditures in research and technology are on a collision course ripe for activist groups to exploit the gap between the science and the ethics of nanotechnology.

Unsurprisingly and unfortunately, "there is a paucity of serious published research into the ethical, legal, and social implications of nanotechnology. As the science leaps ahead, the ethics lags behind. There is danger of derailing nanotechnology if the study of ethical, legal, and social implications does not catch up with the speed of scientific development."[42] While there are dozens of researchers who have begun to think about nanotechnology, they remain paddling to catch the wave about to crest beyond their reach. As a compatriot of mine put it, "at this pace we may have little to offer when time comes to assess nanotechnology critically."

After reviewing citations in scientific databases from 1985 to 2001 on ethics and social implications of nanotechnology, Singer's group in Toronto filed this complaint: "While the number of publications on nanotechnology per se has increased dramatically in recent years, there is very little concomitant increase in publications on the subject of ethical and social implications to be found in the science, technology, and social science literature."[43] STS research has been going on for a few decades or more, so why hasn't nanotechnology found itself the subject of scrutiny?

There have been conferences upon conferences, a handful of them exclusively about SEIN issues. So where is the research? Where are all the findings?

SEIN literature is hardly sufficient to fuel a dialogue between research institutes, granting bodies, and the public on the implications and direction of nanotechnology. Singer fears this gap between scientific and SEIN research "may have devastating consequences including public fear and rejection of nanotechnology without adequate study of its ethical and social implications."[44]

Who Benefits from SEIN Research?

First of all, the government has no serious rationale to hawk a dangerous technology and deny research that draws that conclusion. While it is challenging to debunk findings from studies that cast doubt on the safety of a product or series of products or applications spun off a government initiative, like the NNI, discouraging or repressing research and its findings is immensely counterproductive. The Abu Ghraib prisoner abuse scandal made that case very clear. A sniff of conspiracy by governments becomes grist for media outlets from the *Washington Post* to FOX-TV News. As such, NNI officials like Clayton Teague of the NNCO continue to reiterate, "No one is going to take these studies lightly. They will be taken as important new pieces of information as we proceed to develop appropriate regulatory mechanisms ensuring that the technology is moved forward in a responsible way and health and environmental impacts are given high priority."[45]

SEIN research provides a second way to focus the debate. "If the research community engages quickly to infuse technical data into this debate, the actual risks of engineered nanomaterials will become better defined. Such data will also provide the means to minimize environmental consequences well before a nanotechnology industry is established, leading to more successful and profitable technologies."[46] While Vicki Colvin's argument seems to justify increasing investment into toxicological research, much like what is done at Rice's CBEN, she makes a powerful claim for a broader SEIN agenda.

Third, business and industry benefit as well "in part because of the financial consequences of lawsuits and widespread negative publicity."[47] In 2002, Colvin warned that the

> low investment in environmental nanotechnology research [roughly $500,000 of its $700 million nano research budget[48]] also creates problems for the fledgling nanotechnology industry. The legal and regulatory landscape plays a crucial role in defining the pace for any new technology. Companies investing in the nanotechnology sector need to handicap what this landscape will look like for nanotechnology and plan their business strategies appropriately. Firm data on health impact and a quantitative and general risk assessment model is thus a necessity for commercialization.[49]

On the other hand, bad data can damage commercialization. While poorly researched data can be damaging, well-researched negative data may allow a company to tweak a product, process, or production line to minimize adverse implications. Concerns are real and already are having an impact on the industry. How bad it is for business to market dangerous products in the twenty-first century? According to Kevin Ausman of Rice's CBEN, "It's bad for business if they create a product with liability issues."[50] CNI cofounder Dan Colbert says, "We don't intend to put a product out in the market place until we've determined it's safe."[51] Josh Wolfe put it bluntly: "Nanotube companies risk losing their shirts if nanotubes prove toxic, especially since they're not diversified."[52]

In the short term, it is incontrovertible that recent studies in nanotoxicology are already affecting investment. Steve Jurvetson has voiced some concern over the recent release of nanotoxicology experimental results. A venture capital firm that has stakes in nine nanotechnology companies steers clear of any that raise environmental questions. As Barnaby Feder says, "Until other people's money and research have proven it safe, we'll assume it isn't."[53]

Start-ups are responding as well because they simply cannot handle the risk. "Some smaller nanotechnology start-ups say they simply do not have the resources to push into promising areas that pose health questions."[54] For example, Argonide Nanomaterials of Sanford, Florida, said, "it preferred to focus on filtering products, which are not implanted in the body and thus require much less testing."[55] As such, the research and commercialization agenda of business and industry is already responding to the recent spate of toxicology studies on nanoparticles.

Finally, investors will benefit from more information as well: "Rather than alarm investors, such information (whatever its conclusions) will be reassuring and increase the likelihood that viable nanotechnology products are developed. Perhaps, more important, hard data on the environmental effects of nanomaterials will also go a long way in building the public's trust."[56] With benefits exceeding costs, it remains surprising that toxicology studies on nanoparticles remain generally underfunded.

The Warning

It is important to observe that the immediate reaction to the question, Why societal and ethical research? is mostly, We can't afford another biotechnology disaster! For example, Former Representative Nick Smith (R-MI) provided a typical diatribe:

> Scaremongering tactics of widespread misinformation can be very effective, and in fact often help raise significant amounts of money for the organization, with which they use to attack the science further. This same strategy has been successful in damaging the reputation of biotechnology—delaying research, devel-

opment, and adoption of several safe and beneficial products.... It is important that safe and beneficial nanotechnology innovations do not suffer the problems of emotion and delay that hindered biotechnology applications before them.[57]

This unease is shared by experts in the field as well. For example, outgoing NSF director Rita Colwell warned, "We can't risk making the mistakes that were made with the introduction of biotechnology. It's much too important."[58] Ted Agres, a writer for the *Scientist*, summed up the attitudes of over a hundred participants at an NSF workshop in December 2003: "Scientists, engineers, and government officials must confront head-on the ethical and societal implications of nanoscience and nanotechnology in order to keep the field from falling victim to the obstacles that hampered progress in biotechnology."[59] The assumption is that misinformation leads to a sufficient amount of dread to provoke an irrational response. NanoSig's Bo Varga believes that "if enough misinformation spreads, there's a tremendous potential to generate a social backlash."[60] Actually, the University of Toronto's Joint Centre for Bioethics predicts just such a backlash.[61]

Toxicologists have joined the outcry, but while they seem unwilling to predict a backlash per se, they do forewarn that misinformed consumers might make the commercialization very problematic. Rice's Colvin agrees: "You can make the best technology in the world, but if the public for some reason doesn't buy it, or has misinformed opinions about its consequences, it can tank your technology."[62] One of her colleagues, Emmanuelle Schuler, concurred: "Perception can quickly overtake reality in the court of public opinion, and dominate public acceptance for years to come even when data suggest that the fear was overblown."[63] As Lux Research warned, "stakeholders in the commercialization of nanotechnology have massively undercommunicated on EHS issues to date and risk losing the battle for mindshare by default."[64]

This complaint is amplified not only when media report these concerns but also when taken up by NGOs with an axe to grind. For example, Doug Parr of Greenpeace links nanotechnology to the GMO controversy: "Depending on the development pathway, some aspect of nanotechnology might get a rocky ride, as its social constitution is like that of GMO crops."[65] That linkage carries with it some of the protestations and outright bans the GMO industry experienced in Europe and Asia.

Some public opinion polling has also noted a similar relationship. After running correlation coefficients on the results from *Survey 2001*,[66] W. S. Bainbridge reported that "respondents associated nanotechnology with genetic engineering, and public controversies about genetically modified foods may have serious negative implications for public attitudes toward nanotechnology."[67] While the conclusion is not the necessary consequent of the antecedent, Bainbridge, who has a distinguished relationship with the NSF, does command an important and powerful audience.

This is not to suggest that the concern is universally held. Indeed, at the December 2003 meeting of the NSF, Roco was especially vocal against the quality of the analogy. While Roco might be attempting to neutralize concerns, his cohorts seem to be heeding other voices. Responding to pressure, especially from Congress, SEIN continues to be written into authorization bills dealing with nanotechnology. There are rationales independent of this analogy about why SEIN research has become popularized. Nevertheless, the analogy is one of the most prevalent and important instances of hyperbole in the nanotechnology debate.

Teasing Motives

Why has the NSF among others become so interested in promoting SEIN research as part of both the NNI and NNIN?[68] The NNI at its inception acknowledged "the need to integrate societal studies and dialogues concerning the perceived dangers of nanotechnology with its investment strategy."[69] Indeed, the press release for the NNI expressed support for five kinds of activities, one of which is Ethical, Legal, Societal Implications and Workforce Education and Training. It reads: "Efforts will be undertaken to promote a new generation of skilled workers in the multidisciplinary perspectives necessary for rapid progress in nanotechnology." Then, it adds: "The impact nanotechnology has on society from legal, ethical, social, economic, and workplace preparation perspectives will be studied. The research will help us identify potential problems and teach us how to intervene efficiently in the future on measures that may need to be taken."[70]

Please notice a few things. The primary directive to the activity is "to promote a new generation of skilled workers." Even when the activity begins to address societal and ethical concerns, the sentence ends with "workplace preparation perspectives will be studied." When the research is applied, it seems to be "to intervene efficiently ... on measures that need to be taken." It doesn't take a linguist to read the subtext: nanotechnology will improve the US economy, and we must find a way to integrate it smoothly into the current economic infrastructure. This cynical view is becoming more and more popular. The Committee for the Review of National Nanotechnology Initiative warned, "Some technologists, such as those in the nuclear power and genetically modified foods industries, have ignored these kinds of challenges and suffered the consequences."[71] The report added that we currently do not have a comprehensive and well-established knowledge base on how social and technical systems affect each other in general, let alone for the specific case of nanoscale science and technology.[72] The Committee for the Review of the NNI asserts since "nanoscale science and technology are still in their infancy, a relatively small investment now in examining societal implications has the potential of a big payoff."[73]

To date, SEIN research has not been substantial. Little has been published because SEIN research has been undertaken sparingly and grant support has been underwhelming. The Joint Center for Bioethics filed this reservation: "While there are significant research funds available in the US, these funds are not being used. In 2001, the US NNI allocated $16–28 million to societal implications, but spent less than half that amount. The NSF, responsible for spending $8 million did not fund a single social science project focused on societal implications of nanotechnology."[74] When the NAS reviewed the NNI, they laid some of the blame on NSET. "NSET has not given sufficient consideration to the societal impact and developments in nanoscale science and engineering. Agencies *willing* to engage in assessing societal implication must be given a budgetary incentive to do so. The committee believes that NSET should develop a funding strategy that treats societal implications as a supplement or set-aside to agency core budget requests."[75] They footnoted: "Such a funding strategy is not new. For example, most federal agencies resisted involvement in the SBIR (Small Business Innovation Research) until Congress required agencies to set aside a certain percentage of their budget for the program."[76]

While reviewers seem genuinely concerned with the consequences of nanotechnology, a review of the evaluations of FY 2001 funded programs made two telling observations. First, "the funding committed to societal implications for FY 2001 *appears to range between $16 million and $28 million.*" To explain the range, it reported the documents mixed requests with actual expenditures, and reports of expenditures differed from source to source and sometimes did not reconcile within a source. (No one seems able to explain what this means.) Second, the report impugned the rhetoric of government agencies claiming concern and interests in societal implications:

> Two agencies, NIST and NIH reported no activity or expenditures in the area. DOD indicated that it had made 46 awards under a nano-focused fellowship program within the Defense University Research Initiative on Nanotechnology. The NSF reportedly committed between roughly $8 million and $20 million in budgeted support and expenses for NNI societal implications activities during FY 2001. NSF appears to be the only agency to have engaged in major efforts to study societal implications during 2001.[77]

When Christine Peterson testified to Congress, she added that in 2001 "few proposals were submitted and none were funded."[78] Anecdotally, a group associated with me received a partial grant in 2002 from the NSF funded at less than 10 percent of its request. When queried, the response was that once the technical grants were handed out, there wasn't money left for societal implications. The grant award was hobbled together from unspent funds.

When the NAS reviewed this issue, they offered these observations:

There are a number of reasons both funding strategies failed (i.e. competing with science funding and integrated within science and engineering centers). First, given the differences in goals, knowledge bases, and methodologies, it was probably very difficult for social science group and individual proposals to compete with nanotechnology science and engineering proposals.... Second, studies of societal implications were only one of six optional activities (including international collaboration, shared experimental facilities, systems-level focus, proof-of-concept test beds, and connection to design and development activities) that individual proposals could include.... Finally, NSF's review committees and site visit teams did not include social scientists.[79]

The NAS warned: "Unless things change dramatically during FY 2002, the social implications theme will be simply a fancy title for a relatively straightforward educational initiatives targeted at graduate and undergraduate students."[80] While my colleagues and I were pleased to receive some funding, the case for a *major effort* did not surface until funding in FY 2002. On some level, the substantial increase in SEIN awards evidences a commitment sufficient to associate funds with rhetoric, at least in terms of development. Whether driven by personality or institutional commitment, it seems some of the recommendations by the NAS were heeded.

The NSF estimated the increased investment in societal and ethical, educational, and environmental implications of nanotechnology. In FY 2002, the amount dedicated to societal and educational implications was approximately $30 million. In FY 2003 it would rise to $35 million and to $40 million in FY 2004. Research and development support associated with environmental implications would move from $50 million in FY 2002 to $60 million over the next two years.[81] While the amounts may seem considerable, the first set of data conflated both SEIN and educational outreach. A more detailed analysis reveals some overclaims: "The $4 million the EPA expects to award in 2004 for risk studies is barely measurable against the $847 million in federal money that President Bush proposed for nanotechnology research and development for the 2004 FY."[82]

SEIN AS SYMBOLOGY

"Nanotechnology has a unique opportunity in the history of technology: this could be the first platform technology that introduces a culture of social sensitivity and environmental awareness early in the lifecycle of technology development."[83] However, that assumes the research is done well and policy makers attend to the findings. And it also assumes they are interested in public input. For example, there has been a wave of articles on the toxicity of nanoparticles (see chapter 9). Fears are this could snowball into the kind of publicity that leads to regulation. Josh Wolfe put it best:

Anytime I hear things like wide societal debate in the same context as government decisions (read: regulations) I begin to shudder. The last thing nanotechnology needs is a societal debate and intense government scrutiny. How can you intelligently discuss and regulate something that is still in the discovery and development stage, before it really exists in a practical manufacturing sense? In the case of fullerenes, an overzealous regulator could potentially have used the fish brain damage finding to shoot down buckyball production.[84]

Is SEIN an authentic concern, or is it only window dressing and a public relations effort? The skeptic would argue that it is symbolic. Congresspersons funding social and ethical research are responding to constituent concerns. Others among us would add the corollary that educational outreach exists to belay criticism and to establish an educational workforce able to extend the nanotechnological revolution. Administrators at the NSF may be responding to pressure from politicians who fear nanotechnology might go the way of genetically modified organisms and, especially, foods.[85]

By advancing an agenda to address overclaims and misconceptions, SEIN research might debunk the deceivers and the Neo-Luddites. However, there is sufficient anecdotal evidence to suggest that cynicism is legitimate. For example, some scientific conferences are being used as public relations tools today. Recently, B. McKibben described a germline conference at UCLA in 1998 as cheerleading.[86] He quoted Dr. F. Seitz's comments on a Lounsbery Foundation–sponsored exhibit on genomics at the American Museum of Nature History in New York City as "designed to get people to run the corner."[87]

Along a different vein of concern, "there is also a tendency for those who conduct research about the ethical dimensions of emerging technology to gravitate toward the more comfortable, even trivial questions involved, avoiding issues that might become a focus of conflict. The professional field of bioethics, for example, (which might become, alas, a model for nanoethics) has a great deal to say about many fascinating things, but people in this profession rarely say *no*."[88] Langdon Winner, political scientist from Rensselaer Polytechnic Institute, continued: "Indeed, there is a tendency for career-conscious social scientists and humanists to become a little too cozy with researchers in science and engineering, telling them exactly what they want to hear (or what scholars think the scientists want to hear). Evidence of this trait appears in what are often trivial exercises in which potentially momentous social upheavals are greeted with arcane, highly scholastic rationalizations."[89]

These events and predictions warn us that perception management might be the actual purpose behind worthy prospects, including the SEIN hoopla. On the other hand, others are much less cynical. According to Dr. Mae-Wan Ho, "in contrast to the debate on genetic engineering, where misinformation, denial and obfuscation abound, scientists in this new area are informing the public with admirable clarity and candor, especially in separating hype from reality and in anticipating some of the risks involved."[90] Hopefully, Dr. Ho is correct.

ONGOING RESEARCH

Is there time for SEIN research to have an impact? Ridiculing claims by Colvin and NIOSH's Andrew Maynard that we are at an optimal time to study these problems, the ETC Group disagrees. "In reality, regulators missed both the train and the birth. It's time to acknowledge that nano's life cycle is nearing a mid-life crisis and that environmental, safety and social concerns demand immediate action."[91] Beyond the mixed metaphors, there seems to be time, especially when we can learn from models already in place. For example, workplace safety is under the auspices of a regulatory agency with ongoing research undertaken by a national institute. Standard setting issues have produced a cry for more nanotoxicity studies and increased funding. We are not starting from a standstill but from a healthy power walk.

Fears mount that the naysayers may have the lead in the PR blitz on nanotechnology. Bureaucrats are concerned that the young nanotechnology industry may experience the same reaction that the GMO industry did when it marketed its seeds and products abroad. Still others are anxious that Michael Crichton's book, once released as a movie, will trigger a nightmare of public backlash, and through all of this we hear that "it's not too late to ensure that nanotechnology develops responsibly and with strong public support."[92] "We are at an optimal time to study these problems. We are at the birth of a new market. We can shape this area with knowledge as it develops."[93]

Assuming for a moment that nanotechnology can develop responsibly, the assumption that strong public support can be prompted may be warrantless. Singer, like Roco and company, has a tendency to see SEIN as a safety net of sorts. If erected in a timely fashion, SEIN might catch the nanotechnology mandate should it become wedged off its pedestal by nanomiscreants. As Winner explained to Congress, "late in the process, it does little good to tell those who are unwilling that they are simply being irrational."[94] It is simply too late to tell opponents that they are wrong. The GMO industry discovered the post hoc problem painfully. "The industries that were involved in it really didn't think it was important for them to sort of articulate what they were doing to the general public. Then what happens is there is this backlash, and now you have to sort of go back, deal with the backlash, and then deal with what is people's fundamental misconceptions about what is going on."[95] According to Colvin, "at present, nanotechnology has a very high *wow* index. However, every new technology brings with it a set of societal and ethical concerns that can rapidly turn *wow* into *yuck*." Her prescription: "early research into unintended consequences redirects the *wow-to-yuck* trajectory."[96] Research may uncover toxicological and other health and environmental issues, but translating findings into memes that the public will both understand and willingly defend in debates over nanotechnology is a long stretch. As such, the NSF and other government bodies have provided grants to selected American universities to do an assortment of research projects hoping to affect the *wow-to-yuck* trajectory. "Unfortunately, due in part to unrealistic scenarios like the one in *Prey*, nanotechnology's *yuck* index is rising as people take as fact the fiction of *invisible nanorobots*."[97] To counter the *yuck*, some universities have embraced SEIN research with a vengeance.

AMERICAN UNIVERSITIES

There are two very different ways to examine SEIN issues. Anecdotally, they seem to reflect the duality of thinking about SEIN research. Their difference seems to reveal a secondary philosophy toward it: SEIN is a new discipline that requires fundamental research on some levels and metaresearch on others by groups of humanists and social scientists who are situated within or juxtaposed to nanoscience research teams.

The drawbacks to this bipolar approach include (1) a tendency toward bootstrapping rather than stand-alone SEIN centers and (2) favoring universities that can offer nanoscience studies. The first drawback should have been mitigated by the Center for Nanotechnology in Society. "One of the prime reasons behind the center's creation, observers say, is to try to avoid some of the public skepticism that surrounded the debate over biotech advances such as genetically modified foods, while at the same time dispelling some of the misconceptions the public may already have about nanotechnology."[98] On the other hand, if the center exists merely to facilitate efforts at commercialization, then it will fall short of the stated goals. While the center was not found in President Bush's 2005 budget, it was treated as an NSEC in 2004, though it may not resemble one once the grant money is distributed. The second might be partially resolved by the authorized NNIN, which should allow scientists and engineers, regardless of their physical location, to undertake nanoscience research at one of the nodes of the network. However, this is only speculation, and healthy skepticism remains in order on the part of SEIN researchers and the public.

There have been many NSF awards on societal implications, according to Roco. While meritorious, some seem less concerned with fundamental societal and ethical concerns and are only loosely categorized as SEIN projects. The overinclusion hints at some concern by Roco to convince others that he is on board. In

2001 a career award was given to R. Berne at Virginia to study ethics and belief relating to nanotech, and in the same year, M. L. Lynch from Cornell was given a dissertation research award to research the genesis and practices of the scanning probe microscope. In 2002 one-year seed grants were given to M. Gorman at Virginia and D. Baird at South Carolina to begin on-point research into the philosophical, social, and ethical dimensions of nanotechnology. Also in 2002, J. Yardley at Columbia University received money to design courses on societal implications, while J. J. Richardson from the Potomac Institute received a grant to do a comparative study on shaping science and technology to serve national security. 2003 saw three Nanotechnology Interdisciplinary Research Team (NIRT) grants, two of which are examined in detail below, while the third funded a preliminary study on public opinion at Michigan State. In 2004 another major grant was funded at Michigan State in addition to their earlier polling grant. They were well over $1 million apiece, and it seemed to greatly expand a commitment to study the societal implications of nanotechnology. In 2005 no NIRTs for SEIN were awarded, and the NSEC Center for Nanotechnology in Society (CNS) seems to be in the throes of being carved up into a set of smaller CNSettes.

The NSF recently announced that the CNS will be shared as an informal network bifurcated between teams at the University of California at Santa Barbara (UCSB), which is home to the California NanoSystems Initiative, and another at Arizona State University (ASU) in Tempe. UCSB will receive $5 million over five years, and ASU will receive $6.2 million over the same period. UCSB's center for Information Technology and Society under Bruce Bimber will focus on risk perception and social protest movements, especially regarding global networks' responses to trends in nanoresearch and applications. In addition, they plan to study historical and contemporary contexts and effects on technology diffusion and commercialization. ASU under David Guston and the Consortium for Science, Policy, and Outcomes will work primarily to develop a real-time technology assessment model for nanotechnology around two themes: freedom/privacy/security and human-identity/enhancement-biology. Press releases from both UCSB and ASU insist they will function as the center. Presumably all of this will be resolved in a February 2006 meeting. Smaller awards seemingly associated with the CNS were given to the University of South Carolina ($1.4 million over five years) to continue its work researching the role that images play in communicating all things "nano" and to study the impact of government-promoted nanotechnology on how science and engineering are practiced. Harvard University with UCLA and others ($1.7 million over five years) will develop the NanoConnection to Society NanoIndicator Series, which would add a NanoEthicsBank and a NanoEnvironBank to the current UCLA NanoBank initiative (see below).

A network rather than a center seems problematic for two reasons. First, networks like the NNUN and the NNIN (see chapter 4) make sense only when the sites have user facilities. Second, the call from Congress was for a dedicated center with a national clearinghouse rather than a fractured group of SEIN researchers. While spreading the wealth seems the egalitarian thing to do, it is questionable whether it is good for society. These schools have produced an impressive list of names for their programs and centers, and Harvard-UCLA might win the award for finding creative ways to use "nano" as a prefix. While they could have been included below, it seems more prudent to wait and see what is produced while concentrating on the few programs that have some sort of track record.

UCLA—Nanobank

Lynne Zucker and her colleagues are studying how newly acquired knowledge about nanotechnology makes its way from the laboratory to the marketplace. Lynne Zucker is a sociologist and "studied the growth of the biotech industry since the late 1980s and began tracking nanotech around 2001."[99] One of the major deliverables of the UCLA study will be an extensive database on small start-up firms in the nanotechnology arena and what factors influence how well ideas succeed in the marketplace. Called NanoBank, "it will be a resource for scientists, journalists, policymakers—everyone," Zucker says. "It will help us understand how the knowledge is transmitted, what facilitates that transfer, what blocks it, and what works well."[100] NanoBank is a national, interdisciplinary database comparable to their database developed for biotechnology. In essence, NanoBank will provide free information in an integrated database form. Zucker argues that "NanoBank is simultaneously a data archive, an active site for exchanging papers and ideas for social scientists and ethicists, and a site for interdisciplinary learning across scientific disciplines through the construction of analogies and other methods."[101] The goal is to have the database system launched in the next four years.[102] NanoBank seeks to compile a myriad information and effectively cross-list the data using defining data element links.[103] For example, NanoBank will be the domain of information on scientific articles, patents, firms, universities, national labs, nonprofit interest groups, and so on. Examples of the defining data element links include the name of the person, the discipline of the person, the date or time, the organization, the industry of the organization, science and technology area codes, geolocation, inputs, outputs and success measures, and interdisciplinary convergence.

University of South Carolina—nanoSTS

The other major grant awarded in 2003 went to the University of South Carolina and its nanoScience and Technology Studies program under the direction of Davis Baird, a professor in the philosophy of science. It has a team of humanists and social scientists who are both working side-by-side with scientists. They are situated in the University of South Carolina's NanoCenter and deliver research, curricula, publications, and conferences. One of its goals is to produce a model for a nanoliterate campus in which "a cadre of scientists, engineers, and scholars are used to thinking about the societal and technical problems side-by-side."[104] The grant to South Carolina has allowed Baird and his colleagues to set up an ongoing dialogue among as many points of view as possible. Just as researchers need to consider societal implications from the start, Baird emphasizes, "ethicists and other scholars need to understand what's possible in the lab." Most important, he says, "students who are trained now in the right interdisciplinary setting—will become [the above-mentioned cadre],"[105] hence a major element of the program is education—faculty on faculty, mentoring undergraduate and graduate students, and a highly innovative citizen's school networking nanoscience and SEIN scholars with the public within the NanoCenter. Baird and his colleagues are "investigating how we can go down a better path with nanotechnology" as we prepare the nanotechnology workforce of tomorrow.[106] In 2005 they received another multiyear grant to study imaging, mental modeling, and the reconfiguration of science, technology, and policy as it relates to nanotechnology as part of the CNS NSEC funding initiative.

Michigan State University

In July 2004 Michigan State University received the third major NIRT to study SEIN. They titled their grant "Building Capacity for Social and Ethical Research and Education in Agrifood Technology." With over $1.7 million, Paul Thompson and Brady Deaton from the University of Guelph planned a project with three objectives: (1) derive lessons from the GMO controversy that may be useful to the entire range of researchers in nanotechnology, (2) build a competent team of senior researchers in social and ethical issues associated with the agrifood industry who have collaborated to develop communicative strategies in engineering applications and a junior team in social and economic dimensions of agrifood science, and (3) identify nanotechnology applications in agrifood and develop a proactive SEIN strategy. They plan to have symposia and workshops, to collaborate with a European research team starting a project on public participation, to develop materials for education of stakeholders, and so on.[107] Their first major conference was held in October 2005.

The Nanotechnology Business Alliance—HEITF

Government and academe should not be the only partners in SEIN research. Presumably, it was joined by business and industry, which might want to protect themselves from expensive liability associated with manufacturing and marketing a nanoproduct linked to morbidity and mortality. Independent of any concern beyond short-term profiteering, product liability cases as individual or mass torts can set an industry back for years, in some cases leading to its virtual extinction as happened to the asbestos industry a few decades ago. In addition, industry might simply want to insulate itself from costly and disruptive litigation.

A more cynical, but not wholly unlikely, explanation would be simply that these efforts to encourage SEIN research coming from business and industry might be used to inoculate the industry from outrage produced by a disastrous event. Much like its biomedical function, instilling the public with some expectation of a lesser harm mitigates the intensity of the reaction to a truly horrendous occurrence.

As such, Mark Modzelewski, former executive director of the NbA, said that despite the overwhelming public health and environmental benefits associated with nanotechnology, "the industry should not turn a blind eye to public perceptions and the potential for issues to arise in the future."[108] As such, the NbA jumped onto the bandwagon when it formed the Health and Environmental Issues Task Force (HEITF). HEITF was led by a cross-section of researchers and business leaders in the nanotechnology sector, including Sharon Smith (Lockheed Martin), Josh Wolfe (Lux Capital), Michael Relyea (NYSTAR), Ben Savage (Wasserstein Venture Capital), Nora Savage (EPA), Jack Solomon (Praxair), Scott Rickert (NanoFilm), Jim von Ehr (Zyvex), Cynthia Kuper (Nanobusiness Development Group), Stephen Maebius (Foley & Lardner), and Frank Yang (ISTN).[109]

Its goals were to develop information about the impacts of nanotechnology and to educate public and industry leaders to counter misinformation. Its objectives included networking researchers, foundation, and public interest groups and companies to work on their issues; acting as a repository of research and information relating to health and the environment; acting as a clearinghouse to educate the public, government, and industry leaders on health and environmental effects; developing a Web-based library; and creating a dialogue with the public and industrial, medical, and environmental communities on how nanotech affects health and the environment. It anticipated studies with research institutions, such as the Rice University Center for Biological and Environmental Nanotechnology; the development of standards for manufacturing, workplace conditions, and after-life issues;[110] and program and research efforts to address these concerns.[111] Presumably, "the alliance [would be] organizing discussions with environmental groups, the government and others about environmental issues and how they intersect with nanotechnology."[112]

These were honorable goals as long as the information is objective if that is possible, and if not, it needs to be categorized as subjective speculation. The NbA insisted the HEITF would serve as a resource for a regulatory regime. Modzelewski added, "HEITF will develop standards for manufacturing, workplace safety, and afterlife issues of nanotech" in an effort "to be proactive and prevent problems in the future."[113]

Helping regulators develop criteria to establish roles and regulations is laudable. When agencies find themselves charged to regulate, often they promulgate rules without communicating with regulated entities. They turn to scientists who receive grants to set tolerance levels. However, they have learned it is more productive to ask the entities to be regulated to comment on a viable class of regulatory options, on what constitutes the best available technology, and so on. As one of the goals of NbA and HEITF, business and industry have an incentive to help gauge rules associated with commercialized nanoproducts.

In June 2003 the first meeting of the NbA's HEITF took place in Washington, DC, with intentions "to educate the public and identify any problems before they get out of hand."[114] According to Bo Varga, executive director of NanoSig, "If enough misinformation spreads, there's a tremendous potential to generate a social backlash."[115] Hence HEITF would generate accurate information fostering an informed and legitimate debate, and maybe even defuse the backlash.

HEITF outlined both short-term and long-term goals. In the short term, it sought to gather information and make that information public through its organization and on its Web site.[116] In the long term, it would serve as a standards-producing body, working with similar international organizations to craft appropriate protocols. Moreover, it would provide information in crafting legislation with the US Congress if it finds there are potential hazards to nanotechnology.[117]

Clearly, an agenda with specific goals had been established by the HEITF. However, an analysis of the results of those goals is telling when discussing the efficacy of HEITF. Initially, notice that one of its short-term goals was to develop a Web-based information library that was to be available by September 2003. For example, as of 2005, the Web site http://www.nanobusiness.org contains no comprehensive library. Even if we consider the archives-and-articles section of the NbA Web site, the reader finds that the last uploaded article is from 2002. Nanotechnology has seen incredible advances in the last three years. If the purpose of the database is to be both comprehensive and educative, this lapse constitutes a fundamental problem with the goal of the database. Moreover, the assumption that the archives-and-articles section of the site at the Web-based library is an attempt to spin gold out of straw. Last, even if the archives-and-articles section of the Web site is in fact said library, this knowledge was never released to the media or the public. The front page of the site makes no reference to the archives-and-articles section as a public, user-friendly nanotechnological information resource.

It seems, then, that HEITF did not even begin to fulfill its most basic of goals outlined in July 2003. Why has it not lived up to its promises? One argument may be that the NbA hoped to stave off serious governmental influence over activities involving nanotechnology by taking up the serious health and environmental con-

cerns themselves. The announcement that occurred over a year ago has yet to produce any publicly released findings.

Perhaps if the time frame had not been proposed by HEITF itself, there
would be some semblance of hope that HEITF would undertake real research and
have it released to both governmental and nongovernmental entities alike. Yet
Business Wire reported that the "work [was] to begin immediately,"[118] while *Bio-IT
World* noted that HEITF "will meet soon to begin its work."[119] Thus, when no
noticeable activity has yet to commence, one doubts the words of Ben Savage, who
hypothesized that the HEITF would be "an important catalyst for organizing
intelligent and responsible dialogue between the environmental movement and
the burgeoning global nanotechnology community."[120] To anyone with a healthy
sense of cynicism, it seems that HEITF was just a smoke-and-mirrors campaign
by industry to assuage the fears of both the public and government alike.

Ironically, Modzelewski commented that nanotechnology has yet to have a
major public snafu, but "this however doesn't mean we as an industry should turn
a blind eye to public perceptions and the potential for issues to arise in the
future."[121] Unfortunately, it seems Modzelewski and the NbA have done precisely
what he warns they shouldn't. Turning a blind eye to public perception has fueled
many crises against emerging technologies (genetically modified foods in the EU,
nuclear power in the United States). HEITF's refusal to publish any material or
announce any such intention to do so is a fatal blow to its stated goals of information dissemination and public perception management. The lasting legacy of NbA
HEITF may have more to do with spin and hype than actual scientific research
and advancement.

Winner does caution that business efforts must be counterbalanced with independent efforts.[122] When a business association, qua lobbying and promotional
group, starts to invest in health and safety research, it becomes imperative that
competing groups rise to challenge them to keep them honest. There are too many
incentives to cut corners, underreport findings, emphasize expediency, discredit
criticism, convert objectivity into public relations, and co-opt opposing voices.

A strong case can be made that SEIN research is good for industry. "Firm data
on health impact and a quantitative and general risk assessment model [are] thus
a necessity for commercialization. Indeed, for an industry that is already producing thousands of tons of nanomaterials each year—for applications ranging
from cosmetics to solar cells—the importance of characterizing potential environmental impacts seems evident. Rather than alarm investors, such information
(whatever its conclusions) will be reassuring and increase the likelihood that
viable nanotechnology products are developed."[123] Of course, industry wants to
do its best to avoid controversy and liability. Whether it will be willing to admit
that a current product is unsafe and must be removed from the market remains
unclear, especially given that a removal may constitute culpability. As such, it

behooves the NbA and HEITF to take their goals seriously. Whether it is wise to leave the regulation to the regulated and the direct beneficiaries of the nanoproduct market is a question that must be left unanswered. However, the overall government and industrial record to date suggests we should remain cautious. Is regulation really an option when data is wholly absent? Agencies are already grossly understaffed and underfunded.

STATE OF SEIN

What grade does SEIN deserve? The current spate of SEIN programs at universities and the HEITF do not seem to bode well for the new discipline. While there have been smaller awards associated with SEIN, they have not been particularly impressive in their reach. The problem may be one of vision, as we will see below.

It is fairly important to understand how challenging this research can be. "Societal impact research is very hard to do. It requires teams that predict the future and then decide what those futures might be. The first step, technology forecasting, must be done by nanotechnologists that are closely involved with applications development. The second step requires both social and environmental scientists to evaluate the consequences."[124] Depending on how generous your definition of *scientist* may be, there are some concerns much like those voiced by Representative Dana Rohrabacher (R-CA), who protested at the 2003 hearings: "It sounds like to me you are putting all of the sociology and literature majors in charge of defining the goals of the engineering and science majors."[125] He added, "Am I the only one who is skeptical of the social sciences here? We're injecting bureaucracies into the sciences," and bureaucracies are only good at "transforming pure energy into solid waste.... You'll be giving a forum to the very nuts you are trying to overcome."[126]

When the Twenty-First Century Nanotechnology R&D bill was being debated, it included a call for a comprehensive SEIN center. Lovy called a SEIN center "a philosophy and communications department head's wet dream come true,"[127] and I have little doubt he was describing me—and on one level he is correct. For far too long, scientists have simply ignored the role of the public. Jay Clayton, a professor of English at Duke, noted that "humanists often believe that their only contribution is to object loudly, but that simplifies the ethical and philosophical issues. Technology is part of us."[128]

Too many people think that anyone who specializes in a field outside of the natural sciences is opposed to them. This is hardly true given the large number of science and technology studies programs that have propagated across the United States over the last few decades.

Nonetheless, even among the researchers there are reservations. "Ethics cannot substitute for solid scientific assessments of risk, critics note. And even some of the social scientists who have received those grants say they are skeptical about their roles."[129] Nevertheless, Winner thinks scientists need their help. "Scientists can tell you the knowledge required to make things work. Engineers can tell you how to make them work in practice. What neither of those groups really can do except to perform their own roles as citizens as well, is what these technologies will mean to people when they enter the world of practice [after] they enter the environment."[130]

This is not to suggest that all SEIN research is worthwhile per se. While there are biennial meetings hosted to discuss social and ethical implications of nanotechnology, the Toronto group criticized the 2000 and 2002 meetings for their generalizations and motherhood statements and little else. Though at this early stage, that may seem inevitable and obvious, a December 2003 SEIN meeting at the NSF was still overwhelmingly not specific in its research agendas and recommendations. By and large, stand-alone research teams and creative efforts to integrate humanists and social scientists as part of the research teams at nanocenters in academe, within government laboratories, and in commercial and industrial settings seem to be mostly cosmetic.

This is the vision problem. The primary strategy for most SEIN research can best be described as bootstrapping. For example, at a recent call for competitive proposals for the NNIN, the winning proposal involved a team of research universities where one or a handful of social scientists at major research institutions within the network was to be associated with the science end as well as providing an overall clearinghouse to market nanoscience and nanotechnology to the media and the general public. A competing proposal that would have established a totally independent SEIN research arm that would provide entry and exit SEIN requirements as a cost of participating in the NNIN network as well as embedded humanists and social scientists was rejected.

Unfortunately, bootstrapping as a strategy is incredibly suspect and rightly so. As a strategy for subjecting nanoscience research and development to SEIN considerations, it has serious drawbacks.

First, SEIN as a discipline is much less developed and sophisticated than nanoscience. While SEIN researchers may draw from work in Science and Technology Studies and have models like ELSI, they are struggling over precepts and applications. The capacity to compete for rhetorical space against a well-defined discipline is problematic. It only worsens when the setting for SEIN research is framed by the scientific work being done at the site where the social scientist is associated. Focus is increasingly finite and specialized. SEIN research is retarded by the relative lack of breadth of the established science team. If the team does not research, say, in nanophotonics, then the SEIN researchers not only do not have a sample or subject in that field to study but also may find themselves assigning increasing importance to the science around them out of proportion to its significance.

Second, attaching a small group of SEIN experts to an already established group of scientists places the social scientists at a home field disadvantage. Even when they move into the same building or on the same floor, they are often the first to be relocated when a lab is installed or expanded for "real" science. As outsiders, they are more easily discredited, having their qualifications and expertise challenged by a tight-knit group of scientists who can rhetorically marginalize critics by speaking in another tongue—the increasing abstruse argot of science. The decision-making axiology of an established science team may be codified, and the attached team of social scientists may find their role additive and subtractive only. As a result, negotiating a calculus or algorithm of societal and ethical implications that is fully integrative into *if, how,* and *when choices are made,* becomes seriously impaired.

Third, while texts can be written, rewriting is much more challenging. There is an institutionalized value to the extant that often provides it with a special valence. Institutional momentum is a powerful force in precluding change and reform. Making an argument is much easier when it does not require direct refutation as a prerequisite function. Texts that are inherent have presumption and the burdens of proof necessary to refute them is foreboding, if not overwhelming.

Fourth, adding a group of critics to a well-defined cadre of specialists or experts begs tokenism. Once associated with a SEIN researcher, the nanoscience research team insulates itself from some level of criticism. When public concern crests, the nanoscientists can identify the bootstrapped critic as protector of the public interest. Bootstrapping enables a programmed response. Settings without an independent SEIN voice may use their bootstrapped SEIN team for plausible deniability when societal and ethical challenges occur.

Fifth and finally, the strategy of taking on a group of social scientists from the same institution, like a university, risks seniority problems. Young, untenured social scientists are reticent to speak out and criticize colleagues. Junior faculty will be restricted in what they can accomplish and succeed in their pursuits of tenure and promotion. SEIN research is so novel that most tenure committees would be hard-pressed to qualify publications. Also, nanocenters, often centers of excellence at a university, are important to universities. University centers for nanoresearch are highly prized on campuses, and discord at the hand of a social scientist may not float. Complaints by resident faculty can be discouraged directly, and some researchers avoid this altogether by censoring themselves. Miscreants can be marginalized by less drastic scenarios than the denial of tenure, such as internal allocation of resources, scheduling of duties and obligations, and so on.

The problem is not merely how they are situated; it also includes what they do. What is clearly missing from the research initiatives to date are some of the following: First, the way risk is assessed needs to be reexamined. "It cannot be assumed that the conceptualizations and analytical categories currently available will be able to capture what may prove most distinctive about nanotechnology."[131] Assuming the premise of this complaint by Demos is correct, identifying what those variables are that make current risk assessment appropriate has not been completed. While the team from South Carolina has begun to chip away at some of the visions and imaginaries of technology and questioned the utopist nature of technologically enhanced futurism, it was mostly overlooked in a recent call for proposals of Societal and Ethical NSEC for a more pragmatic one that accepted the extant assumptions that focus more on communicating policies already in effect and assisting commercialization of research from labs.

Second, the entire deficit model in communicating science and technology handicaps thinking about scientific and technological policy decision making. However, an upstream strategy whereby citizens participate in making policy rather than evaluate it after the fact has not been investigated beyond trivial experiments in deliberative polling. How to incorporate forecasting and prediction as variables to weigh as hazards or against hazards has not been undertaken. Balancing situations where one group benefits but bears none of the hazards from a decision against another group that bears the hazards but not the benefits still baffles assessment scholars. How low-probability/high-impact benefits and hazards can be incorporated into a debate of policy evades us as well. We have no algorithm(s) to examine newly emerging technologies beyond appeals to analogies and metaphors.

Third, visions associated with technology need to be challenged. For example, how long should we postpone aging and death? Are the development models that are in place in Western countries appropriate to non-Western ones?

There is also some concern regarding who in the humanist and social-science world are doing SEIN research. According to Carl Batt from Cornell, "Making scientific discoveries, publishing papers, going to meetings to exchange ideas with my colleagues is only part of the process."[132] The NSF works from a set of premises, including perception management, risk communication, mediated argumentation, and so on. The call from the NNI is not to overanalyze, overdetermine, and overstand what is nanotechnology and what is meant by the images and the prosaic constructions of the nanoworld. The SEIN researchers must be expected to come onboard at a steady clip.

One of the greatest problems confronting those engaged in SEIN research is missing the boat entirely—generating a body of "scholarship" that is archival in nature and function. The findings will be quaint, whimsical, and totally irrelevant. As nanoscience and nanotechnology become commonplace and are commercialized and marketed, there will be taxonomists explaining the meaning of the very small, the essential nature of smallness, and the reality attached to this smallness. All the while, nanotechnology will be all around us, and they will have played no role in determining whether the world we find ourselves in is the world in which we want to live. Participation demands the capacity to enter into the debate, but the debate is already upon us. While deep thought is not without value, it will not provide the loci that will provide grapples for arguments. What if that is the goal of the NNI? What if—SEIN is a shell game? What if humanists and social scientists are asked to study SEIN because they are less likely to demand responsible action by government and industry? What if this enterprise is simply prestidigitation? Nanotoxicologists might discover that single-walled carbon nanotubes are toxic. That finding would be unfortunate for the NNI and the race to commercialize. While "thinkers" hypothesize on the meaning of the small, the world transforms around them, forcing them to return to the underenrolled classrooms to teach Hume, Heidegger, and Habermas. Their contribution would be unread articles and books, some on Internet home pages and a few in libraries that still buy paper versions of these text files.

CONCLUSION

While the previous chapter seemed to indicate that environmental considerations are the only SEIN-related concerns, that is not my intention. On the contrary, there are a plethora of considerations, however there is limited time and space. Indeed, there are serious moral and ethical concerns with many nanorelated projects.[133]

SEIN has payoffs. "Such self-examination can pay off big," says Thompson from Michigan State and an advocate of SEIN who points to the information technology industry. "Early on, gadflies and other concerned industry professionals spurred internal debates on societal implications—for example, the loss of privacy. Partly as a result, personal computers and the Internet enjoy widespread public acceptance and adoption." By contrast, he says, "the nuclear industry devoted little discussion to societal questions and has paid a steep price in public acceptance."[134]

While there seems to be some public support for science and technology, science in general, and even nanotechnology in particular, it is dangerous to be lulled by these findings. "The public's enthusiasm for an emerging new technology can easily turn to fear, with grave consequences for commercialization. Emerging technologies do pose risks that are ill-characterized, and the best thing nanoscientists can do—both for the discipline and society—is draw attention to possible risks and study them carefully."[135]

For SEIN research to matter we need a combination of two models: a major center dedicated to SEIN research—which seems highly unlikely (rhetoric to the contrary)—and allied faculty at institutions with active nanoscience research teams. A center would foster and coordinate SEIN research on a macroscopic level, uniting teams made of allied faculty from disparate settings to improve their methodologies and samples. By bringing together researchers in a variety of settings, such as professional meetings, Internet and other teleconferences, and town meetings, a center could improve the exchange of findings and the abduction of subsequent rounds of hypotheses that may need to be tested.

For SEIN research to really matter, work must be done on three levels: (1) business and industry need to associate SEIN considerations in their development and marketing of nanoproducts; (2) centers of scientific and engineering research in nanoscience and nanotechnology should include SEIN experts in all phases of their operations; and (3) a national or international center should coordinate and publicize SEIN bootstrapped centers and business and industrial groups. Unfortunately, the proposed SEIN center will focus on technology transfer and policy recommendations associated with commercialization.

As academics without actual and perceived self-interest, we might be able to shed some light on this emerging technology. Maintaining independence and objectivity is never easy. The lure of federal grants and industrial contracts is awfully attractive, especially in a climate where higher education is being defunded because of tight state budgets. There may be hope in the efforts of the NSF and others to respect SEIN research with financial support. However, the promotional philosophy underlying the NNI and the recent Twenty-First Century Nanotechnology R&D Act suggests attentiveness. Be wary of governments and industries bearing gifts. Researchers need to assure the public that they are acting in the public's best interest by maintaining the highest standards of impartiality. While the rhetoric underlying the trend to fund SEIN can be a favorable omen, it may simply be smoke and mirrors.

CHAPTER 11

A PUBLIC SPHERE
IN NANOSCIENCE
AND TECHNOLOGY
POLICY MAKING

P ublic participation is a basic principle to democracies and constitutional republican governments. Generally, the public participates by voting for a candidate during a public election cycle. Between elections, there are limited opportunities to play a role in the governing process. We can keep appraised of issues, attempt to communicate directly or indirectly with an elected or appointed official, protest and boycott policies and the consequences of policies with which we disagree, and we can actively campaign for or against a candidate based on some policy issue. Predictably enough, these options may not be sufficient in the twenty-first century when communication is nearly instantaneous and policy is more a function of polling and focus groups than insight and reason. Recently, the US government through its flagship agency of science has called upon researchers to consider SEIN of the NNI in nearly all of its manifestations to keep the public informed and to encourage a rational debate on the subject.

This chapter suggests a public sphere in science and technology decision making in the United States may be a pipe dream. At best, professionals may be able to improve both the type and quality of information provided to the public so they may be able to make informed decisions as citizen-consumers, maybe affecting purchasing decisions and voting behavior. However, science and technology seldom, if ever, appear on political agendas beyond local concerns, so it seems unlikely that a national science and technology policy will ever be the subject of a national referendum on its own or a substantive component of a platform for a national political candidate seeking election. As such, the public sphere envisioned in SEIN rhetoric will function predominantly as a consuming public sphere. Consequently, the difference between public relations and public empowerment becomes difficult to discern.

THE CALL

At the December 2003 workshop on the Societal Implication of Nanoscience and Nanotechnology, outgoing NSF Director Rita Colwell offered her rationales for SEIN research in her welcoming remarks:

> As we strive to advance nanoscience and engineering, we must do so benignly and equitably. This will require active involvement with the social sciences and with concepts of managing risk.... Our broader mandate includes informing the public and engaging broad segments of society in decisions on how new capabilities should be applied. We need to anticipate and guide change in order to design the future of our choice, not just one of our making. We want society to *be prepared for*, though not necessarily *control*, the results of far-reaching research. Future generations may well judge our success—and our wisdom—by how well we realize the potential for nanoscience and engineering while avoiding the pitfalls.... We must act quickly as a community to develop the means for measuring and predicting the societal and economic impacts of nanoscale research.[1]

The United States is hardly alone in its concerns. Europe, mostly in terms of the United Kingdom and the European Union, has also begun to study SEIN. The ESRC and the Royal Society began with soliciting studies and publishing results from workshops. However, in 2005, in response to the Royal Society report, the United Kingdom fell short, according to Ann Dowling, chair of the working group on nanotechnology: "The government has not provided new money for the research that will be needed before we can draw up appropriate regulations. It did signal that it will create a group to coordinate research, yet it did not commit any additional money to it."[2] The EC did launch Nanologue to boost nanotechnology discourse. Hans Kastenholz of the Swiss Federal Laboratories for Materials Testing and Research (EMPA) is part of the initiative that also includes Forum for the Future, United Kingdom and Triple Innova, Germany. He commented that "consumer acceptance will be a key issue for nanotechonlogy's future development and thus key for financial markets and venture capitalists. Engaging society in dialogue about the opportunities and potential risks will address and help to mitigate some of these uncertainties."[3]

There have been calls that Japan should get into the act as well. According to the *Yomiuri Shimbun*, "The government therefore is urged to conduct reliable safety assessments to ensure that the nation adopts nanotechnology without risk to humans or the environment, instead of leaving the task of confirming the safety to the technology to Europe and the United States."[4] This call was made within the context of a parallel demand for citizen participation. "Steps should be taken to market nanotechnology products only after a majority of the public has reached a consensus about their safety."[5]

While Colwell claimed the NSF had begun to invest in exploratory grants for SEIN study, it remains to be seen whether research will be forthcoming that empowers citizens to participate in science and technology decision making.

There remains a tenor of elitism in Colwell's remarks among others seeking to inform citizens rather than to actually empower them to participate. There is a significant difference between the two. The first is uni-directional, and the second is transactional. The first involves downward communication to a passive audience. The second is lateral communication, blurring the distinction between expert and inexpert, evidence and opinion, media and medium, and speaker and audience. In addition, the first is linear, and the second is not linear at all. The first is autocratic and technocratic and grounds for healthy suspicion. The second is participatory and fundamentally democratic.

According to the British Association for the Advancement of Science, "The public should be involved in early stages of research and scientists must be prepared to answer questions on what is driving their science."[6]

> What is needed, some scientists say, is substantial public discussion as nanotechnology forges ahead; otherwise activists will exploit people's fears, and the technology will flounder. "I don't mean a PR exercise to explain to people nanotechnology is a good thing," said Peter A. Singer, Director of the University of Toronto's Joint Center for Bioethics. "I mean a real, two-way- balanced discussion, so scientists can share what they're up to and what they think is good—and the public can share its concerns," he said. "A really good dialogue that can shape the way the technology folds out."[7]

There is an additional imperative. "Initially, nanotechnology products are expected to be more expensive than traditional products." According to Frost & Sullivan industry analyst Kirti Timmanagoudar, "Convincing the customer that the added value is worth the additional cost will be the key determinant of the growth of this technology."[8]

THE STATE OF THE PUBLIC

There have been nearly a dozen studies on public perceptions of nanotechnology. One of the first was done in 1989. The Assessing Molecular and Atomic Scale Technologies (MAST) study completed by Futuretrends, Inc., was reported by the Lyndon B. Johnson School of Public Affairs.[9] It asked for definitions of nanotechnology, projected relevant paths, technologies central to its development, wet and dry directions, projections, applications, and barriers. There were only twenty-three respondents, and the findings were not replicated.

Projections regarding nanotechnology generally have been rhetorical flourishes often based on educated guesses. For example, Bill McKibben reported in *Enough*[10] a 1995 *Wired* survey[11] that claimed top scientists projected the arrival of assemblers between 2010 and 2025. Five scientists—Hall, Smalley, Birge, Drexler, and Brenner—made up the sample, and their projections ranged from 2000 to 2025, with Nobel Prize winner Smalley predicting 2000, though he recanted his estimate later. It is unclear what was meant by molecular assembly since while some addressed STM and AFM developments, others referenced grander technologies.

Bainbridge, in "Public Attitudes toward Nanotechnology," inserted three items into a 2001 survey, two using Likert scaling (a three- to ten-item range usually from highly negative to highly positive reactions to a question) and the third asking an open question.[12] There were 3,909 Internet respondents. He reported that nearly 60 percent believed nanotechnology will be beneficial, with only 9 percent believing "our most powerful twenty-first century technologies," including nanotechnology, would threaten humanity.

More recently, the 2001 EC Eurobarometer 55.2 released findings (n=16,029) indicating that EU respondents were uninterested in scientific and technical developments associated with nanotechnology, with over 65 percent indicating they felt they did not understand the "topic" of nanotechnologies.[13] A more recent British study reported, "29 percent of respondents from both the workshops and the survey said they were aware of the term *nanotechnology*" and "68 percent . . . of those who were able to give a definition . . . felt it would improve life. . . . Only 4 percent thought it would make things worse. . . . 13 percent of the workshop respondents said that nanotechnology would make things better or worse depending on how it was used."[14] A late 2004 EC online study (n=749) reported that 80 percent expect nanotechnology to play a role in their lives by 2015, and 75 percent of respondents believed Europe should embark on studies on the risks and societal implications of nanotechnology.[15]

A 2003 American ISTPP National Nanotechnology Survey (n=928) found 65 percent of respondents had no subjective knowledge on nanotechnology, with only 15 percent reporting that bad things might happen from its use.[16] A 2004 study from North Carolina State (n=1,536) reported 52 percent had no familiarity with nanotechnology, and 83 percent heard little to nothing.[17]

Even more interesting is an online survey (n=400) conducted by GolinHarris in late 2004.[18] Likely voters placed a high priority on US leadership in technology. Even though 80 percent could not name a single company that is a leader in nanotechnology, 60 percent felt it was important for state governments to get involved in nanoscience research funding, and 60 percent said the government should increase funding levels for nanotechnology research.

Two sets of findings were released in 2005. The first mostly corroborated earlier data sets, while the second evidenced the importance of biases when citizen-consumers were asked to express their opinions on nanotechnology.

Jane Macoubrie with the Woodrow Wilson Center in DC released her findings from research completed during May and June 2005.[19] A follow-up to the NC State study referenced above, the research was a modified focus group/citizen jury methodology in three cities (n=177). Pretest results indicated that over 80 percent knew little or nothing about nanotechnology. The respondents in the sample then read a prepared document, and they predictably knew more and answered more favorably to nanotechnology (21 percent in the pretest to 49 percent in the post-test), with more expecting benefits to exceed risks and many more rejecting any sort of ban or moratorium. There was additional data on levels of trust in government regulators.

The most recent study by a team of three researchers, Chul-joo Lee, Dietram Scheufele, and Bruce Lewenstein, examines opinion and mediation.[20] They collected their data from a national telephone survey (n=706). Overall, only 25 percent of respondents reported having heard about nanotechnology, and only 16 percent felt somewhat informed. Using a Likert scale as well as a multivariate analysis, the researchers demonstrated that some people who were unaware of nanotechnology still had an opinion about its risks and benefits. For example, "59 percent of those who indicated they are aware of nanotechnology expressed overall support of the emerging technology, compared to 28 percent support from those who were not aware."[21] While more information may seem important, the analysis also suggests other variables are at work. Indeed, the authors conclude biases or intuitive shortcuts are used by citizen-consumers when they craft answers to questions about nanotechnologies.

There seems to be some wiggle room in terms of generating public opinion. In addition, highly negative values are absent at this time. While some Europeans might transfer negative attitudes about GMOs to nanotechnology, that hasn't happened yet, and that may be very good news for nanoscience.

THE CHALLENGE

What role should the public play in science and technology policy decision making? While it might be honorable to claim that the public should participate because we live in a democracy, the question needs to be asked: Are policy makers and the scientific establishment really open to increased public participation? Do the scientific and technological communities really want citizen participation? Are they willing to abide by the consensus brought to the bargaining table by the public? Or do they see citizens as unwashed and unkept, ignorant of science and fearing specters? If so, at what level? Should the public be empowered to referendum science and technology policy? Or is that too much input? Should the public be provided with enough information to understand decisions made? What would the American people do with true access to decision making in science and technology? Provided agency, how would the science and technology public sphere function and what would be the results?

Idealists visualize a functioning public sphere as the precondition to a full-bodied concept of democracy. They advocate a variety of experiments involving citizen groups in panels much like political focus groups.

Realists are concerned a public sphere in science and technology might be an impediment to economic growth and competitiveness. Less fretful about the knowledge and critical capacity of the public, they note that public arguments concerning science and technology are often less than rational, more emotional, fearful, and sometimes grounded in religious bigotry.

Or is this call for action just hype? For skeptics, providing public access would be enough to convince the electorate that they had input and to demobilize criticism and protest. For political handlers, beyond symbolic access, participation would be minimal and would involve the same voices currently shouting from the proverbial sidelines. Instead of screaming from the streets, they would move indoors, buy or borrow suits and ties, and sit behind tables. Extremists would be given a public platform to scream their marginalized complaints, convincing themselves their voices have been heard.

There are some avenues not foreclosed to the public. On one level, the public may affect funding decisions, by either communicating directly with their elected leaders or indirectly at election time.

> Policy-makers especially if they are not themselves technically trained in the appropriate area will be guided in part by assumptions they share with the general public. Decision-makers in government, academia, and industry often have to take into account how the wider public will judge their decisions. Availability of funds to invest in nanotechnology is in part determined by where it stands in the marketplace of ideas including individual investors as well as financial institutions.[22]

As a rule, public opinion has found few venues for expression because America does not have a viable public sphere over science and technology decision making. Nevertheless, producing an informed class of citizen-consumers might shield them from individual and organized groups of discontents who engage falsehoods, exaggeration, and hyperbole to marshal protest, if not boycotts, of commercialized nanoproducts.

DEFINING THE PUBLIC SPHERE

We begin with an idealistic, if not romantic, view of citizenship and politics in the Alexis de Tocqueville,[23] Max Weber,[24] John Dewey,[25] and Hannah Arendt[26] conception of the *public sphere*.

By participating in the general affairs of society, citizens become more fully engaged in other forms of collective life. Functioning as a collective or community, citizens can understand interests competing with their own. As such, agenda setting grew from not only commonalities but also differences protecting interests of a broad spectrum of a society's members.

Antonio Gramsci understood engagement as an alternative to citizens as manipulated objects deployed whimsically by elite preferences:[27] "The nature of political participation and the discursive character of political interaction have a deep significance to our understanding of the state of contemporary democracy."[28] "Without an open and vigorous public sphere the realization of anything resembling authentic democracy is out of the question."[29]

Hundreds of writers have summarized Jürgen Habermas's conception of the public sphere, and each of them focuses on one or more problems associated with it in the current political realities of the late twentieth and very early twenty-first centuries. Habermas's history of the concept of the public sphere is well known. He believed it might be possible to revive the normative content of modernity. While Habermas is associated with the Frankfurt School, which analyzed social relations structured around domination and alienation, he found Karl Marx's theory of labor inadequate to explain democracy in advanced capital societies, though his communication theory reflects some of the same arguments. It remains important to note that his theory on the public sphere excluded the household and the economy. The only two criteria for membership were education and property ownership. For Habermas, "the sphere of commodity exchange and social labor as well as the household and the family relieved from productive functions without distinctions were deemed to belong to the private sphere of civil society."[30] As such, his version of the classical *public sphere* would deny the very nature of the consuming public.

Defining what constitutes the public sphere both enlivens and effaces it. By trying to define it, we locate it or a facsimile. If the characteristics we use to note it are broad and fluid, we will find it. If discrete and finite, we may not discover it at all.

"The problem of the public sphere is to some extent an iatrogenic disease. After all, our deliberations about the public sphere are a part of its reality. Defining the public sphere is both a theoretical problem and an ongoing feature of daily life."[31] Simply put, debating the existence of the public sphere is a characteristic of the sphere itself. Challenging its vitality, its very existence, is the public sphere. As such, even if it doesn't exist, trying to find it may be as productive as locating it.

Architecture of the Public Sphere

The public sphere is partly a place, such as the ancient agora, the town hall, the local church, and even the street corner. Descriptions of the public sphere, especially Habermas's version, generally include four constituents.[32] These largely interrelated elements form the architecture of the public sphere. They are the media, conversation, public opinion, and participation. Paralleling Habermas's listing, Walter Lippman[33] found cause for the collapse of the public sphere in four interrelated phenomena: media corrupting the habit of critical thought, schools and universities failing to educate, government creating a public that excludes groups of people, and corporatism. Lippman also contended that citizens might be expected to do too much, since they must be able to be reflexive about one's own value positions; able to distance oneself from one's convictions and entertain them from the perspective of others; able to live with religious, ethical, and aesthetic incommensurables; and able to accept the multiplicity of values and the clash of the gods in an disenchanted universe. Lippman seemed to understand that a public both underinformed and untrained to handle a copious amount of inconsistent, sometimes incoherent, information constituted a fundamental impediment to a truly functional sphere of sorts.

An effective public sphere in the United States developed after 1945 with the defeat of fascism and associated authoritarian regimes and reached its zenith in the '60s and '70s with the civil rights movement. However, over the last two decades it seems to have shifted in the opposite direction.

Actually, a case can be made that the public sphere may have taken a radical turn for the worse post 9/11 with the "war" on terrorism and the "regime change" initiatives by Bush and Blair. Though Bush was hesitant to invade Iraq until after a case had been made to the public, actions like his continue to limit the agency of the contemporary public sphere to sanctioning or validating decisions rather than participating in their construction.

In the United States, a public sphere of sorts has taken the forms of political parties, labor unions, interest groups, civic associations, political machines, social movements, and ad hoc expressions of political initiatives. This political capacity produces at most a reactionary public sphere probably best illustrated by the recall of Gov. Gray Davis in California and the election of Arnold Schwarzenegger in 2003. The contemporary public sphere is truly immediate and rash, and it is the imprudent reactivity of a public that concerns nanoboosters, as we shall see below.

Status of the Public Sphere

The public sphere as conceived by Dewey,[34] Lippman,[35] Arendt,[36] Habermas,[37] Oskar Negt and Alexander Kluge,[38] and others has always been in trouble from the start. The agora is only a phantasm. It has been consigned to a spiritual half-life, and public opinion is the silhouette of the specter.

On one level, many have argued the public sphere never existed outside the academic machinations of political scientists. "Indeed, democratic citizens as described in these theories seem to live on another planet: they are devoid of race, class, and gender and all the benefits and liabilities associated by Americans with these features. The theories are ethnocentric, racist, and sexist. In general, they concern white male privileged society."[39] Failing to confront the intersectional dynamics in public-sphere theory may help to explain the penchant for contemporary social movements to form around identity rather than issues. The *us versus them* approach is highly vulnerable to sating, misdirection, demobilization, and a host of other disempowerment tactics.

On another level, participation in the public sphere can be problematical because the public sphere may be fictional. According to Seyla Benhabib,

> Influential currents in contemporary political theory, under the guidance of economic models of reasoning in particular, proceed from a methodological fiction. This is the fiction of an individual with an ordered set of coherent preferences.... On complex social and political issues, more often than not, individuals may have views and wishes but no ordered set of preferences, since the latter would imply that they would be enlightened not only about the preferences but about the consequences and relative merits of each of their preferred choices in advance.[40]

Another fiction is that by engaging the public sphere a transformation occurs. More often than not, the public is unprepared to constitute a position from the plethora of opinions in its midst.

Habermas contends that the transformation of the public sphere has been momentous.

> The infrastructure of the public sphere has changed along with the forms of organization, marketing, and consumption of a professionalized book production that operates on a larger scale and is oriented to new strata of readers, and of a newspaper and periodical press whose contents have also not remained the same. It changed with the rise of the electronic mass media, the new relevance of advertising, the increasing fusion of entertainment and information, the greater centralization in all areas, the collapse of the liberal associational life, the collapse of survey-able public spheres on the community level, etc.[41]

For our purposes, we may need to differentiate between categories of public spheres. As such, it may be important to separate the two functions of the public sphere, as Lynch puts it, analytical categories, because the function may affect the overall characteristics of the model:

The efficacy of the public sphere should be broken down into two analytical categories: constraining and enabling. The public sphere as constraint marks the
modified rationalist conception: to what extent does public opinion constrain the
behavior of state actors? Efficacy would be defined as the extent to which a publicly articulated position succeeds in forcing state actors to act contrary to their
interpretation of their interests. The constructivist concepts of efficacy incorporated an enabling dimension: to what extent does participation in public sphere
discourse changes actors' conception of their identity and/or interests?[42]

In terms of science and technology decision making, we seem to be
attempting to discover a constraining public sphere, a challenging variant since by
its nature it must debunk overclaims and misclaims, which can be a difficult mandate given the control of information by policy makers. Revoking policy is always
the greater hurdle since a policy, once adopted, generates its own institutional
momentum and vested interests. Once attacked, it adopts a defensive posture with
both informational and economic resources working to its advantage. The
debunkers find themselves telling stories and hurling epithets against a well-
trained, well-financed militia of sorts. A constraining public sphere labors from a
substantial disadvantage, especially in expertise-dominated fields like science and
technology policy.

While it may be possible to identify an enabling public sphere during election
time, as challengers undertake to dethrone incumbents, the incumbent advantage
is only another bit of proof of the especially daunting challenges an enabling
sphere confronts, and this advantage is made more poignant given the relative
transparency of political information compared with information in science and
technology. While a public demanding the commercialization of newly emerging
technologies might be the fantasy of some proponents of nanotechnology, building such a sphere remains daunting.

SCIENCE AND THE PUBLIC SPHERE

New frictional surfaces, new conflicts, emerge across the terrain of civil society. Science policy making is one of those surfaces. Though the public sphere may be
unhealthy, it might be the only bulwark against the rhetorical extravagances of some
scientific claims. While it may be unknown whether Habermas intended to exclude
science from his discourse ethics, it seems his concern for staged displays would indicate so. It might be inconsistent with his other concerns to allow the colonization of
science discourse, for he wrote, "the development of discourse is dependent on the
proliferation of diverse specialized publics throughout the terrain of society."[43]

Scientific argument has a built-in advantage using a methodology privileging
facts to coerce belief. Dewey claimed that this phenomenon

proceeded from method, from the technique of research and calculation. No one is ever forced by just the collection of facts to accept a particular theory of their meaning, so long as one remains intact some other doctrine by which we can marshal them. Only when the faces are allowed free play for the suggestion of new points of view is any significant conversion of conviction as to its meaning possible. Take away from physical science its laboratory apparatus and its mathematical technique, and the human imagination might run wild in its theories of interpretation.[44]

Absent the presumed objective nature of its methodology, science becomes vulnerable to arguments less refutable by claims of authority.

Public preparation for debunking scientific arguments is sorely deficient. Students studying for undergraduate degrees seldom take more than a few courses in science unless their professional aspirations are in those fields. Universities seldom make much of an effort beyond the core to encourage students to take courses in science and technology. Fewer still demand students in science and technology take course work in liberal arts beyond the core as well.

The media don't pick up the slack. The media in science are worse than journalism in general. Few newspapers and local television stations employ experts in science journalism, and fewer still expose science and technology to even the most superficial of critical reportage. More often than not, journalists will reprint press releases provided to them from industry and government. This brings us to the data: "In 2004, no fewer than 12,343 stories were printed about nanotechnology (that's about 10 times the number of companies actually doing nanotech), up from 7,631 in 2003. In the first two months of 2005 alone nanotechnology has been referenced more than 2,600 times in the popular press."[45] Nathan Tinker of *The Nanotech Company* offered his spin of these numbers: "These stories provide little information helpful for the general reader in understanding nanotechnology or, worse yet, confuse reality and fiction to the point where readers can't tell the difference between Crichton's evil, utterly fictional nanoswarms in *Prey* and the very real and harmless nanoparticles we breath every day."[46] While this may be an overstatement, it is not significantly off the mark.

According to James Curran, this has led to a crisis of legitimacy and citizen disengagement, in turn, leading to a greater "passive dependence on powerful institutions and groups as accredited sources."[47]

The media, especially when covering science and technology, are wholly dependent on the people upon whom they report. This leads to lazy journalism, where "journalists fail to ferret independently for information and evaluate truth from falsehood."[48]

In addition, citizens have lost confidence in the media as sources of reliable information, especially on complex issues. On one level, the advent of the Internet came with proclamations of the virtual town hall, hyperdemocracy, and so on, but these technologies have failed to live up to their expectations. On another note, a spate of accusations and admissions of plagiarism and forgeries by newspeople have not helped their credibility.

Journalism in science and technology is not unlike journalism in any area requiring a finite expertise. Writers reflect the ideology of their sources because of their dependence on them:

> Add to this the simple fact that information is distributed asymmetrically and reflects many socio-economic inequalities and these will be reflected in the structures of the public sphere. A representative mass media is defined in terms of existing structures of power and privilege. The newspersons stress *hard news* and factual reports which actually may disguise their own unconscious reliance of dominant frameworks for selecting and making sense of the new.[49]

This leads to a crisis of legitimacy and citizen disengagement, in turn, leading to a greater "passive dependence on powerful institutions and groups as *accredited* sources."[50]

The media are not limited to news reports. Habermas "neglects both the rhetorical and playful aspect of communication action, which leads to a sharp distinction between information and entertainment and to a neglect of the link between citizenship and theatricality."[51] Media entertainment functions to communicate norms relating to the nature of social relations and stereotypes as a focus of displaced fears. It provides a way of mapping and interpreting society. It provides models for emulation and instruction. Entertainment media communicate social knowledge via axiologies, webs of values, which are easily discernible from programming and personalities included in popular culture venues, such as film and television.

According to Cal-Irvine's Robert May, "many people learn more about science from the media than they do from anywhere else."[52] Hence, a deficient media perpetuate scientific illiteracy. A study from the Cardiff University School of Journalism found "what people know usually corresponded with those aspects of a story that received most persistent coverage."[53] Hence, sexy science and unusual findings are emphasized over less exciting reports, which is one of the reasons that one area of research associated on some level with science received more air time than it deserves—dieting—while global warming seldom makes the evening news at all unless it is linked to a new blockbuster film release.

A 1998 study by researchers at the University of Oldenburg supposed, "Apart from the educational system, the most important and *nearly the only* source of information about sciences are the mass media."[54] As such, the public remains misinformed on issues of science. For example, researchers at the University of Idaho found "the media, in particular television, are the primary and often the only source people will turn to find information about environmental issues."[55] They have a key position in determining the relations between the public and the sub-system of science.[56] Unfortunately, the Oldenburg team reported that especially telegenic topics tend to be emphasized disproportionately to their true relevance, including nanosystems. The team concluded: "more or less all programs are characterized by a basic lack of scientific accuracy... conveying neither adequate popularized scientific knowledge nor adequate social orientation."[57] As such, May added that media constructions in science are generally sensationalized.[58]

Popular culture interpretations of science influence public opinion about science, and those constructs may be wholly inadequate, especially when they are fictionalized. Therefore and unsurprisingly, a book as presumably irrelevant and intrinsically inconsequential as Crichton's *Prey* has captured the attention of legislators, bureaucrats, and industry representatives so much so that NanoScience Technologies recently hired TVA Productions to develop and launch a national media campaign focusing on building awareness for its DNA nanotechnology.[59] Once before, a popular book misdirected a lot of public discourse about nanotechnology. The worry is that *Prey* could become the next *Engines of Creation.*

All this has fostered argument by authority as the default warrant in scientific argument. Even Lippman advocated delegation of authority to specialized and competent experts. Carol Boggs warned, "The myth of professionalized knowledge, expertise, and specialized discourses continues to hold sway, with powerfully disenchanting consequences."[60]

Theorists have argued science should be impervious to challenges. G. W. F. Hegel made the case against science being subject to public scrutiny:

> The sciences, however, are not to be found anywhere in the field of opinion and subjective views, provided of course that they be sciences in other respects; their exposition is not a matter of clever turns of phrase, allusiveness, half-utterances, and semi-reticences, but consist in the unambiguous, determinate, and open expression of their meaning and purpose. It follows that they do not fall under the category of public opinion.[61]

Hegel did not anticipate the new role the public might play in science and technology policy making. Revenue generated by taxes fund much of public science. The taxpayers vote and, as constituents, reject legislators out of touch with the sensibilities of their electorate. In addition, the public can protest and boycott. As you would have thought, the case for some science policy, nanotechnology at least, is being presented to the public for their sanction, maybe their input.

PUBLICS AND COUNTERPUBLICS

In terms of science and technology decision making, we seem to be attempting to discover a constraining public sphere. Failing that, we might want to approach an alternative. The assumption that a single public sphere is preferred in general rather than separate public spheres specific to a set of issues or problems has never been justified. Though a single, all-knowing public sphere has some organizational advantages, its hierarchical nature is foreboding. Dissent may be nurtured where multiple publics are available and squelched by the more vertical construction of dissent of a single spere. Indeed, no single public sphere may be sufficiently expert to address the plethora of questions of policy-making discourse. Maybe something short of the Habermas version of the public sphere is in order.

It is time to distinguish between *the* Public Sphere, a public sphere, and a public or counterpublic. *The* Public Sphere is a historical artifact if it ever existed at all. A public sphere is something that mediates between the private world in which a citizen lives and the public world typified by government. A public and a counterpublic are spaces of discourse, and they are self-creating and self-organized though affected by preexisting forms and channels of communication. They must be more than a list of one's friends. They must include strangers though selected by shared social space. They must be active, not somnolent.

This is where the major conflict exists for the proponents of nanotechnology. Fearing a reactive public that might demand regulation of nanotechnology to a point whereby it might become inefficient to develop nanotechnologies and nano-products within the United States, some fear the business of nanotechnology might prosper abroad. With the translocation of nanotechnology, the United States might lose a competitive edge on two fronts: national security and economic growth and development.

Mihail Roco, though less than some others, seems to think the antiglobaliza-tion and anti–genetically engineered seeds and foods public might relocate their angst to confront nanotechnology. It took no less than six years for the European Commission to grant its first GMO food approval with Syngento's pest-resistant sweet corn.[62] Even so, the secretary general of EuropaBio warned, "This is only the first step on the road to unblocking the approval process. We will have to wait to see whether further approvals, including those for cultivation, are forth-coming."[63] During this period, the domestic biotechnology industry suffered greatly. "A moratorium on sales of most GMO crops in Europe has been costing American corn growers $100 million to $300 million a year in sales."[64]

Whether or not a valid fear, it seems the GMO analogy underpins much of the rhetoric associated with the NNI's SEIN mandates. As such, one of the goals of SEIN research is associated with producing an informed counterpublic to answer the protests and keep the Roco-motion express on track. However, there is also a second call for a public, and the solution might lie in building a counter-public to answer protests and a public to advance commercialization.

Publics and counterpublics are viable alternatives to Habermas's public sphere. Mary Ryan offered some good examples of these rhetorical vehicles.[65] A set of competing counterpublics can compete with Habermas's misogynist bourgeois public sphere: nationalist publics, popular peasant publics, elite women's publics, and working-class publics. In the nineteenth century, these used alternative styles of political behavior and norms of public speech. Women challenged the masculinist notions of the emergent form of class rule. While subaltern voices may not have the resources or even the desire to participate in national and supranational public spheres, publics and counterpublics are more amenable to less fluent advocates or highly fluent men and women of color.

So, too, groups of critics of newly emerging science and technology might see the need to engage new forms of public behavior and speaking to make their case heard. We have seen the beginning of these new forms of behavior, for example, blogs, virtual publications, Web pages, and so on.

Nancy Fraser posits a dual character to these publics and counterpublics: as spaces of withdrawal and regroupment and as bases and training grounds for agitational activities directed toward wider publics.[66] She makes the case that multiple publics provide a plurality of public arenas in which groups with diverse values and rhetorics can participate.[67] While purists might feel that publics and counterpublics are short changed, the assumption is based on the belief there is an actual center stage on which real policy is made in contemporary society. This is no longer the case. There has been a major devolution of policy making. While it may seem that the post–cold war world has left us with only two superpowers, others would be slighted by this remark. Policy is greater than a few military powerhouses, for today's power rests in economies and trade blocs, in transnational agreements and treaties, and in a plethora of private institutions that include national and subnational voices: world health, agriculture, finance, and so on. The world in all of its complexity cannot be addressed by a single sphere; there are too many issues and too many players. Assuming single or major spheres vertically concentrates decision-making authority in a handful of organizations.

This is not to suggest publics and counterpublics are ideal alternatives to the public sphere. Fraser warns that competing spheres often mask and reproduce domination, hence a comprehensive public sphere is preferable to a multiplicity of competing publics.[68] However, the assumption that a single sphere would lift all the masks and end domination is wishful.

This means that while people may speak the same language, people as publics tend to speak better. The plurality tends to diversify the issues debate and entertain consequences sometimes crowded out by speakers who are not facilitated by breadth of interests people, as publics, would consider. The quality-of-life variables for inner-city families led by a single parent may not be noticed by a white male plutocrat who spent most of his life defending corporate clients against mass torts. Publics offer them a place to meet and exchange sensibilities so when the public speaks out it does so with a composite voice.

"The interactive relation postulated in public discourse, in other words, goes far beyond the scale of conversation or discussion to encompass a multigeneric lifeworld organized not just by a relational axis of utterance and response but by potentially infinite axes of citation and characterization."[69] Publics have a deeper database of sensibilities and ideas and their voices are emergent as well; the voice of the public is greater than the sum of the voices of its individual members.

In the United States, there are already multiple publics built around issues and categories or clusters of issues. For example, there are publics associated with the national election cycle in American politics: we have issues-dominated parties, such as the Reform Party or the Green Party; philosophy-dominated parties, such as the Democrats and the Republicans; and we also have publics, such as the senior-citizen-interest publics, dominated by the AARP; labor publics, such as the Teamsters and the AFL-CIO; and even single-issue publics, such as the National Rifle Association. Hence, when we are discussing a public sphere for nanotechnology, we are talking about a group of individuals with competencies and an interest in the future of them.

ABOUT MOVEMENTS

If we expect a public to function as a public sphere responsive to the civil challenges promulgated by the arrival of nanotechnology, it might behoove us to examine current and prospective movements, which may coalesce around this new technology. The links between publics and movements are significant. Initially, much of the literature on public issues uses specific movements to illustrate theses and conclusions. Indeed, some of the more animated public responses to government policies have been provoked and sustained by groups of people acting as a movement—Vietnam antiwar, pro-life vs. pro-choice, and so on. Delineating between the two (publics and movements) is incredibly difficult to do. In addition, once we begin to understand that there exists many publics and counterpublics, the term *movement* seems to fit the bill better, especially if they are defined more loosely.

Movement theory dates back to the '30s when Jerome Davis's *Contemporary Social Movements* was written.[70] A debate over whether a movement needs a formalized organization has never been resolved satisfactorily. However, there is little controversy that movements have a temporal characteristic; they involve sustained activity. They may form a collective action with and without recognized leadership and formal organization over a period of time. Oftentimes, they erupt out of rapid social change with or without social disintegration. They often develop in response to social groups being left behind or excluded from a societal payoff and are sustained when their sentiment base is strong and societal hostility is low. They are related intimately to the mass media that can help set a movement's agenda by ignoring it, hence, the movement decays, or by featuring it, hence, the movement strengthens. As such, movements are very much concerned with communications to the point where a network for communicating with nonmembers and the media is part of its organizational structure. Any concerted anti-nanotechnology public would need these characteristics as well as a counterpublic to oppose them.

Movements decay over time or change radically. These generally involve goal transformation, such as diffusion of goals whereby unattainable goals are replaced by diffuse goals to increase the number of targets or merely to maintain the legitimacy of the movement. It can include a shift in organizational maintenance whereby the primary activity of the organization becomes its maintenance. The third and final transformative process includes the concentration of power in a minority of the members. Lesser processes include coalitions with other organizations, organizational disappearances, factional splits, increased rather than decreased radicalism, and the life careers of members.[71] Movements are flexible and adjust to new circumstances by morphing themselves. As such, the first truly anti-nanotechnology public may morph from a grander goal like anti-GMO or antiglobalization. A viable counterpublic would need a vision sufficiently grander than simply opposing an anti-nanotechnology public to sustain itself as targets shift.

The rhetorical treatment of movements can be traced back seven years earlier than Davis's book.[72] Most of the studies have tended to examine movements historically, some in terms of the role of a leader or group of leaders and their effect or effects on the movement. The following summarizes Doug McAdam and David Snow's compilation on movement theory.[73] Movements are differentiated from social trends such as industrialization, urbanization, and so forth because individuals act as an aggregate rather than a collectivity. Movements are also compared to sentiment pools, which are change-oriented opinions and beliefs that do not involve collective action. They are contrasted with mass migrations such as diasporas but do not include noninstitutional tactical action. Finally, they are compared against interest groups, which are embedded within the mainstream environment and use institutionalized means.

Movements (read: publics) have certain informal characteristics involving collective or joint action, change-oriented goals, some degree of organization, some degree of temporal continuity, and some extrainstitutional collective action. Movements can move from a local phenomenon into a global one and vice versa. At times, local movements may maintain their local focus while serving as a component of a larger global network. Local movements seldom succeed by addressing global implications of issue without anchoring the larger issue to a local phenomenon. In turn, global movements can provide opportunities for local groups seeking to challenge powerful interests at home. The growth of media, especially the Internet, has increased to permeability of movements at both the local and global scales to cross-fertilization by ideology, tactics, and strategies from each other.

Generally speaking, most of the scholarship on movements has reflected on national movements without a transnational character, hence many of their observations seem challenged with increasing globalization. John A. Guidry and his team see the transnational public sphere as a postmodern construct that "undermines the traditional case with which the commonality of that national public sphere can be imagined."[74]

Transnational movements mostly are the product of transformations in transportation and communication technology, not the least of which is twenty-four-hour cable news coverage from CNN, the general inability of states to establish order and mediate change within borders, and the proliferation of intergovernmental institutions and transnational corporations. With such a degree of time and space compression, movements tend to spill over borders, and others exist in a space that does not have geographical restraints at all, existing in a virtual sense. In general, it is safe to claim that the transnational public makes local and global movements transformative.[75]

These characteristics of movements provide us with some guides regarding how we may need to respond to the anti-nanotechnology movement. Any anti-nanotechnology movement will be challenged to maintain its coherence since the foci of its protest will be varied. Nanotechnology affects many product lines and has dozens, if not hundreds, of potential applications. What intersections can be negotiated between media and the agenda and tactics of the movement will determine its vitality. Attracting media attention toward a response strategy would reduce media exposure that protestors desperately need to sustain their activity, sometimes livelihood. We can anticipate local response teams of protestors in league with transnational groups, which might enable boycotts. The transnational nature of an anti-nanotechnology movement might be challenging given what little we know about transnational movements, though it does suggest that an international counterpublic might be in order, and responses by them would need to be tailored at both local and national levels. As local teams attempt to become more national, they lose footing in the community that supports them both conceptually and financially, making them vulnerable. If the membership is anything like the antiglobalization movement, it will be easily distracted with minor victories. According to Jesse Lemisch,

> The wild kids that the media have told us about may in fact be (with important exceptions) deferential, respectful, and with—sad to say—good manners and a poignant longing to love and be loved by their adversaries.... The longing to be at

peace with your adversaries is death to social movements. A generation many of whose members think that forthright debate should be avoided as "towel-snapping," and groove on the benighted and directionless kids of *Blair Witch Project* is a generation which, even at its best, sometimes reflects the worst in its times.[76]

The Anti-GMO Movement

The GMO story: Enter the Flavr Savr tomato and Bt (*Bacillus thuringiensis*) cotton and corn. Some 60 to 75 percent of processed foods—ranging from major soy-based baby formulas to some of the most popular brands of corn chips—include some genetically engineered ingredients.[77] While grains and vegetables enhanced by biotechnology have been common since the mid-'90s in the United States, there have been no ill effects on American consumers. Nonetheless, there is a worldwide neurosis over GMO products. In the United States alone, more than a hundred separate organizations agitate against Frankenfoods and have spent more than $400 million worldwide since 2001 to support protests and boycotts.[78]

The case for the exportation of transgenic crops was made in terms of feeding the hungry much like the case made for the Green Revolution decades earlier. Friends of the Earth reported that acceptance of GMO crops by developing countries as a condition for food aid and food donations was "playing with hunger," exploiting famine to promote US agribusiness.[79] Biotech industry analysts admit that the biotech industry made a mistake in trying to market itself as fighting world hunger and then not delivering.

A secondary locus for complaints has been the issue of farmer-saved seed, which allow farmers to purchase the initial planting and reduce subsequent purchases of seed by saving seed from the previous crop. While Monsanto backed away from this research, residual distrust remained sufficient to empower the protest movement.

Monsanto, holder of patents for GMO corn, soybean, and cotton, was demonized as the greedy, mad scientist of the new millennium. In August 2000, Monsanto released its rice genome research into the public domain and dropped its development of the "terminator" gene to block plants from producing next-generation seed. While it is working on producing golden rice in Asia and a special mustard seed in India (both modified to include vitamin A), virus-resistant potatoes in Mexico and Kenya, and virus-resistant papaya in Hawaii and Southeast Asia, it could not shake off the bad press from its early stumbles.[80]

The hype over nanotechnology may create the same resistance faced by GMOs, especially if the debate over nanotechnology is "contained in the scientific and industrial communities and couched in technical and scientific terms." The scientific community and society may "be repeating the path" taken by genetic modification.[81] According to Colvin's congressional testimony,

The campaign against GMOs was successful despite the lack of sound scientific data demonstrating a threat to society. In fact, I argue that the lack of sufficient public scientific data on GMOs, whether positive or negative, was a controlling factor in the industry's fall from favor. The failure of the industry to produce and share information with public stakeholders left it ill-equipped to respond to GMO detractors. This industry went, in essence, from *wow* to *yuck* to *bankrupt.* There is a powerful lesson here for nanotechnology.[82]

No one will forget the effects.

ANTI-NANOTECHNOLOGY MOVEMENT

Strangely enough, the only mobilized groups in the United States have been Top-less Humans Organized for Natural Genetics (THONG) and another unorganized group that protested the Molecular Foundry groundbreaking ceremony at Berkeley in January 2004.

THONG disrupted a 2004 Chicago meeting of the NanoBusiness Alliance and protested outside Eddie Bauer in May 2005. They have gotten most of the attention because, well, they protest in thongs.

According to Nigel Walker, lead scientist for the National Institute of Environmental Health Sciences, "groups like THONG have succeeded in raising awareness of nanotechnology. And, as it was in the case of the backlash against genetically modified organisms, if the government does not become engaged early on and address concerns, there will be problems with public acceptance after a significant number of products are already on the market. At the end of the day we serve the taxpayer."[83] Whether or not this is true and though some nanocommentators suggest we should be vigilant to avoid public protest, recent protests have hardly shut down the laboratories or commercialization.

According to Colvin, "nanotechnology's legacy will be determined in the months ahead, when nanoscientists and the public confront the inevitable *bad news about nanotechnology.*"[84] Over a year ago, she made this prediction and the *bad news* about nanotechnology doesn't seem to have hurt it much. Of course, in 2003, her voice was joined by others like Glenn Reynolds: "Anti-technology activists have already targeted nanotechnology."[85] Again, we're waiting and all we've seen are the backsides of partially naked people protesting a range of industry practices.

Colvin seems overconcerned that nanotechnology might be subject to the same level of criticism leveled against GMO foods: "The public's enthusiasm for an emerging new technology can easily turn to fear, with grave consequences for commercialization. Emerging technologies do pose risks that are ill-characterized, and the best things nanoscientists can do—both for the discipline and society—is to draw attention to possible risks and study them carefully."[86] She believes that study will produce the kinds of information that the nanotechnology industry can use to mitigate the environmental risks or avoid them altogether. "Legislators are concerned that people from the anti-globalization groups and the anti-genetically modified organisms groups will get on the bandwagon and fuse together into some sort of anti-nano group."[87]

Presumably, the industry has a short window of opportunity. She and others have warned that to follow the misguided path of biotechnology simply is not an option. "Given that the well-organized and influential biotechnology industry has suffered some serious setbacks at the hands of activists, some worry that the embryonic nanotechnology industry will be a pushover."[88]

In addition, "various environmental campaigns groups on the look out for the next big thing have been developing nanotechnology as a new target. With child-like simplicity that would make any spin doctor proud, they have been hammering home the message that nanotechnology is *the next GMO*."[89]

Dismissing this potential anti-nanotechnology movement might be unwise. Even if the fears engendered by the movement are scientifically suspect, this does not mean it will fail to have any impact. "Faced with such patent nonsense, e.g., the *goos*, it is easy to dismiss the anti-nanotech campaign as an irrelevant rant—easy, but unwise. The claims made by groups such as ETC may at times appear irrational, but that doesn't mean they won't have resonances with the public."[90] By packaging counter-nanotechnology claims with some bigger issues about which citizens and scientists may share misgivings, the movement may gain a sympathetic ear. Even without evidence and experimental findings, "the European experience with GMO crops has shown that environmental campaigners are adept at building from vague public misgivings to whip up a storm of protest, even if their arguments rest on shaky scientific grounds."[91] Policy makers are less concerned about whether the basis of protests is predicated on correct science than they are about controlling dissension among their constituencies. Correct science seems to mean a lot to scientists but not to others in contemporary American politics.

The movement's umbrella could be a very large one. We have a situation where antiglobalization protestors have converted in many instances into anti-GMO protestors as well. "Campaigners recently warned that they next great anti-technology battle will be fought not in fields of GMO crops but outside the high tech plants of nanotechnology manufacturers."[92] The leap to anti-nanotechnology protestors is not improbable. Indeed, "many of the activists now calling for strict controls on nanotechnology are veterans of the GMO crop wars."[93]

There are multiple grounds for concern. First, much like the spillover effect from anti-GMO protests abroad, reservations voiced abroad may cross the Atlantic and take root here among the domestic antiglobalists. Many of the domestic groups have strong foreign connections, and national borders have eroded under the forces of free trade agreements and other features of globalization. Nanotechnology will be global in application and will be integrated in products as diverse as pharmaceuticals to technological fixes. These transnational connections make the spillover more likely.

Second, anti-GMO protestors share many of the Luddite-like concerns expressed by the anti-nanotechnology folks. Clare Short says that the protesters are "today's Luddites. . . . Their call [is] to halt historical change and tear down our international institutions."[94] They have found an audience, though their audience seems mostly unconcerned about the radical nature of the alternatives. Short's categorization should make nanotechnology enthusiasts wary at least. If these Luddites can find an alternative, they might be very powerful as a counterpublic.

Third, the antiglobalization and anti-GMO movement needs to redefine itself as sites for protests become fewer and more geographically remote, leaving only virtual protest locations in cyberspace. "The movement needs to ask itself where it goes from here. It must find a role that is not only, as Naomi Klein, chronicler of anti-corporatism, puts it, turning up at international meetings like Deadheads following the Grateful Dead. Whether it burns out or turns into the next big thing will take time to see."[95] Some of these groups have fanatic adherents, infrastructures with overhead and salaried employees. Finding a new locus for their protests means a continuing mission and another round of participation. Having an opportunity to continue to work together might be motivation enough to break into an anti-nanotechnology agenda.

To produce a counterpublic to respond to an antinano movement and a public to promote the commercialization of nanotechnology is a monumental undertaking. The citizen-consumer has been shut out of science and technology decision making for generations. Even if they can be prepared to participate, the next challenge will be to design and establish venues for them in which to participate.

EXPERIMENTS

There is no doubt we need ways to communicate with the public on science and technology. For example, the "way of debating opposing propositions, leading to disagreements over extreme positions needs to be tempered by an approach of engaging around alternatives. We need new forms of participatory democracy to stop antagonistic extremes engaging in battles over science policy."[96] There are experiments for evaluating science policy using deliberation and public consultation. Deliberative polling utilizes a random sample of citizens and exposes them to an issue. In 1988 James Fishkin outlined the format in an article for the *Atlantic Monthly*. In *Voice of the People*, he laid out the rationales and process of deliberative polling.[97] This practice is traceable to an experiment on Channel 4 in England when, in 1993, a group of people came together for a four-hour briefing and discussion on crime. Another illustration would be the National Issues Convention in

Austin on January 18–21, 1996, involving 460 citizens and costing $4 million. It was broadcasted for five hours on PBS and NPR and preceded the Iowa caucuses and New Hampshire primary.

People were randomly selected and polled and offered an all-expense paid trip to Austin to discuss three issues: the economy, the state of the family, and America's role in the post–cold war world. When they arrived, they were assigned a group that designed questions, and then they posed them to policy experts, including presidential candidates, during a live broadcast. As they left, they were polled again. In the end, "they viewed flat tax proposals less favorably, grew more supportive of giving states responsibility for the social safety net, and agreed more strongly that the U.S. should cooperate militarily with other nations to address trouble spots."[98] In addition, participants reported they learned a lot, and some left with a newfound interest in politics. "I never voted in my life, and neither has anybody in my family," said Marty Dorn, a recovering drug addict from Monett, Missouri. "But I'm going to be the first, and I'm going to get all my friends to vote, too."[99] In addition, members of the media reported heavily on this event. "By sponsoring public forums, bringing contending parties together to talk, and making politics so vivid it is tough to ignore, journalists could start recalling the public to the public's important business."[100]

Another prominent example is a Danish innovation designed for debates on technology policy. The consensus conference was pioneered in the 1980s by the Danish Board of Technology to stimulate broad and intelligent social debate on technological issues. Since 1987, it has organized over a dozen conferences on topics ranging from genetic engineering to educational technology, food irradiation, air pollution, human infertility, sustainable agriculture, and the future of private automobiles.[101]

A steering committee is selected and a topic framed. Next, advertisements in local newspapers ask for lay participants. They return a one-page letter describing their backgrounds and reasons to participate. From the replies, a panel of about fifteen people is selected that represent the demographic breath of the population. The panel is controlled for prior knowledge and specific interest. Weekend one: the lay panel with an expert facilitator discusses an expert background paper; the panel generates questions and begins to assemble an expert panel for later. Weekend two: more background readings are discussed and the selection of the expert panel is finalized. The expert panel receives the questions and is asked to prepare easy-to-understand, succinct answers. On the last weekend, which will last four days, both panels assemble with invited parties, including the media and policy makers. On the first day of the last week, experts speak briefly, and the lay group retires and discusses the presentations. On the second day, the lay group cross-examines the experts. For the remainder of day two and throughout day three, the lay group prepares a report. On the fourth day, the experts correct the document for outright factual misstatements. The report may run from fifteen to thirty pages.

According to Richard E. Sclove,

> Conferences that were held in the late 80s influenced the Danish Parliament to pass legislation limiting the use of genetic screening in hiring and insurance decisions, to exclude genetically modified animals from the government's initial biotechnology research and development program, and to prohibit food irradiation for everything except dry spices. Manufacturers are taking heed of the reports that emerge from consensus conferences as well.[102]

Recently, consensus conferences on nanotechnology have been held in Raleigh, North Carolina,[103] and Madison, Wisconsin.[104]

A variation on this model was developed at the Jefferson Center, a Minneapolis-based nonprofit organization. It is called a citizens jury. These juries have been conducted since 1974 and have tackled an array of issues, including science and technology ones, such as solid waste disposal, organ transplantation, environmental risks, and so on. Similar projects have been conducted internationally. They involve three to four months of planning. A citizens' jury involves a randomly selected and demographically sensitive panel of eighteen or so jurors that meets for four or five days. The panel hears from expert witnesses in professionally moderated hearings and then deliberates. On the final day, the jury presents its recommendations. An advisory committee helps the project staff avoid bias.[105] In 2005 the Cambridge University–based Interdisciplinary Research Collaboration (IRC) in Nanotechnology hosted a jury. Presumably, UK NanoJury "represented a unique alliance between university researchers in both social and physical sciences, Greenpeace and a national newspaper to put the debate surrounding the future of nanotechnology on a balanced and informed footing."[106] They will report their conclusions in late 2005.

A working model for a citizen advisory panel is the North Carolina Citizens Technology Forum, and Patrick W. Hamlett of North Carolina State University has an ongoing research program that is exploring nonexpert, public deliberations about technology. He organized two such conferences in North Carolina, one meeting face-to-face, the other meeting on the Internet. Both groups examined genetically modified foods. In the second phase of his research, he is conducting eight more consensus conferences, six that are Internet-only and two that will be "mixed mode": some elements conducted face-to-face, some over the Internet. These groups will examine global warming. In 2002 he received a $150,000 grant from the NSF to organize a national distributed-coordinated consensus-building experiment to examine nanotechnology.

PROBLEM SOLVED?

There are a plethora of problems with deliberative experiments. Decisions do not have to be based on rational warrants; material prerequisites for deliberation are unequally distributed (some citizens are more likely to be listened to); the association between deliberation by citizens and policy making by elite is weak (while the process may convince citizens the procedures are democratic, it may only corroborate current inequities), and so on. While interesting, this debate will wait for another day.

On a grander scale, we need to ask whether these experiments will be effective in producing a counterpublic sufficient to demobilize protestors calling for restrictive regulation, if not outright bans, and sufficient to produce an educated consuming public for nanoproducts.

The assumption that conference attendees will stand up against protestors is ludicrous. The informed are not necessarily the same who would take the initiative to respond to protest, directly or indirectly. There is no "National Deliberation Day" (another Fishkin idea). On the other hand, participants in these conferences might go out and purchase nanoproducts. However, it is dubious that these expensive experiments are needed when the advertising industry has found ways to persuade Americans to buy lots of stuff they do not need.

Simply put, if we want a truly participative public sphere in science and technology (and nanotechnology), these experiments may be insufficient to produce an adequate level of change. At least before they can work, we would need venues available to allow citizens' points of view to impact science and technology decision making. Perhaps citizens need to be appointed to science advisory boards and given the power to review grant proposals to federal agencies. Maybe national referenda on science and technology issues should be placed on election ballots. Or maybe not.

Do the US federal government, the NSF, and others really want citizen participation? According to Glenn Reynolds:

> The real problem isn't distrust of science. It's a distrust of people. And viewed in this light, these fears are likely to be strongest where pessimism about humanity is the strongest. It looks that way, too, with Europe leading the way in throwing some people's only favored invention—the wet blanket—over nanotechnology research. In the more-optimistic U.S., there are concerns, but they haven't yet led to a strong interest in regulating nanotechnology.[107]

Or does the federal government want to seem as if they want it? Is this truly an effort to democratize science and technology decision making or merely another round of fear management? They have almost convinced us it is safer to fly these days with the Transportation Security Administration and an antiterrorism policy within the Department of Homeland Security. Time will tell.

CONCLUSION

The public sphere is not healthy in America. The public sphere associated with science and technology policy making is nonexistent. At best, a weak public sphere might be managed as a public, and the public could be constructed from efforts by SEIN researchers to ask questions, collect and manage data, archive and disseminate thoughtful criticisms, and refute the cranks, technophiles, and technophobes, separating the serious researchers from the publicity seekers and the public relations firms supported by business and industry who claim to be experts but are not objective since they work with significant and substantial self-interest. Building a public will not be easy, though it seems there is some commitment to do so at this time, a least in terms of nanotechnology.

Congregated together from antiglobalization, anti-GMO, and even antiwar groups, an anti-nanotechnology movement has some capacity to slow and frustrate nanotechnology policy making in this decade. Selecting an appropriate response strategy will be critically important. By coupling a response strategy with a bona fide effort to build a public for science and technology policy, we may see the very first steps toward the integration of public variables into the science and technology policy calculus.

While it's uncertain what this may mean to liberal democratic theory, it does open governance at a time when issues like nanotechnology have begun to play a more central role in politics and policy. As biotechnology, computerization, advanced robotics, artificial intelligence, and nanotechnology converge, the definition of what it means to be human will be tested. When debates about who and what we are occur, it becomes important to broaden the participation to secure the sensibilities and the interests of society as a whole since the stakes in this deliberation impact everyone. The new global commons will be the field on which these issues are wrestled and subdued, or else people will find themselves living in an unfamiliar community on a foreign planet.

NOTES

FOREWORD

1. M. C. Roco, "Nanoscale Science and Engineering: Unifying and Transforming Tools," *Journal of the American Institute of Chemical Engineering* 50, no. 5 (2000).

INTRODUCTION

1. Jeff Karoub, "Nanotech goes commercial," *Small Times* (July/August 2003): 10.

2. Peter Binks, "Questions loom large in nanotech's tiny world," *Age*, October 21, 2003, http://www.smh.com.au/articles/2003/10/20/1066631346170.html (accessed October 28, 2003).

3. David Pescovitz, "Nano's got the ways and memes for a viral assault on pop culture," *Small Times* (January/February 2004): 56.

4. "Of boys and ducks, odd ducks and quantum ducks," *Small Times*, April 2005, p. 4.

5. Review by Berube.

6. "Fox catches *Prey*," *IGN Insider*, http://filmforce.ign.com/articles/366/366010p1 .html (accessed September 23, 2003).

7. "Nanotech: Hype yes, bubble no," *Nanotech Report* 1, no. 10 (December 2002): 2.

8. "Summer story stocks," *Nanotech Report* 2, no. 7 (July 2003): 3.

9. Rick Weiss, "For science, nanotech poses big unknowns," *Washington Post*, February 1, 2004, p. A01, http://www.washingtonpost.com/ac2/wp-dyn/A1487-2004Jan31 (accessed February 4, 2004).

10. Ralph Hall, Rep. Texas, Committee on Science House of Representatives, *The*

Societal Implications of Nanotechnology, Hearings before the Committee on Science, House of Representatives, 108 Cong. 1st Sess, Serial No. 108-13, April 9, 2003, p. 12.

11. Ibid.

12. Pat Mooney, "The ETC Century: Erosion, technological transformation and corporate concentration in the 21st century," *Development Dialogue* 1, no. 2 (1999): 69.

13. John Roach, "Nanotech: The tiny science is big and getting bigger," *National Geographic News*, March 24, 1005, http://news.nationalgeographic.com/news/2005/03/0324 _050324_nanotech.html (accessed March 28, 2005).

14. Elhanan Helpman, ed., *General Purpose Technologies and Economic Growth* (Cambridge, MA: MIT Press, 1998).

15. Stephen Baker and Ashton Aston, "The Business of Nanotech," *Business Week*, February 15, 2005.

16. Richard Foster and Sarah Kaplan, *Creative Destruction* (New York: Doubleday, 2001).

17. Michael Mauboussin and Kristen Batholdson, "Big money in thinking small: Nanotechnology—What investors need to know," A Report by Credit Suisse First Boston Equity Research, July 15, 2002, http://www.csfb.com/home/index/index.html (accessed May 7, 2003).

18. Pat Mooney, "The ETC Century: Erosion, technological transformation and corporate concentration in the 21st century," *Development Dialogue* 1, no. 2 (1999): 75.

19. "Rage against the machines," *SciFi*, December 2003, p. 59.

20. This term seems to have first appeared in 1974. Norio Taniguchi's "On the basic concept of nanotechnology" appeared in the *Proceedings of the International Conference on Production Engineering*, Tokyo. He defined nanotechnology as "the production technology to get extra high accuracy and ultra fine dimensions, i.e., the preciseness and fineness of the order of 1 nm (nanometer), 10^{-9} in length" (p. 18).

Drexler attempted to explain the scale involved in nanotechnology: "From the 30 micron scale of a fairly typical human cell to the 0.3 nanometer scale of a typical atom is a factor of 100,000 in linear dimension, or a factor of 1,000,000,000,000,000 in volume. This is the difference between a mountain and a marble. Atoms make up molecules, which make up cells, with nanocomputers far larger than typical molecules, and yet far smaller than typical cells." See Drexler, "Nanotechnology: Since engines," *Foresight Background*, no. 6 (1989): 2.

Taniguchi's view of nanotechnology refers to tiny machines and their components either ground down to nanoscale or involving fine resolution lithography techniques, faster silicon devices, integrated optoelectrical systems, and nano-engineering process technologies. See Tim Studt, "Chip advances means faster and smaller IC devices," *R&D Magazine*, May 1991, pp. 72–74.

21. Arthur R. Von Hippel, *Molecular Science and Molecular Engineering* (New York: MIT Press, 1959).

22. Though Drexler popularized the concept of nanotechnology, the concept of microlevel or molecular-level machinery is easily traceable to Richard Feynman and a Christmas 1959 talk he gave at an American Physical Society meeting. See Richard Feynman, "There's plenty of room at the bottom," *Engineering and Science*, 1960, pp. 22–36.

23. Adam Keiper, "The nanotechnology revolution,"*New Atlantis: A Journal of Technology and Society,* Summer 2003, http://www.thenewatlantis.com/archive/2–/keiper-print.htm (accessed September 12, 2003).

24. Ibid.

25. Ibid.

26. Ibid.

27. Ibid.

28. SEIN. Pronounced *sane*, a convenient homonym, it refers to "societal and ethical implications of nanotechnology." "Such areas include environment and safety, equity, and potential conflict of interest arising from the interactions among government, industry, and universities, and intellectual property ownership." See W. Zhou, "Ethics of nanobiology at the frontline," *Santa Clara Computer and High Technology Law Review* 19 (2003): 485.

CHAPTER 1

1. N. Katherine Hayles, "Connecting the quantum dots: Nanotechscience and culture," in *Nanoculture: Implications of the New Technoscience,* N. Hayles, ed. (Bristol, UK: Intellect Books, 2004), p. 11.

2. Kevin Kelleher, "Nanotech: Dissecting what it is, and isn't," *Street.com,* March 29, 2005, http://www.thestreet.com/tech/kevinkelleher/10215012.html (accessed March 29, 2005).

3. Pat Mooney, "Nanotechnology: Nature's tool box," *ABC Radio National,* November 14, 2004, http://www.abc.net.au/rn/talks/bbing/stories/s1241931.htm (accessed December 13, 2004).

4. Bruce Fraser, "Interpretation of novel metaphors," in *Metaphor and Thought,* 2nd ed., Andrew Ortony, ed. (Cambridge: Cambridge University Press, 1993), p. 334.

5. "Portfolio plays in nanotech," *Nanotech Report* 2, no. 4 (April 2003): 2.

6. These terms were drawn from Glenn Reynolds, "Forward to the future: Nanotechnology and regulatory policy," a Pacific Research Institute Paper, November 2002, p. 1, http://www.pacificresearch.org/pub/sab/techno/forward_to_nanotech.pdf (accessed October 3, 2003).

7. Royal Academy of Engineers and Royal Society, *Nanoscience and Nanotechnologies: Opportunities and Uncertainties,* 2004, http://www.nanotec.org.uk/finalReport.htm (accessed July 30, 2004), p. 1.

8. Christopher J. Preston, "The promise and theatre of nanotechnology: Can environmental ethics guide US?" *Hyle* 11, no. 1, 2005, http://www.hyle.org/jounral/issues/11-1/preston.htm (accessed May 2, 2005).

9. Glenn Reynolds, "Forward to the future: Nanotechnology and regulatory policy," A Pacific Research Institute Paper, November 2002, http://www.pacificresearch.org/pub/sab/techno/forward_to_nanotech.pdf (accessed October 3, 2003), p. 2.

10. "Nano name game," *Forbes/Wolfe Nanotech Report* 3, no. 2 (February 2004): 1.

11. Charles Morgan, "Nanotech insiders—You don't know what you're missing," November 23, 2004, e-mail communication.

12. "Nano name game," *Forbes/Wolfe Nanotech Report* 3, no. 2 (February 2004): 1.

13. Ibid.

14. Barnaby Feder, "Concerns that nanotech label is overused," *New York Times*, April 12, 2004, http://www.nytimes.com/2004/04/12/technology/12phone.html?ex=1397188800&en=20db9db23c360970&ei=5007&partner=USERLAND (accessed May 2, 2004).

15. Ibid.

16. Andrew Dietderich, "Nanotechnology will grow slowly, be a tough sell, panel says," *Crains Detroit*, June 15, 2005, http://www.crainsdetroit.com/cgi-bin/news.pl?newId=6240&print=Y (accessed June 16, 2005).

17. Josh Wolfe, "Nano-zilla strikes Tokyo," *Forbes.com*, April 20, 2005, http://www.forbes.com/2005/04/20/cz_jw_0420soapbox_inl_print.html (accessed April 25, 2005).

18. Andrew Dietderich, "Nanotechnology will grow slowly, be a tough sell, panel says," *Crains Detroit*, June 15, 2005, http://www.crainsdetroit.com/cgi-bin/news.pl?newId=6240&print=Y (accessed June 16, 2005).

19. David Forman, "Best laid plans," *Small Times* (July/August 2004), p. 29.

20. Thiemo Lang, "The nano-savvy investor: Making sense of the public markets," *Small Times* (March 2005): 27.

21. "Much ado about almost nothing," *Economist*, March 18, 2004, http://www.economist.com/science/displayStory.cfm?story_id=2521232 (accessed April 5, 2004).

22. "Making it big in nanotech," *Red Herring*, September 15, 2004, http://www.redherring.com/article.aspx?a=10836&hed=Making+it+big+in+nanotech&sector=Industries&subsector=%09%09%09%09%09%09%09Computing (accessed June 17, 2005).

23. Kevin Maney, "Scared of nano-pants? Hey, you may be onto something," *USA Today*, June 21, 2005, http://www.usatoday.com/tech/columnist/kevinmaney/2005-06-21-nano-pants_x.htm (accessed July 14, 2005).

24. Michael E. Davey, "RS20589: manipulating molecules—The National Nanotechnology Initiative," *CRS Report for Congress*, September 20, 2000, http://www.ncseonline.org/NLE/CRSreports/Sceince/st-48.cfm?&cfd=14971140&cftoken=12149058 (accessed July 19, 2004).

25. Ron Wyden, *H.R. 766, Nanotechnology Research and Development Act of 2003*, Hearings before the Committee on Science, House of Representatives, March 19, 2003, p. 18.

26. W. S. Bainbridge, "Sociocultural meanings of nanotechnology: Research methodologies," an unpublished paper from the 2nd Workshop on Societal Implications of Nanoscience and Nanotechnology (Arlington, VA: NSF, December 3–5, 2003).

27. Rustum Roy, "Giga science and society," *Materials Today* 5, no. 12 (2002): 72, http://www.materialstoday.com/pdfs_5_12/opinion.pdf (accessed October 4, 2003).

28. Ibid.

29. Richard Weaver, "Ultimate terms in contemporary rhetoric," in *Language Is Sermonic: Richard M. Weaver on the Nature of Rhetoric*, ed. Richard L. Johannesen, Rennard Strickland, and Ralph T. Eubanks (Baton Rouge: Louisiana State Univ. Press, 1970): 87–112.

30. Department of Trade and Industry, *New Dimensions for Manufacturing: UK Strategy for Nanotechnology*, Report of the UK Advisory Group on Nanotechnology Applications (United Kingdom: Department of Trade and Industry, June 2002) (accessed October 5, 2003), p. 46.

31. William G. Schultz, "Nanotechnology under the scope," *Chemical and Engineering News*, December 4, 2002, http://pubs.acs.org/cen/today/dec4.html (accessed October 5, 2003).

32. Arie Rip, "Technology assessment as part of the co-evolution of nanotechnology and society: The thrust of the TA program in NanoNed, appendix 2: NanoNed proposal, 12 February 2003," a paper contributed to the Conference on *Nanotechnology in Science, Economy and Society*, Marburg, Germany, January 13–15, 2005.

33. "Gartner says no new major IT innovation before 2005," *Press Release*, 2002, http://banners.noticiasdot.com/termometro/boletines/docs/ti/gartner/2002/gartner _Before_2005.pdf (accessed July 25, 2004).

34. See *Gartner Home Page*, 2004, http://www4.gartner.com/research/special _reports/hype_cycle/hc_special_report.jsp (accessed July 23, 2004).

35. Both embedded PowerPoint slides are from a presentation by David Schehr, Research Director, Gartner Financial Services in 2002. See http://www.nicsa.org/ Archives/download/TF04_Schehr.ppt (accessed July 26, 2004).

36. "Gartner says no new major IT innovation before 2005," press release, 2002, http://banners.noticiasdot.com/termometro/boletines/docs/ti/gartner/2002/gartner _Before_2005.pdf (accessed July 25, 2004).

37. Phil Wainwright, "Weekly review: Web services journey will be long," *ASPNews*, October 29, 2002, http://www.aspnews.com/analysis/analyst_cols/print.php/11275 _1490151 (accessed July 23, 2004).

38. Andrew Briney, "Hype, hype, hooray," *InfoSecurity*, July 2003, http://infosecurity mag.techtarget.com/2003/jul/note.shtml (accessed July 23, 2004).

39. Patrick McGee, "Nanotech, but not in a nanosecond," *Wired News*, November 30, 2003, http://www.wired.com/news/technology/0,1282,48737,00.html?tw=wn_story _related (accessed July 27, 2004).

40. Dan Gillmor, "Big breakthroughs can come in small packages," *SiliconValley.com*, February 16, 2002, http://www.siliconvalley.com/mld/siliconvalley/business/columnists/ 2685956.htm (accessed July 27, 2004).

41. Rex Crum, "Talking nanotech with Mike Honda," *CBS.Marektwatch.com*, March 17, 2004, http://cbs.marketwatch.com/news/print_story.asp?print=1&guid={A637E67A -A5D7-41B- (accessed April 5, 2004).

42. Jim von Ehr, transcript, "Imaging and imagining conference," March 4, 2004, unpublished.

43. Philip Bond, "Converging technologies and competitiveness," *Converging Technologies for Improving Human Performance: Nanotechnology, Biotechnology, Information Technology and Cognitive Science*, ed. Mihail C. Roco and William Sims Bainbridge (Arlington, VA: National Science Foundation, June 2002), p. 30.

44. "Don't believe the hype," *Nature* 424, no. 6946 (July 17, 2003): 237.

45. While originally tagged to Zyvex's Web site, it has either been removed or was mistagged by Greenpeace. Hence, Alexander Huw Arnall, *Future Technologies, Today's Choices*. Its secondary title is more enlightening: *Nanotechnology, Artificial Intelligence and Robotics; A technical, political and institutional map of emerging technologies* (London: Greenpeace Environmental Trust, July 2003): p. 37.

46. ETC Group, "Nanotech news in living colour: An update on white papers, red flags, green goo, grey goo (and red herrings)," *Communique* 85 (May/June 2004): 5.

47. Chris Phoenix, "Nanotechnology: Nature's tool box," *ABC Radio National*, November 14, 2004, http://www.abc.net.au/rn/talks/bbing/stories/s1241931.htm (accessed December 13, 2004).

48. K. Eric Drexler, *Engines of Creation* (New York: Anchor Press, 1986), p. 172.

49. Chris Phoenix and Eric Drexler, "Safe exponential manufacturing," *Nanotechnology* 15 (2004): 870.

50. "Don't believe the hype," *Nature* 424, no. 6946 (July 17, 2003): 237.

51. Stephen Strauss, "A far-fetched theory that won't come unstuck," *GlobeandMail.com*, July 24, 2004, http://www.theglobeandmail.com/servlet/ArticleNews/TPStory/LAC/20040724/GOO24/TPScience/ (accessed July 26, 2004).

52. Adam Keiper, "The nanotechnology revolution," *New Atlantis: A Journal of Technology and Society*, Summer 2003, http://www.thenewatlantis.com/archive/2/keiper print.htm (accessed September 12, 2003).

53. Center for Responsible Nanotechnology, "Grey Goo Is a Small Issue," 2003, http://www.crnano.org/BD-Goo.htm (accessed December 15, 2003).

54. Ibid.

55. Amanda Armstrong, "Nanotechnology: Nature's tool box," *ABC Radio National*, November 14, 2004, http://www.abc.net.au/rn/talks/bbing/stories/s1241931.htm (accessed December 13, 2004).

56. David Reid, "Nanotechnology pioneer slays *grey goo* myths," *EurekaAlert!*, June 8, 2004, http://www.eurekaalert.org/pub_releases/2004-06/iop-nps060704.ph (accessed July 9, 2004).

57. Chris Phoenix and Eric Drexler, "Safe exponential manufacturing," *Nanotechnology* 15 (2004): 869.

58. Ibid., p. 871.

59. Paul Rincon, "Nanotech guru turns back on goo," *BBC News*, June 9, 2004, http://newsvote.bbc.co.uk/mpapps/pagetools/print/news.bbc.co.uk/1/hi/sci/tech/3788673.stm (accessed June 10, 2004).

60. Stephen Strauss, "A far-fetched theory that won't come unstuck," *GlobeandMail.com*, July 24, 2004, http://www.theglobeandmail.com/servlet/ArticleNews/TPStory/LAC/20040724/GOO24/TPScience/ (accessed July 26, 2004).

61. Ibid.

62. Sherwood Boehlert, Opening Statement from the Hearing on Nano Consequences, April 9, 2003, http://www.house.gov/science/hearings/full03/apr09/boehlert.htm (accessed January 7, 2004).

63. Glenda Chui, "Who's afraid of nanotechnology," *Mercury News*, September 16, 2003, http://www.bayarea.com/mld/mercurynews/living/health/6783577.htm (accessed October 2, 2003).

64. "Nanotechnology: Public debate takes off," *EurActiv.com Portal*, July 31, 2003, http://www.euractiv.com/cgi-bin/cgint.exe/1268169-86?14&1015=7&1004=1506022 (accessed October 17, 2003).

65. Kate Marshall, "Future present: Nanotechnology and the scene of risk," *Nanocultures: Implication of the New Technoscience*, N. K. Hayles, ed. (Bristol, England: Intellect Books, 2004), pp. 154 and 155.

66. David Reid, "Nanotechnology pioneer slays *grey goo* myths," *EurekaAlert!*, June 8, 2004, http://www.eurekaalert.org/pub_releases/2004-06/iop-nps060704.ph (accessed July 9, 2004).

67. National Science and Technology Council, *Nanotechnology: Shaping the World Atom by Atom* (Washington, DC: U.S. GPO, n.d.), p. 2.

68. Graham Collins, "Shamans of small," *Scientific American* 285, no. 3 (September 2001): 85.

69. Ibid.

70. Ibid.

71. Ibid.

72. Sonia Miller, "Hype or hope?" *New York Law Journal* 233, no. 108 (June 7, 2005).

73. Kelly Kordzik, "Prey tell what nano is? Nope," *Small Times*, (January/February 2003): 8.

74. Geoff Brumfield, "A little knowledge," *Nature* 424, no. 6946 (July 17, 2003): 248.

75. Douglas Brown, "Perception may be nano's biggest enemy, leaders tell Congress," *Small Times*, April 10, 2003, http://www.smalltimes.com/print_doc.cfm?doc_id=5809 (accessed April 24, 2003).

76. Steve Crosby, "Study nano and the environment now, because what we can't see scares us," *Small Times* (January/February 2003): 4.

77. Kelly Kordzik, "Prey tell what nano is? Nope," *Small Times* (January/February 2003): 8.

78. Mike Roco, "Crichton preys on fear, but real nano man's a fan," *Small Times* (January/February 2003): 7.

79. ETC Group, "Nanotech News in Living Colour: An Update on White Papers, Red Flags, Green Goo, Grey Goo (and Red Herrings)," *Communiqué* 85 (May/June 2004): 7.

80. Ibid.

81. "Nanotech: Hype yes, bubble no," *Nanotech Report* 1, no. 10 (December 2002): 1.

82. Ibid.

83. Jack Mason, "As nanotech grows, leaders grapple with public fear and perception," *Small Times*, May 20, 2004, http://www.smalltimes.com/document_display.cfm ?document_id=7926 (accessed July 11, 2004).

84. Barry Glassner, *The Culture of Fear: Why Americans Are Afraid of the Wrong Things: Crime, Drugs, Minorities, Teen Moms, Killer Kids, Mutant Microbes, Plan Crashes, Road Rage and So Much More* (New York: Basic Books, 1999).

85. "ISTPP 2003 National Nanotechnology Survey," George Bush School of Government and Public Service, Texas A&M University, and "Public perception about nanotechnology: Risks, benefits and trust," North Carolina State University, 2004, unpublished.

86. M. C. Roco, "The U.S. National Nanotechnology Initiative after three years (2001–2003): Setting new goals for a responsible nanotechnology," *Abstracts of the Workshop on Societal Implications of Nanoscience and Nanotechnology*, December 3–5, 2003, pp. 6–7.

87. Kelly Kordzik, "Prey tell what nano is? Nope—," *Small Times* (January/February 2003): 8.

88. "Nanotech industry review and outlook," *Nanotech Report* 2, no. 1 (January 2003): 3.

89. Stephen Herrera, "Fear seems not to be a factor in America despite the nanomonsters under the bed," *Small Times* (January/February 2003): 45.

90. Steve Fuller, *Philosophy, Rhetoric, and the End of Knowledge: The Coming of Science and Technology Studies* (Madison: University of Wisconsin Press, 1993), p. 234.

91. John Burnham, *How Superstition Won and Science Lost* (New Brunswick, NJ: Rutgers University Press, 1988).

92. Steve Fuller, *Philosophy, Rhetoric, and the End of Knowledge: The Coming of Science and Technology Studies* (Madison: University of Wisconsin Press, 1993), p. 236.

93. Ibid., p. 235.

94. Ibid.

95. John Brockman, "The emerging third culture: Scientists who publish for general audiences," *Whole Earth Review*, June 22, 1992, p. 16.

96. C. P. Snow, *The Two Cultures: and a Second Look* (Cambridge: Cambridge University Press, 1963).

97. MAST, Policy Research Project on Anticipating Effects of New Technologies, *Assessing Molecular and Atomic Scale Technologies* (MAST), Austin, TX: Lyndon B. Johnson School of Public Affairs, University of Texas at Austin, 1989, p. 26.

98. John Brockman, "The emerging third culture: Scientists who publish for general audiences," *Whole Earth Review*, June 22, 1992, p. 16.

99. Ibid.

100. John Brockman, *The Third Culture* (New York: Simon & Schuster, 1995), p. 17.

101. Ibid., p. 18.

102. Ibid., p. 19.

103. Ibid., p. 26.

104. Ibid.

105. Ibid.

106. Albert E. Moyer, *A Scientist's Voice in American Culture: Simon Newcomb and the Rhetoric of the Scientific Methods* (Berkeley and Los Angeles: University of California Press, 1992), p. 8.

107. Sonia Miller, "Law in a new frontier," *Abstracts of the Workshop on Societal Implications of Nanoscience and Nanotechnology*, December 3–5, 2003, p. 86.

CHAPTER 2

1. Reference to an anonymous ghostwriter later identified as Feynman. James Gleick, *Genius: The Life and Science of Richard Feynman* (New York: Vintage Books, 1992), p. 164.

2. Ibid., p. 68.

3. Ibid., p. 366.

4. Ibid., pp. 11–13.

5. Ibid., p. 145.

6. Ibid., p. 166.

7. Ibid., p. 9.

8. K. Eric Drexler, "Nanotechnology: From Feynman to funding," *Bulletin of Science, Technology & Society* 24, no. 1 (February 2004), http://www.metamodern.com/d/04/00/FeynmanToFunding.pdf (accessed July 26, 2004).

9. James Gleick, *Genius: The Life and Science of Richard Feynman* (New York: Vintage Books, 1992), p. 14.

10. Ibid., p. 356.

11. Adam Keiper, "The nanotechnology revolution," *New Atlantis: A Journal of Technology and Society*, Summer 2003, http://www.thenewatlantis.com/archive/2/keiper print.htm (accessed September 12, 2003).

12. Richard P. Feynman, "There's plenty of room at the bottom," *The Pleasure of Finding Things Out: The Best Short Works of Richard P. Feynman*, ed. Jeffrey Robbins (Cambridge, MA: Perseus Books, 1999), p. 118.

13. Richard Feynman, "Infinitesimal machinery," *Journal of Microelectromechanical Systems* 2, no. 1 (March 1993): 5.

14. Ibid.

15. Ibid.

16. For an accurate verbatim experience, the entire speech is on videotape and can be borrowed from Caltech. The version in the *Journal of Microelectromechanism Systems* is heavily edited and reorganized, including bold headers where none appeared in the speech. In addition, the charm, excitement, and animation in Feynman's performance is lost in the transcription.

17. Freeman Dyson, foreward to Richard P. Feynman, *The Pleasure of Finding Things Out: The Best Short Works of Richard P. Feynman*, ed. Jeffrey Robbins (Cambridge, MA: Perseus Books, 1999), p. ix.

18. Jeffrey Robbins, editor's introduction to Richard P. Feynman, *The Pleasure of Finding Things Out: The Best Short Works of Richard P. Feynman*, ed. Jeffrey Robbins (Cambridge, MA: Perseus Books, 1999), p. xiii–xiv.

19. Richard P. Feynman, "The Pleasure of Finding Things Out," *The Pleasure of Finding Things Out: The Best Short Works of Richard P. Feynman*, ed. Jeffrey Robbins (Cambridge, MA: Perseus Books, 1999), p. 2.

20. Ibid., p. 21.

21. James Gleick, *Genius: The Life and Science of Richard Feynman* (New York: Vintage Books, 1992), p. 355.

22. Danny Hills, *No Ordinary Genius: The Illustrated Richard Feynman* (New York: W. W. Norton, 1994), p. 179.

23. Jagdish Mehra, *The Beat of a Different Drum: The Life and Science of Richard Feynman* (Oxford: Clarendon Press, 1994), p. 446.

24. Richard P. Feynman, *The Meaning of It All: Thoughts of a Citizen-Scientist* (Reading, MA: Perseus Books, 1998), p. 6–7.

25. Ibid.

26. Ibid., p. 23.

27. Ibid., p. 25.

28. Ibid.

29. Ibid., p. 26.

30. Richard P. Feynman, "The cargo cult science: The 1974 CalTech commencement address," *The Pleasure of Finding Things Out: The Best Short Works of Richard P. Feynman*, ed. Jeffrey Robbins (Cambridge, MA: Perseus Books, 1999), p. 208.

31. Richard P. Feynman, "The value of science," *The Pleasure of Finding Things Out:*

The Best Short Works of Richard P. Feynman, ed. Jeffrey Robbins (Cambridge, MA: Perseus Books, 1999), p. 148.

32. Richard P. Feynman, *The Meaning of It All: Thoughts of a Citizen-Scientist* (Reading, MA: Perseus Books, 1998), pp. 26–27.

33. James Gleick, *Genius: The Life and Science of Richard Feynman* (New York: Vintage Books, 1992), p. 125.

34. Richard Feynman, "Infinitesimal machinery," *Journal of Microelectromechanical Systems* 2, no. 1 (March 1993): 4.

35. Jagdish Mehra, *The Beat of a Different Drum: The Life and Science of Richard Feynman* (Oxford: Clarendon Press, 1994), p. 446.

36. Richard P. Feynman, "The value of science," *The Pleasure of Finding Things Out: The Best Short Works of Richard P. Feynman*, ed. Jeffrey Robbins (Cambridge, MA: Perseus Books, 1999), pp. 141–42.

37. Though Drexler popularized the concept of nanotechnology, the concept of microlevel or molecular-level machinery is easily traceable to Richard Feynman and a Christmas 1959 talk he gave at an American Physical Society meeting. See Richard Feynman, "There's plenty of room at the bottom," *Engineering and Science*, 1960, pp. 22–36.

38. Ibid.

39. Eutactic means having atomically precise structure. See Russell Mills, "Steps toward nanotechnology," *Foresight Update* 16 (July 1, 1993): 5.

40. Christine Peterson, "Nanotechnology: Evolution of a concept," *Journal of the British Interplanetary Society*, October, 1, 1992, p. 396.

41. "Nanotechnology's power brokers," *Nanotech Report* 2, no. 3 (March 2003): 3.

42. Christine Peterson, "Nanotechnology: evolution of a concept," *Journal of the British Interplanetary Society*, October, 1, 1992, p. 396.

43. CS 404: Nanotechnology and Exploratory Engineering. Most of Drexler's biography is drawn from Peterson, "Nanotechnology: Evolution," p. 376. As wife and colleague, Christine Peterson offers a unique insight into Drexler's work on nanotechnology.

44. Dallas Brother, "The road to Lilliput is paved with good intention: The Nanotechnology Conference Interviews, November, 1992," *MONDO 2000*, May 1992, p. 100.

45. Ed Regis, *Nano, the Emerging Science of Nanotechnology: Remaking the World—Molecule by Molecule* (New York: Little, Brown, 1995).

46. Gary Stix, "Waiting for breakthroughs," *Scientific American* 274, no. 4 (April 1996): 96.

47. Ibid., p. 98.

48. Ralph Merkle, "A response to *Scientific American*'s new story: Trends in nanotechnology," 1996, http://www.foresight.org/SciAmDebate/SciAmResponse.html (accessed July 21, 2004).

49. "His fees . . . reach $5,000 for an appearance." See Timothy Beardsley, "Nanofuture: How much fun would it be to live forever," *Scientific American* 262, no. 1 (January 1990): 15.

50. Adam Keiper, "The nanotechnology revolution," *New Atlantis: A Journal of Technology and Society*, Summer 2003, http://www.thenewatlantis.com/archive/2/keiper print.htm (accessed September 12, 2003).

51. Lev Nazrozov, "Will Drexler save the West from nano annihilation," *NEWSMAX .COM*, December 19, 2003, http://216.26.163.62/2003/lev12_19.html (accessed December 29, 2003).

52. James M. Pethoukoukis, "Drexler speaks!—Or at least writes," *U.S. News*, September 4, 2003. http://www.usnews.com/usnews/nycu/tech/nextnews/archive/next030904.htm (accessed September 12, 2003).

53. Nancy M. Nelson, "Technology: Metacomputers, nanotechnology, electronic publishing," *Information Today* 8 (April 1991): 13; and Bill McKibben, "More than enough," *Ecologist*, May 22, 2003, http://www.theecologist.org/archive_article.html?artilce=400 (accessed September 2, 2003).

54. *EXTRO3* was sponsored by the Extropy Institute, an institution devoted to the use of technology to overcome human limits, including extending life space, augmenting intelligence, gaining access to space, and achieving control over human biology. See "Recent events: Extropy institute conference," *Foresight Update* 30, September 1, 1997, http://www.foresight.org/Updates/Updates30/update30.5.html (accessed July 26, 2004).

55. K. Eric Drexler, Transcript, "Imaging and Imagining Conference," March 3, 2004, unpublished.

56. Ibid.

57. Glenn Reynolds, "A Tale of Two Nanotechs," *Tech Central Station*, January 28, 2004, http://www.techcentralstation.com/012804A.html (accessed July 11, 2004).

58. Howard Lovy, "Nano re-created in business's image: Is this the best of all futures," *Small Times*, January 23, 2004, http://www.smalltimes.com/document_display.cfm?document_id=7279 (accessed July 11, 2004).

59. Lawrence Lessig, "Stamping out good science," *Wired* 12, no. 7 (July 2004) http://www.wired.com/wired/archive/12.07/view.html?pg=5 (accessed July 9, 2004).

60. Christine Peterson, "Societal implications of nanotechnology," *Federal Document Clearing House*, April 9, 2003, http://web.lexis_nexis.com/congcomp/document?_m=b3afd9dc2b8d55cdb425cc4c2fd2c2b (accessed April 24–25, 2003).

61. Simon Smith, "Dismissing Drexler is bad for business," *Betterhumans.com*, February 26, 2004, http://www.betterhumans.com/Print/article.aspx?articleID=2004-02-26-1 (accessed March 2, 2004).

62. Ibid.

63. Ibid.

64. Ibid.

65. Ibid.

66. Richard Jones, "The future of nanotechnology," *NanotechWeb.org*, August 2004, http://nanotechweb.org/articles/feature/3/8/1/1 (accessed August 9, 2004).

67. Simon Smith, "Dismissing Drexler is bad for business," *Betterhumans.com*, February 26, 2004, http://www.betterhumans.com/Print/article.aspx?articleID=2004-02-26-1 (accessed March 2, 2004).

68. N. Katherine Hayles, "Connecting the quantum dots: Nanotechscience and culture," in *Nanoculture: Implications of the New Technoscience*, ed. N. Hayles (Bristol, UK: Intellect Books, 2004), p. 12.

69. K. Eric Drexler, Transcript, "Imaging and Imagining Conference," March 3, 2004, unpublished.

70. Cynthia Selin, "Expectations in the emergence of nanotechnology," September 25, 2002, p. 7, http://web.cbs.dk/departments/mpp/forskerskolen/calendar/Selin.pdf (accessed May 2, 2004).

71. Neal Lane, personal e-correspondence, March 29, 2004, unpublished.

72. Gary Stix, "Little big science," *Scientific American,* September 16, 2001, http://www.ruf.rice.edu/~rau/phys600/stix.htm (accessed May 2, 2004).

73. T. Kary, "Nanotech: More science than fiction," *ZD net Special,* February 11, 2002, http://zdnet.com/2100-1103-833739.html (accessed May 3, 2004).

74. Adam Keiper et al., "The nanotech schism: High-tech pants or molecular revolution," *New Atlantis,* Winter 2004, http://www.thenewatlantis.com/archive/4/soa/nanotechprint.htm (accessed April 5, 2004).

75. K. Eric Drexler, Transcript, "Imaging and Imagining Conference," March 3, 2004, unpublished.

76. Adam Keiper et al., "The nanotech schism: High-tech pants or molecular revolution," *New Atlantis,* Winter 2004, http://www.thenewatlantis.com/archive/4/soa/nanotechprint.htm (accessed April 5, 2004).

77. Lawrence Lessig, "Stamping out good science," *Wired* 12, no. 7 (July 2004) http://www.wired.com/wired/archive/12.07/view.html?pg=5 (accessed July 9, 2004).

78. Chris Phoenix, "Of chemistry, nanobots, and policy," *Center for Responsible Nanotechnology,* December 2003, http://crnano.org/Debate.htm (accessed December 2, 2003).

79. Adam Keiper et al., "The nanotech schism: High-tech pants or molecular revolution," *New Atlantis,* Winter 2004, http://www.thenewatlantis.com/archive/4/soa/nanotechprint.htm (accessed April 5, 2004).

80. N. Katherine Hayles, "Connecting the quantum dots: Nanotechscience and culture, *Nanoculture: Implications of the New Technoscience,* ed. N. Hayles (Bristol, UK: Intellect Books, 2004), p. 13.

81. K. Eric Drexler, Transcript, "Imaging and Imagining Conference," March 3, 2004, unpublished.

82. Ed Regis, "The incredible shrinking man," *Wired* 12, no. 10 (October 2004), http:// wired.com/wired/archive/12.10/drexler.html (accessed July 20, 2005).

83. Jack Mason, "As nanotech grows, leaders grapple with public fear and perception," *Small Times,* May 20, 2004, http://www.smalltimes.com/document_display.cfm?document_id=7926 (accessed July 11, 2004).

84. Howard Lovy, "Nanotechnology has reached a crossroads somewhere between hypothesis and hype," *Small Times* (July/August 2003): 6.

85. K. E. Drexler, personal e-correspondence, December 6, 2003.

86. Whitesides received an Alfred P. Sloan Fellowship in 1968, the American Chemical Society (ACS) Award in Pure Chemistry in 1975, the Harrison Howe Award (Rochester Section of the ACS) in 1979, an Alumni Distinguished Service Award (California Institute of Technology) in 1980, the Remsen Award (ACS, Maryland Section) in 1983, an Arthur C. Cope Scholar Award (ACS) in 1989, the James Flack Norris Award (ACS, New England Section) in 1994, the Arthur C. Cope Award (ACS) in 1995, the Defense Advanced Research Projects Agency Award for Significant Technical Achievement in 1996, the Madison Marshall Award (ACS) in 1996, the National Medal of Science in 1998, the Sierra Nevada Distinguished Chemist Award (Sierra Nevada Section of the ACS), the Wallac Oy Innovation Award in High Throughput Screening (the Society for Biomolecular Screening) in 1999, the Award for Excellence in Surface Science (the Surfaces in Biomaterials Foundation) in

1999, and the Von Hippel award (Materials Research Society) in 2000. He is a member of the American Academy of Arts and Sciences, the National Academy of Sciences, and the American Philosophical Society. He is also a Fellow of the American Association for the Advancement of Science and the New York Academy of Science, a foreign fellow of the Indian National Science Academy, and an Honorary Fellow of the Chemical Research Society of India.

87. "Harvard's George Whitesides on nanotechnology: 'A word, not a field,'" http://www.sciencewatch.com/july-aug2002/sw_july-aug2002_page3.htm (accessed June 30, 2004).

88. "Small Times Magazine: 2003 Innovator of the Year: Finalists," *Small Times* (November/December 2003): 33.

89. George Whitesides, "The new biochemphysicist: An interview by David Ewing Duncan with George Whitesides," *Discover* (December 2003): 24.

90. G. M. Whitesides, "The once and future nanomachine," *Scientific American* 285, no. 3 (September 2001): 78.

91. Ibid., p. 81. Drexler et al. respond that molecular bearings can run dry, bearings with any low-static friction should be possible, and near-frictionless sliding of nested carbon nanotubes has been demonstrated. See K. E. Drexler, D. Forrest, R. A. Freitas, J. S. Halls, N. Jacobstein, T. McKendree, R. Merkle, and C. Peterson, *A Debate about Assemblers: Many Future Nanomachines: A Rebuttal to Whiteside's Assertion That Mechanical Molecular Assemblers Are Not Workable and Not a Concern*, 2001, http://www.imm.org/SciAmDebate2/whitesides.html (accessed July 18, 2004).

92. G. M. Whitesides, "The once and future nanomachine," *Scientific American* 285, no. 3 (September 2001): 81. Drexler et al. offer chemical power, light power, acoustic energy, radionuclide batteries, and a handful of other sources. See K. E. Drexler, D. Forrest, R. A. Freitas, J. S. Halls, N. Jacobstein, T. McKendree, R. Merkle, and C. Peterson, *A Debate about Assemblers: Many Future Nanomachines: A Rebuttal to Whiteside's Assertion That Mechanical Molecular Assemblers Are Not Workable and Not a Concern*, 2001, http://www.imm.org/SciAm Debate2/whitesides.html (accessed July 18, 2004).

93. Ibid. Drexler et al. suggest storing information in polymers and broadcasting information and directions. See K. E. Drexler, D. Forrest, R. A. Freitas, J. S. Halls, N. Jacobstein, T. McKendree, R. Merkle, and C. Peterson, *A Debate about Assemblers: Many Future Nanomachines: A Rebuttal to Whiteside's Assertion That Mechanical Molecular Assemblers Are Not Workable and Not a Concern*, 2001, http://www.imm.org/SciAmDebate2/whitesides.html. (accessed July 18, 2004).

94. Ibid.

95. Ibid. Drexler claims *pincers* or *jaws* are imprecise and misleading metaphors, and his response to both the *fat* and *sticky* fingers problem are covered in the next section. See K. E. Drexler, D. Forrest, R. A. Freitas, J. S. Halls, N. Jacobstein, T. McKendree, R. Merkle, and C. Peterson, *A Debate about Assemblers: Many Future Nanomachines: A Rebuttal to Whiteside's Assertion That Mechanical Molecular Assemblers Are Not Workable and Not a Concern*, 2001, http://www.imm.org/SciAmDebate2/whitesides.html (accessed July 18, 2004).

96. Ibid., p. 82. Drexler answers that neutrophils, lymphocytes, and monocytes manage well enough. This is a medical and engineering problem rather than a fundamental

problem with physical limits. See K. E. Drexler, D. Forrest, R. A. Freitas, J. S. Halls, N. Jacobstein, T. McKendree, R. Merkle, and C. Peterson, *A Debate about Assemblers: Many Future Nanomachines: A Rebuttal to Whiteside's Assertion That Mechanical Molecular Assemblers Are Not Workable and Not a Concern*, 2001, http://www.imm.org/SciAmDebate2/whitesides.html (accessed July 18, 2004).

97. Ibid. Responding, Drexler states that water molecules would be much smaller than nanobots and cells and much larger than a nanosub and would tumble along in the bloodstream, and nanosubs might select a general rather than a specific destination. See K. E. Drexler, D. Forrest, R. A. Freitas, J. S. Halls, N. Jacobstein, T. McKendree, R. Merkle, and C. Peterson, *A Debate about Assemblers: Many Future Nanomachines: A Rebuttal to Whiteside's Assertion That Mechanical Molecular Assemblers Are Not Workable and Not a Concern*, 2001, http://www.imm.org/SciAmDebate2/whitesides.html (accessed July 18, 2004).

98. K. E. Drexler, *Engines of Creation* (New York: Anchor Press, 1986), p. 105.

99. G. M. Whitesides, unpublished lecture, NanoVentures 2003, Richardson, TX, February 28, 2003.

100. G. M. Whitesides, "The once and future nanomachine," *Scientific American* 285, no. 3 (September 2001): 81. Drexler et al. respond that molecular bearings can run dry, bearings with any low-static friction should be possible, and near-frictionless sliding of nested carbon nanotubes has been demonstrated. See K. E. Drexler, D. Forrest, R. A. Freitas, J. S. Halls, N. Jacobstein, T. McKendree, R. Merkle, and C. Peterson, *A Debate about Assemblers: Many Future Nanomachines: A Rebuttal to Whiteside's Assertion That Mechanical Molecular Assemblers Are Not Workable and Not a Concern*, 2001, http://www.imm.org/SciAmDebate2/ whitesides.html (accessed July 18, 2004).

101. George Whitesides, "The new biochemphysicist: An interview by David Ewing Duncan with George Whitesides," *Discover* (December 2003): 24.

102. G. M. Whitesides, "The once and future nanomachine," *Scientific American* 285, no. 3 (September 2001): 83. Even Drexler has backed away from this scenario, though in 2001 he foresaw no technical problems per se, citing examples from biotechnology research on DNA, conceptual designs and modeling, and Japanese robot manufacturing. See K. E. Drexler, D. Forrest, R. A. Freitas, J. S. Halls, N. Jacobstein, T. McKendree, R. Merkle, and C. Peterson, *A Debate about Assemblers: Many Future Nanomachines: A Rebuttal to Whiteside's Assertion That Mechanical Molecular Assemblers Are Not Workable and Not a Concern*, 2001, http://www.imm.org/SciAmDebate2/whitesides.html (accessed July 18, 2004). In 2004, Drexler and Chris Phoenix merely argue self-replication is not a necessary characteristic of molecular manufacturing without addressing feasibility per se. See Chris Phoenix and Eric Drexler, "Safe exponential manufacturing," *Nanotechnology* 15 (2004): 869–72.

103. G. M. Whitesides, Workshop on Societal Implications of Nanoscience and Nanotechnology, December 3–5, 2003, unpublished lecture.

104. G. M. Whitesides, "Nanotechnology: Art of the possible," *Technology Review* 101, no. 6 (November/December 1998): 85.

105. G. M. Whitesides, "The once and future nanomachine," *Scientific American* 285, no. 3 (September 2001): 83.

106. K. E. Drexler, "Letter to Whitesides," personal e-correspondence, May 20, 2003.

107. Ibid.

108. R. Baum, "Nanotechnology: Drexler and Smalley make the case for and against molecular assemblers," *Chemical and Engineering News* 81, no. 48 (December 1, 2003), http:// pubs.acs.org/cen/coverstory/8148/8148counterpoint.html (accessed December 2, 2003).

109. "Nanotech's power elite: 2004," *Nanotech Report* 3, no. 3 (March 2004): 2.

110. Candace Stuart, "He has a Nobel, a company and Washington's ear," *Small Times* (November/December 2003): 37.

111. "Nanotechnology's power brokers," *Nanotech Report* 2, no. 3 (March 2003): 1.

112. "Nanotech's power elite: 2004," *Nanotech Report* 3, no. 3 (March 2004): 2.

113. "Nanomaterials: A chemical reaction," *Nanotech Report* 3, no. 1 (2004): 1.

114. "Nanotechnology in brief," *NanotechWeb*, July 9, 2004, http://nanotechweb.org/ article/news/3/7/5 (accessed July 10, 2004).

115. Frank D. Roylance, "Tiny tubes and big dreams," *Baltimore Sun*, May 10, 2004, p. 11A.

116. "Carbon Nanotechnologies, Inc. announces the allowance of a U.S. patent for coating single-walled carbon nanotubes," *NanoInvestorNews*, July 13, 2004, http://www .nanoin- vestornews.com/modules.php?name=News&file=print&sid=3064 (accessed July 19, 2004).

117. Ibid.

118. William Schulz, "Creating a national nanotechnology effort," *CENEAR—Special Report* 78, no. 42 (October 16, 2000): 39–42, http://pub.acs.org/cen/nanotechnology/7842/ print/7842government.html (accessed June 15, 2004).

119. Howard Lovy, "Clash of the nanotech titans," *Howard Lovy's Nanobot*, December 1, 2003, http://nanobot.blogspot.com/2003_11_30_nanobot_archive.html (accessed December 2, 2003).

120. Ibid.

121. ETC Group, "Nanotech takes a giant step down," March 6, 2002, http:// etcgroup.org/article.asp?newsid=305 (accessed May 21, 2003).

122. R. Baum, "Nanotechnology: Drexler and Smalley make the case for and against molecular assemblers," *Chemical and Engineering News* 81, no. 48 (December 1, 2003), http:// pubs.acs.org/cen/coverstory/8148/8148counterpoint.html (accessed December 2, 2003).

123. R. Kurzweil, "The Drexler-Smalley Debate on Molecular Assembly," *KurzweilAI .net*, December 1, 2003, http://www.kruzweilai.net/articles/art0604.html (accessed December 2, 2003).

124. K. E. Drexler, "Drexler open letter," 2003, http://pubs.acs.org/cen/coverstory/ 8148/8148counterpoint.html (accessed December 2, 2003).

125. K. E. Drexler, "Drexler writes Smalley open letter on assemblers," Posted *Nano- dot*, April 20, 2003, http://www.foresight.org/NanoRev/Letter.html (accessed May 15, 2003).

126. Richard Smalley, "Smalley responds," in Rudy Baum, "Nanotechnology: Drexler and Smalley make the case for and against molecular assemblers," *Chemical and Engineering News* 81, no. 48 (December 1, 2003), http://pubs.acs.org/cen/coverstory/8148/8148counterpoint.html (accessed December 2, 2003).

127. Ibid.

128. Chris Phoenix, "Of chemistry, nanobots, and policy," December 2003, *Center for Responsible Nanotechnology*, http://crnano.org/Debate.htm (accessed December 2, 2003).

129. K. Eric Drexler, "Drexler counters," in Rudy Baum, "Nanotechnology: Drexler and Smalley make the case for and against molecular assemblers," *Chemical and Engineering News* 81, no. 48 (December 1, 2003), http://pubs.acs.org/cen/coverstory/8148/8148 counterpoint.html (accessed December 2, 2003).

130. Richard Smalley, "Smalley concludes," in Rudy Baum, "Nanotechnology: Drexler and Smalley make the case for and against molecular assemblers," *Chemical and Engineering News* 81, no. 48 (December 1, 2003), http://pubs.acs.org/cen/coverstory/8148/8148counterpoint.html (accessed December 2, 2003).

131. Ray Kurzweil, "The Drexler-Smalley debate on molecular assembly," *KurzweilAI.net*, December 1, 2003, http://www.kruzweilai.net/articles/art0604.html (accessed December 2, 2003).

132. Chris Phoenix, "Of chemistry, nanobots, and policy," December 2003, *Center for Responsible Nanotechnology*, http://crnano.org/Debate.htm (accessed December 2, 2003).

133. R. Baum, "Nanotechnology: Drexler and Smalley make the case for and against molecular assemblers," *Chemical and Engineering News* 81, no. 48 (December 1, 2003), http://pubs.acs.org/cen/coverstory/8148/8148counterpoint.html (accessed December 2, 2003).

134. K. Eric Drexler, "Drexler writes Smalley open letter on assemblers," Posted Nanodot, April 20, 2003, http://www.foresight.org/NanoRev/Letter.html (accessed May 15, 2003).

135. Richard Smalley, "Smalley concludes," in Rudy Baum, "Nanotechnology: Drexler and Smalley make the case for and against molecular assemblers," *Chemical and Engineering News* 81, no. 48 (December 1, 2003), http://pubs.acs.org/cen/coverstory/8148/8148counterpoint.html (accessed December 2, 2003).

136. Howard Lovy, "Nano re-created in business's image: Is this the best of all futures," *Small Times,* January 23, 2004, http://www.smalltimes.com/document_display.cfm?document_id=7279 (accessed July 11, 2004).

137. Chris Phoenix, "Letters," *CENEAR* 82 (2004): 6–8, http://pubs.acs.org/isubscribe/journals/cen/82/i04/print/8204lett.html (accessed February 4, 2004).

138. "Nobel winner Smalley responds to Drexler's challenge: Fails to defend national nanotech policy," *Foresight Institute Press Center,* December 1, 2003, http://www.foresight.org/press.html (accessed December 2, 2003).

139. Zac Goldsmith, "Discomfort and Joy: Bill Joy interview," *Ecologist,* September 22, 2000, http://www.theecologist.org/archive_article.html?article=188&category=84 (accessed March 22, 2004).

140. "Interview with Bill Joy," *Weekly Edition: The Best of NPR News,* March 18, 2000, http://Lexis-Nexis.com/.

141. "Small science, big questions," *SNH.com.au,* November 27, 2003, http://www.smh.com.au/articles/2003/11/16/10768917669906.html (accessed November 25, 2003).

142. Virginia Postrel, "Joy, to the world," *Reasononline,* June 2000, http://reason.com/0006/co.vp.joy.shtml (accessed July 27, 2004).

143. Bill Joy, "Why the future doesn't need us," *Wired* 8, no. 4 (April 2000), http://www .wired.com/wired/archive/8.04/joy.html (accessed March 22, 2004).

144. Zac Goldsmith, "Discomfort and Joy: Bill Joy interview," *Ecologist,* September 22, 2000, http://www.theecologist.org/archive_article.html?article=188&category=84 (accessed March 22, 2004).

145. Ibid.

146. Jon Gertner, "Proceed with caution," *New York Times Magazine*, June 6, 2004, p. 34.

147. Ibid., p. 32.

148. Ibid., p. 36.

149. Virginia Postrel, "Joy, to the world," *Reasononline*, June 2000, http://reason .com/0006/co.vp.joy.shtml. (accessed July 27, 2004).

150. Joel Garreau, "From Internet scientist, a preview of extinction," *Washington Post*, March 12, 2000, http://www2.gol.com/users/coynerhm/from_internet_scientist_htm (accessed July 28, 2004).

151. Virginia Postrel, "Joy, to the world," *Reasononline*, June 2000, http://reason.com/ 0006/co.vp.joy.shtml. (accessed July 27, 2004).

152. Ibid.

153. Jon Gertner, "Proceed with caution," *New York Times Magazine*, June 6, 2004, p. 34.

154. Ibid.

155. John Markoff, "Technologist gives his peers a dark warning," *New York Times on the Web*, March 13, 2000, http://www.nytimes.com/library/tech/00/03/biztech/articles/ 13joy.html (accessed July 27, 2004).

156. "Can Zac save the planet," *Guardian*, November 7, 2003, http://www .guardian.co.uk/gmdebate/Story/0,2763,835010,00.html (accessed July 21, 2003).

157. "HARDtalk: Zac Goldsmith," *BBC Press Releases*, July 10, 2002, http://www.bbc .co.uk/print/pressoffice/pressreleases/stories/2002/10_october/07/hardtalk_ (accessed July 26, 2003).

158. "Can Zac save the planet," *Guardian*, November 7, 2003, http://www .guardian.co.uk/gmdebate/Story/0,2763,835010,00.html (accessed July 21, 2003).

159. "Young, gifted and Zac," *Observer*, April 13, 2003, http://observer.guardian .co.uk/magazine/story/0,11913,935831,00.html (accessed July 21, 2003).

160. "Can Zac save the planet," *Guardian*, November 7, 2003, http://www .guardian.co.uk/gmdebate/Story/0,2763,835010,00.html (accessed July 21, 2003).

161. Ibid.

162. P. Brown, "Printers pulp Monsanto edition of Ecologist," *Guardian*, September 29, 1998, http://www.csdl.tamu.edu/FLORA/328Fall98/Reading14.html (accesssed July 21, 2003).

163. "Young, gifted and Zac," *Observer*, April 13, 2003, http://observer.guardian .co.uk/magazine/story/0,11913,935831,00.html (accessed July 21, 2003).

164. Zac Goldsmith, "The seeds of discord," *Sunday Telegraph*, March 2, 2002, http://millennium-debate.org/suntel3mar4.htm (accessed July 21, 2003).

165. Ibid.

166. Ibid.

167. Ibid.

168. Michael Brunton, "Little worries: Invasion of the nanobots? Critics warn of a 'gray goo' future as nanotechnology enters the marketplace," *Time Europe*, May 4, 2003, http://www.time.com/time/europe/magazine/article/0,13005,901030512-449458,00.html (accessed July 22, 2004).

169. "So where is the hero who can save the planet before it's too late," *BIT@E: Monthly Bulletin on Armenian Information Technology Market*, May 2003, p. 24.

170. ETC reported that Prince Charles "hasn't talked with, [but] he did order several copies of *The Big Down*." See ETC Group, "Nanotech and the precautionary prince," *Genotype*, May 2, 2003, http://www.etcgroup.org/article.asp?newsid=397 (accessed May 21, 2003).

171. "Charles fears science could kill the Earth," *Floyd Report*, April 27, 2003, http://www.floydreport.com/view_article.php?lid=209 (accessed July 21, 2003).

172. "HRH the Prince of Wales: Menace in the minutiae," *Independent.co.uk*, July 11, 2004, http://argument.independent.co.uk/commentators/story.jsp?story=539977 (accessed July 12, 2004).

173. "Prince Charles warns of dangers of the breakthrough science of nanotechnology," Reuters, July 10, 2004, http://nanotechwire.com/news.asp?nid=925&ntid=&pg=i (accessed July 11, 2004).

174. "HRH the Prince of Wales: Menace in the minutiae," *Independent.co.uk*, July 11, 2004, http://argument.independent.co.uk/commentators/story.jsp?story=539977 (accessed July 12, 2004).

175. Geoffrey Lean, "Hundreds of firms using nanotech in food," *Independent.co.uk*, July 18, 2004, http://news.independent.co.uk/uk/environment/story.jsp?story=542140 (accessed July 19, 2004).

176. "Prince Charles warns of dangers of the breakthrough science of nanotechnology," Reuters, July 10, 2004, http://nanotechwire.com/news.asp?nid=925&ntid=&pg=i (accessed July 11, 2004).

177. "HRH the Prince of Wales: Menace in the minutiae," *Independent.co.uk*, July 11, 2004, http://argument.independent.co.uk/commentators/story.jsp?story=539977 (accessed July 12, 2004).

178. "Will Prince Charles et al diminish the opportunities of developing countries in nanotechnology?" NanoTechWeb, January 28, 2004, http://www.nanotechweb.org/articles/society/3/1//1/1 (accessed February 18, 2005).

179. "Prince Charles warns of dangers of the breakthrough science of nanotechnology," Reuters, July 10, 2004, http://nanotechwire.com/news.asp?nid=925&ntid=&pg=i (accessed July 11, 2004).

180. "Blair is the enemy of the Greens," *Observer*, August 18, 2002, http://observer.guardian.co.uk/worldview/story/0,11581,776678,090.html (accessed July 21, 2003).

181. Zac Goldsmith, "Time's up for the global economy," *Global Agenda Magazine*, n.d., http://www.globalagendamagazine.com/default.asp?page=2&SID=3&theme=3 (accessed July 21, 2003).

CHAPTER 3

1. Rick Weiss, "For science, nanotech poses big unknowns," *Washington Post*, February 1, 2004, p. A01, http://www.washingtonpost.con/ac2/wp-dyn/A1487-2004Jan31 (accessed February 4, 2004).

2. Harold Brubaker, "Nanotech finds fans in industry," *Sun Herald*, April 30, 2004, http://www.sunherald.com/mld/thesunherald/business/8554757.htm (accessed May 3, 2004).

3. Jay Lindquist, "In the spotlight," *NanoBusiness News*, February 3, 2004.

4. William G. Schulz, "Nanotechnology under the scope," *Chemical and Engineering News*, December 4, 2002, http://pubs.acs.org/cen/today/dec4.html (accessed October 5, 2003).

5. Scott Burnell, "Interview: Sen. Wyden eyes nanotech," *United Press International*, September 22, 2002, http://www.upi.com/view.cfm?StoryID=20020920-030036-4843r (accessed July 26, 2004).

6. Ron Wyden, "Wyden Chairs First Senate Nanotechnology Hearing: 'Small Science' could change the way Americans live, work, treat disease," September 17, 2002, http://wyden.senate.gov/media/2002/09172002_nanotech.html (accessed July 26, 2004).

7. Scott Burnell, "Interview: Sen. Wyden eyes nanotech," *UPI Science News*, September 22, 2002, http://www.upi.com/view.cfm?StoryID=20020920-030036-4843r (accessed July 26, 2004).

8. Ibid.

9. Ibid.

10. Ron Wyden, Committee on Science in the House of Representatives, *H.R. 766, Nanotechnology Research and Development Act of 2003*, Hearings before the Committee on Science in the House of Representatives, March 19, 2003, p. 18.

11. Scott Burnell, "Interview: Sen. Wyden eyes nanotech," *UPI Science News*, September 22, 2002, http://www.upi.com/view.cfm?StoryID=20020920-030036-4843r (accessed July 26, 2004).

12. George Allen, Committee on Science in the House of Representatives, *H.R. 766, Nanotechnology Research and Development Act of 2003*, Hearings before the Committee on Science in the House of Representatives, March 19, 2003, pp. 19 and 21.

13. Juliana Gruenwald, "Congress decides to study first and regulate later," *Small Times*, May/June 2004, p. 12.

14. Susan R. Morissey, "Harnessing nanotechnology," *Government and Policy* 82, no. 16 (April 14, 2004): 30–33, http://pubs.acs.org/cen/nlw/print/8216gov1.html (accessed June 15, 2004).

15. "Nanotech's power elite: 2004," *Nanotech Report* 3, no. 3 (March 2004): 1.

16. Ibid.

17. Doug Brown, "Changing of the guard," *Small Times*, January/February 2003, p. 14.

18. "Biotechnology in the Tobacco Belt: Danville Weighs In," *NanoInvestorNews*, April 12, 2005, http://www.nanoinvestornews.com/modules.php?name=Content&pa=showpage&pid=19 (accessed June 17, 2005).

19. Juliana Gruenwald, "Senator forming caucus to keep nanotech issues on forefront," *Small Times*, April 8, 2004, http://www.smalltimes.com/print_doc.cfm?doc_id=7697 (accessed April 13, 2004).

20. Susan R. Morissey, "Harnessing nanotechnology," *Government and Policy* 82, no. 16 (April 14, 2004): 30–33, http://pubs.acs.org/cen/nlw/print/8216gov1.html (accessed June 15, 2004).

21. Ron Wyden, Committee on Science in the House of Representatives, *H.R. 766,*

Nanotechnology Research and Development Act of 2003, Hearings before the Committee on Science in the House of Representatives, March 19, 2003, p. 17.

22. Ron Wyden, "Q&A," *Small Times*, January/February 2004, p. 10.

23. Ron Wyden, Committee on Science in the House of Representatives, *H.R. 766, Nanotechnology Research and Development Act of 2003*, Hearings before the Committee on Science in the House of Representatives, March 19, 2003, p. 17.

24. Susan R. Morissey, "Harnessing nanotechnology," *Government and Policy* 82, no. 16 (April 14, 2004): 30–33, http://pubs.acs.org/cen/nlw/print/8216gov1.html (accessed June 15, 2004).

25. "Nanotech's power elite: 2004," *Nanotech Report* 3, no. 3 (March 2004): 1.

26. "Senate bill: $103 million for Oregon defense work," *Bend.com New Sources*, June 23, 2004, http://www.bend.com/news/ar_view%5E3Far_id%5E3D16350.htm (accessed July 10, 2004).

27. Juliana Gruenwald, "Congress decides to study first and regulate later," *Small Times*, May/June 2004, p. 11.

28. John Spooner, "Companies look forward to nanotechnology," *Monsters and Critics*, May 24, 2005, http://news.monstersandcritics.com/business/printer_1002047.php (accessed May 25, 2005).

29. Susan R. Morissey, "Harnessing nanotechnology," *Government and Policy* 82, no. 16 (April 14, 2004): 30–33, http://pubs.acs.org/cen/nlw/print/8216gov1.html (accessed June 15, 2004).

30. J. Wakefield, "Science's political bulldog," *Scientific American* 290, no. 5 (May 2004): 52.

31. National Science and Technology Council, *Nanotechnology: Shaping the World Atom by Atom* (Washington, DC: U.S. GPO, 1999), p. 1.

32. Neal Lane, "Testimony before the Senate Appropriations Committee," *Federal News Service*, May 7, 1998, http://web.lexis-nexis.com/congroup/document (accessed September 8, 2003).

33. Ibid.

34. Ibid.

35. Ibid.

36. "Dr. Neal Lane to return to Rice University," press release, December 15, 2000, http://www.ostp.gov/html/001215.html (accessed July 10, 2004).

37. "Thinking small: Mike Roco," *Nanotech Report* 1, no. 5 (July 2002): 5.

38. Ibid.

39. "Nanotech's power elite: 2004," *Nanotech Report* 3, no. 3 (March 2004): 1.

40. "Nanotechnology's power brokers," *Nanotech Report* 2, no. 3 (March 2003): 1.

41. Mihail Roco, introduction to *Societal Implications of Nanoscience and Nanotechnology*, ed. M. C. Roco and W. S. Bainbridge (New York: Kluwer, 2001), p. 1.

42. A. Leo, "Get ready for your nano future," *Technology Review*, May 4, 2001, http://www.technologyreview.com/articleswo_leo050401.asp (accessed October 5, 2003).

43. "Thinking small: Mike Roco," *Nanotech Report* 1, no. 5 (July 2002): 5.

44. Ibid.

45. Ibid.

46. A. Leo, "Get ready for your nano future," *Technology Review*, May 4, 2001, http://www.technologyreview.com/articleswo_leo050401.asp (accessed October 5, 2003).

47. "Nanotech's power elite: 2004," *Nanotech Report* 3, no. 3 (March 2004): 2.

48. William Schulz, "Creating a national nanotechnolgoy effort," *CENEAR—Special Report* 78, no. 42 (October 16, 2000): 39–42, http://pub.acs.org/cen/nanotechnology/7842/print/7842government.html (accessed June 15, 2004).

49. A. Leo, "Get ready for your nano future," *Technology Review*, May 4, 2001, http://www.technologyreview.com/articleswo_leo050401.asp (accessed October 5, 2003).

50. Howard Lovy, "How to fight nanotech misinformation in two easy words: Honesty, imagination," *Small Times*, May/June 2004, p. 20.

51. Jeff Karoub, "Taking some initiative in D.C., a Small Times Q&A with Thomas Kalil," *Small Times*, March 4, 2005, http://smalltimes.org/document_display.cfm?section_id=97&document_id=8892 (3 June 2005).

52. Josh Wolfe, "Decoding future nanotech investment success," *Forbes/Wolfe Nanotech Report*, October 10, 2002, http://www.forbes.com/2002/10/10/1010soapbox_print.html (June 3, 2005).

53. Jeff Karoub, "Taking some initiative in D.C., a Small Times Q&A with Thomas Kalil," *Small Times*, March 4, 2005, http://smalltimes.org/document_display.cfm?section_id=97&document_id=8892 (June 3, 2005).

54. Jeff Karoub, "Nano's Road to the Future," *Small Times* 5, no. 1, (January/February 2005): 25.

55. Ibid., p. 26.

56. Ibid.

57. Tom Kalil, "Next steps for the National Nanotechnology Initiative," *Nanotechnology Law and Business Journal* 1, no. 1, (2004): article 7.

58. "Tom Kalil Biography," April 3, 2003, http://www.eecs.berkeley.edu/IPRO/IBMday03/bio/kalil.html (June 3, 2005).

59. Adeel Iqbal, "Scientific Revolution Tinier than an Atom," *Daily Californian* 6 (February 2004), http://www.mindfully.org/Technology/2004/Scientific-Revolution-Nanotech6feb04.htm (accessed June 3, 2005).

60. In Realis, *A Critical Investor's Guide to Nanotechnology*, February 2002, p. 10.

61. R. Weiss, "Some see solution, some potential dangers," *Washington Post*, February 10, 2004, http://newsday.com/news/printedition/health/ny-dsspdn3663472feb10,0,5417935,pri (accessed February 16, 2004).

62. R. Hapanowicz, "NanoCommerce 2003 conference overview part 1," December 10, 2003, http://news.nanoapex.com/modules.php?name=News&file=article&sid=4080 (accessed December 15, 2003).

63. Susan R. Morissey, "Harnessing nanotechnology," *Government and Policy* 82, no. 16 (April 14, 2004): 30–33, http://pubs.acs.org/cen/nlw/print/8216gov1.html (accessed June 15, 2004).

64. Candace Stuart, "He has a Nobel, a company and Washington's ear," *Small Times*, November/December 2003, p. 36.

65. Susan R. Morissey, "Harnessing nanotechnology," *Government and Policy* 82, no. 16 (April 14, 2004): 30–33, http://pubs.acs.org/cen/nlw/print/8216gov1.html (accessed June 15, 2004).

66. "Thinking small: Phillip J. Bond," *Nanotech Report* 2, no. 7 (July 2003): 51.

67. R. Hapanowicz, "NanoCommerce 2003 conference overview part 1," December 10, 2003, http://news.nanoapex.com/modules.php?name=News&file=article&sid=4080 (accessed December 15, 2003).

68. "Thinking small: Phillip J. Bond," *Nanotech Report* 2, no. 7 (July 2003): 51.

69. Liza Porteus, "Psyching up kids for tech could help U.S. jobs," *FoxNews*, July 5, 2004, http://www.foxnews.com/story/0,2933,123562,00.html (accessed July 9, 2004).

70. "Thinking small: Phillip J. Bond," *Nanotech Report* 2, no. 7 (July 2003): 51.

71. Ibid.

72. Susan R. Morissey, "Harnessing nanotechnology," *Government and Policy* 82, no. 16 (April 14, 2004): 30–33, http://pubs.acs.org/cen/nlw/print/8216gov1.html (accessed June 15, 2004).

73. Jack Mason, "As nanotech grows, leaders grapple with public fear and perception," *Small Times*, May 20, 2004, http://www.smalltimes.com/document_display.cfm?document_id=7926 (accessed July 11, 2004).

74. "Thinking small: Phillip J. Bond," *Nanotech Report* 2, no. 7 (July 2003): 51.

75. "Parables, predictions, and wisdom of crowds," *Forbes/Wolfe blog*, May 31, 2004, http://www.forbeswolfe.com/ (accessed June 15, 2004).

76. Susan R. Morissey, "Harnessing nanotechnology," *Government and Policy* 82, no. 16 (April 14, 2004): 30–33, http://pubs.acs.org/cen/nlw/print/8216gov1.html (accessed June 15, 2004).

77. Jack Mason, "As nanotech grows, leaders grapple with public fear and perception," *Small Times*, May 20, 2004, http://www.smalltimes.com/document_display.cfm?document_id=7926 (accessed July 11, 2004).

78. Howard Lovy, "How to fight nanotech misinformation in two easy words: Honesty, imagination," *Small Times*, May/June 2004, p. 20.

79. Ibid.

80. "Beware the nanobubble," *Nanotech Report* 2, no. 6 (June 2003): 3.

81. "2003 nanotech product guide," *Nanotech Report* 2, no. 7 (July 2003): 1.

82. "Decoding future nanotech investment success," *Nanotech Report* 1, no. 7 (September 2002): 1.

83. Ibid.

84. W. Blanpied, "Inventing US Science Policy," *Physics Today* 51, no. 2 (1998): 34–40.

85. Ibid.

86. Ibid.

87. Ibid.

88. Ibid.

89. Christine M. Matthews, "U.S. National Science Foundation: An overview," *CRS Report for Congress*, July 1, 1998, http://www.ncseonline.org/NLE/CRSreports/Science/st-6.cfm?&CFID=14969499&CFTOKEN=65710512 (accessed on July 20, 2004).

90. Ibid.

91. Ibid.

92. In his February 2002 budget submission to Congress, President Bush outlined a management agenda for making government more focused on citizens and results, which includes expanding Electronic Government—or E-Government. E-Government uses improved Internet-based technology to make it easy for citizens and businesses to interact

with the government, save taxpayer dollars, and streamline citizen-to-government communications. T. Smith, "National Science Foundation in the FY 2004 budget," *AAAS Report XXVIII: Research & Development FY 2004*, 2004, http://www.aaas.org/spp/rd/04pch7.htm (accessed on July 20, 2004).

93. Ibid.

94. Ibid.

95. Ibid.

96. Christine M. Matthews, "U.S. National Science Foundation: An overview," *CRS Report for Congress*, July 1, 1998, http://www.ncseonline.org/NLE/CRSreports/Science/st-6.cfm?&CFID=14969499&CFTOKEN=65710512 (accessed on July 20, 2004).

97. Ibid.

98. T. Smith, "National Science Foundation in the FY 2004 budget," *AAAS Report XXVIII: Research & Development FY 2004*, 2004, http://www.aaas.org/spp/rd/04pch7.htm (accessed on July 20, 2004).

99. Ibid.

100. Ibid.

101. Ibid.

102. "National Science Foundation," The Office of Management and Budget, The Executive Office of the President, 2004, http://www.whitehouse.gov/omb/budget/fy2005/nsf.html (accessed July 19, 2004).

103. Ibid.

104. Ibid.

105. Ibid.

106. R. Jones, "FY 2005 National Science Foundation Budget request," *AIP Bulletin of Science Policy News*, February 4, 2004, http://www.aip.org/fyi/2004/011.html (accessed on July 19, 2004).

107. Ibid.

108. Ibid.

109. R. Terra, "National Nanotechnology Initiative in FY2001 budget," *Foresight Update*, http://www.foresight.org/Updates/Update40/Update40.1.html (accessed on July 19, 2004).

110. "Megabucks for Nanotech," *Scientific American*, September 2001, p. 8.

111. "Science Scope," *Science* 287, no. 5457, issue 25 (February 25, 2000), http://www.sciencemag.org/content/vol287/issue5457/s-scope.shtml.

112. Ibid.

113. Ibid.

114. Ibid.

115. D. Brown, "Nation depends on more money for nano, advocates tell Senate," *Small Times*, 2002, http://www.smalltimes.com/document_display.cfm?&document_id=3829 (accessed on July 21, 2004).

116. "National Science Foundation," The Office of Management and Budget, The Executive Office of the President, 2004, http://www.whitehouse.gov/omb/budget/fy2005/nsf.html (accessed July 19, 2004).

117. "National Science Foundation," 2002, http://www.aecom.yu.edu/ogs/Guide/NSFInfo.htm (accessed July 21, 2004).

118. R. Terra, "National Nanotechnology Initiative in FY2001 budget," *Foresight Update*, http://www.foresight.org/Updates/Update40/Update40.1.html (accessed on July 19, 2004).

119. "$1 billion for National Nanotechnology Initiative," 2004, http://www.azonano.com/news_old.asp?newsID=50 (accessed on July 20, 2004).

120. "NIH 'soft landing' turns hard in 2005," *The AAAS R&D Funding Update on R&D in the FY 2005 NIH Budget*, February 20, 2004, http://www.aaas.org/spp/rd/nih05p.htm (accessed on July 21, 2004).

121. Ibid.

122. Ibid.

123. J. Weisman, "2006 cuts in domestic spending on table," *Washington Post*, May 27, 2004, p. A1.

124. T. Smith, "National Science Foundation in the FY 2004 budget," *AAAS Report XXVIII: Research & Development FY 2004*, 2004, http://www.aaas.org/spp/rd/04pch7.htm (accessed on July 20, 2004).

125. M. Davey, "RS20589: Manipulating molecules: The National Nanotechnology Initiative," *CRS Report for Congress*, September 20, 2000, http://www.ncseonline.org/NLE/CRSreports/Science/st-48.cfm?&CFID=14991140&CFTOKEN=13149058 (accessed July 19, 2004).

126. C. Schulz, "Crafting a national nanotechnology effort," *Chemical and Engineering News* 78, no. 42 (October 16, 2000), http://pubs.acs.org/cen/nanotechnology/7842/print/7842government.html (accessed on June 15, 2004).

127. Ibid.

128. Ibid.

129. M. Davey, "RS20589: Manipulating molecules: The National Nanotechnology Initiative," *CRS Report for Congress*, September 20, 2000, http://www.ncseonline.org/NLE/CRSreports/Science/st-48.cfm?&CFID=14991140&CFTOKEN=13149058 (accessed July 19, 2004).

130. Ibid.

131. D. Brown, "Nation depends on more money for nano, advocates tell Senate," *Small Times*, 2002, http://www.smalltimes.com/document_display.cfm?&document_id=3829 (accessed on July 21, 2004).

132. Ibid.

133. Ibid.

134. Daniel S. Greenberg, *Science, Money, and Politics* (Chicago: University of Chicago Press, 2001).

135. Ibid.

136. "President Bush's FY 2005 budget request: How does science fare?" 2004, http://www.chemistry.org/portal/a/c/s/1/feature_pol.html?DOC=policymakers%5Cpol_capcon_feb04.html (accessed on June 23, 2004).

137. Barbara Goldstein, "Nanotechnology is BIG at NIST," 2003, http://www.nist.gov/public_affairs/nanotech.htm (accessed July 14, 2004).

138. M. C. Roco, "National Nanotechnology Investment in the FY 2005 budget request," *AAAS Report on U.S. R&D in FY 2005*, Washington, D.C., March 2004 (preprint).

139. Ibid.

140. "NIST launches advanced measurement laboratory to support nanotech," 2004, http://www.linuxelectronics.com/article.php?story=20040622083043387 (accessed on June 30, 2004).

141. Ibid.

142. Barbara Goldstein, "Nanotechnology is BIG at NIST," 2003, http://www.nist .gov/public_affairs/nanotech.htm (accessed July 14, 2004).

143. Thomas Cellucci, "Fiscal year 2005 NIST budget: Views from industry," 2004, http://www.zyvex.com/News/Testimony042404_F.html (accessed on May 3, 2004).

144. Ibid.

145. Ibid.

146. Ibid.

147. Ibid.

148. "The Advanced Technology Program: A progress report on the impacts of an industry-government technology partnership," 1996, http://www.atp.nist.gov/atp/ repcong/repcong.htm (accessed June 23, 2004).

149. Ibid.

150. Ibid.

151. Douglas Brown, "Another year, another threat to the Advanced Technology Program," *Small Times*, February 6, 2003, http://www.smalltimes.com/document_display.cfm ?document_id=5456 (accessed June 23, 2004).

152. Ibid.

153. "ATP's Project Portfolio in Nanotechnology," 2004, http://www.nist.gov/public _affairs/atp_nanotech.htm (accessed June 23, 2004).

154. Ibid.

155. Ibid.

156. Audrey Leath, "Advanced Technology Program: A funding retrospective," *American Institute of Physics Bulletin of Science Policy News*, 2000, http://www.aip.org/fyi/ 2000/fy00.080.htm (accessed June 23, 2004).

157. Ibid.

158. Douglas Brown, "Another year, another threat to the Advanced Technology Program," *Small Times*, February 6, 2003, http://www.smalltimes.com/document_display.cfm ?document_id=5456 (accessed June 23, 2004).

159. Debra M. Bryant, "Overview Bush administration FY 2005 budget recommendations for information technology," The Ferguson Group, March 5, 2004, http://www.njtc .org/publicpolicy/FY%202005%20IT%20Budget%20Rec.doc (accessed June 23, 2004).

160. Douglas Brown, "Another year, another threat to the Advanced Technology Program," *Small Times*, February 6, 2003, http://www.smalltimes.com/document_display.cfm ?document_id=5456 (accessed June 23, 2004).

161. Ibid.

162. Ibid.

163. Ibid.

164. Juliana Gruenwald, "Despite House's okay, little time left to pass nanotech bill this year," *Small Times*, July 12, 2004, http://www.smalltimes.com/document_display.cfm ?document_id=8159 (accessed July 19, 2004).

165. Candace Stuart, "Rationing resources through fed's drought," *Small Times*, April 2005, p. 26.

166. Mihail Roco, introduction to *Societal Implications of Nanoscience and Nanotechnology*, ed. M. C. Roco and W. S. Bainbridge (New York: Kluwer, 2001), p. 2.

167. R. Meeks, "President's Budget Includes Modest Increase for R&D in FY 2004," *National Science Foundation InfoBrief*, October 2003.

168. J. Uldrich and D. Newberry, *The Next Big Thing Is Really Small: How Nanotechnology Will Change the Future of Your Business* (New York: Crown Business, 2003), pp. 66–67.

169. R. Smalley, "Nanotechnology: The state of nano-science and its prospects for the next decade," Hearings before the Subcommittee on Basic Research of the House Committee on Science, 104th Cong., June 22, 1999, http://www.house.gov/science/smalley_062299.htm (accessed April 26, 2004).

170. Mihail Roco, Introduction to *Societal Implications of Nanoscience and Nanotechnology*, ed. M. C. Roco and W. S. Bainbridge (New York: Kluwer, 2001), p. 2.

171. Mihail Roco, "The US National Nanotechnology Initiative after three years (2001–2003): Setting new goals for responsible nanotechnology," Societal Implications of Nanoscience and Nanotechnology Workshop, December 3, 2003, p. 1.

172. NanoInvestorNews.com, 2003, http://www.nanoinvestornews.com/modules.php?name=facts_figures&op=sho&im=locfunding/usa (accessed August 29, 2004).

173. J. Uldrich and D. Newberry, *The Next Big Thing Is Really Small: How Nanotechnology Will Change the Future of Your Business* (New York: Crown Business, 2003), pp. 66–67.

174. T. Henderson, "Bayh-Dole Act of 1980 opened the doors for researchers to profit from inventions," *Small Times*, July 15, 2001, http://www.smalltimes.com/document_display.cfm?document_id=1427 (accessed April 26, 2004).

175. National Institute of Mental Health, "Technology transfer legislation summary," August 2000, http://intramural.nimh.nih.gov/techtran/legislation.htm (accessed June 10, 2003).

176. Gary Stix, "Razing the tollbooths: A call for restricting patents on basic biomedical research," *Scientific American*, April 2003, http://www.sciam.com/article.cfm?articleID=00092FBB-6BDC-1E61-A98A809EC5880105 (accessed April 26, 2004).

177. Glenn Fishbine, *The Investor's Guide to Nanotechnology and Micromachines* (New York: George Wiley and Sons, 2002), p. 40.

178. J. Uldrich and D. Newberry, *The Next Big Thing Is Really Small: How Nanotechnology Will Change the Future of Your Business* (New York: Crown Business, 2003), pp. 66–67.

179. Glenn Fishbine, *The Investor's Guide to Nanotechnology and Micromachines* (New York: George Wiley and Sons, 2002), p. 40.

180. The American Society of Mechanical Engineers, *Monthly Report for the White House Office of Science and Technology Policy*, June 2003, http://www.asme.org/gric/fellows/reports/mackint_09-02.html (accessed June 10, 2003).

181. Ibid.

182. M. C. Roco, "The U.S. National Nanotechnology Initiative after three years (2001–2003): Setting new goals for a responsible nanotechnology," *Abstracts of the Workshop on Societal Implications of Nanoscience and Nanotechnology*, December 3–5, 2003, p. 1.

183. Glenn Fishbine, *The Investor's Guide to Nanotechnology and Micromachines* (New York: George Wiley and Sons, 2002), p. 43.

184. Ibid.

185. Office of Basic Energy Sciences, Office of Science, *Nanoscale Science, Engineering and Technology in the Department of Energy: Research Directions and Nanoscale Science Research Centers*, March 2004, p. 16.

186. Ibid., p. 4.

187. Jamie Dinkelacker, "Policy Watch," *Foresight Update*, no. 18 (April 15, 1994): 8–9.

188. Richard Russell, Committee on Science in the House of Representatives, *H.R. 766, Nanotechnology Research and Development Act of 2003*, Hearings before the Committee on Science in the House of Representatives, March 19, 2003, p. 86.

189. "Nanotech's new Apollo mission: Energy," *Nanotech Report* 2, no. 6 (June 2003): 1.

190. J. Uldrich and D. Newberry, *The Next Big Thing Is Really Small: How Nanotechnology Will Change the Future of Your Business* (New York: Crown Business, 2003), pp. 66–67.

191. T. McCarthy, "Molecular nanotechnology and the world system," January 9, 1996, http://www.bcf.usc.edu/~tmccarth/main.html (accessed June 14, 2004).

192. R. Mullen, "DOE science 2004 budget: A few winners," *New Technology Week*, 2003, p. 1.

193. J. Uldrich and D. Newberry, *The Next Big Thing Is Really Small: How Nanotechnology Will Change the Future of Your Business* (New York: Crown Business, 2003), p. 115.

194. ETC Group: The Action Group on Erosion, Technology, and Concentration, *From Genomes to Atoms: The Big Down, Atomtech and Technologies Converging at the Nano-scale* (Winnipeg, Manitoba: ETC Group, 2003), p. 61.

195. Phil Beradelli, "Interview: Marburger defends R&D policies," *United Press International*, June 23, 2004, http://www.upi.com/view.cfm?StoryID=20040622-07343506586r (accessed July 9, 2004).

196. National Nanotechnology Initiative Grand Challenge Workshop, *Nanoscience Research for Energy Needs* (March 16–18 2004): v.

197. Doug Brown, "Bush's proposed budget makes nanotechnology a top priority," *Small Times*, February 5, 2003, http://www.smalltimes.com/document_display.cfm?document_id=5449 (accessed April 26, 2004).

198. Ibid.

199. William Schulz, "Creating a national nanotechnology effort," *CENEAR—Special Report* 78, no. 42 (October 16, 2000): 39–42, http://pub.acs.org/cen/nanotechnology/7842/print/7842government.html (accessed June 15, 2004).

200. "Nanotech on the front lines," *Nanotech Report* 1, no. 9 (November 2002): 1.

201. "DOD lacks database to determine foreign dependence, GAO says," *Aerospace Daily* 170, no. 4 (April 6, 1994): 30.

202. "Media Watch," *Foresight Update*, no. 19 (September 15, 1994): 9–11.

203. "Nanotech on the front lines," *Nanotech Report* 1, no. 9 (November 2002): 1.

204. Jamie Dinkelacker, "Policy watch," *Foresight Update*, no. 18 (April 15, 1994): 8–9.

205. Glenn Fishbine, *The Investor's Guide to Nanotechnology and Micromachines* (New York: George Wiley and Sons, 2002), p. 46.

206. S. Cobb, "International news," *Foresight Update*, no. 10 (October 30, 1990): 4–5.

207. Jamie Dinkelacker, "Policy watch," *Foresight Update*, no. 18 (April 15, 1994): 8–9.

208. "Nanotech relief for G.I. Joe," *Nanotech Report* (April 2004): 3.

209. "Nanotech on the front lines," *Nanotech Report* 1, no. 9 (November 2002): 1.

210. The Institute of Nanotechnology, "Latest on nanotechnology in the United

States: MIT wins army contract," March 2003, http://www.nano.org.uk/thisweek71.htm (accessed June 11, 2003).

211. "Nanotech on the front lines," *Nanotech Report* 1, no. 9 (November 2002): 1.

212. Ibid.

213. Phil Copeland, "Future warrior exhibits super powers," *AFIS Defenselink* (July 27, 2004), http://www.defenselink.mil/news/Jul2004/n07272004_2004072705.html (accessed July 30, 2004).

214. J. Uldrich and D. Newberry, *The Next Big Thing Is Really Small: How Nanotechnology Will Change the Future of Your Business* (New York: Crown Business, 2003), p. 69.

215. United States Army Logistics Management College, *Thinking Small: Technologies That Can Reduce Logistics Demand*, January 2000, http://www.almc.army.mil/alog/issues/MarApr00/MS523.htm (accessed June 13, 2003).

216. "Nanotech on the front lines," *Nanotech Report* 1, no. 9 (November 2002): 1.

217. Glenn Fishbine, *The Investor's Guide to Nanotechnology and Micromachines* (New York: George Wiley and Sons, 2002), p. 110.

218. J. Uldrich and D. Newberry, *The Next Big Thing Is Really Small: How Nanotechnology Will Change the Future of your Business* (NY: Crown Business, 2003), p. 139.

219. Ibid.

220. "DARPA funds collaborative quantum computer center," *NanoDot*, February 14, 2001, http://nanodot.org/article.pl?sid=01/09/14/1313244 (accessed June 27, 2004).

221. "Nanotech on the front lines," *Nanotech Report* 1, no. 9 (November 2002): 2.

222. Ibid.

223. M. Meyer, "Socio-economic research on nanoscale science and technology: A European overview and illustration," *Societal Implications of Nanoscience and Nanotechnology*, ed. M. C. Roco and W. S. Bainbridge (New York: Kluwer, 2001), p. 221.

224. Ibid., p. 222.

225. Dean Takahashi, "Creating a sticky situation," *CentreDaily.com*, December 25, 2004, http://www.centredaily.com/mld/centredaily/business/technology/10497914.htm (accessed January 12, 2005).

226. John Gartner, "Mobile Army requires solar soldiers," *Technology Review*, May 16, 2005, http://www.technologyreview.com/articles/05/05/wo/wo_051605gartner.asp (accessed July 19, 2005).

227. "Infectech, Inc., approves name change to NanoLogix and enters hydrogen production, biodefense, medical diagnostics and cancer drug markets," *NanoInvestorNews*, April 7, 2005, http://www.nanoinvestornews.com/modules.php?name=News&file=print&sid=4195 (accessed April 14, 2005).

228. "Nano-engineered powders tackle toxic chemicals," *PhysOrg.com*, April 28, 2005, http://www.physorg.com/printnews.php?newsid=3914 (accessed May 10, 2005).

229. "Nanotech on the front lines," *Nanotech Report* 1, no. 9 (November 2002): 1–2.

230. Kevin Rayburn, "Researchers look to nanotechnology to improve military missile defense," *Louisville Cardinal Online*, January 25, 2005, http://www.louisvillecardinal.com/vnews/display.v/ART/2005/01/25/41f7c4d1450e8 (accessed July 19, 2005).

231. "MIT scientists improve explosives detection," *NewsNanoApex*, April 21, 2005, http://news.nanoapex.com/modules.php?name=News&file=article&sid=5602 (accessed July 19, 2005).

232. J. Uldrich and D. Newberry, *The Next Big Thing Is Really Small: How Nanotechnology Will Change the Future of Your Business* (New York: Crown Business, 2003), p. 69.

233. Ibid.

234. W. Knight, "Military robots get swarm intelligence," *New Scientist*, April 25, 2003, http://www.newscientist.com/news/news.jsp?id=ns99993661 (accessed May 13, 2003).

235. "Nanotech on the front lines," *Nanotech Report* 1, no. 9 (November 2002): 1.

236. Barnaby Feder, "Defense Department expands nanotechnology research," *New York Times*, April 8, 2003, http://www.siliconvalley.com/mld/siliconvalley/news/5585217 .htm (accessed August 5, 2004).

237. Ibid.

238. Ibid.

239. Mihail Roco, Introduction to *Societal Implications of Nanoscience and Nanotechnology*, ed. M. C. Roco and W. S. Bainbridge (New York: Kluwer, 2001), p. 10.

240. Hongda Chen, "U. S. Department of Agriculture Research on Nanotechnology," NNI: From Vision to Commercialization, March 31–April 2, 2004.

241. J. Uldrich and D. Newberry, *The Next Big Thing Is Really Small: How Nanotechnology Will Change the Future of Your Business* (New York: Crown Business, 2003), p. 73.

242. Robert Cowan, "A tiny robot swarm—Fiction no longer," *Christian Science Monitor*, April 7, 2005, http://www.csmonitor.com/2005/0407/p14s01-stct.html (accessed April 7, 2005).

243. "Nanotechnology's Homeland Security potential to be explored," *Space Daily*, December 11, 2003, http://www.spacedaily.com/news/nanotech-03zzr.html (accessed August 10, 2004).

244. Candace Stuart, "Small tech provides some answers to a nation questioning its safety," *Small Times*, September 11, 2004, http://www.smalltimes.com/document_display .cfm?section_id=45&document_id=4586 (accessed August 10, 2004).

CHAPTER 4

1. "Decoding future nanotech investment success," *Nanotech Report* 1, no. 7 (September 2002): 1.

2. Robert F. Service, "Nanotechnology grows up," *Science* 304 (June 18, 2004): 1733.

3. "C60 market monitor for May 2004," May 16, 2004, http://www.nanoinvestor news.com/modules.php?name=News&file=article&sid=2792 (accessed May 20, 2004).

4. *Partnership in Nanotechnology*, From NSF Grantees Conference, January 29–30, 2001, Arlington, VA: NSF.; *Proceedings of the Joint NSF-NIST Conference on Nanoparticles: Synthesis, Processing into Functional Nanostructures, and Characterization*, May 12–13, 1997, Arlington, VA: NSF; *Proceedings of the Joint NSF-NIST Conference on Ultrafine Particle Engineering*, May 25–27, 1994, Arlington, VA: NSF.

5. *NNUN Abstracts 2002*, http://nsf.cornell.edu/nnun/2002NNUNinfo.pdf (accessed July 13, 2004).

6. See Sandip Tiwari in Department of Trade and Industry and Office of Science and Technology, *New Dimensions for Manufacturing: A UK Strategy for Nanotechnology*, http://

www.oft.osd.mil/library/library_files/document_134_nanotechnologyreport.pdf
(accessed June 21, 2004), p. 21.

7. *NNUN Abstracts 2002.* (2002): 3 http://nsf.cornell.edu/nnun/2002NNUNinfo
.pdf. (accessed July 13, 2004).

8. Ibid.

9. Neal Lane, "Testimony before the Senate Appropriations Committee," *Federal
News Service,* May 7, 1998, http://web.lexis-nexis.com/congroup/document (accessed September 8, 2003).

10. Department of Trade and Industry and Office of Science and Technology, *New
Dimensions for Manufacturing: A UK Strategy for Nanotechnology,* http://www.oft.osd.mil/
library/library_files/document_134_nanotechnologyreport.pdf (accessed June 21, 2004),
p. 21.

11. William Schulz, "Creating a national nanotechnology effort," *CENEAR—Special
Report* 78, no. 42 (October 16, 2000): 39–42, http://pub.acs.org/cen/nanotechnology/7842/
print/7842government.html (accessed June 15, 2004).

12. National Science and Technology Council, Committee on Technology, Subcommittee on Nanoscale Science, Engineering, and Technology (NSET), *National Nanotechnology Initiative: Research and Development Supporting the Next Industrial Revolution, Supplement
to the President's FY 2004 Budget* (Arlington, VA: National Nanotechnology Coordination
Office, August 29, 2003), p. 2.

13. "Decoding future nanotech investment success," *Nanotech Report* 1, no. 7 (September 2002): 1.

14. "Overview of ATP," September 10, 2003, http://www.atp.nist.gov/atp/
overview.htm (accessed January 3, 2004).

15. National Science and Technology Council, Committee on Technology, Subcommittee on Nanoscale Science, Engineering, and Technology (NSET), *National Nanotechnology Initiative: Research and Development Supporting the Next Industrial Revolution, Supplement
to the President's FY 2004 Budget* (Arlington, VA: National Nanotechnology Coordination
Office, August 29, 2003), p. 3.

16. Scott Burnell, "Interview: Sen. Wyden eyes nanotech," *UPI Science News,* September 22, 2002, http://www.upi.com/view.cfm?StoryID=20020920-030036-4843r (accessed July 26, 2004).

17. "President's Council of Advisors on Science and Technology," *PCAST Home Page,*
June 22, 2004, http://www.ostp.gov/PCAST/pcast.html (accessed July 11, 2004).

18. National Science and Technology Council, Committee on Technology, Subcommittee on Nanoscale Science, Engineering, and Technology (NSET), *National Nanotechnology Initiative: Research and Development Supporting the Next Industrial Revolution, Supplement
to the President's FY 2004 Budget* (Arlington, VA: National Nanotechnology Coordination
Office, August 29, 2003), p. 6.

19. "Nanotechnology Work Plan, March 3, 2003," Committee on Science in the
House of Representatives, *H.R. 766, Nanotechnology Research and Development Act of 2003,*
Hearings before the Committee on Science in the House of Representatives, March 19,
2003, p. 90.

20. ETC Group: The Action Group on Erosion, Technology, and Concentration,

From Genomes to Atoms: The Big Down, Atomtech and Technologies Converging at the Nano-scale (Winnipeg, Manitoba: ETC Group, 2003), p. 61.

21. Competitive process involving the submission of university grant proposals to study a specific area of nanotechnology.

22. Glenn Fishbine, *The Investor's Guide to Nanotechnology and Micromachines* (New York: George Wiley and Sons, 2002), p. 19.

23. ETC Group: The Action Group on Erosion, Technology, and Concentration, *From Genomes to Atoms: The Big Down, Atomtech and Technologies Converging at the Nano-scale* (Winnipeg, Manitoba: ETC Group, 2003), p. 61.

24. Michael Davey, "Manipulating molecules: The National Nanotechnology Initiative," *CRS Report for Congress*, In Committee on Science in the House of Representatives, *H.R. 766, Nanotechnology Research and Development Act of 2003*, Hearings before the Committee on Science in the House of Representatives, March 19, 2003, p. 94.

25. Ibid., p. 95.

26. Committee on Science in the House of Representatives, *H.R. 766, Nanotechnology Research and Development Act of 2003*, Hearings before the Committee on Science in the House of Representatives, March 19, 2003, p. 6.

27. Susan R. Morissey, "Harnessing nanotechnology," *Government and Policy* 82, no. 16 (April 14, 2004): 30–33, http://pubs.acs.org/cen/nlw/print/8216gov1.html (accessed June 15, 2004).

28. Candace Stuart, "Rationing resources through fed's drought," *Small Times*, April 2005, p. 25.

29. Susan R. Morissey, "Harnessing nanotechnology," *Government and Policy* 82, no. 16 (April 14, 2004): 30–33, http://pubs.acs.org/cen/nlw/print/8216gov1.html (accessed June 15, 2004).

30. Candace Stuart, "President's advisors recommend NNI branch out," *Small Times*, May 19, 2005, http://www.smalltimes.com/document_display.cfm?document_id=9253 (accessed July 19, 2005).

31. Susan R. Morissey, "Harnessing nanotechnology," *Government and Policy* 82, no. 16 (April 14, 2004): 30–33, http://pubs.acs.org/cen/nlw/print/8216gov1.html (accessed June 15, 2004).

32. Committee on Science in the House of Representatives, *H.R. 766, Nanotechnology Research and Development Act of 2003*, Hearings before the Committee on Science in the House of Representatives, March 19, 2003, p. 14.

33. "Decoding future nanotech investment success," *Nanotech Report* 1, no. 7 (September 2002): 2.

34. National Science and Technoloy Council, Committee on Technology, The Interagency Working Group of Nanoscience, Engineering and Technology, "Nanotechnology—Shaping the World Atom by Atom," September 1999, http://www.wtec.org/loyola/nano/IWGN.Public.Brochure/ (accessed January 3, 2004).

35. K. Eric Drexler, "Nanotechnology: From Feynman to funding," a preprint.

36. Ibid.

37. Lev Nazrozov, "Will Drexler save the West from nano annihilation," *NEWSMAX .COM*, December 19, 2003, http://216.26.163.62/2003/lev12_19.html (accessed December 29, 2003).

38. William J. Clinton, "Remarks on Science and Technology Investments," California Institute of Technology, Pasadena, CA, January 21, 2000, http://www.columbia .edu/cu/osi/nanopotusspeech.html (accessed August 7, 2003).

39. William J. Clinton, State of the Union Address, Joint Session of Congress, Washington, DC, January 27, 2000, http://archives.cnn.com/2000/ALLPOLITICS/stories /01/27/sou.trans/ (accessed November 4, 2005).

40. Adam Keiper, "The nanotechnology revolution," *New Atlantis: A Journal of Technology and Society*, Summer 2003, http://www.thenewatlantis.com/archive/2/keiper print.htm (accessed September 12, 2003).

41. Ibid.

42. Ibid.

43. National Science and Technology Council, Committee on Technology, Subcommittee on Nanoscale Science, Engineering, and Technology (NSET), *National Nanotechnology Initiative: Research and Development Supporting the Next Industrial Revolution, Supplement to the President's FY 2004 Budget* (Arlintgton, VA: National Nanotechnology Coordination Office, August 29, 2003), p. 3.

44. "Nanotech industry review and outlook," *Nanotech Report* 2, no. 1 (January 2003): 3.

45. "Nanotechnology's power brokers," *Nanotech Report* 2, no. 3 (March 2003): 1.

46. Juliana Gruenwald, "After the bill: Bush's budget a bit short," *Small Times*, March/April 2004, p. 10.

47. Ibid.

48. Ibid.

49. *National Nanotechnology Infrastructure Network* (Ithaca, NY: Cornell University).

50. Juliana Gruenwald, "After the bill: Bush's budget a bit short," *Small Times*, March/April 2004, p. 10.

51. "Trouble in Nanoland," *Economist*, December 5, 2002, http://www.economist .com/science/PrinterFriendly.cfm?Story_IS=1477445 (accessed October 5, 2003).

52. Mihail Roco, "Nanoscale Science and Engineering: Unifying and Transforming Tools," *AIChE Journal* 50, no. 5 (May 2004): 893.

53. Mark Modzelewski, "Nanotech industry can help groundbreaking bill fulfill its promise through collaboration, public policy," *Small Times*, January/February 2004, p. 12.

54. Charles Choi, "Nano Bill promises real results," *TechNewsWorld*, December 4, 2003, http://www.technewsworld.com/perl/story/32298.html (accessed December 9, 2003).

55. "U.S. Senate Passes Nano Bill," NbA press release, November 19, 2003.

56. Rex Crum, "Talking nanotech with Mike Honda," *CBS.Marketwatch.com*, March 17, 2004, http://cbs.marketwatch.com/news/print_story.asp?print=1&guid={A637E67A -A5D7-41B- (accessed April 5, 2004).

57. Committee on Science in the House of Representatives, *H.R. 766, Nanotechnology Research and Development Act of 2003*, Hearings before the Committee on Science in the House of Representatives, March 19, 2003, p. 11.

58. Charles Choi, "Nano Bill promises real results," *TechNewsWorld*, December 4, 2003, http://www.technewsworld.com/perl/story/32298.html (accessed December 9, 2003).

59. David Berube, ibid.

60. Michael Davey, "Manipulating molecules: The National Nanotechnology Initiative," *CRS Report for Congress*, In Committee on Science in the House of Representatives, *H.R. 766, Nanotechnology Research and Development Act of 2003*, Hearings before the Committee on Science in the House of Representatives, March 19, 2003, p. 97.

61. David Berube quoted by Jeff Karoub, "Ethics Center a Small Obstacle as Senate Nears Nano Bill Passage," *Small Times*, November 6, 2003, http://www.smalltimes.com/document_display.cfm?document_id=6912 (accessed November 18, 2003).

62. Declan McCullagh, "House earmarks billions for nanotech," *CNET News.com*, May 7, 2003, http://news.com.com/2100-1028_3-1000408.html?tag=prntfr (accessed July 14, 2004).

63. Juliana Gruenwald, "PCAST will advise Bush on nanotech," *Small Times*, July 20, 2004, http://www.smalltimes.com/document_display.cfm?document_id=8171 (accessed July 26, 2004).

64. "U.S. Senate Passes Nano Bill," NbA press release, November 19, 2003.

65. Ibid.

66. Barnaby Feder, "It's a tiny new world," *New York Times*, December 22, 2003, p. C1.

67. Phil Beradelli, "Interview: Marburger defends R&D policies," *United Press International*, June 23, 2004, http://www.upi.com/view.cfm?StoryID=20040622-07343506586r (accessed July 9, 2004).

68. Glenn Fishbine, *The Investor's Guide to Nanotechnology and Micromachines* (New York: George Wiley and Sons, 2002), p. xi.

69. "Semiconductor Industry Association Says U.S. Could Lose Race for Nanotechnology Leadership," *NanoTech Wire*, March 16, 2005, http://nanotechwire.com/news.asp?nid=1735 (accessed July 14, 2005).

70. M. Mauboussin and K. Bartholdson, *Big Money in Thinking Small: Nanotechnology— What Investors Need to Know* (Boston, MA: Credit Suisse Equity Research/First Boston, May 2003).

71. Ibid., p. 15.

72. "EU-US-Japan Government Spending," *NanoInvestorNews*, July 27, 2004, http://www.nanoinvestornews.com/modules.php?name=Facts_Figures&op=sho&im=fund_euusjapan9702 (accessed July 27, 2004).

73. Glenn Fishbine, *The Investor's Guide to Nanotechnology and Micromachines* (New York: George Wiley and Sons, 2002), p. 59.

74. Sudhir Chowdhary, "Govt to commercialize nanotechnology products," *Financial Express*, August 19, 2004, http://www.financialexpress.com/fe_full_story.php?content_id=66304 (accessed May 11, 2005).

75. Josh Wolfe, *FORBES/WOLFE Nanotech Weekly Insider*, December 10, 2004, e-mail communication.

76. "U.S. risks losing nano lead," *PhysOrg.com*, July 6, 2005, http://www.physorg.com/printnews.php?newsid=4963 (accessed July 8, 2005).

77. "Developing global nanotech," *Red Herring*, April 12, 2005, http://www.redherring.com/PrintArticle.aspx?a=11765&sector=Industries (accessed April 18, 2005).

78. "Nanoscience, Nanotechnology: The state of play in the Union's NMP priority," *EUROPA*, January 7, 2005, http://europa.eu.into/comm/research/headlines/news/article_05_01_07_en.html (accessed January 12, 2005).

79. A joint report from France's Academics of Science and Technology recently studied France's efforts. According to Philippe Nozieres, one of the report authors, "The money is very dispersed and there is no control over the way it is spent. In France there are many excellent labs, but they are very dispersed and don't integrate very well into a larger project." See Genevieve Oger, "French academy urges government to invest in small tech," *Small Times*, April 30, 2004, http://www.smalltimes.com/document_display.cfm?document _id=7787 (accessed May 3, 2004).

80. "Nanotechnology news in brief," *NanoTechWeb*, April 15, 2005, http://nanotechweb.org/articles/news/4/4/11?alert=1 (accessed April 18, 2005).

81. "Nanoscience, Nanotechnology: The state of play in the Union's NMP priority," *EUROPA*, January 7, 2005, http://europa.eu.into/comm/research/headlines/news/article _05_01_07_en.html (accessed January 12, 2005).

82. M. Mauboussin and K. Bartholdson, *Big Money in Thinking Small: Nanotechnology— What Investors Need to Know* (Boston, MA: Credit Suisse Equity Research/First Boston, May 2003), p.18.

83. Glenn Fishbine, *The Investor's Guide to Nanotechnology and Micromachines* (New York: George Wiley and Sons, 2002), p. 62.

84. C. Rowan, "It's Ours to Lose—An Analysis of EU Nanotechnology Funding and the Sixth Framework Program," 2002, http://www.nanoeurope.org/docs/European%20 Nanotech%20Funding.pdf (accessed on April 26, 2004).

85. C. Rowan, "It's Ours to Lose—An Analysis of EU Nanotechnology Funding and the Sixth Framework Program," 2002, http://www.nanoeurope.org/docs/European%20 Nanotech%20Funding.pdf (accessed on April 26, 2004), p. 9.

86. "A new EU strategy in the field of nanotechnology," *Welcome Europe.com*, May 12, 2004, http://www.welcomeurope.com/news_info.asp?idnews=1519 (accessed May 20, 2004).

87. European Union, "Third European Report on Science and Technology Indicators," March 2003, http://www.cordis.lu/indicators/third_report.htm (accessed April 26, 2004), p. 2.

88. The National Technology Transfer Center (NTTC) is a full-service technology-management center, providing access to federal technology information, knowledge management and digital learning services, technology assessment, technology marketing, assistance in finding strategic partners, and electronic-business development services. The NTTC fosters relationships with federal clients, showcases technologies, and facilitates partnerships between clients and U.S. industry. See http://www.nttc.edu/ (accessed August 10, 2004).

89. NanoInvestorNews.com, 2003, http://www.nanoinvestornews.com/modules .php?name=facts_figures&op=sho&im=locfunding/eu (accessed on August 29, 2004).

90. "A new EU strategy in the field of nanotechnology," *Welcome Europe.com*, May 12 2004, http://www.welcomeurope.com/news_info.asp?idnews=1519 (accessed May 20, 2004).

91. "EU-Government Funding for Nanotechnology," *NanoInvestorNews*, July 26, 2004, http://www.nanoinvestornews.com/modules.php?name=Facts_Figures&op=sho &im=locfunding/eu (accessed July 26, 2004).

92. C. Rowan, "It's Ours to Lose—An Analysis of EU Nanotechnology Funding and

the Sixth Framework Program," 2002, http://www.nanoeurope.org/docs/European%20 Nanotech%20Funding.pdf (accessed on April 26, 2004), p. 6.

93. Glenn Fishbine, *The Investor's Guide to Nanotechnology and Micromachines* (New York: George Wiley and Sons, 2002), p. 60.

94. European Union, *Third European Report on Science and Technology Indicators*, March 2003, http://www.cordis.lu/indicators/third_report.htm (accessed April 26, 2004), p. 2.

95. M. Kelly, "European Nanotech Depends Too Much on Government Handouts, Expert Says," *Small Times*, November 30, 2001, http://www.smalltimes.com/document _display.cfm?document_id=2656 (accessed April 26, 2004).

96. Genevieve Oger, "European Union turns to research projects in hopes of building up its sagging economy," *Small Times*, January/February 2004, p. 20.

97. Ibid.

98. Althea Lipsett, "Plenty more," *Guardian Unlimited*, June 22, 2004, http:// education.guardian.co.uk/egweekly/story/0,5500,1244084,00.html (accessed July 9, 2004).

99. Victoria Knight, "EU unveils business-friendly research strategy," *Dow Jones Newswires*, April 6, 2005, e-communication.

100. "European Union earmarks billions to prepare for a nanofuture," *Small Times*, July 6, 2004, http://www.smalltimes.com/print_doc.cfm?doc_id=8144 (accessed July 12, 2004).

101. European Union, "Towards a European Strategy for Nanotechnology," 2004, ftp://ftp.cordis.lu/pub/nanotechnology/docs/nano_com_en.pdf (accessed July 13, 2004).

102. "EU, CEOs call for nanotechnology push," *CNN.com*, June 29, 2004, http:// www.cnn.com/2004/TECH/biztech/06/29/nanotechnology.europe.reut/ (accessed August 4, 2004).

103. John Ryan, "Nanotechnology; separating fact, fiction, hype and hope," *Physics-Web*, August 2004, http://physicsweb.org/article/world/17/8/2 (accessed August 9, 2004).

104. "2nd Annual International Nanoscience Conference to Be Held in Grenoble, France," May 18, 2004, http://news.nanoapex.com/modules.php?name=News&file =article&sid=4545 (accessed May 25, 2004).

105. Antonio Correia and Pastora Martinex Samper, "Nanobiotechnology and the PHANTOMS Network," *Applied Nanoscience* 1, no. 1 (2004): 13–14.

106. Ibid.

107. Cientifica Report, *The Nanotechnology Opportunity Report*, 2003, http://nanotech now.com/nanotechnology-opportunity-report.htm (accessed April 26, 2004), p. 8.

108. Committee on Science and Technology House of Commons, *Science and Technology— Fifth Report*, March 22, 2004, http://www.publications.parliament.uk/pa/cm200304/cmselect /cmstech/56/5606.htm (accessed November 3, 2005), p. 58.

109. Ibid., p. 59.

110. Ibid.

111. "Surface Technology Systems' investment in UCL aims to place UK at forefront of nanotechnology research," *NanoInvestorNews*, February 25, 2005, http://www.nano investornews.com/modules.php?name=News&file=print&side=4084 (accessed February 28, 2005).

112. Phil Beradelli, "Interview: Marburger defends R&D policies," *United Press Inter-*

national, June 23, 2004, http://www.upi.com/view.cfm?StoryID=20040622-07343506586r (accessed July 9, 2004).

113. "Asian governments open their wallets for nanotech," *Nanotech Report* 2, no. 1 (January 2003): 3.

114. C. Rowan, "It's Ours to Lose—An Analysis of EU Nanotechnology Funding and the Sixth Framework Program," 2002, http://www.nanoeurope.org/docs/European%20 Nanotech%20Funding.pdf (accessed on April 26, 2004), p. 8.

115. Ibid.

116. Foreign Press Center—Japan, "Nanotechnology in Japan: Present Situation and Future Outlook," March 2001, http://www.fpcj.jp/e/gyouji/br/2001/010309.html (accessed June 13, 2003).

117. Ibid.

118. Ivan Meakin, Ministry for Education, Science, and Technology, "Nanotechnology in Japan—A Guide to Public Spending in FY2002," 2002.

119. Glenn Fishbine, *The Investor's Guide to Nanotechnology and Micromachines* (New York: George Wiley and Sons, 2002), p. 64.

120. D. Jones, "Nanotechnology Bill Gets Bipartisan Support from Capital Hill," *Inside Energy*, May 12, 2003, http://www.external.ameslab.gov/news/headlines/ ie030512.pdf (accessed April 26, 2004), p. 7.

121. Tanya Sienko, Asian Technology Information Program, "Overview of Nanotechnology Efforts in Japan," 1996.

122. Advertisement, "Nanotech Report—Investing today" 2004, 4.

123. D. Swinbanks, "MITI Aims at Small Research," *Nature* 349 (February 7, 1991a): 449.

124. Ibid.

125. Ibid., p. 90.

126. D. Jones, "Nanotechnology Bill Gets Bipartisan Support from Capital Hill," *Inside Energy*, May 12, 2003, http://www.external.ameslab.gov/news/headlines/ie030512 .pdf (accessed April 26, 2004), p. 7.

127. P. Hansson, "Nanotechnology: Prospects and Policies," *Futures*, October 1991, pp. 849–59.

128. Paul Kallender, "Japan boosts nano budget and industrial cooperation," *Small Times*, March/April 2004, p. 12.

129. Strategy of Nanotechnology Research and Development, *Nanotech Process Foundry web page*, January 26, 2004, http://www.sanken.osaka-u.ac.jp/labs/foundry/en/project.html (accessed August 5, 2004).

130. Paul Kallender, "Japan boosts nano budget and industrial cooperation," *Small Times*, March/April 2004, p. 13.

131. Paul Kallender, "Asia-Pacific governments invest in labs and research centers," *Small Times*, January/February 2004, p. 21.

132. Paul Kallender, "China and Taiwan are sinking some serious cash into nanotech," *Small Times*, November/December 2003, p. 10.

133. "Japanese firm, state of New Mexico sign deal to commercialize technology," *Small Times* n.d., http://www.smalltimes.com/print_doc.cfm?doc_id=8066 (accessed July 9, 2004).

134. Paul Kallender, "Japan's nano program encourages interdisciplinary coopera-

tion," *Small Times*, June 22, 2004, http://www.smalltimes.com/document_display.cfm ?document_id=8079 (accessed July 10, 2004).

135. "Asian nanotech fever running hot," *Nanotech Report* 2, no. 1 (January 2003): 2.

136. Ibid.

137. Ibid.

138. Ibid.

139. Ibid.

140. "Toray's nano-breakthrough," *Intel@tex*, October 24, 2004, http://www.inteletex .com/NewsDetail.asp?PubId=&NewsId=3406 (accessed November 1, 2004).

141. M. Mauboussin and K. Bartholdson, *Big Money in Thinking Small: Nanotechnology— What Investors Need to Know* (Boston, MA: Credit Suisse Equity Research/First Boston, May 2003), p. 17.

142. "China tops the world in nanopapers," *People's Daily Online*, June 10, 2005, http:// english.people.com.cn/200506/10/print20050610_189642.html (accessed June 14, 2005).

143. "Developing global nanotech," *Red Herring*, April 12, 2005, http://www.red herring.com/PrintArticle.aspx?a=11765&sector=Industries (accessed April 18, 2005).

144. Paul Kallender, "Japans builds on its nanopillars," *Small Times*, July/August 2003, p. 18.

145. Liz Kalaugher, "Smoking grass leads to nanotubes," *PhysicsWeb*, June 13, 2005, http://physicsweb.org/articles/news/9/6/8/1 (accessed June 13, 2005).

146. K. Kwuscha, "China Gaining Momentum in Nanotech R and D and Business," *Asia Pacific Nanotech Weekly*, November 3–5, 2002, http://www.nanoworld.jp/apnw/ articles/library/pdf/3.pdf (accessed April 26, 2004), and Paul Kallender, "China and Taiwan are sinking some serious cash into nanotech," *Small Times*, November/December 2003, p. 10.

147. Fu Jing, "Nanotech needs major capital injection," *China Daily*, May 20, 2004, http://www.chinadaily.com.cn/english/doc/12004-05/20/content_332367.htm (accessed May 25, 2004).

148. M. Mauboussin and K. Bartholdson, *Big Money in Thinking Small: Nanotechnology— What Investors Need to Know* (Boston, MA: Credit Suisse Equity Research/First Boston, May 2003), p. 17.

149. Paul Kallender, "Asia-Pacific governments invest in labs and research centers," *Small Times*, January/February 2004, p. 20.

150. Paul Kallender, "Japan boosts nano budget and industrial cooperation," *Small Times*, March/April 2004, p. 13.

151. Paul Kallender, "China and Taiwan are sinking some serious cash into nanotech," *Small Times*, November/December 2003, p. 10.

152. K. Kwuscha, "China Gaining Momentum in Nanotech R and D and Business," *Asia Pacific Nanotech Weekly*, November 3–5, 2002, http://www.nanoworld.jp/apnw/ articles/library/pdf/3.pdf (accessed April 26, 2004).

153. "Asian governments open their wallets for nanotech," *Nanotech Report* 2, no. 1 (January 2003), p. 2.

154. J. Lin-Liu, "China, Emboldened by Breakthroughs, Sets Out to Become Nanotech Power," *Small Times*, December 18, 2001, http://www.smalltimes.com/section _display.cfm?section_id=51&summary=1&startpos=821 (accessed April 26, 2004).

155. Paul Kallender, "China and Taiwan are sinking some serious cash into nanotech," *Small Times,* November/December 2003, p. 10.

156. Liang Yu, "Nanotech Shrouded in Doubt," *News.Nano.Apex,* July 30, 2001, http://news.nanoapex.com/modules.php?name=News&file=article&sid=560 (accessed July 27, 2004).

CHAPTER 5

1. Richard Siegel, Interagency Working Group of Nanoscience, Engineering, and Technology, "Executive Summary," *Nanostructure Science and Technology: A Worldwide Study* (Loyola College, MD: WTEC, September 1999), p. xviii.

2. Ibid., p. xix.

3. Ibid., p. xxi.

4. Ibid., p. xxiii.

5. Interagency Working Group of Nanoscience, Engineering, and Technology, *Nanotechnology: Shaping the World Atom by Atom* (Washington, DC: NSTC/CT, 1999), p. 8.

6. Ibid., p. 3.

7. Ibid., p. 97.

8. Ibid., p. 13.

9. Ibid., pp. 14–16.

10. Ibid., p. 18.

11. Ibid., p. 21.

12. Ibid., p. 24.

13. Ibid., p. 27.

14. Ibid., p. 9.

15. Ibid., p. 17.

16. Ibid., pp. 47–48.

17. Ibid., p. 54.

18. Ibid., pp. 1–3.

19. Ibid., p. 17.

20. Ibid., p. 48.

21. Ibid., p. 28.

22. National Science and Technology Council, *National Nanotechnology Initiative: Research and Development Supporting the Next Industrial Revolution* (Washington, DC: NSTC/CT, 2003).

23. Ibid., p. 2.

24. Mike Roco, "National Nanotechnology Investment in the FY 2005 Budget Request," *AAAS Report on US R&D in FY 2005* (preprint), March 2004, pp. 256–57.

25. Ibid., p. 257.

26. Ibid., p. 264.

27. Mihail Roco, "Nanoscale Science and Engineering: Unifying and Transforming Tools," *AIChE Journal* 50, no. 5 (May 2004): 890–97.

28. "Who We Are," 2003, http://www.esrc.ac.uk/esrccontent/aboutesrc/aboutus.asp (accessed September 15, 2003).

29. Ibid.

30. "Our Mission," 2003, http://www.esrc.ac.uk/esrccontent/esrcgen/display.gen/mission.asp (accessed September 15, 2003).

31. Ibid.

32. Steven Wood, Richard Jones, and Alison Geldart, *The Social and Economic Challenges of Nanotechnology* (Swindon, UK: Economic and Social Research Council, 2003), http://www.esrc.ac.uk/esrccontent/DownloadDocs/Nanotechnology.pdf (accessed January 3, 2004), p. 5.

33. Ibid.

34. Ibid., p. 39.

35. Ibid., p. 12.

36. Ibid., p. 40.

37. Ibid., p. 41.

38. Ibid., p. 42.

39. George M. Whitesides, "The once and future nanomachine," *Scientific American*, September 16, 2001.

40. Richard E. Smalley, "Of chemistry, love and nanobots," *Scientific American*, September 16, 2001, pp. 76–77.

41. Philip Ball, "Natural strategies for the molecular engineer," *Nanotechnology* 13 (2002): R15–R28.

42. Steven Wood, Richard Jones, and Alison Geldart, *The Social and Economic Challenges of Nanotechnology* (Swindon, UK: Economic and Social Research Council, 2003), http://www.esrc.ac.uk/esrccontent/DownloadDocs/Nanotechnology.pdf (accessed January 3, 2004), p. 25.

43. "Small science, big questions," *SNH.com.au*, November 27, 2003, http://www.smh.com.au/articles/2003/11/16/10768917669906.html (accessed November 25, 2003).

44. "U.K. recognizes importance of perception," *Howard Lovy's NanoBot*, September 30, 2003, http://nanobot.blogspot.com/2003_09_28_nanobot_archive.html (accessed October 15, 2003).

45. Ibid.

46. "Royal Society working group on best practice in communicating the results of new scientific research to the public," October 31, 2003, http://www.royalsoc.ac.uk/news/comm/htm (accessed November 24, 2003).

47. "Nanotechnology and nanoscience—Progress report," September 2, 2003, http://www.nanotec.org.uk/ReportSep03.htm.

48. Royal Society and Royal Academy of Engineering, *Nanotechnology: Views of scientists and engineers*, December 5, 2003, http://www.nanotec.org.uk/workshopOct03.htm (accessed June 25, 2004).

49. "Nanotechnology and nanoscience—Progress report," September 2, 2003, http://www.nanotec.org.uk/ReportSep03.htm.

50. J. Carroll, "Untitled," 2003, http://www.nanotec.org.uk/evidence/49John Carroll.htm (accessed March 30, 2004).

51. J. Wiggins, "Untitled," 2003, http://www.nanotec.org.uk/evidence/65aJason Wiggins.htm (accessed March 30, 2004).

52. Royal Society and Royal Academy of Engineering, *Nanotechnology: Views of scientists and engineers*, December 5, 2003, http://www.nanotec.org.uk/workshopOct03.htm (accessed June 25, 2004), pp. 6 and 11.

53. V. Alakson, "Nanotechnology call for views," 2003, http://www.nanotec.org.uk/evidence/62aForumForTheFuture.htm (accessed March 30, 2004).

54. See National Consumer Council, "Untitled," 2003, http://www.nanotec.org.uk/evidence/45NNCSubl.htm (accessed March 30, 2004).

55. P. Healey, "Nanotechnology: Social, ethical, legal and economic issues—A STAGE input," 2003, http://www.nanotec.org.uk/evidence/93bAltmannGobrudd.htm (accessed March 30, 2004).

56. M. Gannett, "Untitled," 2003, http://www.nanotec.org.uk/evidence/85aMartin Gannett.htm (accessed March 30, 2004).

57. T. Harper, "Oral evidence—Tim Haper, Cientifica," 2003, http://www.nanotec.org.uk/evidence/oralHarperTim.htm (accessed March 30, 2004).

58. R. Balmer, "Hopes and concerns of nanotechnology," 2003, http://www.nanotec.org.uk/evidence/90BalmerALDES.htm (accessed March 30, 2004).

59. J. Gimzewski, "Untitled," 2003, http://www.nanotec.org.uk/evidence/53aGSub.htm (accessed March 30, 2004).

60. R. Brazil, "Nanotechnology—The issues," 2003, http://www.nanotec.org.uk/evidence/78aRSC.htm (accessed March 30, 2004).

61. P. Dobson, "Nanotechnology," October 2003, http://www.nanotec.org.uk/evidence/58aPeterDobson.htm (accessed March 30, 2004).

62. Novartis Foundation, "Untitled," 2003, http://www.nanotec.org.uk/evidence/86DerekChadwick.htm (accessed March 30, 2004).

63. S. Wood and R. Jones, "Report of oral evidence session with Professor Stephen Wood (SW) & Professor Richard Jones (RJ)," 2003, http://www.nanotec.org.uk/evidence/oralJones&Wood.htm (accessed March 30, 2004).

64. Kristen Kulinowski, "Oral evidence—Dr. Kristen Kulinowski, Center for Biological and Environmental Nanotechnology, Rice University, Houston TX (USA)," 2003, http://www.nanotec.org.uk/evidence/oralKulinowskiDrKristen.htm (accessed March 30, 2004).

65. G. Fenton, "Untitled," 2003, http://www.nanotec.org.uk/evidence/54Gary Fenton.htm (accessed March 30, 2004).

66. Warwick University—Nanosystems Group, "Untitled," 2003, http://www.nanotec.org.uk/evidence/89WarwickUniversity.htm (accessed March 30, 2004).

67. B. Evans, "Untitled," October 2003, http://www.nanotec.org.uk/evidence/56BarryEvans.htm (accessed March 30, 2004).

68. D. Rickerby, "Societal and policy aspects of the introduction of nanotechnology in healthcare," 2003, http://www.nanotec.org.uk/evidence/46bSocietalAndPolicyAspects.htm (accessed March 30, 2004).

69. F. Albertario and J. Snape, "Untitled," 2003, http://www.nanotec.org.uk/evidence/79aFabioAlbertario.htm (accessed March 30, 2004).

70. R. Hanson, "Untitled," 2003, http://www.nanotec.org.uk/evidence/73Robin Hanson.htm (accessed March 30, 2004).

71. B. Wang, "Untitled," 2003, http://www.nanotec.org.uk/evidence/83Brian Wang.htm (accessed March 30, 2004).

72. J. Altmann, "Military uses of nanotechnology—Risks and proposals for precautionary action," 2003, http://www.nanotec.org.uk/evidence/93bAltmannGobrudd.htm (accessed March 30, 2004).

73. CRN, "Invited commentary on Royal Society Nanotechnology Workshop," December 2003, http://www.crnano.org/RSWorkshop1.htm (accessed December 8, 2003).

74. Ibid.

75. Ibid.

76. BMRB Social Research, *Nanotechnology: Views of the general public, Quantitative and qualitative research carried out as part of the nanotechnology study*, BMRB International Report 45101666, 2004, http://www.nanotec.org.uk/Market%20Research.pdf (accessed June 25, 2004), p. 3.

77. Ibid., p. 4.

78. Ibid., p. 19.

79. Ibid., p. 36.

80. Ibid., p. 44.

81. Committee on Science and Technology in the House of Commons, *Science and Technology—Fifth Report*, March 22, 2004, http://www.publications.parliament.uk/pa/cm200304/cmselect/cmstech/56/5606.htm (accessed November 3, 2005), p. 2.

82. Genevieve Oger, "Government panel displeased with British support for nano," *Small Times*, May/June 2004, p. 12.

83. Department of Trade and Industry and Office of Science and Technology, *New Dimensions for Manufacturing: A UK Strategy for Nanotechnology*, Report of the UK Advisory Group on Nanotechnology Applications submitted to Lord Sainsbury, http://www.oft.osd.mil/library/library_files/document_134_nanotechnologyreport.pdf (accessed June 21, 2004), p. 3.

84. Ibid., p. 7.

85. Ibid., p. 8.

86. Committee on Science and Technology in the House of Commons, *Science and Technology—Fifth Report*, March 22, 2004, http://www.publications.parliament.uk/pa/cm200304/cmselect/cmstech/56/5606.htm (accessed November 3, 2005), p. 3.

87. Ibid.

88. Ibid.

89. Ibid., pp. 13–14.

90. Ibid., p. 72.

91. Melanie Reynolds, "UK Government Rules Out Nano Fab Plans," *Electronics Weekly*, June 30, 2004, http://nanotechwire.com/news.asp?nid=900&ntid=&pg=2 (accessed July 9, 2004).

92. Committee on Science and Technology in the House of Commons, *Science and Technology—Fifth Report*, March 22, 2004, http://www.publications.parliament.uk/pa/cm200304/cmselect/cmstech/56/5606.htm (accessed November 3, 2005), p. 3.

93. Ibid., p. 9.

94. Ibid., p. 56.

95. Ibid., p. 72.

96. W. Lance Haworth and Ezio Andreta, *European Commission. 3rd Joint EC-NSF Workshop on Nanotechnology. 31 January–1 February 2002. Nanotechnology Revolutionary Opportunities and Societal Implications*, ed. Mihail Roco and Renzo Tomellini (Luxembourg: Office for Official Publications of the European Communities, 2002), p. 5.

97. Ibid., p. 13.

98. Ibid.

99. Ibid., p. 16.

100. Ibid.

101. Ibid., p. 19.

102. IUPAC, http://www.iupac.org/dhtml_home.html (accessed June 21, 2004).

103. ACS, http://www.chemistry.org/portal/a/c/s/1/home.html/ (accessed June 21, 2004).

104. CAS, http://www.cas.org (accessed June 21, 2004).

105. EC's Community Health and Consumer Protection Directorate, *Nanotechnologies: A Preliminary Risk Analysis on the Basis of a Workshop Organized in Brussels on 1–2 March 2004*, 2004, http://europa.eu.int/comm/health/ph_risk/events_risk_en.htm (accessed June 20, 2004), p. 11.

106. Ibid., p. 24.

107. Ibid., p. 36.

108. A. Arnall and D. Parr, "Moving the nanoscience and technology (NST) debate forwards: Short-term impacts, long-term uncertainty and the social constitution," EC's Community Health and Consumer Protection Directorate, *Nanotechnologies: A Preliminary Risk Analysis on the Basis of a Workshop Organized in Brussels on 1–2 March 2004*, 2004, http://europa.eu.int/comm/health/ph_risk/events_risk_en.htm (accessed June 20, 2004), p. 44.

109. Ibid., p. 45. The nano-optimism remarked was backcited to "Trouble in nanoland," *Economist*, December 5, 2002.

110. Ibid., pp. 44–45.

111. Vicki Colvin, "Engineered nanomaterials and risks: One perspective," EC's Community Health and Consumer Protection Directorate, *Nanotechnologies: A Preliminary Risk Analysis on the Basis of a Workshop Organized in Brussels on 1–2 March 2004*, 2004, http://europa.eu.int/comm/health/ph_risk/events_risk_en.htm (accessed June 20, 2004), pp. 48–49.

112. A. Hett, "Nanotechnology—From the insurers' perspective," EC's Community Health and Consumer Protection Directorate, *Nanotechnologies: A Preliminary Risk Analysis on the Basis of a Workshop Organized in Brussels on 1–2 March 2004*, http://europa.eu.int/comm/health/ph_risk/events_risk_en.htm (accessed June 20, 2004), pp. 103–104.

113. Ibid., p. 103.

114. O. Renn, "Public perception of nanotechnology," EC's Community Health and Consumer Protection Directorate, *Nanotechnologies: A Preliminary Risk Analysis on the Basis of a Workshop Organized in Brussels on 1–2 March 2004*, http://europa.eu.int/comm/health/ph_risk/events_risk_en.htm, citing Mihail Roco and William Bainbridge, eds., *Societal implication of nanoscience and nanotechnology* (Kluwer, Doredrecht, the Netherlands: 2004) (accessed June 20, 2004), p. 123.

115. Ibid.

116. Ibid.

117. Royal Academy of Engineers and Royal Society, *Nanoscience and nanotechnologies:*

Opportunities and uncertainties, 2004, http://www.nanotec.org.uk/finalReport.htm (accessed July 30, 2004), p. 46.

118. Ibid., p. 51.

119. Ibid., p. 56.

120. Ibid., p. 72.

121. Ibid., p. 77.

122. Ibid., p. 109.

123. A. Hett, *Nanotechnology: Small matter, many unknowns* (Zurich, Switzerland: Swiss Reinsurance, 2004), p. 6.

124. M. Mauboussin and K. Bartholdson, *Big money in thinking small: Nanotechnology— What investors need to know*, May 7, 2003, p. 3.

125. Ibid., p. 4.

126. Ibid., p. 7.

127. Ibid., p. 9.

128. Ibid., p. 14.

129. Ibid., p. 28.

130. Ibid.

131. "Free report tells you—5 nanotech stocks to buy now," 2004.

132. Ibid.

133. "Nanotech: Hype yes, bubble no," *Nanotech Report* 1, no. 10 (December 2002): 1.

134. Advertisement, "Nanotech Report—Investing today," (2004): 3.

135. Ibid., p. 4.

136. Ibid., p. 5.

137. Ibid., p. 7.

138. "Nanotech: Hype yes, bubble no," *Nanotech Report* 1, no. 10 (December 2002): 1.

139. Advertisement, "Nanotech Report—Investing today," (2004): 2.

140. Ibid., p. 3.

141. "Bush and the nano pretenders," *Nanotech Report* 2, no. 12 (December 2003): 3.

142. David Forman, "News analysis: 2004 finds 'nano' crossing into speculative space," *Small Times*, December 31, 2003, http://www.smalltimes.com/document_display.cfm?document_id=7155 (accessed January 6, 2004).

143. "Bush and the nano pretenders," *Nanotech Report* 2, no. 12 (December 2003): 3.

144. Ibid.

145. "Beware the nanobubble," *Nanotech Report* 2, no. 6 (June 2003): 3.

146. Ibid.

147. Ibid., p. 1.

148. "Nano name game," *Nanotech Report* 3, no. 2 (February 2004): 1.

149. Ibid., p. 2.

150. "Portfolio plays in nanotech," *Nanotech Report* 2, no. 4 (April 2003): 2.

151. "A new world is born," *Nanotech Report* 1, no. 1 (March 2002): 2.

152. "NWNR wins award for editorial excellence!" *Nanotech Report* 2, no. 6 (June 2003): 1.

153. "2003 nanotech product guide," *Nanotech Report* 2, no. 7 (July 2003): 1.

154. Anna Salleh, "Nanotech risk put on insurance agenda," *News in Science*, June 7, 2004, http://www.abc.net.au/science/news/stories/s1124943.htm (accessed July 26, 2004).

155. A. Hett, *Nanotechnology: Small matter, many unknowns* (Zurich, Switzerland: Swiss Reinsurance, 2004) pp. 22–24.

156. Ibid., p. 22.

157. Ibid., p. 20.

158. Ibid., p. 22.

159. Ibid., p. 8.

160. Ibid., p. 13.

161. Anna Salleh, "Nanotech risk put on insurance agenda," *News in Science*, June 7, 2004, http://www.abc.net.au/science/news/stories/s1124943.htm (accessed July 26, 2004).

162. A. Hett, *Nanotechnology: Small matter, many unknowns* (Zurich, Switzerland: Swiss Reinsurance, 2004) p. 30.

163. Ibid., p. 33.

164. Ibid.

165. Ibid., p. 41.

166. Ibid., p. 5.

167. Ibid., p. 37.

168. Ibid., p. 45.

169. Ibid., p. 46.

170. Ibid., p. 47.

171. Ibid.

CHAPTER 6

1. M. Roco, introduction to *Societal Implications of Nanoscience and Nanotechnology*, ed. M. C. Roco and W. S. Bainbridge (New York: Kluwer, 2001), pp. 3–4.

2. "Nanotechnology: Tiny Hope or Big Hype?" *CNN Technology*, March 15, 2004, p. 1.

3. Ibid.

4. M. Roco, introduction to *Societal Implications of Nanoscience and Nanotechnology*, ed. M. C. Roco and W. S. Bainbridge (New York: Kluwer, 2001), p. 5.

5. Stephen Baker and Ashton Aston, "The Business of Nanotech," *Business Week*, February 15, 2005.

6. Ibid.

7. Sean Murdock, "From the Director," *NanoBusiness News*, e-mail communication, October 29, 2004.

8. Matthew Nordan, "Nanomyths and Nanotruths," *NanoBusiness News*, October 29, 2004, e-mail communication, October 29, 2004.

9. "Nanotechnology news in brief," *NanoTechWeb*, September 23, 2004, http://nano techweb.org/articles/news/3/9/15 (accessed October 4, 2004).

10. "Stanford software brings precision and practicality to nanotechnology," *PhysOrg.com*, May 24, 2005, http://www.physorg.com/printnews.php?newsid=4243 (accessed May 25, 2005); and Carl Wherrett and John Yelovich, "The Players and Pretenders of Nanotech," *Motley Fool*, August 30, 2004, http://www.fool.com/news/commentary/2004/commentary04083001.htm (accessed on September 7, 2004).

11. Charles Choi, "Nano World: Nanocatalysts for oil, drugs," March 25, 2005,

http://www.wpherald.com/storyview.php?StoryID=20050325-123219-4868r (accessed July 12, 2005).

12. Carl Wherrett and John Yelovich, "The Tiny Next Big Thing," *Motley Fool*, November 24, 2004, http://www.fool.com/news/commentary/2004/commentary 04112405.htm?source=EDNWFT (accessed June 17, 2005).

13. "Nanotechnology news in brief," *NanoTechWeb*, December 3, 2004, http://nano techweb.org/articles/news/3/12/4 (accessed December 13, 2004).

14. David Rotman, "Magnetic-Resonance Force Microscopy," "10 Emerging Tech-nologies," *Technology Review*, May 2005, http://ww.technologyreview.com/articles/05/05/issue/feature_emerging.asp?p=0 (accessed April 19, 2005).

15. "Innovative fountain pen writes on the nanoscale," *EurekaAlert!* April 25, 2005, http//www.eurekaalert.org/pub_releases/2005-04/nu-ifp042695.php (accessed May 2, 2005).

16. Charles Choi, "Ten overlooked nano firms," *Washington Times*, May 9, 2005, http://washingtontimes.com/upi-breaking/20050506-011337-9236r.htm (accessed July 3, 2005).

17. Howard Wolinsky, "High-tech tweezers enable nano-assembly lines," *Chicago Sun Times*, October 3, 2004, http://www.suntimes.com/special_sections/innovate/cst-fin-cia 05arryx.html (accessed July 8, 2005).

18. "Good Vibrations in the Nanoworld," *NewsNanoApex*, September 28, 2004, http://news.nanoapex.com/modules.php?name=News&file=article&sid=5059 (ac-cessed July 8, 2005).

19. Charles Choi, "Ten overlooked nano firms," *Washington Times*, May 9, 2005, http://washingtontimes.com/upi-breaking/20050506-011337-9236r.htm (accessed July 3, 2005).

20. "Nanotechnology new in brief," October 1, 2004, http://nanotechweb.org/articles/news/3/10/2 (accessed October 4, 2004).

21. "Nano-stamping makes its mark," *NanoTechWeb*, June 6, 2005, http://nanotechweb.org/articles/news/4/6/7?alert=1 (accessed June 13, 2005).

22. Belle Dume, "Superlens could image nanoscale with light," *NanoTechWeb*, May 12, 2005, http://nanotechweb.org/articles/news/4/5/6?alert=1 (accessed May 23, 2005).

23. James Tyrell, "EUV microscope explores nanoscale," June 27, 2005, http://nanotechweb.org/articles/news/4/6/13?alert=1 (accessed July 5, 2005).

24. "Nanorods could lead to superlenses," *NanoTechWeb*, April 21, 2005, http://nanotechweb.org/articles/news/4/4/14?alert=1 (accessed April 25, 2005).

25. "Oil Worth Its Weight in Gold in Directed Nanomachining," *NewsNanoApex*, May 25, 2005, http://news.nanoapex.com/modules.php?name=News&file=article&sid=5649 (accessed July 8, 2005).

26. "Nanotechnology news in brief," *NanoTechWeb.org*, March 11, 2005, e-mail com-munication (accessed March 17, 2005).

27. Carl Wherrett and John Yelovich, "The Players and Pretenders of Nanotech," *Motley Fool* (August 30, 2004), http://www.fool.com/news/commentary/2004/commentary 04083001.htm (accessed on September 7, 2004).

28. "A Few Steps Closer to Nanoscale Photonic Technology," *PhysOrg.com*, May 20, 2005, http://www.physorg.com/news4198.html (accessed July 7, 2005).

29. M. Roco, introduction to *Societal Implications of Nanoscience and Nanotechnology*, ed. M. C. Roco and W. S. Bainbridge (New York: Kluwer, 2001), p. 6.

30. "Stanford software brings precision and practicality to nanotechnology,"

PhysOrg.com, May 24, 2005, http://www.physorg.com/printnews.php?newsid=4243 (accessed May 25, 2005).

31. "Nanotechnology news in brief," *NanoTechWeb*, December 10, 2004, http://www.nanotechweb.org/articles/news/3/12/7 (accessed December 13, 2004).

32. Claudia Hume, "The Outer Limits of Miniaturization," *Chemical Specialties*, September 2000, p. 48.

33. Charles Choi, "Nano World: Nanocatalysts for oil, drugs," *World Peace Herald*, March 25, 2005, http://www.wpherald.com/storyview.php?StoryID=20050325-123219 -4868r (accessed July 8, 2005).

34. Robert Service, "Color-Changing Nanoparticles Offer a Golden Rule for Molecules," *Science*, May 20, 2005, e-communication.

35. Peter Weiss, "Falling into Place: Atom mist yields nanobricks and mortar," *Science News Online* 166, no. 11 (September 11, 2004), http://www.sciencenews.org/articles/20040911/ fob5.asp (accessed July 8, 2005).

36. John Gartner, "Military reloads with nanotech," *Technology Review*, January 21, 2005, http://www.technologyreview.com/articles/05/01/wo/wo_gartner012105.asp (accessed July 8, 2005).

37. "Industry Veteran to Validate QuantumSphere Inc.'s Nanopowders," *NanoInvestorNews*, April 26, 2005, http://www.nanoinvestornews.com/modules.php?name =News&file=article&sid=4238 (accessed July 8, 2005).

38. Carl Wherrett and Josh Yelovich, "Nanophase finally ready?" *Motley Fool*, November 22, 2004, http://www.fool.com/Server/FoolPrint.asp?File=/news/mft/2004/ mft04111212.htm (accessed November 15, 2004).

39. Kevin Maney, "Nanotechnology's everywhere," *USA Today*, June 1, 2005, http://www.usatoday.com/tech/columnist/kevinmaney/2005-05-31-nanotech_x.htm (accessed July 14, 2005).

40. "Cyclics Sees Sizeable Growth in Nano Technology; Cyclics' Plastic Supports Nano-composite Materials," *NanoInvestorNews*, March 16, 2005, http://www.nanoinvestor news.com/modules.php?name=News&file=article&sid=4144 (accessed July 8, 2005).

41. "NaturalNano to Be Honored as First Nano Company Exploiting Natural Formed Nanotubes at Nanobusiness 2005," *NanoInvestorNews*, May 23, 2005, http://www.nano investornews.com/modules.php?name=News&file=article&sid=4312 (accessed July 8, 2005).

42. Kevin Maney, "Nanotechnology's everywhere," *USA Today*, June 1, 2005, http://www.usatoday.com/tech/columnist/kevinmaney/2005-05-31-nanotech_x.htm (accessed July 14, 2005).

43. "Industrial Nanotech Launches Nansulate Translucent Product Line," *NanoInvestorNews*, May 23, 2005, http://www.nanoinvestornews.com/modules.php?name =News&file=print&sid=4311 (accessed May 24, 2005).

44. "Paint to help clean and purify bad air," *Shanghai Daily*, November 12, 2004, http://english.eastday.com/eastday/englishedition/metro/userobject1ai710823.html (accessed December 14, 2004).

45. Alan Osborn, "Nanomaterials help stop bullets," *Plastic and Rubber Weekly*, November 23, 2004, http://www.prw.com/main/newsdetails.asp?id=3444 (accessed May 31, 2005).

46. Charles Choi, "Ten overlooked nano firms," *Washington Times*, May 9, 2005, http://washingtontimes.com/upi-breaking/20050506-011337-9236r.htm (accessed July 3, 2005).

47. Ibn Campusino, "Nuts, knots and vertex spirals—Three aspects of fullerenes," *Sunday Times*, February 27, 2005, http://www.timesofmalta.com/core/print_article.php ?id=179331 (accessed February 28, 2005).

48. Jeff Sturgeon, "Big steps of nanotech," *Roanoke Times*, http://www.roanoke.com/business/18857.html (accessed July 8, 2005).

49. "Masters of the Flame: Industrial Production of Fullerenes Becomes a Reality," *Nano-C*, October 2003, http://nano-c.com/pdf/MastersoftheFlame.pdf (accessed July 11, 2005).

50. Candace Stuart, "Worker Safety Rises to the Fore," *Small Times*, January/February 2005, p. 19.

51. Liz Kalaugher, "Carbon nanotubes fill up with magnetic nanoparticles," *Nano-TechWeb*, April 1, 2005, http://nanotechweb.org/articles/news/4/4/1?alert=1 (accessed April 3, 2005).

52. "Laboratory grows world record length carbon nanotube," *NewsNanoApex*, September 15, 2004, http://news.nanoapex.com/modules.php?name=News&file=article &sid=5007 (accessed July 7, 2005).

53. "Nanotubes and Energy—Hype or Hope?" *NanoTechWire*, April 18, 2005, http://nanotechwire.com/news.asp?nid=1839&ntid=133&pg=1 (accessed July 11, 2005).

54. "DNA Makes Nanotube Transistors," *Technology Review*, December 3, 2004, http://www.technologyreview.com/articles/04/12/rnb_120304.asp (accessed December 13, 2004).

55. "For Cheap Nanotubes, Just Add Water," *Betterhumans*, November 18, 2004, http://betterhumans.com/News/news.aspx?articleID=2004-11-18-3 (accessed November 29, 2004).

56. Phil Schewe, James Riordan, and Ben Stein, "Carbon nanowires," *Physics News Update* 635, no. 3 (May 1, 2003), http://www.aip.org/enews/physnews/2003/split/535-3.html (accessed September 22, 2004).

57. "Futuristic smart yarns on the horizon," *NewsNanoApex*, November 19, 2004, http://news.nanoapex.com/modules.php?name=News&file=article&sid=5270 (accessed July 7, 2005).

58. "Researchers create nanotubes that change colors, form nanocarpet, and kill bacteria," *NewsNanoApex*, September 25, 2004, http://news.nanoapex.com/modules.php ?name=News&file=print&sid=5050 (accessed September 27, 2004).

59. "Multipurpose Nanocables Invented," *NewsNanoApex*, November 18, 2004, http://news.nanoapex.com/modules.php?name=News&file=article&sid=5261 (accessed July 7, 2005).

60. "Nanotech advances makes carbon nanotubes more useful," *Innovations Report*, April 12, 2005, http://www.innovations-report.com/html/reports/physics_astronomy/report-42923.html (accessed July 19, 2005).

61. Roger Bishop, "Don't Fall in the Knowledge Gap," *Small Times*, May 9, 2005, http://www.smalltimes.com/document_display.cfm?document_id=9166 (accessed May 9, 2005).

62. David Vink, "Dow and GE use nanocomposites for car parts," *Plastic & Rubber Weekly*, March 16, 2005, http://www.prw.com/main/newsdeatils.asp?id=3823 (accessed May 10, 2005).

63. Tim Moran, "Mirror, Mirror, on the Fenders," *New York Times*, June 13, 2005, http://www.nytimes.com/2005/06/13/automobiles/13TECH.html (accessed June 16, 2005).

64. "Johns Manville goes nano," *Inteletex*, June 15, 2005, http://www.inteletex.com/NewsDetail.asp?PubId=&NewsId=3969 (accessed July 1, 2005).

65. Matt Marshall, "Turning to tech for cleaner water," *Mercury News*, May 30, 2005, http://www.mercurynews.com/mld/mercurynews/business/11773272.htm (accessed July 11, 2005).

66. Charles Choi, "Ten overlooked nano firms," *Washington Times*, May 9, 2005, http://washingtontimes.com/upi-breaking/20050506-011337-9236r.htm (accessed July 3, 2005).

67. Charles Choi, "Nano World: Water, water everywhere nano," *World Peace Herald*, March 18, 2005, http://www.wpherald.com/print.php?StoryID=20050318-112217-1110r (accessed July 12, 2005).

68. Ibid.

69. Ronald Bailey, "Nanotechnology: Hell or Heaven?" *Reasononline*, October 27, 2004, http://www.reason.com/rb/rb102794.shtml (accessed November 1, 2004).

70. "U.S.-Russian Nano-Filter Enters Space Technology Hall of Fame," *NanoInvestorNews*, April 13, 2005, http://www.nanoinvestornews.com/modules.php?name=News&file=print&sid=4209 (accessed April 18, 2005).

71. "Small wonders," *Economist*, December 29, 2004, http://www.economist.com/printededition/PrinterFriendly.cfm?Story_ID=3494722 (accessed January 12, 2005).

72. M. B. Owens, "Nanotech feeding into industry," *City Paper Online*, September 3, 2004, http://www.nashvillecitypaper.com/index.cfm?section_id=10&screen=news&news_id=35522 (accessed July 19, 2005).

73. Roger Bishop, "Don't Fall in the Knowledge Gap," *Small Times*, May 9, 2005, http://www.smalltimes.com/document_display.cfm?document_id=9166 (accessed May 9, 2005).

74. Charles Choi, "Ten overlooked nano firms," *Washington Times*, May 9, 2005, http://washingtontimes.com/upi-breaking/20050506-011337-9236r.htm (accessed July 3, 2005).

75. "ObjectSoft Corp., to Be Renamed Nanergy Corp., First Nanotechnology Produce Released for Manufacture, Details Capital Structure," *NanoInvestorNews*, March 17, 2005, http://www.nanoinvestornews.com/modules.php?name=News&file=article&sid=4148 (accessed July 8, 2005).

76. "Evident Technologies Awards NSF Grant to Develop Advanced Quantum-Dot Based Anti-Counterfeiting Materials," *PR Newswire*, January 12, 2005, http://biz.yahoo.com/prnews/050112/nyw093_1.html (accessed February 7, 2005).

77. "*Novapure* appointed as Canadian master distributor for *Green Millennium* photocatalytic coatings: New nanotechnology created to address airborne and surface contaminants," *NanoInvestorNews*, April 26, 2005, http://www.nanoinvestornews.com/modueles.php?name=News&file=print&sid=4239 (accessed May 2, 2005).

78. "Three Nanotech Leaders, Ecology Coatings, NanoDynamics and MetaMateria, Partner to Provide New High-Performance Liquid Technology," *NanoInvestorNews*, March 16, 2005, http://www.nanoinvestornews.com/modules.php?name=News&file=article&sid=4142 (accessed July 8, 2005).

79. Michael Kanellos, "Nanotech company aims to put paint in the past," *CNET News*, April 11, 2005, http://news.com.com/Nanotech+company+aims+to+put+paint+in+the+past/2100-7337_3-5660745.html (accessed July 8, 2005).

80. "Nanotechnology news in brief," *NanoTechWeb*, February 11, 2005, http://nanotechweb.org/articles/news/4/2/8?alert=1 (accessed February 14, 2005).

81. "World's First Commercial Nanotechnology-Based Solid Lubricant Declared Non-Toxic," *NanoTechWire*, April 7, 2005, http://nanotechwire.com/news.asp?nid=1808 &ntid=126&pg=1 (accessed July 11, 2005).

82. Victoria Griffith, "DNA wires herald biological machines," *Financial Times*, March 4, 2005, e-mail communication.

83. "Nanotechnology's Impact on Products: Cancer Treatment Gets Reinvented, Automobiles Get Incrementally Improved," *NanoInvestorNews*, March 30, 2005, http://www.nanoinvestornews.com/modules.php?name=News&file=article&sid=4173 (accessed July 5, 2005).

84. M. Roco, introduction to *Societal Implications of Nanoscience and Nanotechnology*, ed. M. C. Roco and W. S. Bainbridge (New York: Kluwer, 2001), p. 10.

85. ETC Group, *Down on the Farm: The Impact of Nano-Scale Technologies on Food and Agriculture*, November 2004, http://www.etcgroup.org/article.asp?newsid=485 (accessed July 13, 2005).

86. "Breakthrough suggests nanotech applications for food safety," *Food and Production Daily*, March 22, 2005, http://www.foodproductiondaily.com/news/news-ng.asp?n=58894 -breakthrough-suggests-nanotech (accessed July 11, 2005).

87. Cooperative State Research, Education and Extension Service, U.S. Department of Agriculture, *Nanoscale Science and Engineering for Agriculture and Food Systems*, September 2003, http://www.nseafs.cornell.edu/web.roadmap.pdf (accessed July 13, 2005).

88. Rajai Atalla, James Beecher, Robert Caron, et al., U.S. Forest Service, *Nanotechnology for the Forest Products Industry: Vision and Technology Roadmap*, April 6, 2005, http://www.fpl.fs .fed.us/highlighted-research/nanotechnology/forest-products-nano-technology.pdf (accessed July 12, 2005).

89. "Nanotechnology Can Play Vital Role in Forest Products Industries: Technology Roadmap Released," *NanoInvestorNews*, April 5, 2005, http://www.nanoinvestornews.com/ modules.php?name=name=News&file=rpint&sid=4189 (accessed April 14, 2005).

90. Rajai Atalla, James Beecher, Robert Caron, et al., U.S. Forest Service, *Nanotechnology for the Forest Products Industry: Vision and Technology Roadmap*, April 6, 2005, p. 18, http://www .fpl.fs.fed.us/highlighted-research/nanotechnology/forest-products-nano-technology.pdf (accessed July 12, 2005).

91. Ibid., p. 14.

92. Mark Hachman, "Better eating through nanotech," *ExtremeTech*, June 29, 2005, http://www.extremetech.com/article2/0,1697,1832810,00.asp (accessed July 12, 2005).

93. "Nanoscale technology: The future of food safety," *Food & Drink*, May 17, 2005, http://www.foodanddrinkeurope.com/news/printNewsBis.asp?id=60041 (accessed July 12, 2005).

94. Ibid.

95. ETC Group—The Action Group on Erosion, Technology, and Concentration, *From Genomes to Atoms: The Big Down, Atomtech and Technologies Converging at the Nano-Scale* (Winnipeg, Manitoba: ETC Group, 2003), p. 50.

96. "Nanotechnology sales increase to 687.5m in 2004," *Food Production Daily*, May 27, 2005, http://www.foodproductiondaily.com/news/printNewsBis.asp?id=60283 (accessed May 27, 2005).

97. "Nanotechnology begins to make presence felt," *Food & Production Daily*, January 10, 2005, http://www.fooodproductiondaily.com/news/printNewsBis.asp?id=57174 (accessed January 12, 2005).

98. Mark Hachman, "Better eating through nanotech," *ExtremeTech*, June 29, 2005, http://www.extremetech.com/article2/0,1697,1832810,00.asp (accessed July 12, 2005).

99. "Nanotechnology sales increase to 687.5m in 2004," *Food Production Daily*, May 27, 2005, http://www.foodproductiondaily.com/news/printNewsBis.asp?id=60283 (accessed May 27, 2005).

100. Ibid.

101. "Industry meets academia to discuss nanofood," *Food and Production Daily*, June 17, 2005, http://www.foodproductiondaily.com/news/printNewsBis.asp?id=60733 (accessed June 17, 2005).

102. "NanoFood," *ScienCentral News*, June 2, 2005, http://www.sciencentral.com/articles/view.php3?article_id=218392560 (accessed July 3, 2005).

103. "Nanoscale technology: The future of food safety," *Food & Drink*, May 17, 2005, http://www.foodanddrinkeurope.com/news/printNewsBis.asp?id=60041 (accessed July 12, 2005).

104. Matt Marshall, "Sciences of the small produces huge savings," *San Jose Mercury News*, March 29, 2005, e-mail communication.

105. Ibid.

106. Moore's Law refers to the chip industry's ability to double transistor density every 18 months.

107. "Fujitsu touts carbon nanotubes for chips," *IDG News Service*, March 5, 2005, e-mail communication.

108. "Worldwide Nanotech IC Market Projected to Reach $172 Billion in 10 Years," *NanoInvestorNews*, December 23, 2004, http://nanoinvestornews.com/modules.php?name=News&file=article&sid=3757 (accessed July 7, 2005).

109. Charles Choi, "Nano world: Chipping away at chip size," *World Peace Herald*, November 19, 2004, http://www.wpherald.com/print.php?StoryID=20041119-125452-9930r (accessed July 6, 2005).

110. Liz Kalaugher, "Painting nanowires into circuits," *NanoTechWeb*, April 29, 2005, http://nanotechweb.org/articles/news/4/4/18?alert=1 (accessed May 2, 2005).

111. "Nanotechnology news in brief," *NanoTechWeb*, October 24, 2004, http://nanotechweb.org/articles/news/3/10/21 (accessed November 1, 2004).

112. John Spooner, "Intel May Combine Silicon with Carbon Nanotubes," *Extreme-Nano*, May 29, 2005, http://www.extremenano.com/article/Intel+May+Combine+Silicon+with+Carbon+Nanotubes+/153034_1.aspx (accessed July 6, 2005).

113. Carl Wherrett and John Yelovich, "The Players and Pretenders of Nanotech," *Motley Fool* (August 30, 2004, http://www.fool.com/news/commentary/2004/commentary04083001.htm (accessed on September 7, 2004).

114. John Spooner, "Intel Looks at Nanotechnology," June 6, 2005, http://www.extremenano.com/article/Intel+Looks+to+Nanotechnology/153279_1.aspx (accessed July 3, 2005).

115. Carl Wherrett and John Yelovich, "The Tiny Next Big Thing," *Motley Fool*,

November 24, 2004, http://www.fool.com/news/commentary/2004/commentary 04112405.htm?source=EDNWFT (accessed June 17, 2005).

116. Lawrence Gassman, "It's about the numbers, stupid!" *NanoTechWeb*, June 14, 2005, http://nanotechweb.org/articles/column/4/6/1/1 (accessed June 24, 2005).

117. "Three Technologies Tapped for Scientific American 50," *NewsNanoApex*, November 25, 2004, http://news.nanoapex.com/modules.php?name=News&file=article &sid=5292 (accessed July 6, 2005).

118. "Ultrathin Carbon Speeds Circuits," *Technology Review*, November 9, 2004, http://www.technologyreview.com/articles/04/11/rnb_110904.asp (accessed November 15, 2004).

119. Mike Martin, "Nano Fabric May Make Computers Thinner," *CIO Today*, November 29, 2004, http://www.cio-today.com/story.xhtml?story_id=28525 (accessed November 29, 2004).

120. Tom Krazit, "HP technology lets future chips live with mistakes," *IT World*, June 13, 2005, http://www.itworld.com/Comp/1982/050608hpresearch/ (accessed July 3, 2005).

121. United Press International, "Study: A molecule is, in fact, a transistor," June 7, 2005, http://www.sciencedaily.com/upi/index.php?feed=Science&article=UPI-1-20050607-17385000-bc-us-transistor.xml (accessed July 3, 2005).

122. "New concept for single molecule transistor," *PhysOrg.com*, July 9, 2005, http://www.physorg.com/news4345.html (accessed July 11, 2005).

123. ETC Group—The Action Group on Erosion, Technology, and Concentration, *From Genomes to Atoms: The Big Down, Atomtech and Technologies Converging at the Nano-Scale* (Winnipeg, Manitoba: ETC Group, 2003), p. 50.

124. Carl Wherrett and John Yelovich, "The Players and Pretenders of Nanotech," *Motley Fool*, August 30, 2004, http://www.fool.com/news/commentary/2004/commentary 04083001.htm (accessed on September 7, 2004).

125. K. Chang, "IBM Creates a Tiny Circuit out of Carbon," *New York Times*, August 27, 2001, pp. 12A/14A.

126. Lawrence Gasman, "Opportunities in the Emerging Nanostorage Market," *NanoMarkets White Paper*, July 2004, p. 4.

127. Lawrence Gassman, "It's about the numbers, stupid!" *NanoTechWeb*, June 14, 2005, http://nanotechweb.org/articles/column/4/6/1/1 (accessed June 24, 2005).

128. Lawrence Gasman, "Commercialization Opportunities for Nanoelectronics," *NanoBusiness News*, June 3, 2005, e-mail communication.

129. Charles Choi, "Nanotechnology-Based Data Storage on Rise," *E-Commerce Times*, September 12, 2004, http://www.ecommercetimes.com/story/Nanotechnology-Based -Data-Storage-on-Rise-36505.html (accessed July 6, 2005).

130. Gregory T. Huang, "10 Emerging Technologies: Universal Memory," *Technology Review*, May 2005, http://www.technologyreview.com/articles/05/05/issue/feature_ emerging.asp?p=0 (accessed April 19, 2005).

131. Charles Choi, "Nanotechnology-Based Data Storage on Rise," *E-Commerce Times*, September 12, 2004, http://www.ecommercetimes.com/story/Nanotechnology-Based -Data-Storage-on-Rise-36505.html (accessed July 6, 2005).

132. Will Knight, "Nano-levers point to futuristic gadgets," *New Scientist*, June 24, 2005, http://www.newscientist.com/article.ns.?id=dn7577&print=true (accessed July 1, 2005).

133. "Nano-Grating DVDs could store 100 times more," *PhysOrg.com*, May 24, 2005, http://www.physorg.com/printnews.php?newsid+4249 (accessed June 16, 2005).

134. "Nano Down Under: The Sci/Tech Saga," *Best of the NanoWeek*, January 25, 2005, e-mail communication.

135. Charles Choi, "Nanotechnology-Based Data Storage on Rise," *E-Commerce Times*, September 12, 2004, http://www.ecommercetimes.com/story/Nanotechnology-Based -Data-Storage-on-Rise-36505.html (accessed July 6, 2005).

136. ETC Group—The Action Group on Erosion, Technology, and Concentration, *From Genomes to Atoms: The Big Down, Atomtech and Technologies Converging at the Nano-Scale* (Winnipeg, Manitoba: ETC Group, 2003), p. 60.

137. Carl Wherrett and John Yelovich, "What's the intelligence on *Nanosys?*" *Motley Fool*, December 21, 2004, http://www.fool.com/Server/FoolPrint.asp?File=/news/ mft/2004/mft04122105.htm (accessed January 12, 2005).

138. Carl Wherrett and John Yelovich, "A giant leap toward IPO" *Motley Fool*, January 26, 2005, http://www.fool.com/Server/FoolPrint.asp?File=/news/mft/2004/mft05012609 .htm (accessed January 31, 2005).

139. Charles Choi, "Nanotech ready for big changes soon," *TechNewsWorld*, September 10, 2004, http://www.technewsworld.com/sotry/Nanotech-ready-for-Big-changes -Soon-36523.html (accessed September 22, 2004).

140. Carl Wherrett and John Yelovich, "The Players and Pretenders of Nanotech," *Motley Fool*, August 30, 2004, http://www.fool.com/news/commentary/2004/commen-tary04083001.htm (accessed on September 7, 2004).

141. Carl Wherrett and John Yelovich, "The Tiny Next Big Thing," *Motley Fool*, November 24, 2004, http://www.fool.com/news/commentary/2004/commentary 04112405.htm?source=EDNWFT (accessed June 17, 2005).

142. Jennifer Johnston, "Tiny molecule could shrink computers, phones and iPods," *Sunday Herald*, http://www.sundayherald.com/print49868 (accessed May 24, 2005).

143. Monya Baker, "From the Lab: Nanotechnology," *Technology Review*, June 2005, http://www.technologyreview.com/articles/05/06/issues/ftl_nano.asp?p=0 (accessed May 24, 2005).

144. "Case Researchers Grow Carbon Nanotubes in Lab Using Faster, Cheaper Means," *NewsNanoApex*, April 20, 2005, http://news.nanoapex.com/modules.php?name =News&file=article&sid=5597 (accessed July 7, 2005).

145. Liz Kalaugher, "Growing bent carbon nanotubes," *NanoTechWeb*, April 14, 2005, http://nanotechweb.org/articles/news/4/4/10?alert=1 (accessed April 18, 2005).

146. Jack Uldrich, "Something Small, Something Blue," *Motley Fool*, April 21, 2005, http://www.fool.com/news/commentary/2005/commentary05042112.htm (accessed July 7, 2005).

147. "Optics demo does quantum logic," *TRN*, April 6/13, 2005, http://www.trnmag .com/Stories/2005/040605/Optics_demo_does_quantum_logic_Brief_040605.html (accessed July 7, 2005).

148. "Nanotechnology news in brief," *NanoTechWeb*, February 11, 2005, http://-nanotechweb.org/articles/news/4/2/8?alert=1 (accessed February 14, 2005).

149. "Nanomechanical memory cell could catapult efforts to improve data storage,"

NewsNanoApex, October 1, 2004, http://news.nanoapex.com/modules.php?name=News &file=article&sid=5068 (accessed July 6, 2005).

150. Barnaby Feder, "At IBM, a Tinier Transistor Outperforms Its Silicon Cousins," *New York Times,* May 20, 2002, pp. 1C/9C.

151. "Carbon Nanotube Electronics Will Lead to $3.6 Billion in Business Opportunities," *NanoInvestorNews,* May 4, 2005, http://www.nanoinvestornews.com/modules.php ?name=News&file=article&sid=4257 (accessed July 6, 2005).

152. Doug Tsuruoka, "Nanotechnology Niche Called Quantum Dots Carries Big Possibilities," *Investor's Business Daily,* October 11, 2004, http://www.investors.com/ editorial/tech01.asp?view=1 (accessed October 11, 2004).

153. "Nanotubes Cut Copper Creep," *ECN Asia,* May 21, 2005, http://www .ecnasiamag.com/article.asp?id=1968 (accessed May 24, 2005).

154. Stephen Baker and Ashton Aston, "The Business of Nanotech," *Business Week,* February 15, 2005, http://www.businessweek.com/magazine/content/05_07/b3920001 _mz001.htm (accessed July 6, 2005).

155. "Nanotubes enter flat-panel display market," *NanoTechWeb,* May 23, 2005, http://nanotechweb.org/articles/news/4/5/11?alert=1 (accessed July 6, 2005), and Stephen Shankland, "Motorola builds a nanotube-based display," *CNET.com,* http:// earthlink.com/Motorola+builds+nanotube-based+display/2100-7337_3-5698503.html (accessed July 6, 2005).

156. "Three Technologies Tapped for Scientific American 50," *NewsNanoApex,* November 25, 2004, http://news.nanoapex.com/modules.php?name=News&file=article &sid=5292 (accessed July 6, 2005).

157. "Nanotech firm makes paper and ink quality displays," *SiliconPublic.com,* February 15, 2005, http://www.siliconrepublic.com/news/news.nv?story=single4444 (accessed February 24, 2005).

158. "New nanomaterial technology used in flat panel displays," *Silicon Valley North News,* February 25, 2005, http://www.siliconvalleynorth.com/home/newsFpPDjilvDU 20050224_pf.html (accessed February 28, 2005).

159. Ibid.

160. "Nanotubes respond to gas attacks," *PhysicsWeb,* January 6, 2005, http://physics-web.org/articles/news/9/1/3/1 (accessed January 12, 2005).

161. Belle Dume, "Nanotube sensor detects nerve agents," *NanoTechWeb,* November 14, 2003, http://nanotechweb.org/articles/news/2/11/9/1 (accessed April 3, 2005).

162. Tim Stephens, "Nanotechnology grant will support work on new sensor technology," *Currents Online,* May 9, 2005, http://currents.ucsc.edu/04-05/05-09/schmidt.asp (accessed July 7, 2005).

163. "Nanotechnology news in brief," *NanoTechWeb.org,* March 11, 2005, e-mail communication (accessed March 17, 2005). Recently, it won a $500,000 grant from the NSF to create a collaborative development kit of its nanoelectronic detection devices. See "Nanotechnology news in brief," *NanoTechWeb,* April 15, 2005, http://nanotechweb.org/articles/ news/4/4/11?alert=1 (accessed April 18, 2005).

164. Charles Choi, "Ten overlooked nano firms," *Washington Times,* May 9, 2005, http:// washingtontimes.com/upi-breaking/20050506-011337-9236r.htm (accessed July 3, 2005).

165. Carl Wherrett and John Yelovich, "The Players and Pretenders of Nanotech,"

Motley Fool, August 30, 2004, http://www.fool.com/news/commentary/2004/commentary 04083001.htm (accessed on September 7, 2004).

166. "CDT plans $40.25 million IPO as it seeks bigger share of FPD market," *Data-Monitor,* August 14, 2004, http://www.datamonitor.com/~56a0c9fcc784419a93c16c4ff 8487f82~/industries/news/artic (accessed April 14, 2005).

167. "Tiny superconductors withstand stronger magnetic fields," *I-Newswire,* February 6, 2005, http://i-newswire.com/goprint5430.html (accessed February 8, 2005).

168. "Nanotechnology leads to discovery of super superconductors," *NewsNanoApex,* September 12, 2004, http://news.nanoapex.com/modules.php?name=News&file=article &sid=5000 (accessed July 7, 2005).

169. Bill Robinson and Natasha Starkell, "FutureTech," *Gateway to Russia,* December 16, 2004, http://www.gateway2russia.com/st/art_260273.php (accessed July 7, 2005).

170. "Nanotechnology news in brief," *NanoTechWeb,* May 27, 2005, http://nanotech web.org/articles/news/4/5/16/1 (accessed July 5, 2005).

171. "Genetic Engineering News Reports on Nanotech in Biotech and Medicine," *NewsNanoApex,* March 17, 2005, http://news.nanoapex.com/modules.php?name=News &file=article&sid=5560 (accessed July 5, 2005).

172. *FORBES/WOLFE Nanotech Weekly Insider,* October 15, 2004, e-mail communication.

173. Josh Wolfe, "On the Ground Floor of Nanotechnology," *Forbes.com,* November 24, 2004, http://www.forbes.com/2004/11/24/cz_jw_1124soapbox_print.html (accessed November 29, 2004).

174. Ravi Chandrasekaran, John Miller, and Michael Gertner, "Detecting Molecules: The Commercialization of Nanosensors," *Nanotechnology Law & Business Journal* 2, no. 1, (2005): 10.

175. "New Report Says Nano-Enabled Drug Discovery Market to Reach $1.3 Billion by 2009," *NanoInvestorNews,* March 7, 2005, http://www.nanoinvestornews.com/modules .php?name=News&file=article&sid=4109 (accessed July 5, 2005).

176. "ORNL Nanoscience Center *Jump Starts* Medical Compound Device," *Newswise,* May 24, 2005, http://www.newswise.com/p/articles/news/512110/ (accessed May 31, 2005).

177. "Biophan CEO Pens Column for Nanotech Briefs Magazine, Discussed Convergence of Nanotechnology and Biotechnology," *Business Wire,* March 4, 2005, http://home .businesswire.com/portal/site/google/index.jsp?ndmViewId=news_view&newsId=200503 03005031&newsLang=en (accessed March 10, 2005).

178. Sean Murdock, "From the director," *NanoBusiness News,* June 3, 2005, e-mail communication.

179. "NIH awards Emory and Georgia Tech $10 million for partnerships in cancer nanotechnology," *Medical News Today,* October 7, 2004, http://www.medicalnewstoday .com/medicalnews.php?newsid=14551 (accessed July 5, 2005).

180. M. Roco, introduction to *Societal Implications of Nanoscience and Nanotechnology,* ed. M. C. Roco and W. S. Bainbridge (New York: Kluwer, 2001), p. 7.

181. Charles Choi, "Ten overlooked nano firms," *Washington Times,* May 9, 2005, http://washingtontimes.com/upi-breaking/20050506-011337-9236r.htm (accessed July 3, 2005).

182. Josh Wolfe, "On the Ground Floor of Nanotechnology," *Forbes.com,* November 24,

2004, http://www.forbes.com/2004/11/24/cz_jw_1124soapbox_print.html (accessed November 29, 2004).

183. "Nanotechnology news in brief," *NanoTechWeb*, February 11, 2005, http://nano techweb.org/articles/news/4/2/8?alert=1 (accessed February 14, 2005).

184. "Asthma and diabetes sufferers could benefit nano-particle research," *Medical Research News*, May 1, 2005, http://www.news-medical.net/?id=9658 (accessed May 10, 2005).

185. Carl Wherrett and John Yelovich, "The Tiny Next Big Thing," *Motley Fool*, November 24, 2004, http://www.fool.com/news/commentary/2004/commentary 04112405.htm?source=EDNWFT (accessed June 17, 2005).

186. "Nanotechnology to revolutionize drug delivery," *in-Pharma Technologist.com*, July 3, 2005, e-mail communication.

187. "Liquidia Technologies Announces Breakthrough," *NanoInvestorNews*, June 22, 2005, http://www.nanoinvestornews.com/modules.php?name=News&file=article&sid =4396 (accessed July 12, 2005).

188. Carl Wherrett and John Yelovich, "The Players and Pretenders of Nanotech," *Motley Fool*, August 30, 2004, http://www.fool.com/news/commentary/2004/commentary 04083001.htm (accessed on September 7, 2004).

189. "Nanotechnology to revolutionize drug delivery," *in-Pharma Technologist.com*, July 3, 2005, e-mail communication.

190. Jeffrey M. Perkel, "The Ups and Downs of Nanobiotech," *Scientist*, August 30, 2004, p. 16.

191. Josh Wolfe, "Nanotechnology's Disruptive Future," *Forbes.com*, October 21, 2004, http://www.forbes.com/investmentnewsletters/2004/10/21/cz_jw_1021soapbox.html (accessed July 11, 2005).

192. "Nanomedicines, medical devices now in development total 152," *NewsNanoApex*, January 5, 2005, http://news.nanoapex.com/modules.php?name=News&file=article&sid =5410 (accessed July 5, 2005).

193. Stephen Baker and Ashton Aston, "The Business of Nanotech," *Business Week*, February 15, 2005, http://www.businessweek.com/magazine/content/05_07/b3920001 _mz001.htm (accessed July 5, 2005)

194. Carl Wherrett and John Yelovich, "A Giant Leap for Nano," *Motley Fool*, January 18, 2005, http://www.fool.com/News/mft/2005/mft05011811.htm (accessed July 5, 2005).

195. Liz Kalaugher, "Silicon nanowires promote bone growth," *NanoTechWeb*, April 21, 2005, http://www.nanotechweb.org/articles/news/4/4/13/1 (accessed July 5, 2005).

196. "Aligned nanotubes accommodate bone," *Technology Review*, December 9, 2004, http://www.technologyreview.com/articles/04/12/rnb_120904.asp (accessed December 14, 2004).

197. Robert Service, "Nanofibers Seed Blood Vessels," *Science*, April 1, 2005, e-mail communication.

198. "Nano-bumps could help repair clogged blood vessels," *PhysOrg.com*, April 20, 2005, http://www.physorg.com/news3807.html (accessed July 5, 2005).

199. M. Roco, introduction to *Societal Implications of Nanoscience and Nanotechnology*, ed. M. C. Roco and W. S. Bainbridge (New York: Kluwer, 2001), p. 7.

200. Stephen Baker and Ashton Aston, "The Business of Nanotech," *Business Week*,

February 15, 2005, http://www.businessweek.com/magazine/content/05_07/b3920001 _mz001.htm (accessed July 5, 2005).

201. Liz Kalaugher, "Nanocrystals highlight DNA mutations," May 10, 2005, http://nanotechweb.org/articles/news/4/5/5?alert=1 (accessed May 23, 2005).

202. "Revolutionary nanotechnology illuminates brain cells at work," *PhysOrg.com*, May 30, 2005, http://www.physorg.com/news4321.html (accessed July 8, 2005).

203. Paul Rincon, "Test could spot Alzheimer's early," *BBC News*, November 12, 2004, http://news.bbc.co.uk/1/hi/sci/tech/4003593.stm (accessed July 5, 2005).

204. LGC Limited, "Products and Services: Research and Development," http://www.lgc.co.uk/research_labonachip.asp (accessed July 5, 2005).

205. Ravi Chandrasekaran, John Miller and Michael Gertner, "Detecting Molecules: The Commercialization of Nanosensors," *Nanotechnology Law & Business Journal* 2, no. 1, (2005): 11.

206. "Acrongenomics Inc. Further Develops its NanoJETA Platform," *Nano-InvestorNews*, March 14, 2005, http://www.nanoinvestornews.com/modules.php?name =News&file=article&sid=4134 (accessed July 5, 2005).

207. "Infectech, Inc., Approves Name Change to NanoLogix and Enters Hydrogen Production, BioDefense, Medical Diagnostics and Cancer Drug Markets," *Nano-InvestorNews*, April 7, 2005, http://www.nanoinvestornews.com/modules.php?name=News &file=print&sid=4195 (accessed April 14, 2005).

208. Belle Dume, "Nanodevices target viruses," *PhysicsWeb*, October 4, 2004, http://physicsweb.org/articles/news/8/10/6/1 (accessed October 11, 2004).

209. "Purdue proves concept of using nano-materials for drug discovery," *Innovations Report*, February 18, 2005, http://www.innovations-report.com/html/reports/life _sciences/report-40607.html (accessed February 24, 2005).

210. "Acrongenomics Inc. Further Develops its NanoJETA Platform," *NanoInvestor-News*, March 14, 2005, http://www.nanoinvestornews.com/modules.php?name=News &file=article&sid=4134 (accessed July 5, 2005).

211. "Nanotechnology news in brief," October 1, 2004, http://nanotechweb.org/articles/ news/3/10/2 (accessed October 4, 2004).

212. "Nanotech Method Detects Respiratory Syncytial Virus," *Medical News Today*, June 12, 2005, http://www.medicalnewstoday.com/medicalnews.php?newsid=25998 (accessed July 3, 2005).

213. Sean Murdock, "From the director," *NanoBusiness News*, June 3, 2005, e-mail communication.

214. Cory Johnson, "Nanotech investors curb their enthusiasm," *MSNBC.com*, December 16, 2004, http://msnbc.com/id/6713189/print/1/displaymode/1098 (accessed January 12, 2005).

215. "Nanotechnology news in brief," *NanoTechWeb*, May 27, 2005, http://nanotech web.org/articles/news/4/5/16/1 (accessed July 5, 2005).

216. "Nanotechnology to detect tumors," *My DNA News*, April 29, 2005, http://www .mydna.com/resources/news/200504/news_20050429_tumde.html (accessed July 5, 2005).

217. Kevin Maney, "Nanotechnology's everywhere," *USA Today*, June 1, 2005, http://www.usatoday.com/tech/columnist/kevinmaney/2005-05-31-nanotech_x.htm (accessed July 14, 2005).

218. Cory Johnson, "Nanotech investors curb their enthusiasm," *MSNBC.com*, December 16, 2004, http://msnbc.com/id/6713189/print/1/displaymode/1098 (accessed January 12, 2005).

219. "Tiny bundles seek and destroy breast cancer cells," *NewsNanoApex*, May 24, 2005, http://news.nanoapex.com/modules.php?name=News&file=article&sid=5646 (accessed July 5, 2005).

220. "Study: Nano therapy kills tumors, extends life," November 11, 2004, http://smalltimes.org/document_display.cfm?section_id=53&document_id=8429 (accessed July 5, 2005).

221. Elizabeth Sabrio, "Research team develops gel to grow blood vessels," *Daily Northwestern*, April 8, 2005, http://www.dailynorthwestern.com/vnews/display.v/ART/2005/04/08/42561f86ce59e (accessed July 5, 2005).

222. "Nanoparticles Silence Childhood Cancer," *Betterhumans*, April 19, 2005, http://www.betterhumans.com/News/news.aspx?articleID=2005-04-19-3 (accessed April 20, 2005).

223. "The Search for a Kinder, Gentler Chemotherapy," *NewsNanoApex*, September 12, 2004, http://news.nanoapex.com/modules.php?name=News&file=article&sid=5003 (accessed July 5, 2005).

224. Jack Uldrich, "Releasing Nanotech's Potential?" *Motley Fool*, May 23, 2005, http://www.fool.com/News/mft/2005/mft05052318.htm (accessed July 5, 2005).

225. Charles Choi, "Ten overlooked nano firms," *Washington Times*, May 9, 2005, http://washingtontimes.com/upi-breaking/20050506-011337-9236r.htm (accessed July 3, 2005).

226. Nicole Dyer, "Tumor-killing Nanoshells," *Popular Science*, October 2004, http://www.popsci.com/popsci/medicine/article/0,20967,703146,00.html (accessed October 11, 2004).

227. M. Roco, introduction to *Societal Implications of Nanoscience and Nanotechnology*, ed. M. C. Roco and W. S. Bainbridge (New York: Kluwer, 2001), p. 7.

228. Carl Wherrett and John Yelovich, "The Players and Pretenders of Nanotech," *Motley Fool*, August 30, 2004, http://www.fool.com/news/commentary/2004/commentary04083001.htm (accessed on September 7, 2004).

229. "First Pure Magnetic NanoPowders Advance Medical Diagnostics," *NanoInvestorNews*, March 22, 2005, http://www.nanoinvestornews.com/modules.php?name=News&file=article&sid=4153 (accessed July 5, 2005).

230. "Kyoto Univ. research trap hydrogen in fullerene," *RedNova News*, January 14, 2005, http://www.rednova.com/modules/news/tools/php?tool=print&id=119143 (accessed February 7, 2005).

231. "Nano-Probes Allow Inside Look at Cell Nuclei," *NewsNanoApex*, March 18, 2005, http://news.nanoapex.com/modules.php?name=News&file=article&sid=5563 (accessed July 5, 2005).

232. Genevieve Oger, "The Nanoscience behind Beauty Is Serious Business at L'Oreal," *Small Times*, December 26, 2002, http://www.smalltimes.com/document_display.cfm?document_id=5236 (accessed July 5, 2005).

233. "Nanotechnology news in brief," *NanoTechWeb*, February 11, 2005, http://nanotechweb.org/articles/news/4/2/8?alert=1 (accessed February 14, 2005).

234. Alan Cane, "Sunscreen to save sun lovers' skin," *Financial Times*, April 22, 2005, e-mail communication.

235. Mike Langberg, "Envisioning a big future in nanotechnology," May 20, 2005, http://www.kentucky.com/mld/kentucky/business/technology/11696029.htm (accessed July 5, 2005).

236. "Malvern's Zatasizer nano is used to optimize topical skin care formulation," *NanoInvestorNews*, June 8, 2005, http://www.nanoinvestornews.com/modules.php?name =News&file=article&sid=4364 (accessed July 3, 2005).

237. "Hosokawa Powder Technology Research Institute Develops Innovative Hairgrowth Technology Using Nanotechnology," April 16, 2005, *NanoInvestorNews*, http://www.nanoinvestornews.com/modules.php?name=News&file=article&sid=4210 (accessed July 5, 2005).

238. "Nanotechnology news in brief," *NanoTechWeb*, December 3, 2004, http://nano techweb.org/articles/news/3/12/4 (accessed December 13, 2004).

239. "Argonide Unveils a New Water Filter with 20 Times the Capacity of Conventional Filters," *NanoInvestorNews*, March 10, 2005, http://www.nanoinvestornews.com/ modules.php?name=News&file=article&sid=4133 (accessed July 5, 2005).

240. Charles Choi, "Nano World: Water, water everywhere nano," *Washington Times*, March 18, 2005, http://www.washtimes.com/upi-breaking/20050317-095517-1543r.htm (accessed March 22, 2005).

241. Francesca Hopkins, "Nanotechnology Used to Study Environment," *Daily Californian*, January 19, 2005, http://www.dailycal.org/particle.php?id=17297 (accessed January 28, 2005).

242. John K. Borchardt, "Nanotech shows promise for cheaper Superfund cleanup," *USA Today*, February 10, 2005, http://www.usatodaycom/tech/news/nano/2005-02-10 -nano-iron-cleanup_x.htm (accessed February 15, 2005).

243. "Tiny particles could solve billion-dollar problem," *EurekaAlert!* February 23, 2005, http://www2.eurekalert.org/pub_releases/2005-02/ru-tpc022305.php (accessed February 23, 2005).

244. John K. Borchardt, "Nanotech shows promise for cheaper Superfund cleanup," *USA Today*, February 10, 2005, http://www.usatoday.com/tech/news/nano/2005-02-10- nano-iron-cleanup_x.htm (accessed February 15, 2005).

245. Ibid.

246. "Iron nanoparticles may be effective in cleaning up carbon tetrachloride in contaminated groundwater," *NewsNanoApex*, January 14, 2005, http://news.nanoapex.com/ modules.php?name=News&file=article&sid=5444 (accessed July 3, 2005).

247. "Nanoscale Iron Available for Environmental Remediation Projects," *NanoInvestorNews*, April 20, 2005, http://www.nanoinvestornews.com/modules.php?name =News&file=article&sid=4227 (accessed July 8, 2005).

248. "Tiny particles could solve billion-dollar problem," *EurekaAlert!* February 23, 2005, http://www2.eurekalert.org/pub_releases/2005-02/ru-tpc022305.php (accessed February 23, 2005).

249. Wayne Curtis, "The Methuselah Report: Living to be 120 might be attainable, but is it desirable?" *AARP Bulletin Online*, July–August 2004, http://www.aarp.org/ bulletin/yourhealth/Articles/a2004-07-07-methuselah.html (accessed July 12, 2004).

250. Will Knight, "Drugs delivered by robots in the blood," *New Scientist*, October 1, 2004, http://www.newscientist.com/news/news.jsp?id=ns99996474 (accessed October 4, 2004).

251. Liz Kalaugher, "Nanotube transistor interacts with cell membrane," *NanoTechWeb*, April 19, 2005, http://www.nanotechweb.org/articles/news/4/4/12/1 (accessed July 5, 2005).

252. "Nanomaterials break out of laboratory into marketplace," *NewsNanoApex*, October 3, 2004, http://news.nanoapex.com/modules.php?name=News&file=article&sid=5077 (accessed July 5, 2005).

253. "Nanotechnology in brief," *NanoTechWeb*, September 10, 2004, http://nanotechweb.org/articles/news/3/9/7 (accessed September 22, 2004).

254. Roland Pease, "Brush up on your nanotechnology," *BBC News*, June 12, 2005, http://news.bbc.co.uk/1/hi/sci/tech/4085214.stm (accessed July 3, 2005).

255. Liz Kalaugher, "Carbon nanotubes head for brain repair," *NanoTechWeb*, May 26, 2005, http://nanotechweb.org/articles/news/4/5/14/1 (accessed July 5, 2005).

256. "Research and Markets: Opportunities for Nano-Engineered Solutions in the Electrical Power Industry Worldwide," *NanoInvestorNews*, March 24, 2005, http://www.nanoinvestornews.com/modules.php?name=News&file=article&sid=4162 (accessed July 11, 2005).

257. "Nanotubes juice super batteries," *TRN*, March 23/30, 2005, http://www.trnmag.com/Stories/2005/032305/Nanotubes_juice_super_batteries_Brief_032305.html (accessed July 19, 2005).

258. Michael Kanellos, "Alternative energies are looking good again," *CNET News.com*, July 12, 2004, http://news.com.com/Energy+heats+up+high+tech/2009-7337_3-5263772.html (accessed August 4, 2004).

259. M. Roco, introduction in *Societal Implications of Nanoscience and Nanotechnology*, ed. M. C. Roco and W. S. Bainbridge (New York: Kluwer, 2001), p. 8.

260. "New research may advance the nanoelectronics field," *NewsNanoApex*, October 19, 2004, http://news.nanoapex.com/modules.php?name=News&file=article&sid=5153 (accessed July 6, 2005).

261. Liz Kalaugher, "Electrodes trap bacteria for nanoscale assembly," *NanoTechWeb*, March 22, 2005, http://nanotechweb.org/articles/news/4/3/10/1 (accessed March 28, 2005).

262. Liz Kalaugher, "Bacteria sport nanowire hairs," *NanoTechWeb*, June 29, 2005, http://nanotechweb.org/articles/news/4/6/14?alert=1 (accessed July 5, 2005).

263. "Discovery captures, converts heat," *PhysOrg.news*, April 6, 2005, http://www.physorg.com/news3613.html (accessed July 11, 2005).

264. Many of these come from "Rick Smalley's top 5 nano-energy solutions," *Nanotech Report* 2, no. 6 (June 2003): 2

265. John Barratt, "Innovators who view the fossil fuel industry as a fossil industry miss out on opportunities," *Small Times*, April 26, 2005, http://www.smalltimes.com/print_doc.cfm?doc_id=9013 (accessed June 16, 2005).

266. Ibid.

267. Susan Shor, "Toshiba Puts Nano-Material into Fast-Charging Battery," *TechNews World*, March 30, 2005, http://www.technewsworld.com/story/hardware/41889.html (accessed April 4, 2005).

268. "Nanotechnology News in Brief," *NanoTechWeb*, April 22, 2005, http://nanotechweb.org/articles/news/4/4/15?alert=1 (accessed April 25, 2005).

269. Testimony before the Research Subcommittee of the Committee on Science, U.S. House of Representatives, 109 Cong., 1st sess., May 18, 2005, http://www.house.gov/science/hearings/research05/May18/Kennedy%20testimony.pdf (accessed July 11, 2005).

270. "Nano pyramids boost fuel cells," *TRN Magazine*, May 4/11, 2005, http://www.trnmag.com/Stories/2005/050405/Nano_pyramids_boost_fuel_cells_Brief_050405.html (accessed July 11, 2005).

271. "Big Hopes for New Hydrogen Storage Material," *Renewable Energy Access*, May 17, 2005, http://renewableenergyaccess.com/rea/news/story?id=29928 (accessed May 24, 2005).

272. "Nano pyramids boost fuel cells," *TRN Magazine*, May 4/11, 2005, http://www.trnmag.com/Stories/2005/050405/Nano_pyramids_boost_fuel_cells_Brief_050405.html (accessed July 11, 2005).

273. "Nanotechnology new in brief," October 1, 2004, http://nanotechweb.org/articles/news/3/10/2 (accessed October 4, 2004).

274. Dean Takahashi, "Creating a sticky situation," *CentreDaily.com*, December 25, 2004, http://www.centredaily.com/mld/centredaily/business/technology/10497914.htm (accessed January 12, 2005).

275. Josh Wolfe, "Nanotechnology's Disruptive Future," *Forbes.com*, October 21, 2004, http://www.forbes.com/investmentnewsletters/2004/10/21/cz_jw_1021soapbox.html (accessed July 11, 2005).

276. Paul Carlstrom, "As solar gets smaller, its future gets brighter: Nanotechnology could turn rooftops into a sea of power-generating stations," *SFGate.com*, July 11, 2005, http://www.sfgate.com/cgi-bin/article.cgi?f=/c/a/2005/07/11/BUG7IDL1AF1.DTL&hw=solar&sn=001&sc=1000 (accessed July 12, 2005).

277. "Nanotechnology news in brief," *NanoTechWeb*, January 21, 2005, http://nanotechweb.org/articles/news/4/1/12?alert=1 (accessed January 28, 2005).

278. "Enerl Nanotech Unit Prepares to Commercialize Revolutionary Battery Electrodes," *NanoInvestorNews*, February 15, 2005, http://www.nanoinvestornews.com/modules.php?name=News&file=print&sid=4039 (accessed July 11, 2005).

279. "McDaniel to Lead Enerl Nanotech Business," *PRNewswire*, June 14, 2005, http://www.prnewswire.com/cgi-bin/stories.pl?ACCT=109&STORY=/www/story/06-14-2005/0003870514&EDATE (accessed July 13, 2005).

280. Josh Wolfe, "Nanotechnology's Disruptive Future," *Forbes.com*, October 21, 2004, http://www.forbes.com/investmentnewsletters/2004/10/21/cz_jw_1021soapbox.html (accessed July 11, 2005).

281. "NanoHorizons Patent Cost and Efficiency Breakthrough for Solar Cells and Organic LEDs," *NanoInvestorNews*, June 8, 2005, http://www.nanoinvestornews.com/modules.php?name=News&file=article&sid=4363 (accessed July 3, 2005).

282. "Nanotechnology news in brief," *NanoTechWeb*, April 8, 2005, http://nanotechweb.org/articles/news/4/4/7?alert=1 (accessed April 14, 2005).

283. Charles Choi, "Ten overlooked nano firms," *Washington Times*, May 9, 2005, http://washingtontimes.com/upi-breaking/20050506-011337-9236r.htm (accessed July 3, 2005).

284. Liz Kalaugher, "Vesicles roll up for rechargeable batteries," *NanoTechWeb*, June 29, 2005, http://nanotechweb.org/articles/news/4/6/15?alert=1 (accessed July 5, 2005).

285. Erika Jonietz, "10 Emerging Technologies: Quantum Wires," *Technology Review*, May 2005, http://www.technologyreview.com/articles/05/05/issue/feature_emerging.asp ?p=0 (accessed April 19, 2005).

286. "Konarka and Evident Technologies form Joint Research Program to Develop Ultra High Performance Power Plastic," *NanoInvestorNews*, March 23, 2005, http://www .nanoinvestornews.com/modules.php?name=News&file=article&sid=4157 (accessed July 11, 2005).

287. "U.S. Army Taps Konarka for $1.6 Million Renewable Energy Program," May 4, 2005, http://www.nanoinvestornews.com/modules.php?name=News&file=article&sid =4258 (accessed July 11, 2005).

288. "Energy start-ups bank on nanotechnology," *NanoTechWeb*, May 3, 2005, http://nanotechweb.org/articles/news/4/5/2?alert=1 (accessed May 9, 2005).

289. "Nanotubes crank out hydrogen," *Fuel Cell Today*, January 27, 2005, http://www .fuelcelltoday.com/FuelCellToday/IndustryInformation/IndustryInformationExternal/ NewsDisplayArticle/0,1602,5504,00.html (accessed July 19, 2005).

290. "Platinum key to new nano solar energy technique," *Platinum Today*, March 21, 2005, http://www.platinum.matthey.com/media_room/1111399203.html (accessed March 28, 2005).

291. "Nanotechnologists' New Plastic Can See in the Dark," *Science Daily*, January 10, 2005, http://www.sciencedaily.com/releases/2005/01/050110112209.htm (accessed May 24, 2005).

292. "Fuel Cell News: The Scoop," *Best of the NanoWeek*, January 25, 2005, e-mail communication.

293. "Nanotechnology news in brief," January 7, 2005, http://nanotechweb.org/ articles/news/4/1/4 (accessed January 12, 2005).

294. James Mackintosh, "GM and Toyota in talk on hydrogen cars," *Financial Times*, May 12, 2005, e-mail communication.

295. "First Pure Magnetic Nanopowders Advance Medical Diagnostics," March 22, 2005, http://www.nanoinvestornews.com/modules.php?name=News&file=article&sid =4153 (accessed July 11, 2005).

296. Michael Kanellos, "Nanoparticles for energy, explosions," *CNET*, http://news .com.com/Nanoparticles+for+energy%2+explosions/2100-7337_3-5421090.html (accessed October 27, 2004).

297. Charles Choi, "Nano World: Nanocatalysts for oil, drugs," *Washington Times*, March 25, 2005, http://washingtontimes.com/upi-breaking/20050317-123506-4921r.htm (accessed July 11, 2005).

298. "Nanotechnology news in brief," *NanoTechWeb*, February 11, 2005, http:// nanotechweb.org/articles/news/4/2/8?alert=1 (accessed February 14, 2005).

299. Candace Stuart, "Oxonica's deal with bus fleet puts company on right road," *Small Times*, February 7, 2005, http://www.smalltimes.com/document_display.cfm ?document_id=8742 (accessed February 8, 2005).

300. "Acta's breakthrough fuel cell technology wins Summit Medi-Chem contract,"

Fuel Cell Today, May 21, 2005, http://www.fuelcelltoday.com/FuelCellToday/Industry Information/IndustryInformationExternal/NewsDisplayArticle/0,1602,5992,00.html (accessed July 11, 2005).

301. "Green Plus Helps Taxi Drivers in London Lower Emissions and Save Fuel," *NewsNanoApex*, November 18, 2004, http://news.nanoapex.com/modules.php?name =News&file=article&sid=5262 (accessed July 11, 2005).

302. Nicole Wallace, "New Project Studies Nanotechnology and Policy," *Chronicle of Philanthropy*, May 26, 2005, http://philanthropy.com/premium/articles/v17/i16/ 16002902.htm, e-communication.

303. "Easton Sport Apply Carbon Nanotube Technology to Baseball," *NetComposites*, May 19, 2005, http://www.netcomposites.com/news.asp?2974 (accessed May 19, 2005).

304. The ETC Group offers a list of nanoproducts that it claims come from an unofficial EPA document. See http://www.etcgroup.org/documents/nanoproducts _EPA.pdf.

305. Stephen Baker and Ashton Aston, "The Business of Nanotech," *Business Week*, February 15, 2005.

306. Michael Kanellos, "Nanoparticles for energy, explosions," *CNETNews.com*, http://news .com.com/Nanoparticles+for+energy%2+explosions/2100-7337_3-5421090 .html (accessed October 27, 2004).

307. Josh Wolfe, "Nanotechnology's Disruptive Future," *Forbes.com*, October 21, 2004, http://www.forbes.com/investmentnewsletters/2004/10/21/cz_jw_1021soapbox.html (accessed July 11, 2005).

308. Meredith Dodge, "The cleanest underwear in town," *Taipei Times*, June 23, 2005, p. 15.

309. "New study: Nanotechnology poised to revolutionize tech, manufacturing markets," *NanoInvestorNews*, October 30, 2004, http://www.nanoinvestornews.com/modules .php?name=News&file=article&sid=3669 (accessed July 20, 2005).

310. Carl Wherrett and John Yelovich, "Is Nanotechnology for Real?" *Motley Fool* (July 23, 2004), http://www.fool.com/news/commentary/2004/commentary 04072305.htm (accessed July 26, 2004).

311. Advertisement, "Nanotech Report—Investing today," (2004): 9–10.

CHAPTER 7

1. EmTech Research, *Small Times* (May/June 2005): 14.

2. Robert F. Service, "Nanotechnology Grows Up," *Science* 304 (June 18, 2004): 1733.

3. Ann M. Thayer, "Nanotech investing," *Chemical and Engineering News*, May 2, 2005, http://pubs.acs.org/cen/coverstory/83/print/8318nanotech.html (accessed May 6, 2005).

4. In Realis, *A Critical Investor's Guide to Nanotechnology* (February 2002): 1.

5. Stacy Lawrence, "Nanotech grows up," *Technology Review*, May 23, 2005, http:// www.technologyreview.com/articles/05/06/issue/datamine.asp?p=0 (accessed May 24, 2005).

6. Lux Research, "Industry at a Glance," *San Francisco Chronicle* (February 1, 2004), http://sfgate.com/cgi-bin/article.cgi?f=/chronicle/a/2004/02/01/BUGCJ4J1D51.DTL (accessed July 11, 2004).

7. ETC Group, "Nanotech News in Living Colour: An Update on White Papers, Red Flags, Green Goo, Grey Goo (and Red Herrings)," *Communiqué* 85 (May/June 2004): 1.

8. "Lux Research Releases *The Nanotech Report 2004* Key Findings," August 16, 2004, e-mail communication.

9. Charles Choi, "Nanotubes, buckyballs surprise investors," *United Press International*, May 19, 2004, http://www.upi.com/view.cfm?StoryID=20040518-011947-3488r (accessed May 25, 2004).

10. David Forman, "As the Analyst: Four Micro/Nano Experts Show What Life's Like Inside Market Research," *Small Times*, April 7, 2005, http://www.smalltimes.com/print_doc.cfm?doc_id=8983 (accessed April 14, 2005).

11. Alan Shalleck, "Where Are All the New Products?" *NanoInvestorNews*, March 28, 2005, http://www.nanoinvestornews.com/modules.php?name=Content&pa=showpage &pid=18 (accessed March 28, 2005).

12. Howard Lovy, "Can Nano Create New Markets," *NanoMarkets White Paper* (April 2004).

13. Ibid.

14. Ibid.

15. Ibid.

16. Glenda Chui, "Who's afraid of nanotechnology?" *Mercury News*, September 16, 2003, http://www.bayarea.com/mld/mercurynews/living/health/6779281.htm?template =content (accessed October 17, 2003).

17. "Nanotech is Wall Street's latest love," *Seattle Post-Intelligencer*, July 13, 2004, http://seattlepi.nwsource.com/business/aptech_story.asp?category=1700&slug =Nanotechnology%20Hype (accessed July 19, 2004).

18. "The Nanotechnology Industry, an estimated $961 million for FY 2004," *NanoInvestorNews*, July 15, 2004, http://www.nanoinvestornews.com/modules.php?name=News &file-Print&sid=3-72 (accessed July 19, 2004).

19. David Forman, "Nanotech rides a rising tide," *Small Times* (March/April 2004): 18–19.

20. David Forman, "Nano takes to the Street: Second IPO ready to go," *Small Times* (July 21, 2004), http://www.smalltimes.com/document_display.cfm?section_id =51&document _id=8175 (accessed July 26, 2004).

21. Robert F. Service, "Nanotechnology Grows Up," *Science* 304 (June 18, 2004): 1733.

22. Stacy Lawrence, "Nanotech grows up," *Technology Review*, May 23, 2005, http://www.technologyreview.com/articles/05/06/issue/datamine.asp?p=0 (accessed May 24, 2005).

23. Tiffany Kary, "Is small the next big thing?" *CNET News.com*, February 11, 2002, http://news.com.com/2100-1001-833691.html (accessed July 27, 2004).

24. Ibid.

25. "Bingaman Calls for DARPA Head to Resign over Terrorism Futures Idea," *ABQ Journal* (July 29, 2003), http://www.abqjournal.com/news/metro/112110nm07-29-03.htm (accessed April 26, 2004).

26. "Nanotechnology Facts and Figures," *NanoInvestorNews*, January 26, 2004, http://www.nanoinvestornews.com/modules.php?name=Facts_Figures&op=sho&im =dbloc/countriespie (accessed June 25, 2004).

27. C. Cerf and V. Navasky, *The Experts Speak: The Definitive Compendium of Authoritative Misinformation* (New York: Pantheon Books, 1984), p. x.

28. Ibid.

29. B. Laird, "Policy study compares nanotechnology funding in the United States and Japan," *Foresight Update* 18 (April 15, 1994): 2.

30. Ralph Merkle, "A response to Scientific American's new story: Trends in nanotechnology," 1996, http://www.foresight.org/SciAmResponse.html (accessed August 29, 2004).

31. Jay Lindquist, "In the Spotlight," *NanoBusiness News* (February 3, 2004).

32. Jamie Dinkelacker, "Policy Watch," *Foresight Update* 18 (April 15, 1994): 8.

33. Cientifica, *The Nanotechnology Opportunity Report*, 2003, 9, http://nanotechnow.com/nanotechnology-opportunity-report.htm (accessed April 26, 2004).

34. "NanotechnologyInvestment.com Exclusive Interview with Nanosys Inc.," *MarketWire*, November 10, 2004, http://www.marketwire.com/mw/release_html_b1?release_id=75744 (accessed July 12, 2005).

35. NanoMarkets, "What will the future nanotech industry look like?" *NanoMarkets White Paper* (March 2004): 5.

36. Tiffany Kary, "Nanotech: More Science Than Fiction," *CNET News.com*, February 11, 2002, http://zdnet.com.com/2102-1103_2-833739.html (accessed May 3, 2002).

37. Howard Lovy, "Business has redefined nano in its own image; will that vision bring us the best of all futures?" *Small Times*, January/February 2004, p. 14.

38. Jeff Karoub, "Nano tool market is no small change," *Small Times*, July/August 2003, p. 20.

39. "European Free Alliance in the European Parliament," June 11, 2003, http://www.greens-efa.org/en/agenda/detail.php?id=1105&lg=en (accessed July 30, 2003).

40. Tiffany Kary, "Nanotech: More Science Than Fiction," *CNET News.com*, February 11, 2002, http://zdnet.com.com/2102-1103_2-833739.html (accessed May 3, 2002).

41. Michael J. Mauboussin and Kristen Bartholdson, "Big Money in Thinking Small: Nanotechnology—What Investors Need to Know" (Credit Suisse Equity Research/First Boston, May 7, 2003), p. 19.

42. Mihail Roco, "Nanoscale Science and Engineering: Unifying and Transforming Tools," *AIChE Journal* 50, no. 5 (May 2004): 893.

43. Michael J. Mauboussin and Kristen Bartholdson, "Big Money in Thinking Small: Nanotechnology—What Investors Need to Know" (Credit Suisse Equity Research/First Boston, May 7, 2003), pp. 19–23.

44. Jon Van, "Nanotechnology gets $3.7 billion federal backing," November 26, 2003, http://www.menafn.com/qn_news_story.asp?StoryId=Cp8we0eicveiTtKfotlrfq0G (accessed December 2, 2003).

45. "Nanotech industry review and 2004 outlook," *Nanotech Report* 3, no. 1 (January 2004): 4.

46. Stephen Baker and Ashton Aston, "The Business of Nanotech," *Business Week*, February 15, 2005.

47. Ed Moran, "Hype of hope," a publication of Deloitte & Touche LLP, Summer 2002, p. 1.

48. M. Mauboussin and K. Bartholdson, "Big Money in Thinking Small: Nanotechnology—What Investors Need to Know" (Credit Suisse Equity Research/First Boston: Boston, MA, 2003).

49. "Corporate nanotechnology investments are at risk of being wasted," *NanoInvestorNews.com* January 14, 2005, http://www.nanoinvestornews.com/modules.php?name=News&file=article&sid=3961 (accessed January 28, 2005).

50. M. Mauboussin and K. Bartholdson, "Big Money in Thinking Small: Nanotechnology—What Investors Need to Know" (Credit Suisse Equity Research/First Boston: Boston, MA, 2003), p. 19.

51. Ibid., p. 20.

52. Stephen Baker and Ashton Aston, "The Business of Nanotech," *Business Week*, February 15, 2005.

53. Michelle Rama, "Nanotech: Mini-science, mega-potential," *Boston.com* (July 25 2004), http://www.boston.com/business/technology/articles/2004/07/25/nanotech_mini_science_mega_potential/ (accessed July 26, 2004).

54. Ibid.

55. Arnaud Paris, "Micro- and nanosystems in biology; from innovative applications to market," *Applied Nanoscience* 1, no. 1 (2004): 3–6.

56. Josh Wolfe, "Nanotech Insiders: Nov 12 (Waves, Genes and Happiness)," *Forbes/Wolfe Nanotech Insider*, November 12, 2004, e-mail communication, November 12, 2004.

57. M. Mauboussin, and K. Bartholdson, "Big Money in Thinking Small: Nanotechnology—What Investors Need to Know, (Credit Suisse Equity Research/First Boston: Boston, MA, May 7, 2003), p. 18.

58. John Cook, "Venture Capital: Nanotech may take years to hit it big," *Seattle Post-Intelligencer*, August 6, 2004, http://seattlepi.nwsource.com/venture/185099_vc06.html (accessed August 9, 2004).

59. "Investing in nano: Venture capitalists tell their side of the story," *Small Times*, March 7, 2005 (accessed August 25, 2005).

60. Ann M. Thayer, "Nanotech investing," *Chemical and Engineering News*, May 2, 2005, http://pubs.acs.org/cen/coverstory/83/print/8318nanotech.html (accessed May 6, 2005).

61. Ibid.

62. Charles Choi, "Nanotech ready for big changes soon," *TechNewsWorld*, September 10, 2004, http://www.technewsworld.com/story/Nanotech-ready-for-Big-changes-Soon-36523.html (accessed September 22, 2004).

63. Ibid.

64. Ibid.

65. David Forman, "VC forecast: Slow, steady, boring," *Small Times* (November/December 2003): 8.

66. Ann M. Thayer, "Nanotech investing," *Chemical and Engineering News*, May 2, 2005, http://pubs.acs.org/cen/coverstory/83/print/8318nanotech.html (accessed May 6, 2005).

67. Barnaby Feder, "Tiny ideas coming of age," *New York Times*, October 24, 2004, http://www.nytimes.com/2004/10/24/weekinreview/24fede.html (accessed October 27, 2004).

68. Ann M. Thayer, "Nanotech investing," *Chemical and Engineering News*, May 2, 2005, http://pubs.acs.org/cen/coverstory/83/print/8318nanotech.html (accessed May 6, 2005).

69. M. Mauboussin and K. Bartholdson, "Big Money in Thinking Small: Nanotechnology—What Investors Need to Know," (Credit Suisse Equity Research/First Boston, Boston, MA, May 7, 2003), p. 18.

70. David Forman, "Nanotech rides a rising tide," *Small Times* (March/April 2004): 19–20.

71. Cory Johnson, "Nanotech investors curb their enthusiasm," *MSNBC.com*, December 16, 2004, http://msnbc.com/id/6713189/print/1/displaymode/1098 (accessed January 12, 2005).

72. Nanotechnology moves other technology to new levels. Credit Suisse Equity Research labels it a general-purpose technology. The term was invented by Ethanan Helpman. GTPs include steam engines, electricity, and railroads and have been the basis for major economic revolutions.

73. Thiemo Lang, "The nano-savvy investor: Making sense of the public markets," *Small Times* (March 2005): 27.

74. Waqar Qureshi interviewed by Jeff Karoub, "From the front: True tales from industry veterans," *Small Times* (January/February 2005): 8.

75. Ed Moran, "Hype of hope," A publication of Deloitte & Touche LLP (Summer 2002): AV1.

76. Waqar Qureshi interviewed by Jeff Karoub, "From the front: True tales from industry veterans," *Small Times* (January/February 2005): 10.

77. Alan Marty, Committee on Science in the House of Representatives, *H.R. 766, Nanotechnology Research and Development Act of 2003*, Hearings before the Committee on Science in the House of Representatives, March 19, 2003, p. 60.

78. Ibid., p. 57.

79. Michael J. Mauboussin and Kristen Bartholdson, "Big Money in Thinking Small: Nanotechnology—What Investors Need to Know" (Credit Suisse Equity Research/First Boston, Boston, MA, May 7, 2003), p. 18.

80. Candace Stuart, "States should look beyond borders for true nature of small tech clusters," *Small Times* (March/April 2004): 7.

81. "Nano stock skepticism spells opportunity," *Nanotech Report* 2, no. 10 (October 2003): 3.

82. Neil Aronson, "Nanotech's market potential," *Boston Business Journal* (July 16, 2004), http://www.bizjournals.com/boston/stories/2004/07/19/editorial3.html (accessed July 26, 2004).

83. "The Nanotechnology Industry, an estimated $961 million for FY 2004," *NanoInvestorNews*, July 15, 2004, http://www.nanoinvestornews.com/modules.php?name=News&file-Print&sid=3-72 (accessed July 19, 2004).

84. Steve Jurvetson, "On the record: Nanotechnology—Unlocking the smallest secrets," *San Francisco Chronicle*, February 1, 2004, http://sfgate.com/cgi-bin/article.cgi?file=/chronicle/archive/2004/02/01/BUGCJ4J2761.DTL&type=business (accessed July 11, 2004).

85. "Lux Research Releases *The Nanotech Report 2004* Key Findings," August 16, 2004, e-mail communication.

86. David Forman, "'04 nano funding report: Less money but record number of

rounds," *Small Times*, February 2, 2005, http://www.smalltimes.com/print_doc.cfm?doc
_id=8744 (accessed February 8, 2005).

87. David Forman, "Channeling nano's funding future," *Small Times* (January/February 2005): 13.

88. David Forman, "'04 nano funding report: Less money but record number of rounds," *Small Times*, February 2, 2005, http://www.smalltimes.com/print_doc.cfm?doc
_id=8744 (accessed February 8, 2005).

89. Ann M. Thayer, "Nanotech investing," *Chemical and Engineering News*, May 2, 2005, http://pubs.acs.org/cen/coverstory/83/print/8318nanotech.html (accessed May 6, 2005).

90. "Market Watch," *Nature* 235 (2005): 273.

91. Ann M. Thayer, "Nanotech investing," *Chemical and Engineering News*, May 2, 2005, http://pubs.acs.org/cen/coverstory/83/print/8318nanotech.html (accessed May 6, 2005).

92. M. Mauboussin and K. Bartholdson, "Big Money in Thinking Small: Nanotechnology—What Investors Need to Know" (Credit Suisse Equity Research/First Boston, Boston, MA, May 2003), p. 21.

93. Jeff Karoub, "HP Official: Ignorance and Greed Could Spoil Nanotech's Credibility [Electronic Version]," *Small Times Magazine* (2001): 1.

94. Michael Boland, "Nanosys IPO Withdrawal; What Does It Means for the Industry?" *Innovation World TechConnect*, August 6, 2004, http://web2.innovationworld.net/techconnect/ (accessed August 9, 2004).

95. David Forman, "Nano start-ups skip across chasm," *Small Times* (April 2005): 21.

96. Ann M. Thayer, "Nanotech investing," *Chemical and Engineering News*, May 2, 2005, http://pubs.acs.org/cen/coverstory/83/print/8318nanotech.html (accessed May 6, 2005).

97. "Asian nanotech fever running hot," *Nanotech Report* 2, no. 1 (January 2003): 1.

98. Ibid.

99. Ibid.

100. Ibid.

101. Ibid., p. 2.

102. "Top ten investment banks," *Nanotechnology Law & Business Review* 2, no. 2 (2005): no page (electronic subscription).

103. Ann M. Thayer, "Nanotech investing," *Chemical and Engineering News*, May 2, 2005, http://pubs.acs.org/cen/coverstory/83/print/8318nanotech.html (accessed May 6, 2005).

104. Ibid.

105. Ibid.

106. "Merrill's nano index: Gilded list or fool's gold?" *Nanotech Report* 3, no. 6 (June 2004): 3.

107. Merrill Lynch, "Comment: Nanotechnology—Introducing the Merrill Lynch Nanotech Index" (April 8, 2004).

108. Thiemo Lang, "The nano-savvy investor: Making sense of the public markets," *Small Times* (March 2005): 27.

109. Stephen Baker and Ashton Aston, "The Business of Nanotech," *Business Week*, February 15, 2005.

110. "Merrill Lynch Creates *Nanotech Index* to Track Evolving Industry," *Merrill Lynch Press Release*, April 1, 2004, http://www.ml.com/about/press_release/04012004-1
_nanotech_index_pr.htm (accessed June 3, 2004).

111. "Merrill's nano index: Gilded list or fool's gold?" *Nanotech Report* 3, no. 6 (June 2004): 3.

112. Mike Langberg, "Nanotech IPO may lead to a new bubble," *Mercury News,* June 18, 2004, http://www.mercurynews.com/mld/mercurynews/business/technology/895361.htm (accessed July 9, 2004).

113. "Lux Nanotech Index Launches on American Stock Exchange," *Nano-InvestorNews.com,* March 18, 2005 http://www.nanoinvestornews.com/modules.php?name=News&file=print&sid=4150 (accessed March 22, 2005).

114. Ibid.

115. Ann M. Thayer, "Nanotech investing," *Chemical and Engineering News,* May 2, 2005, http://pubs.acs.org/cen/coverstory/83/print/8318nanotech.html (accessed May 6, 2005).

116. Thiemo Lang, "The nano-savvy investor: making sense of the public markets," *Small Times* (March 2005): 27.

117. Ibid.

118. Ann M. Thayer, "Nanotech investing," *Chemical and Engineering News,* May 2, 2005, http://pubs.acs.org/cen/coverstory/83/print/8318nanotech.html (accessed May 6, 2005).

119. Marcelo Prince, "Nanotech start-ups: Wall St.'s new fad," *DelawareOnline.com,* July 18, 2004, http://cgi.delawareonline.com/cgi-bin/advprint/print.cgi (accessed August 4, 2004).

120. "Nanophase Announces Second Quarter 2004 Results," *NanoInvestorNews.com,* July 22, 2004, http://www.nanoinvestornews.com/modules.php?name=News&file=article&sid=3102 (accessed July 26, 2004).

121. David Forman, "Nano takes to the Street: Second IPO ready to go," *Small Times,* July 21, 2004, http://www.smalltimes.com/document_display.cfm?section_id=51&document_id=8175 (accessed July 26, 2004).

122. Josh Wolfe, "On the Ground Floor of Nanotechnology," *Forbes.com,* November 24, 2004, http://www.forbes.com/2004/11/24/cz_jw_1124soapbox_print.html (accessed November 29, 2004).

123. "Nanosys selected for *Red Herring Top 100 Innovators* Award," *NanoInvestorNews,* December 7, 2004, http://www.nanoinvestornews.com/modules.php?name=News&file=print&sid=3833 (accessed December 14, 2004).

124. "Nanosys files to go public," *Reuters,* April 22, 2004, http://www.reuters.com/financeNewsArticle.jhtml?type=bondsNews&storyID=4914432 (accessed May 3, 2004).

125. Michael Kanellos, "Nanosys prices IPO, looks ahead," *CNET News.com,* July 15, 2004, http://news.com.com/Nanosys+prices+IPO,+looks+ahead/2100-1001_3-5271016.html (accessed July 29, 2004).

126. Mark Brandt and Drew Harris, "Talk loudly or be stealthy about new nanotechnology," *Nanotechnology Law & Business Journal* 2, no. 2 (2005): no page (accessed electronically).

127. John Boudreau, "Nanosys pulls IPO," *Mercury News,* August 5, 2004, http://www.siliconvalley.com/mld/siliconvalley/9324713.htm (accessed August 5, 2004).

128. Marcelo Prince, "Nanotech start-ups: Wall St.'s new fad," *DelawareOnline.com,* July 18, 2004, http://cgi.delawareonline.com/cgi-bin/advprint/print.cgi (accessed August 4, 2004).

129. John Boudreau, "Nanosys pulls IPO," *Mercury News,* August 5, 2004, http://www.siliconvalley.com/mld/siliconvalley/9324713.htm (accessed August 5, 2004).

130. David Forman, "Best laid plans," *Small Times* (July/August 2004): 34.

131. Ibid.

132. Ibid., p. 31.

133. Michael Boland, "Nanosys IPO withdrawal; What Does It Means for the Industry?" *Innovation World TechConnect*, August 6, 2004, http://web2.innovationworld.net/techconnect/ (accessed August 9, 2004).

134. Mike Langberg, "Nanotech IPO may lead to a new bubble," *Mercury News,* June 18, 2004, http://www.mercurynews.com/mld/mercurynews/business/technology/8953611.htm (accessed July 9, 2004).

135. Michael Kanellos, "Nanosys prices IPO, looks ahead," *CNET News.com*, July 15, 2004, http://news.com.com/Nanosys+prices+IPO,+looks+ahead/2100-1001_3-5271016.html (accessed July 29, 2004).

136. Carl Wherrett and John Yelovich, "Nano's Banner Company?" *Motley Fool*, August 4, 2004, http://www.fool.com/News/mft/2004/mft04080307.htm (accessed August 9, 2004).

137. "Nanosys awarded U.S. Defense Department contract to develop flexible solar cells," *NanoInvestorNews.com*, August 19, 2004, http://www.nanoinvestornews.com/modules.php?name=News&file=article&sid=3274 (accessed August 25, 2004).

138. Paul La Monica, "No-no for nanotech," *CNNMoney*, August 5, 2004, http://money.cnn.com/2004/08/05/technology/techinvestor/lamonica/ (accessed August 9, 2004).

139. David Forman, "Industry tries to make sense of Nanosys IPO withdrawal," *Small Times*, August 5, 2004, http://www.smalltimes.com/print_doc.cfm?doc_id=8202 (accessed August 9, 2004).

140. Josh Wolfe, "Nanotech Insider," *Nanotech Weekly Newsletter* (August 6, 2004), e-mail correspondence.

141. "Nanosys Puts IPO on hold," *Scientist*, August 30, 2004, p. 18.

142. David Forman, "After Nanosys IPO withdrawal: CDT to brave NASDAQ," *Small Times*, September 14, 2004, http://www.smalltimes.com/print_doc.cfm?doc_id=8281 (accessed September 22, 2004).

143. Paul La Monica, "No-no for nanotech," *CNN Money*, August 5, 2004, http://money.cnn.com/2004/08/05/technology/techinvestor/lamonica/ (accessed August 9, 2004).

144. Mike Langberg, "Nanotech IPO may lead to a new bubble," *Mercury News*, June 18, 2004, http://www.mercurynews.com/mld/mercurynews/business/technology/8953611.htm (accessed July 9, 2004).

145. Josh Wolfe, "On the Ground Floor of Nanotechnology," *Forbes.com*, November 24, 2004, http://www.forbes.com/2004/11/24/cz_jw_1124soapbox_print.html (accessed November 29, 2004).

146. David Forman, "Industry tries to make sense of Nanosys IPO withdrawal," *Small Times*, August 5, 2004, http://www.smalltimes.com/print_doc.cfm?doc_id=8202 (accessed August 9, 2004).

147. "Future Nanotech IPOs—The A-list," *Nanotech Report* (June 2004): 1.

148. Thiemo Lang, "The nano-savvy investor: Making sense of the public markets," *Small Times* (March 2005): 27.

149. David Forman, "Nano start-ups skip across chasm," *Small Times* (April 2005): 22.

150. Ann M. Thayer, "Nanotech investing," *Chemical and Engineering News*, May 2, 2005, http://pubs.acs.org/cen/coverstory/83/print/8318nanotech.html (accessed May 6, 2005).

151. David Forman, "Nano start-ups skip across chasm," *Small Times* (April 2005): 21.

152. Ibid.

153. Jae Kim, "Dawn of the Nani-Age," *Nanalyze*, November 1, 2004, http://www.nanalyze.com/articles/foresight_oct_04/printalr.aspx (accessed November 8, 2004).

154. Stephen Baker and Ashton Aston, "The Business of Nanotech," *Business Week*, February 15, 2005.

155. Ann M. Thayer, "Nanotech investing," *Chemical and Engineering News*, May 2, 2005, http://pubs.acs.org/cen/coverstory/83/print/8318nanotech.html (accessed May 6, 2005).

156. Ibid.

157. "Nano-Tex Secures $35 Million Series—A Round to Drive Development, Marketing, Global Expansion of Fabric Innovations," *NanoInvestorNews*, March 8, 2005, http://www.nanoinvestor news.com/modules.php?name=News&file=print&sid=4117 (accessed March 17, 2005).

158. Current products include *Resist Spills*, a liquid repellant; *Coolest Comfort*, moisture wicking for synthetics and wrinklefree cotton; *Resists Static*, the first permanent antistatic treatment for synthetics; and *Repels and Releases Stains*, an antistain treatment. See ibid.

159. Ibid.

160. Ann M. Thayer, "Nanotech investing," *Chemical and Engineering News*, May 2, 2005, http://pubs.acs.org/cen/coverstory/83/print/8318nanotech.html (accessed May 6, 2005).

161. See David Forman, "After Nanosys IPO withdrawal: CDT to brave NASDAQ," *Small Times*, September 14, 2004, http://www.smalltimes.com/print_doc.cfm?doc_id=8281 (accessed September 22, 2004).

162. Ibid.

163. Ann M. Thayer, "Nanotech investing," *Chemical and Engineering News*, May 2, 2005, http://pubs.acs.org/cen/coverstory/83/print/8318nanotech.html (accessed May 6, 2005).

164. S. Ehrrera and L. Aragon, "Small Worlds: Nanotechnology wins over mainstream venture capitalists [Electronic Version]," *Red Herring* (2001): 1.

165. "In nanotechnology, Josh Wolfe is at the door," *Red Herring*, January 28, 2002, http://www.redherring.com/Article.aspx?a=4359 (accessed July 27, 2004).

166. Ibid.

167. Ibid.

168. S. Ehrrera and L. Aragon, "Small Worlds: Nanotechnology wins over mainstream venture capitalists [Electronic Version]," *Red Herring* (2001): 1.

169. J. Uldrich and D. Newberry, *The Next Big Thing Is Really Small: How Nanotechnology Will Change the Future of Your Business* (New York: Crown Business, 2003), p. 71.

170. Joshua Wolfe, personal interview, May 27, 2003.

171. Ibid.

172. J. Henderson, "Nanotechnology," *Small Times Magazine* (2003): 1.

173. F. Mason, "Nanotechnology in the 21st Century," *Small Times Magazine* (2003): 3.

174. Tiffany Kary, "Nanotech: More Science than Fiction," *CNET News.com*, February 11, 2002, http://zdnet.com.com/2102-1103_2-833739.html (accessed May 3, 2002).

175. Ibid.

176. Ibid.

177. J. Henderson, "Nanotechnology," *Small Times Magazine* (2003): 1.

178. David Pescovitz, "Plenty of G's and a tone of Gee-Whiz," *Small Times* (January/February 2004): 33.

179. Ibid., p. 30.

180. *Forbes/Wolfe Nanotech Report* 3, no. 3 (March 2004): 2.

181. David Pescovitz, "Plenty of G's and a tone of Gee-Whiz," *Small Times* (January/February 2004): 30.

182. J. Henderson, "Nanotechnology," *Small Times Magazine* (2003): 1.

183. David Pescovitz, "Plenty of G's and a tone of Gee-Whiz," *Small Times* (January/February 2004): 31.

184. Ibid., p. 32.

185. Ibid., p. 34.

186. David Pogue, "Explaining Nanotechnology," *New York Times*, February 3, 2005, http://www.nytimes.com/2005/02/03/technology/circuits/03POGUE-EMAIL .html (accessed February 3, 2005).

187. Ann M. Thayer, "Nanotech investing," *Chemical and Engineering News*, May 2, 2005, http://pubs.acs.org/cen/coverstory/83/print/8318nanotech.html (accessed May 6, 2005).

188. David Pescovitz, "Plenty of G's and a tone of Gee-Whiz," *Small Times* (January/February 2004): 31.

189. M. Calvey, "Jurvetson Pins Big Hopes on Tiny Nanomachines," *San Francisco Business Times* (2001): 1.

190. "Big potential in TINY," *Nanotech Report* 1, no. 3 (2002): 1.

191. See David Forman, "Nano start-ups skip across chasms," *Small Times*, April 2005, http://www.smalltimes.com/document_display.cfm?document_id=9025 (accessed April 25, 2005). Forman cites Nantero CEO Greg Schmegal, who claims Nanotero's nanotube-based/nonvolatile RAM (NRAM) integrated into a standard CMOS manufacturing environment "is most likely to lead to an initial product that competes with the super-fast memory built right into computer processors."

192. "Beware the nanobubble," *Nanotech Report* 2, no. 6 (June 2003): 3.

193. Much of this history comes from the Silicon Investor Web page, http://www .siliconinvestor.com/stocktalk/subject.gsp?subjectid=52700 (accessed June 27, 2004).

194. "Earnings season hits the nanosphere," *Nanotech Report* 2, no. 11 (2003): 4.

195. Ibid.

196. "Nanotech's power elite: 2004," *Nanotech Report* 3, no. 3 (March 2004): 4.

197. "Big potential in TINY," *Nanotech Report* 1, no. 3 (2002): 1.

198. "Harris & Harris Group Announced Closing of Follow-on Offering and Full Exercise of Overallotment Option," *NanoInvestorNews*, July 7, 2004, http://www.nanoinvestornews. com/modules.php?name=News&file=print&sid=3033 (accessed July 12, 2004).

199. "Beware the nanobubble," *Nanotech Report* 2, no. 6 (June 2003): 3.

200. Ann M. Thayer, "Nanotech investing," *Chemical and Engineering News*, May 2, 2005, http://pubs.acs.org/cen/coverstory/83/print/8318nanotech.html (accessed May 6, 2005).

201. Ibid.

202. "Nanotechnology in Brief," *NanotechWeb*, July 1, 2004, http://nanotechweb.org/ articles/news/3/7/1 (accessed July 11, 2004).

203. Ann M. Thayer, "Nanotech investing," *Chemical and Engineering News*, May 2, 2005, http://pubs.acs.org/cen/coverstory/83/print/8318nanotech.html (accessed May 6, 2005).

204. M&A refers to M&A Online, a resource for browsing and posting companies for sale, see www.maol.com.

205. Modzelewski's biographical material is from an NbA Web page, http://www.nanobusiness.org/mark.html (accessed July 14, 2004).

206. "Beware the nanobubble," *Nanotech Report* 2, no. 6 (June 2003): 3.

207. Ibid., p. 4.

208. "Forbes Ranks NanoBusiness Alliance Director as One of Nanotech Industry's Leading Power Brokers NanoBusiness Association," *NanoTechWire*, March 18, 2004, http://nanotechwire .com/news.asp?nid=778&ntid=116&pg=2 (accessed June 25, 2004).

209. "Nanotech's power elite: 2004," *Nanotech Report* 3, no. 3 (March 2004): 1.

210. "Nanotechnology's power brokers," *Nanotech Report* 2, no. 3 (March 2003): 2.

211. "Thinking small: Mark Modzelewski," *Nanotech Report* 1, no. 10 (December 2002): 5.

212. Ibid.

213. Ibid.

214. Jack Mason, "NanoBusiness Alliance leader steps down, AtomWorks chief takes over," *Small Times*, May 18, 2004, http://www.smalltimes.com/print_doc.cfm?doc_id=7915 (accessed May 18, 2004).

215. Ibid.

216. Ibid.

217. "About Lux Research," *Lux Research Homepage*, July 4, 2004, http://www.luxresearchinc.com/ (accessed July 11, 2004).

218. Jack Mason, "NanoBusiness Alliance leader steps down, AtomWorks chief takes over," *Small Times*, May 18, 2004, http://www.smalltimes.com/print_doc.cfm?doc_id=7915 (accessed May 18, 2004).

219. Ibid.

220. "NanoBusiness Alliance and working-in-nanotechnology.com Join to Advance Nanotech Career Opportunities," *NanoInvestorNews*, May 19, 2004, http://www.nanoinvestornews.com/modules.php?name=News&file=articles&sid=2908. (accessed May 25, 2004).

221. "Nanotech Leaders Advance Policy Agenda on Capitol Hill; Congress and Administration Urged to Bolster Funding for Nanotech R&D, Ensure American Leadership," *NanoInvestorNews.com*, February 8, 2005, http://www.nanoinvestornews.com/modules.php?name=News&file=print&sid=4024, accessed February 15, 2005.

222. Michael J. Mauboussin and Kristen Bartholdson, "Big Money in Thinking Small: Nanotechnology—What Investors Need to Know" (Boston: Credit Suisse Equity Research/First Boston, May 7, 2003), pp. 26–35.

223. Matthew Nordan, "Nanomyths and Nanotruths," *NanoBusiness News*, October 29, 2004, e-mail communication, October 29, 2004.

224. Ibid.

CHAPTER 8

1. Zach Goldsmith, "Blair is the enemy of the Greens," *Observer*, August 12, 2003, http://observer.guardian.co.uk/worldview/story/0,11581,776678,090.html (accessed July 20, 2005).

2. Jonathan Oliver, "Nightmare of grey goo," *Mail*, April 27, 2003.

3. ETC Group, *The Big Down: From Genomes to Atoms, Atomtech: Technologies Converging at the Nano-Scale*, January 2003, http://www.etcgroup.org/documents/TheBigDown.pdf (accessed September 5, 2003), p. 26; "How safe is nanotechnology? Japan needs to find out soon," *Yomiuri Shimbun*, December 1, 2003, *Small Times*, http://www.smalltimes.com/document_display.cfm?document_id=7021 (accessed December 9, 2003).

4. The correct citation is Andrzej Hucko, Hubert Lange, Ewa Calko, Hanna Grubek-Jaworska, and Pawel Droszcz, "Physiological Testing of Carbon Nanotubes: Are They Asbestos Like," *Fullerene Science and Technology* 9, no. 2 (2002): 251–54.

5. Jessica Gorman, "Taming high-tech particles: Cautious steps into the nanotech future," *Science News* 161, no. 13, March 30, 2002, http://www.sciencenws.org/20020330/bob8.asp (accessed October 5, 2003).

6. ETC Group, "Nanotech News in Living Colour: An Update on White Papers, Red Flags, Green Goo, Grey Goo (and Red Herrings)," *Communiqué* 85 (May/June 2004).

7. "To smooth and speed the growth of an effective foresight movement, we can try to develop an effective framework for organizing—call it the 'foresight metagroup,' a group of groups. A metagroup of this sort would help develop organizations to prepare for the coming breakthroughs, and would help build an understanding of nanotechnology and AI by focusing critical discussion. It would help us build a broad movement by fostering diversity." See "What is the Foresight Institute," *Foresight Background* no. 3, 1987, p. 5.

8. Ibid., p. 6.

9. Ibid.

10. Drexler, Peterson, and him.

11. Jamie Dinkelacker, "Interview: Jim Bennett," *Foresight Update* 14 (July 15, 1992): 12.

12. Ibid.

13. Christine Peterson, "Taking action: Making a difference in nanotechnology," *Foresight Update* 17 (December 15, 1993): 17–18.

14. Ibid., p. 14.

15. Ibid., p. 15.

16. Ibid.

17. Ibid.

18. Ibid.

19. Ibid.

20. Ibid.

21. "For Every Book You Buy Online, Amazon Donates to Foresight," *Foresight Update* 29 (August 30, 1998), http://www.foresight.org/Updates/Updates34/update34.1.html (accessed July 26, 2004).

22. Christine Peterson, "Taking action: Making a difference in nanotechnology," *Foresight Update* 17 (December 15, 1993): 15.

23. Dr. Dinkelacker stepped down to become vice president for marketing at the American Information Exchange in 1993. See "New advisers named," 1993, p. 11.

24. Jamie Dinkelacker, "Executive Director notes," *Foresight Update* 14 (July 15, 1992): 14.

25. "Foresight Perspective & Policy," *Foresight Update* 21 (June 1, 1995): 16.

26. "When a reader views any site through the window of a CritLink server, any word, phrase, or text block may be highlighted by diamond brackets of different colors that signify agreement, comment, disagreement, or a query. Clicking on the highlighted phrase allows the reader to view the full comment, which of course may itself contain many links to other comments." See "Foresight Unveils Backlink Mediator to Provide Other Half of the Web," *Foresight Update* 31 (December 15, 1997), http://www.foresight.org/Updates/Updates31/update31.3.html (accessed July 26, 2004).

27. Christine Peterson, "Thank you," *Foresight Update* 33 (May 30, 1998), http://www.foresight.org/Updates/Updates33/update33.5.html (accessed July 26, 2004).

28. "First Feynman Prize in nanotechnology awarded: Caltech chemistry researcher Charles Musgrave receives $5000 prize," *Foresight Update* 17 (December 15, 1993): 1.

29. Lewis Phelps, "Progress toward molecular manufacturing detailed by Foresight Conference speakers," *Foresight Update* 23 (November 30, 1995): 1–2.

30. "Foresight Institute offers $250,000 Feynman Grand Prize for major advances in molecular nanotechnology," *Foresight Update* 24 (April 15, 1996): 1.

31. Biography of Philippe van Nedervelde, 2004, http://www.foresight.org/FI/VanNedervelde.html (accessed July 26, 2004).

32. Christine Peterson, "Inside Foresight: Paradigm Shift in Progress," *Foresight Update* 29 (June 30, 1997), http://www.foresight.org/Updates/Updates29/update29.2.html (accessed July 26, 2004).

33. "About the Institute for Molecular Manufacturing," a flyer, n.d., n.p.

34. "IMM to fund molecular manufacturing," *Foresight Update* 12 (August 1, 1991): 1.

35. Kathleen Shatter recently replaced Steigler as IMM's executive director. Steigler left to pursue an advanced degree. See "New faces...," 1992, p. 15.

36. "New faces at IMM," *Foresight Update* 14 (July 15, 1992): 15.

37. James Bennett, "Funding nanotechnology ourselves," *Foresight Update* 16 (July 1, 1992): 11.

38. Mike Langberg, "Nanotechnology needs vision," *Calgary Herald,* January 7, 1993, p. B7.

39. Drexler, undated solicitation letter, n.p.

40. Gary Stix, "Waiting for breakthroughs," *Scientific American* 274, no. 4 (April 1996): 96.

41. "New advisers named," *Foresight Update* 15 (February 15, 1993), p. 11.

42. "Interview: Nanotechnologist Ralph Merkle: Molecules and Computers—Modeling, Design and Visualization," *Foresight Update* 17 (December 15, 1993): 9.

43. James Bennett, "Funding nanotechnology ourselves," *Foresight Update* 16 (July 1, 1992): 11.

44. Announcement flyer.

45. Ibid.

46. Ibid.

47. James Bennett, "The politics of technology in the United States," *Prospect in*

Nanotechnology: Toward Molecular Manufacturing, ed. Markus Krummenacker and James Lewis (New York: John Wiley & Sons, 1995), p. 229.

48. Jamie Dinkelacker, "Interview: Jim Bennett," *Foresight Update* 14 (July 15, 1992): 13.

49. Ibid.

50. James Bennett, "The politics of technology in the United States," *Prospect in Nanotechnology: Toward Molecular Manufacturing*, Markus Krummenacker & James Lewis, eds. (New York: John Wiley & Sons, Inc., 1995), p. 238.

51. Ibid., p. 237.

52. Sander Olson, "1st Foresight Conference on Advanced Nanotechnology A Success," *NewsNanoApex*, October 31, 2004, http://news.nanoapex.com/mopdules.php??name =News&file=print&sid=5201 (accessed November 1, 2004).

53. It can be downloaded from http://www.nanoengineer-1.com/mambo/index.php ?option=com_content&task=view&id=33&Itemid=2.

54. Foresight Nanotech Institute, "Foresight Nanotech Institute Adopts New Mission," May 23, 2005, http://www.foresight.org/cms/press_center/113 (accessed May 27, 2005).

55. Ibid.

56. Ibid.

57. Cientifica Blog, "A Self Assembled Steak Sandwich and an Intelligent Dessert in Every Home?" http://www.cientifica.com/archives/000586.html (accessed May 27, 2005).

58. "Gateway founder gives $250k to Foresight; Institute developing nano roapmap," *Small Times*, June 21, 2005, http://smalltimes.com/documnet_display.cfm?documnet_id =9401 (accessed July 1, 2005).

59. See http://www.incipientposthuman.com/nyta.htm.

60. "benefits—risks—solution—urgency—politics," http://www.crnano.org/index .html (accessed July 28, 2003).

61. Recently Phoenix coauthored an article with Drexler (Chris Phoenix and Eric Drexler, "Safe exponential manufacturing," *Nanotechnology* 15 [2004]: 870), and Treder has written for *Future Briefs*, which is a publication of new Global Initiatives, Inc., a group out of Bethesda, Maryland.

62. "Inside CRN," http://responsiblenanotechnology.org/inside.htm (accessed July 28, 2003).

63. "CRN Research: Current Results—Bootstrapping a Nanofactory," http:// crnano.org/bootstrap.htm (accessed July 28, 2003).

64. Ibid.

65. "NSF misses the point on nanotechnology," *NewsNanoApex*, September 8, 2004, http://news.nanoapex.com/modules.php?name=News&file=print&sid=4992 (accessed September 20, 2004).

66. Chappell Brown, "Nanofactories a few years away from realization?" *Electronic Engineering Times*, March 21, 2005, p. 46.

67. Chris Phoenix and Mike Treder, "Three Systems of Ethics: A Proposed Application for Effective Administration of Molecular Nanotechnology," October 2003, http:// crnano.org/systems.htm (accessed October 22, 2003).

68. "Patchwork Regulation of Nanotech Could Be Grave Danger," February 15, 2003, http://responsiblenanotechnology.org/PR-patchwork.htm (accessed July 28, 2003).

69. Ibid.

70. "Published Papers," http://responsiblenanotechnology.org/papers.htm (accessed July 28, 2003).

71. "CRN Research: Current Results—Overview," http://responsiblenanotechnology.org/overview.htm. (accessed July 28, 2003).

72. "Managing Magic," http://responsiblenanotechnology.org/magic.htm (accessed July 28, 2003).

73. "CRN Research: Current Results—The Need for Early Development," http://responsiblenanotechnology.org/early.htm (accessed July 28, 2003).

74. Ibid.

75. Ibid.

76. "CRN Research: Current Results—No Simple Solutions," http://responsible nanotechnology.org/solutions.htm (accessed July 28, 2003).

77. "CRN Research: Current Results—A Solution That Balances Many Interests," http://responsiblenanotechnology.org/early.htm. (accessed July 28, 2003).

78. For more information on EMS, see Gregory R. Ganger and David F. Nagle, "Enabling Dynamic Security Management of Networked Systems via Device-Embedded Security," December 2000, http://www.pdl.cmu.edu/PDL-FTP/Storage/CMU-CS-00 -174.pdf (accessed August 19, 2003).

79. "CRN Research: Current Results—A Solution That Balances Many Interests," http://responsiblenanotechnology.org/early.htm. (accessed July 28, 2003).

80. "CRN Research: Current Results—Possible Technical Restrictions," http:// responsiblenanotechnology.org/restrictions.htm (accessed July 28, 2003).

81. "CRN Research: Current Results—A Solution That Balances Many Interests," http://responsiblenanotechnology.org/early.htm (accessed July 28, 2003).

82. "CRN Research: Current Results—Possible Technical Restrictions," http:// responsiblenanotechnology.org/restrictions.htm (accessed July 28, 2003).

83. "CRN Research: Current Results—A Solution That Balances Many Interests," http://responsiblenanotechnology.org/early.htm (accessed July 28, 2003).

84. Ibid.

85. "Issues, Positions—and Urgency—in Nanotechnology Policy," http:// responsiblenanotechonlogy.org/positions.htm (accessed July 28, 2003).

86. "CRN Research: Current Results—Overview," http://responsiblenano technology.org/overview.htm (accessed July 28, 2003).

87. Ibid.

88. Chris Phoenix and Mike Treder, "Applying the Precautionary Principle to Nanotechnology," January 2003, http://responsiblenanotechnology.org/precautionary.htm (accessed July 28, 2003).

89. Ibid.

90. Ibid.

91. "CRN Research: Current Results—Overview," http://responsiblenanotechnology .org/overview.htm (accessed July 28, 2003).

92. Chris Phoenix and Mike Treder, "Applying the Precautionary Principle to Nanotechnology," January 2003, http://responsiblenanotechnology.org/precautionary.htm (accessed July 28, 2003).

93. "CRN Research: Current Results—Overview," http://responsiblenanotechnology.org/overview.htm (accessed July 28, 2003).

94. Ibid.

95. "Managing Magic," http://responsiblenanotechnology.org/magic.htm (accessed July 28, 2003).

96. Ibid.

97. "CRN Research: Current Results—No Simple Solutions," http://responsible-nanotechnology.org/solutions.htm (accessed July 28, 2003).

98. "CRN Research: Current Results—Dangers of Molecular Nanotechnology," http://responsiblenanotechnology.org/dangers.htm (accessed July 28, 2003).

99. "CRN Research: Current Results—The Need for International Development," http://responsiblenanotechnology.org/early.htm (accessed July 28, 2003).

100. "CRN Research: Current Results—Dangers of Molecular Nanotechnology," http://responsiblenanotechnology.org/dangers.htm (accessed July 28, 2003).

101. "CRN Research: Current Results—A Solution That Balances Many Interests," http://responsiblenanotechnology.org/early.htm (accessed July 28, 2003).

102. Blue goo is called that because of the typical color of police uniforms. See Tihamer Toth-Fejel "Grey Goo Begone," discussion, Sci.nanotech Discussion List, available ftp: PLANCHET.RUTGERS.EDU, April 26, 1995. A variation is called security fog. See Peter Merel "Security fog," discussion, Sci.nanotech Discussion List, available ftp: PLANCHET.RUTGERS.EDU, June 14, 1995. Another involved response teams: NERT: Nanotechnology Emergency Response Teams. They are mostly blue goo.

103. "CRN Research: Current Results—The Need for International Development," http://responsiblenanotechnology.org/early.htm (accessed July 28, 2003).

104. Ibid.

105. Mike Treder, "CRN Offers Qualified Endorsement of Greenpeace Nanotech Report," June 30, 2003, http://responsiblenanotechnology.org/PR-Greenpeace.htm (accessed August 17, 2003).

106. "A Technical Commentary on Greenpeace's Nanotechnology Report," September 2003, http://crnano.org/Greenpeace.htm (accessed September 4, 2003).

107. "CRN and Students Team Up to Tackle Nanotechnology Issues," *CRN Press Release*, June 21, 2004, e-mail communication (accessed June 21, 2004).

108. See http://crnano.typepad.com/crnblog/2004/04/transnational_r.html.

109. See http://www.law.asu.edu/Apps/Faculty/Faculty.aspx?Individual_ID=6.

110. Jeremy Smith and Tom Wakeford, "Who's in Control," *Ecologist*, May 22, 2003, http://www.theecologist.org/archive_article.html?article=405 (accessed September 2, 2003).

111. Geoff Brumfield, "A Little Knowledge," *Nature* 424, no. 6946 (July 17, 2003): 246.

112. ETC Comments, *Environmental Futures: Nanotechnology and the Environment*, December 2, 2003, http://www.environmentalfutures.org/Images/nanoetccomments.pdf (accessed July 19, 2005).

113. Waldemar Ingdahl, "Nitpicking nanotechnology," *Tech Center Station*, January 24, 2004, http://www.techcentralstation.com/012704C.html (accessed April 18, 2004).

114. Pat Mooney and Olle Nordberg, "Introduction to the ETC Century," in "The ETC Century: Erosion, technological transformation and corporate concentration in the 21st century," *Development Dialogue* (1999): 5–7.

115. "RAFI Becomes ETC Group," *News Release*, September 4, 2001.

116. Michael Brunton, "Invasion of the nanobots? Critics warn of a 'gray goo' future as nanotechnology enters the marketplace," *Time Europe*, 2003, http://www.time.com/time/europe/magazine/printout/0,13155,901030512-449458,00.html (accessed July 26, 2003).

117. Jasper Gerard, "Charles gets in a wee tizz over nanotechnology," *Sunday Times*, April 27, 2003, p. 17.

118. ETC Group, *The Big Down: From Genomes to Atoms, Atomtech: Technologies Converging at the Nano-Scale*, January, 2003, http://www.etcgroup.org/documents/TheBigDown.pdf (accessed September 5, 2003).

119. "Opposition to nanotechnology," *New York Times*, August 19, 2002, http://www.mainescience.org/news/2002/08h_nanotech.html (accessed July 30, 2003).

120. Ibid.

121. Philip Shropshire, "The Big Letdown," *Betterhumans*, March 3, 2003, http://www.betterhumans.com/Features/Columns/Red_Hour_Orgy/column.aspx?articleID=2003-03-02-4 (accessed July 27, 2004).

122. Pat Mooney, "The ETC Century: Erosion, technological transformation and corporate concentration in the 21st century," *Development Dialogue* (1999): 91.

123. ETC Group, *The Big Down: From Genomes to Atoms, Atomtech: Technologies Converging at the Nano-Scale*, January 2003, http://www.etcgroup.org/documents/TheBigDown.pdf (accessed September 5, 2003), p. 31.

124. Pat Mooney, "The ETC Century: Erosion, technological transformation and corporate concentration in the 21st century," *Development Dialogue* (1999): 5–7.

125. A, C, T , G, and now "F" thus increase the potential diversity (or destructiveness) of life. As evidence, they refer experiments on proteins. For example, Rice researchers have been studying F-active, a filamentous protein that might serve as biosensor; its conductivity might make it a possible alternative to silicon nanowires. See ETC Group, *The Big Down: From Genomes to Atoms, Atomtech: Technologies Converging at the Nano-Scale*, January 2003, http://www.etcgroup.org/documents/TheBigDown.pdf (accessed September 5, 2003), p. 30.

126. "Nanotech un-gooed! Is the grey/green goo brouhaha the industry's second blunder," *ETC Group Communique* 80 (July/August 2003), http://www.etcgroup.org/article.asp?newsid=399 (accessed May 21, 2003).

127. Philip Shropshire, "The Big Letdown," *Betterhumans*, March 3, 2003, http://www.betterhumans.com/Features/Columns/Red_Hour_Orgy/column.aspx?articleID=2003-03-02-4 (accessed July 27, 2004).

128. ETC Group, *The Big Down: From Genomes to Atoms, Atomtech: Technologies Converging at the Nano-Scale*, January 2003, http://www.etcgroup.org/documents/TheBigDown.pdf (accessed September 5, 2003), p. 72.

129. Ibid., p. 6.

130. Waldemar Ingdahl, "Nitpicking nanotechnology," *Tech Center Station*, January 24, 2004, http://www.techcentralstation.com/012704C.html (accessed April 18, 2004).

131. "No small matter," *Red Herring*, December 16, 2002, http://www.redherring.com/Article.aspx?f=articles%2farchive%2finsider%2f2002%2f12%2f (accessed October 5, 2003).

132. Ibid.

133. ETC Group, "Nanotech and the Precautionary Prince," *Genotype*, May 2, 2003, http://www.etcgroup.org/article.asp?newsid=397 (accessed May 21, 2003).

134. ETC Group, *The Big Down: From Genomes to Atoms, Atomtech: Technologies Converging at the Nano-Scale*, January 2003, http://www.etcgroup.org/documents/TheBigDown.pdf (accessed September 5, 2003), p. 71.

135. Timothy Mullaney and Specer Ante, "Info Wars," *Business Week*, June 5, 2000, p. 107.

136. Pat Mooney, "The ETC Century: Erosion, technological transformation and corporate concentration in the 21st century," *Development Dialogue* (1999): 80–81.

137. ETC Group, *The Big Down: From Genomes to Atoms, Atomtech: Technologies Converging at the Nano-Scale*, January 2003, http://www.etcgroup.org/documents/TheBigDown.pdf (accessed September 5, 2003), p. 67.

138. Pat Mooney, "The ETC Century: Erosion, technological transformation and corporate concentration in the 21st century," *Development Dialogue* (1999): 30–36.

139. Only a small portion of *The ETC Century* examines nanotechnology. See Pat Mooney, "The ETC Century: Erosion, Technological Transformation and Corporate Concetration in the 21st Century," *Development Dialogue* (1999): 43–52.

140. Ibid., p. 45.

141. Ibid., p. 47.

142. Ibid., p. 52.

143. Ibid., p. 110.

144. Ibid., p. 118.

145. "Nanotech un-gooed! Is the grey/green goo prouhaha the industry's second blunder," *ETC Group Communiqué* 80 (July/August 2003), http://www.etcgroup.org/article.asp?newsid=399 (accessed May 21, 2003).

146. ETC Group, "Nanotech News in Living Colour: An Update on White Papers, Red Flags, Green Goo, Grey Goo (and Red Herrings)," *Communiqué* 85 (May/June 2004): 1.

147. Ibid.

148. Ibid., p. 2.

149. Ibid.

150. See http://www.etcgroup.org/documents/nanoproducts_EPA.pdf.

151. ETC Group, "Nanotech News in Living Colour: An Update on White Papers, Red Flags, Green Goo, Grey Goo (and Red Herrings)," *Communiqué* 85 (May/June 2004): 2.

152. Lee-Anne Broadhead and Sean Howard, "The heard of darkness," *Resurgence* 221 (November/December 2003), http://resurgence.gn.apc.org/issues/broadhead221.htm (accessed November 4, 2003).

153. Pat Mooney, "The ETC Century: Erosion, Technological Transformation and Corporate Concetration in the 21st Century," *Development Dialogue* (1999): 29.

154. See http://www.etcgroup.org/article.asp?newsid=486.

155. See http://www.etcgroup.org/article.asp?newsid=509.

156. Ole PeterGalaasen, "Nanotechnology: The regulation and oversight of nanotechnology," *Plausible Futures*, March 4, 2005, http://www.plausiblefutures.com/index.php?id=232445&cat=5911&printable=1 (accessed March 31, 2005).

157. Glenn Reynolds, "Inside Foresight: Greenpeace and Nanotechnology," *Foresight Update* 52 (August 31, 2003), http://www.foresight.org/Updates/Updates52/update52.2.html (accessed July 26, 2004).

158. Ibid.

159. Ibid.

160. Howard Lovy, "The Greenpeace Report, Part II: NanoWars," *Howard Lovy's NanoBot*, July 25, 2003, http://nanobot.blogspot.com/2003_07_01_nanobot_archive.html #105905157013774164 (accessed July 27, 2004).

161. Bruce Goldfarb, "New task force focuses on nanotechnology health effects," *NanoBiotech News*, August 13, 2003, p. 3.

162. Mark Modzelewski, "The Greenpeace Report, Part II: NanoWars," *Howard Lovy's NanoBot*, July 25, 2003, http://nanobot.blogspot.com/2003_07_01_nanobot_archive.html #105905157013774164 (accessed July 27, 2004).

163. James M. Pethokoukis, "Turning green over nanotech," *USNEWS.com*, July 31, 2003, http://www.usnews.com/usnews/nycu/tech/nextnews/archive/next030731.htm (accessed August 7, 2003).

164. Doug Parr, foreward to *Future Technologies, Today's Choices*. Its secondary title is more enlightening: *Nanotechnology, Artificial Intelligence and Robotics; A technical, political and institutional map of emerging technologies* (London: Greenpeace Environmental Trust, July 2003), p. 5.

165. Ibid., pp. 5–6.

166. Ibid.

167. Glenn Reynolds, "Greenpeace and Nanotechnology," *Tech Central Station*, July 30, 2003, http://www.techcentralstation.com/073003B.html (accessed September 2, 2003).

168. "Small science, big questions," *SNH.com.au*, November 27, 2003, http://www.smh.com.au/articles/2003/11/16/10768917669906.html (accessed November 25, 2003).

169. Clive Cookson, "Nanotech safety to be reviewed," *Financial Times*, February 26, 2005, http://news.ft.com/home/us (accessed July 19, 2005).

CHAPTER 9

1. Glenda Chui, "Who's Afraid of Nanotechnology," *Mercury News*, September 16, 2003, http://www.bayarea.com/mld/mercurynews/living/health/6783577.htm (accessed October 2, 2003).

2. William G. Schulz, "Nanotechnology under the Scope," *Chemical and Engineering News*, December 4, 2002, http://pubs.acs.org/cen/today/dec4.html (accessed October 5, 2003).

3. Alan Leo, "Get ready for your nano future," *Technology Review*, May 4, 2001, http://www.technologyreview.com/articleswo_leo050401.asp (accessed October 5, 2003).

4. Kevin Maney, "Scared of nano-pants? Hey, you may be onto something," *USA Today*, June 21, 2005, http://www.usatoday.com/tech/columnist/kevinmaney/2005-06-21 -nano-pants_x.htm (accessed July 14, 2005).

5. Pat Mooney, "Nanotechnology: Nature's tool box," *ABC Radio National*, November 14, 2004, http://www.abc.net.au/rn/talks/bbing/stories/s1241931.htm (accessed December 13, 2004).

6. J. Bradford DeLong, "Semi-daily journal of economist Brad DeLong: Fair and balanced almost every day," December 3, 2003, http://www.j-bradford-delong.net .movable_type/2003_archives/002838.html (accessed December 8, 2003).

7. Ibid.

8. Michael J. Mauboussin and Kristen Bartholdson, "Big Money in Thinking Small: Nanotechnology—What Investors Need to Know" (Credit Suisse Equity Research/First Boston, May 7, 2003), p. 13.

9. ETC Group, "Nanotech News in Living Colour: An Update on White Papers, Red Flags, Green Goo, Grey Goo (and Red Herrings)," *Communiqué* 85 (May/June 2004): 3.

10. Wei Zhou, "Ethics of nanobiology at the frontline," *Santa Clara Computer and High Technology Law Review* 19 (2003): 486.

11. "Professor calls for a new branch of learning," *Eureka Alert!* August 30, 2004, http://www.eurekalert.org/pub_releases?2004-08/uoe-pcf083004.php (accessed September 7, 2004).

12. Carolyn Aldred, "Nanotechnology raises concerns about health risks," *Crain's Detroit Business*, November 9, 2004, e-communication.

13. "Nanotech advances need more safety screening-study," *ABC News,* June 14, 2005, http://abcnews.go.com/Technology/print?id=848241 (accessed June 15, 2005).

14. Candace Stuart, "Worker safety rises to the fore," *Small Times* (January/February 2005): 18–20.

15. "Nanotechnology to revolutionize drug delivery," *In-Pharma Technologist.com,* July 3, 2005, e-communication.

16. ETC Group, *The Big Down: From Genomes to Atoms, Atomtech: Technologies Converging at the Nano-Scale*, January 2003, http://www.etcgroup.org/documents/TheBigDown.pdf (accessed September 5, 2003): 25.

17. S. Parkinson, "Nanotechnology," 2003, http://www.nanotec.org.uk/evidence/76aScientistsforGlobalResponsibility.htm (accessed March 30, 2004).

18. Juliana Gruenwald, "Controversial study points to need for more federal research," *Small Times*, April 1, 2004, http://www.smalltimes.com/document_display.cfm?document_id=7658 (accessed April 5, 2004).

19. ETC Group, "Nanotech News in Living Colour: An Update on White Papers, Red Flags, Green Goo, Grey Goo (and Red Herrings)," *Communiqué* 85 (May/June 2004): 5.

20. Steve Crosby, "Study nano and the environment now, because what we can't see scares us," *Small Times* (January/February 2003): 4.

21. Juliana Gruenwald, "Controversial study points to need for more federal research," *Small Times*, April 1, 2004 http://www.smalltimes.com/document_display.cfm?document_id=7658 (accessed April 5, 2004).

22. Peter H. M. Hoet, Abderrahim Nemmar, and Benoit Nemery, "Health impact of nanomaterials," *Nature* 22, no. 1 (January 2004): 19, http://www.nature.com/cgi-taf/DynaPage.taf?file=/nbt/journal/v22/n1/full/nbt0104-19.html (accessed January 12, 2004).

23. Jeffrey M. Perkel, "The Ups and Downs of Nanobiotech," *Scientist*, August 30, 2004, p. 17.

24. ETC Group, "No Small Matter! Nanotech Particles Penetrate Living Cells and Accumulate in Animal Organs," *Communiqué* 76 (May/June 2002), July 23, 2002, http://www.etcgroup.org/article.asp?newsid=356 (accessed May 21, 2003). There was a peculiar press release that claimed "an exception to the conventional scientific notion that objects small enough to be measured in nanometers behave according to different rules than larger objects." See "Is Small Different? Not Necessarily Say George Tech Researchers," *News-*

NanoApex, July 13, 2004. The researchers were studying compression of a carbon nanospring. See M. Poggi et al., "Measuring the Compression of a Carbon Nanospring," *NanoLetters* 4, no. 6 (2004): 1009–16. The release adds: "The findings suggest there may be other nano materials that behave in ways similar to their macroscale counterparts." The researchers draw no such conclusion relating to health and environmental concerns.

25. National Science and Technology Council, *Nanotechnology: Shaping the World Atom by Atom* (Washington, DC: U.S. GPO, 1999), p. 2.

26. Ibid., p. 5.

27. Ibid.

28. A. Nemmar, M. Hoet, B. Vanquickenbone, D. Dinsdale, M. Thomeer, M. Hoy-laerts, H. Vanbilloen, L. Mortelmans, and B. Nemery, "Passage of inhaled particle into the blood circulation of humans," *Circulation* 105, no. 411 (2002), http://circ.ahajournals .org/cgi/content/full/105/4/411 (accessed February 11, 2004). This finding was also found by G. Oberdörster, "Pulmonary effect of inhaled ultrafine particles," *International Archives of Occupational and Environmental Health* 74 (2001): 1–8, and K. Donaldson, V. Stone, A. Seaton et al, *Environmental Health Perspectives* 109, supp. 4 (2001): 523–27.

29. G. Oberdörster, J. Ferin, and B. Lehnert, "Correlation between particle size, in vivo particle persistence, and lung injury," *Environmental Health Perspectives* 102, supp. 5 (1994): 173–79.

30. Steve Jurvetson, "On the record: Nanotechnology—Unlocking the smallest secrets," *San Francisco Chronicle*, February 1, 2004, http://sfgate.com/cgi-bin/article.cgi?file =/chronicle/archive/2004/02/01/BUGCJ4J2761.DTL&type=business (accessed July 11, 2004).

31. Vicki Colvin, "The potential environmental impact of engineered nanomate-rials," *Nature Biotechnology* 21, no. 10 (October 2003), http://www.nature.com/cgi-af/Dyna Page.taf?file=/nbt/journal/v21/n10/full/nbt875.html (accessed March 9, 2004): 1167.

32. Kristen Kulinowski, personal communication, August 28, 2004.

33. Ernie Hood, "Nanotechnology: Looking As We Leap," *Environmental Health Per-spectives* 112, no. 13 (September 2004): A745.

34. Jessica Gorman, "Taming high-tech particles: Cautious steps into the nanotech future," *Science News* 161, no. 13 (March 30, 2002), http://www.sciencenews.org/20020330/ bob8.asp (accessed October 5, 2003).

35. Vicki Colvin, "Responsible nanotechnology: Looking beyond the good news," *Eureka Alert! In Context*, November 1, 2002, http://www.eurekaalert.org/context.php?context =nano&show=essays&essaydate=1102 (accessed October 5, 2003).

36. Vicki Colvin in Susan Morrissey, "Paving the Way for Nanotech," *Chemical and Engineering News* 82, no. 24, June 8, 2004, http://pubs.acs.org/cen/news/8223/8223earlygov .html (accessed June 15, 2004).

37. Catherine Clabby, "Area has role in nano-testing," *NewsObserver.Com*, May 23, 2005, http://newsobserver.com/news/story/2436846p-8841418c.html (accessed May 24, 2005).

38. Carl Wherrett and John Yelovich, "Commercializing nanotechnology," *Motley Fool*, March 2, 2004, http://www.fool.com/Server/FoolPrint.sdp?File=/ news/commen-tary/2—4/commentary040 (accessed March 18, 2004).

39. Vicki Colvin, "The potential environmental impact of engineered nanoma-

terials," *Nature Biotechnology* 21, no. 10 (October 2003), http://www.nature.com/cgi-af/Dyna Page.taf?file=/nbt/journal/v21/n10/full/nbt875.html (accessed March 9, 2004): 1166.

40. ETC Group, "Green goo: Nanobiotechnology comes alive," *Communiqué* 77 (January/February 2003): 2.

41. ETC Group, "No small matter II: The case for a global moratorium," *Occasional Paper Series* 7, no. 1 (April 2003): 2.

42. Ibid., p. 3.

43. Ibid.

44. Vicki Colvin in Susan Morrissey, "Paving the Way for Nanotech," *Chemical and Engineering News* 82, no. 24, June 8, 2004, http://pubs.acs.org/cen/news/8223/8223earlygov .html (accessed June 15, 2004).

45. "What will the future nanotech industry look like," *NanoMarkets White Paper*, March 2004, p. 6.

46. Eric Berger, "Study raises concerns over buckyballs," *Houston Chronicle*, March 29, 2004, http://www.chron.com/cs/CDA/printstory.mpl/health/244472494 (accessed April 5, 2004).

47. Business Communications Company, press release, February 3, 2003, http:// www .bccresearch.com/editors/RGB-245R.html (accessed May 1, 2004).

48. "Nanotech un-gooed! Is the grey/green goo brouhaha the industry's second blunder," *ETC Group Communiqué* 80 (July/August 2003), http://www.etcgroup.org/ article.asp?newsid=399 (accessed May 21, 2003).

49. "Nanotubes: Directions and Technologies," a press release of the Business Communications Company, Inc., February 2003, http://www.bccresearch.com/editors/RGB-245R.html (accessed August 17, 2004).

50. Eric Berger, "It's going to be a giant business," *Houston Chronicle*, March 4, 2004, http://www.chron.com/cs/CDA/printstory.mpl/business/2432521 (accessed March 18, 2004).

51. Ibid.

52. Ibid.

53. Charles Choi, "Nanotubes, buckyballs surprise investors," *United Press International*, May 19, 2004, http://www.upi.com/view.cfm?StoryID=20040518-011947-3488r (accessed May 25, 2004).

54. ETC Group, "No Small Matter! Nanotech Particles Penetrate Living Cells and Accumulate in Animal Organs," *Communiqué* 76 (May/June 2002), http://www.etcgroup .org/article.asp?newsid=356 (accessed May 21, 2003).

55. Vicki Colvin, "The potential environmental impact of engineered nanomaterials," *Nature Biotechnology* 21, no. 10 (October 2003), http://www.nature.com/cgi-af/Dyna Page.taf?file=/nbt/journal/v21/n10/full/nbt875.html (accessed March 9, 2004): 1167.

56. James Wilsdon and Rebecca Willis, "Will nanotechnology go the GM way?" *Hindu*, September 9, 2004, http://www.thehindu.com/thehindu/seta/2004/09/09/stories/ 2004090900031400.htm (accessed September 22, 2004).

57. Barnaby Feder, "As uses grow, tiny materials' safety is hard to pin down," *Silicon-Investor*, November 3, 2003, http://www.siliconinvestor.com/stocktalk/msg.gsp?msgid =19459234 (accessed December 9, 2003).

58. Vicki Colvin, "The potential environmental impact of engineered nanomate-

rials," *Nature Biotechnology* 21, no. 10 (October 2003), http://www.nature.com/cgi-af/Dyna Page.taf?file=/nbt/journal/v21/n10/full/nbt875.html (accessed March 9, 2004): 1167.

59. Andrew Bridges, "Tiniest particles of matter don't behave, raising nanotech concerns," *USA Today*, April 2, 2004, http://www.usatoday.com/tech/news/2004-04-02-nano -flaw_x.htm (accessed April 5, 2004).

60. Jessica Gorman, "Taming High-Tech Particles: Cautious steps into the nanotech future," *Science News Online* 161, no. 13 (March 30, 2002), http://www.sciencenews.org/articles/ 20020330/bob8.asp (accessed July 18, 2005).

61. Geoff Brumfiel, "A Little Knowledge," *Nature* 424, no. 6946 (July 17, 2003): 246–47.

62. Helene F. Lecoanet and Mark R. Wiesner, "Velocity effects on Fullerene and Oxide Nanoparticle Deposition in Porous Media," *Environmental Science and Technology* 38 (2004): 4377–82.

63. J. D. Fortner, D. Y. Lyon, C. M. Sayes, et al., "C_{60} in Water: Nanocrystal Formation and Microbial Response," *Environmental Science and Technology*, April 28, 2005, http:// pubs.acs.org/cgi-bin/jcen?esthag/asap/html/es048099n.html (accessed June 3, 2005).

64. Helene F. Lecoanet and Mark R. Wiesner, "Velocity Effects on Fullerene and Oxide Nanoparticle Deposition in Porous Media," *Environmental Science and Technology* 38 (2004): 4377–82.

65. Alexander Huw Arnall, *Future Technologies, Today's Choices*. Its secondary title is more enlightening: *Nanotechnology, Artificial Intelligence and Robotics; A technical, political and institutional map of emerging technologies* (London: Greenpeace Environmental Trust, July 2003), p. 36.

66. Yuichi Shibata, "How safe is nanotechnology," *Daily Yomiuri*, November 28, 2003, http://www.yomiuri.co.jp/newse/20031128wo71.htm (accessed December 2, 2003).

67. Doug Brown, "Nano Litterbugs? Experts See Potential Pollution Problems," *Small Times* (March/April 2002), http://www.smalltimes.com/document_display.cfm ?document_id=3266 (accessed July 30, 2003).

68. Ibid.

69. Alexander Huw Arnall, *Future Technologies, Today's Choices*. Its secondary title is more enlightening: *Nanotechnology, Artificial Intelligence and Robotics; A technical, political and institutional map of emerging technologies* (London: Greenpeace Environmental Trust, July 2003), p. 36.

70. ETC Group, "Nanotech News in Living Colour: An Update on White Papers, Red Flags, Green Goo, Grey Goo (and Red Herrings)," *Communiqué* 85 (May/June 2004): 5.

71. Andrew Bridges, "Tiniest particles of matter don't behave, raising nanotech concerns," *USA Today*, April 2, 2004, http://www.usatoday.com/tech/news/2004-04-02-nano -flaw_x.htm (accessed April 5, 2004).

72. C. Sayes, J. D. Fortner, W. Guo, W., et al., "The differential cytotoxicity of water-soluble fullerenes," *Nano Letters* 4, no. 10 (2004): 1881–87.

73. J. D. Fortner, D. Y. Lyon, C. M. Sayes, et al., "C_{60} in Water: Nanocrystal Formation and Microbial Response," *Environmental Science and Technology*, April 28, 2005, http:// pubs.acs.org/cgi-bin/jcen?esthag/asap/html/es048099n.html (accessed June 3, 2005).

74. Ibid.

75. Vicki Colvin, "Responsible nanotechnology: Looking beyond the good news,"

Eureka Alert! In Context, November 1, 2002, http://www.eurekaalert.org/context.php?context =nano&show=essays&essaydate=1102 (accessed October 5, 2003).

76. Geoff Brumfiel, "Nanotechnology: A Little Knowledge," *Nature* 424, no. 6946 (July 17, 2003): 247.

77. Alexander Huw Arnall, *Future Technologies, Today's Choices.* Its secondary title is more enlightening: *Nanotechnology, Artificial Intelligence and Robotics; A technical, political and institutional map of emerging technologies* (London: Greenpeace Environmental Trust, July 2003), p. 40.

78. Josh Wolfe, "Nanotech vs. environmentalists," *Forbes,* September 16, 2003, http://www.forbes.com/2003/09/16/cz_jw_0916soapbox.html (accessed October 17, 2003).

79. Ibid.

80. K. Smith, S. Kim, J. Recendez, et al., "Airborne particles of the California Central Valley alter the lungs of healthy adult rats," *Environmental Health Perspectives* 111, no. 7 (June 2003): 902–907

81. S. Gavett, N. Haykal-Coates, L. Copeland, J. Heinrich, and M. Gilmour, "Metal composition of ambient P[M. sub. 2.5] influences severity of allergic airways disease in mice," *Environmental Health Perspectives* 111, no. 12 (September 2003): 1471–78.

82. B. Li, C. Sioutas, A. Cho, et al., "Ultrafine particulate pollutants induce oxidative stress and mitochondrial Damage," *Environmental Health Perspectives* 114, no. 4 (April 2003): 455–61.

83. R. Vincent, P. Kumarathasan, G. Goegan, et al., "Inhalation toxicology of urban ambient particulate matter: Acute cardiovascular effects in rats," Respiratory Response Health Effects Institute Statement, Synopsis of Research Report 104, October 2002: 5–54 and 55–62, and G. Wellenius, B. Coull, J. Godleski, et al., "Inhalation of concentrated ambient air particles exacerbates myocardial ischemic in conscious dogs," *Environmental Health Perspectives* 111, no. 4 (April 2003): 402–409.

84. L. Calderon-Garcidernas, B. Azzarelli, H. Acuna, et al., "Air pollution and brain damage," *Toxicologic Pathology* 30, no. 30 (2002): 373–89.

85. Jocelyn Kaiser, "Mounting Evidence Indicts Fine-Particle Pollution," *Science,* 307–5717, March 25, 2005, pp. 1858–61, http://www.sciencemag.org/cgi/content/full/307/5717/1858a (accessed May 17, 2005).

86. Geoffrey Lean, "Hundreds of firms using nanotech in food," *Independent.co.uk,* July 18, 2004, http://news.independent.co.uk/uk/environment/story.jsp?story=542140 (accessed July 19, 2004).

87. Doug Brown, "Nano Litterbugs? Experts See Potential Pollution Problems," *Small Times* (March/April 2002), http://www.smalltimes.com/document_display.cfm ?document_id=3266 (accessed July 30, 2003).

88. Jessica Gorman, "Taming high-tech particles: Cautious steps into the nanotech future," *Science News* 161, no. 13 (March 30, 2002), http://www.sciencenews.org/20020330/bob8.asp (accessed October 5, 2003).

89. Rick Weiss, "Some see solution, some potential dangers," *Washington Post,* February 10, 2004, http://newsday.com/news/printedition/health/ny-dsspdn3663472feb10 ,0,5417935,pri (accessed February 16, 2004).

90. Kristen Kulinowski, personal communication, August 28, 2004.

91. Peter H. M. Hoet, Abderrahim Nemmar, and Benoit Nemery, "Health impact of nanomaterials," *Nature* 22, no. 1 (January 2004): 19, http://www.nature.com/cgi-taf/Dyna Page.taf?file=/nbt/journal/v22/n1/full/nbt0104-19.html (accessed January 12, 2004).

92. David B. Warheit, Brett R. Lawrence, Kenneth L. Reed, and Thomas R. Webb, "Comparative pulmonary toxicity assessment of single-wall carbon nanotubes in rats," *Toxicological Sciences* 77, no. 1 (2004): 117.

93. Barnaby Feder, "As uses grow, tiny materials' safety is hard to pin down," *Silicon-Investor*, November 3, 2003, http://www.siliconinvestor.com/stocktalk/msg.gsp?msgid =19459234 (accessed December 9, 2003).

94. A. Nemmar, M. Hoet, B. Vanquickenbone, D. Dinsdale, M. Thomeer, M. Hoylaerts, H. Vanbilloen, L. Mortelmans, and B. Nemery, "Passage of inhaled particle into the blood circulation of humans," *Circulation* 105, no. 411 (2002), http://circ.ahajournals.org/ cgi/content/full/105/4/411 (accessed February 11, 2004).

95. Ibid.

96. A. Nemmar, M. Hoylaerts, P. Hoet, D. Dinsdale, T. Smith, H. Xu, J. Vermylen, and B. Nemery, "Ultrafine particles affect experimental thrombosis in an in vivo hamster model," *American Journal of Respiratory and Critical Care Medicine* 166 (2002): 998–1004, http://ajrccm.atsjournals.org/cgi/content/full/166/7/998 (accessed June 10, 2004).

97. A. Nemmar, M. Hoylaerts, P. Hoet, J. Vermylen, and B. Nemery, "Size effect of intratracheally instilled particles on pulmonary inflammation and vascular thrombosis," *Toxicology and Applied Pharmacology* 1 (January 2003): 38–45.

98. G. Oberdörster, Z. Sharp, V. Atudorei, A. Elder, R. Gelein, A. Lunts, W. Kreyland, and C. Cox, "Extrapulmonary translocation of ultrafine carbon particles following whole-body inhalation exposure of rats," *Journal of Toxicology and Environmental Health, Part A* 65 (2002): 1532.

99. J. Schwartz, "Air pollution and daily mortality: A review and meta-analysis," *Environmental Research* 64 (1994): 36–52, and D. Bates, "Health indicates of the adverse effects of air pollution: The question of coherence," *Environmental Research* 59 (1992): 226–349.

100. A. Seaton, W. MacNee, K. Donaldson, and D. Godden, "Particular air pollution and acute health effects," *Lancet* 345 (January 1995): 176.

101. Yihong Zhang, Theodore C. Lee, Benedicte Guillemin, Ming-Chih Yu, and William N. Rom, "Enhances IL-1, and Tumor Necrosis Factor-Release and Messenger RNA Expression in Macrophages from Idiophatic Pulmonary Fibrosis and after Asbestos Exposure," *Journal of Immunology* 150, no. 9 (May 1, 1993): 4188–96.

102. ETC Group, "No Small Matter! Nanotech Particles Penetrate Living Cells and Accumulate in Animal Organs," *Communiqué* 76 (May/June 2002).

103. Andrzej Hucko, Hubert Lange, Ewa Calko, Hanna Grubek-Jaworska, and Pawel Droszcz, "Physiological Testing of Carbon Nanotubes: Are They Asbestos Like," *Fullerene Science and Technology* 9, no. 2 (2002): 253.

104. Jessica Gorman, "Taming High-Tech Particles," *Science News* 61, no. 13 (March 30, 2002), http://www.sciencenews.org/20020330/bob8.asp (accessed July 30, 2003).

105. ETC Group, *The Big Down: From Genomes to Atoms, Atomtech: Technologies Converging at the Nano-Scale*, January 2003, http://www.etcgroup.org/documents/TheBigDown.pdf. (accessed September 5, 2003): 25.

106. S. Takenaka, E. Karg, C. Roth, A. Schulz, U. Heinzmann, P. Schramel, and J. Heyder, "Pulmonary and systemic distribution of inhaled intrafine silver particle in rats," 109, supp. 4 (August 2001): 547.

107. A phagocyte is a cell, such as a white blood cell, that engulfs and absorbs waste material, harmful microorganisms, or other foreign bodies in the bloodstream and tissues.

108. A macrophage is any of the large phagocytic cells of the reticuloendothelial system.

109. S. Takenaka, E. Karg, C. Roth, A. Schulz, U. Heinzmann, P. Schramel, and J. Heyder, "Pulmonary and systemic distribution of inhaled intrafine silver particle in rats," *Environmental Health Perspectives Supplements* 109, no. 4 (August 2001): 550.

110. Chiu-Wing Lam, John T. James, Richard McCluskey, and Robert L. Hunter, "Pulmonary toxicity of single-wall carbon nanotubes in mice 7 and 90 days after intratracheal instillation," *Toxicological Sciences* 77, no. 1 (2004): 126–34.

111. Sarah Graham, "Nanotech: It's not easy being green," *Scientific American.com*, July 28, 2003, http://www.sciam.com/print_version.cfm?articlelID=00077C33-511E-1F20-B8E 780A8418 (accessed March 18, 2004).

112. Geoff Brumfiel, "Nanotechnology: A Little Knowledge," *Nature* 424, no. 6946 (July 17, 2003): 247.

113. Rick Weiss, "Some see solution, some potential dangers," *Washington Post*, February 10, 2004, http://newsday.com/news/printedition/health/ny-dsspdn3663472feb 10,0,5417935,pri (accessed February 16, 2004).

114. Chiu-Wing Lam, John T. James, Richard McCluskey, and Robert L. Hunter, "Pulmonary toxicity of single-wall carbon nanotubes in mice 7 and 90 days after intratracheal instillation," *Toxicological Sciences* 77, no. 1 (2004): 131.

115. Chiu-Wing Lam, John T. James, J. N. Latch, R. F. Hamilton, and A. Holian. "Pulmonary toxicity of simulated lunar and Martian dusts in mice. II. Biomarkers of acute responses after intratracheal instillation," *Inhalation Toxicology* 14 (2002): 917–28.

116. Chiu-Wing Lam, John T. James, Richard McCluskey, and Robert L. Hunter, "Histopathological study of single-walled carbon nanotubes in mince 7 and 90 days after instillation into the lungs," 225th ACS National Meeting, New Orleans, LA, March 23–27, 2003, http://oasys2.confex.com/acs/225nm/techprogram/P595657.htm (accessed December 1, 2003).

117. Chiu-Wing Lam, John T. James, Richard McCluskey, and Robert L. Hunter, "Pulmonary toxicity of single-wall carbon nanotubes in mice 7 and 90 days after intratracheal instillation," *Toxicological Sciences* 77, no. 1 (2004): 133.

118. David B. Warheit, Brett R. Lawrence, Kenneth L. Reed, and Thomas R. Webb, "Comparative pulmonary toxicity assessment of single-wall carbon nanotubes in rats," *Toxicological Sciences* 77, no. 1 (2004): 117–25.

119. David B. Warheit, Brett R. Lawrence, Kenneth L. Reed, and Thomas R. Webb, "Pulmonary-toxicity-screening studies with single-wall carbon nanotubes," the 225th ACS National Meeting New Orleans, LA, March 23–27, 2003, http://oasys2.confex.com/acs/ 225nm/techprogram/P595346.htm (accessed December 1, 2003).

120. Rick Weiss, "Some see solution, some potential dangers," *Washington Post*, February 10, 2004, http://newsday.com/news/printedition/health/ny-dsspdn3663472feb10 ,0,5417935,pri (accessed February 16, 2004).

121. Barnaby Feder, "As uses grow, tiny materials' safety is hard to pin down," *Silicon Investor*, November 3, 2003, http://www.siliconinvestor.com/stocktalk/msg.gsp?msgid=19459234 (accessed December 9, 2003).

122. Sarah Graham, "Nanotech: It's not easy being green," *Scientific American.com*, July 28, 2003, http://www.sciam.com/print_version.cfm?articleIID=00077C33-511E-1F20-B8E780A8418 (accessed March 18, 2004).

123. Barnaby Feder, "As uses grow, tiny materials' safety is hard to pin down," *Silicon Investor*, November 3, 2003, http://www.siliconinvestor.com/stocktalk/msg.gsp?msgid=19459234 (accessed December 9, 2003).

124. David B. Warheit, Brett R. Lawrence, Kenneth L. Reed, and Thomas R. Webb, "Comparative pulmonary toxicity assessment of single-wall carbon nanotubes in rats," *Toxicological Sciences* 77, no. 1 (2004): 124.

125. Gunter Oberdörster, "Effects and fate of inhaled ultrafine particles," 225th ACS National Meeting, New Orleans, LA, March 23–27, 2003, http://oasys2.confex.com/acs/225nm/techprogram/P598970.htm (accessed December 1, 2003).

126. Gunter Oberdörster, "Emerging concepts in nanoparticle toxicology," EC's Community Health and Consumer Protection Directorate, *Nanotechnologies: A Preliminary Risk Analysis on the Basis of a Workshop Organized in Brussels on 1–2 March 2004*, 2004, http://europa.eu.int/comm/health/ph_risk/events_risk_en.htm. (accessed June 20, 2004), p. 115.

127. Ibid., p. 116.

128. Ibid., p. 116.

129. "Trouble in Nanoland," *Economist*, December 5, 2002, http://www.economist.com/science/PrinterFriendly.cfm?Story_IS=1477445 (accessed October 5, 2003).

130. Peter H. M. Hoet, Abderrahim Nemmar, and Benoit Nemery, "Health impact of nanomaterials," *Nature* 22, no. 1 (January 2004): 19, http://www.nature.com/cgi-taf/DynaPage.taf?file=/nbt/journal/v22/n1/full/nbt0104-19.html (accessed January 12, 2004).

131. Michael Brunton, "Little worries: Invasion of the nanobots," *Time Europe* 161, no. 19, May 12, 2003, http://www.time.com/time/europe/magazine/article/0,13005,901030512-449458,00.html (accessed June 21, 2004).

132. R. Dunford, A. Salinaro, L. Cai, N. Serpone, S. Horikoshi, H. Hidaka, and J. Knowld, "Chemical Oxidation and DNA Damage Catalyzed by Inorganic Sunscreen Ingredients," *FEBS (Federation of European Biochemical Societies) Letters* 418 (1887): 1–2, 87–90.

133. A. Huczko and H. Lange, "Carbon nanotubes: Experimental evidence for a null risk of skin irritation and allergy," *Fullerene Science and Technology* 9, no. 2 (2001): 247–50.

134. Anne Shvedova, Vincent Castronova, Elena Kisin, Diane Schwegler-Berry, Asley Murray, Vadim Gandelsman, Andrew Maynard, and Paul Baron, "Exposure to Carbon Nanotubes Material: Assessment of Nanotube Cytotoxicity Using Human Keratinocyte Cells," *Journal of Toxicology and Environmental Health, Part A* 66 (2003): 1909–26.

135. Nancy A. Monteiro-Riviere, Robert J. Nemanich, Alfred O. Inman, Yunyu Y. Wang, and Jim E. Riviere, "Multi-walled carbon nanotube interactions with human epidermal keratinocytes," *Toxicology Letters* 163 (2005): 377–84.

136. Andrew Maynard, Paul Baron, Michael Foley, Anna Shvedova, Elena Kisin, and Vincent Castranova, "Exposure to Carbon Nanotube Material: Aerosol Release during the Handling of Unrefined Single-Walled Carbon Nanotube Material," *Journal of Toxicology and Environmental Health, Part A* 67 (2004): 87–107.

137. Vicki Colvin, "Responsible nanotechnology: Looking beyond the good news," *Eureka Alert! In Context*, November 1, 2002, http://www.eurekaalert.org/context.php?context =nano&show=essays&essaydate=1102 (accessed October 5, 2003).

138. Vicki Colvin, "The potential environmental impact of engineered nanomaterials," *Nature Biotechnology* 21, no. 10 (October 2003), http://www.nature.com/cgi-af/Dyna Page.taf?file=/nbt/journal/v21/n10/full/nbt875.html (accessed March 9, 2004): 1168.

139. Barnaby Feder, "As uses grow, tiny materials' safety is hard to pin down," *Silicon Investor*,November 3, 2003, http://www.siliconinvestor.com/stocktalk/msg.gsp?msgid= 19459234 (accessed December 9, 2003).

140. Jessica Gorman, "Taming high-tech particles: Cautious steps into the nanotech future," *Science News* 161, no. 13 (March 30, 2002), http://www.sciencenews.org/20020330/ bob8.asp (accessed October 5, 2003).

141. Ibid.

142. Ibid.

143. ETC Group, *The Big Down: From Genomes to Atoms, Atomtech: Technologies Converging at the Nano-Scale*, January 2003, http://www.etcgroup.org/documents/TheBigDown.pdf (accessed September 5, 2003): 25.

144. Doug Brown, "Nano Litterbugs? Experts See Potential Pollution Problems," *Small Times* (March/April 2002), http://www.smalltimes.com/document_display.cfm ?document_id=3266 (accessed July 30, 2003).

145. ETC Group, "No Small Matter! Nanotech Particles Penetrate Living Cells and Accumulate in Animal Organs," *Communiqué* 76 (May/June 2002), http://www.etcgroup.org /article.asp?newsid=356 (accessed May 21, 2003).

146. M. Jenkins, "Antioxidants and Free Radicals," 1996, http://www.rice.edu/— jenky/sports/antiox.html (accessed June 3, 2005).

147. Bice Fubini, "Surface reactivity in the pathogenic response to particulates," *Environmental Health Perspectives* 105, supp. 5 (September 1997): 1013–20.

148. Andrew Maynard, Paul Baron, Michael Foley, Anna Shvedova, Elena Kisin, and Vicent Castranova, "Exposure to Carbon Nanotube Material: Aerosol Release during the Handling of Unrefined Single-Walled Carbon Nanotube Material," *Journal of Toxicology and Environmental Health, Part A* 67 (2004): 87–107.

149. Vicki Colvin, "Sustainability for Nanotechnology," *Scientist* 26–27.

150. A. Maynard, P. Baron, M. Foley, A. Shvedova, E. Kisin, and V. Castronova, "Exposure to carbon nanotube material: Aerosol release during the handling of unrefined single-walled carbon nanotube material," *Journal of Toxicology and Environmental Health, Part A* 67 (2004): 87–107.

151. J. Landemann, H. Weigmann, C. Rickmeyer, et al., "Penetration of titanium dioxide microparticles in a sunscreen formulation into the horny later and the follicle orifice," *Skin Pharmacology and Applied Skin Physiology* (1999): 247–56.

152. M. Tan, C. Commens. L. Burnett, and P. Snitch, "A pilot study on the percutaneous absorption of microfine titanium dioxide from sunscreens," *Australasian Journal of Dermatology* 37 (1996): 185–87.

153. S. Tinkle, J. Antonini, B. Rich, et al., "Skin as a route of exposure and sensitization in chronic beryllium disease," *Environmental Health Perspectives* 111, no. 9 (July 2003): 1202–208.

154. C. V. Howard, "Nano-particle and toxicity," April 2, 2003, Annex to ETC Group, "No small matter II: The case for a global moratorium," *Occasional Paper Series* 7, no. 1 (April 2003). Howard cites W. G. Kreyling et al., "Translocation of ultrafine insoluble iridium particles from lung epithelium to extrapulmonary organs is size dependent but very low," *Journal of Toxicology and Environmental Health A* 65, no. 20 (October 25, 2002): 1513–30, and Gunter Oberdörster et al., "Ultrafine particles in the urban air: To the respiratory tract and beyond," *Environmental Health Perspectives* 110, no. 8 (August 2002): A440–441.

155. M. Gumbleton, "Caveolae as potential macromolecule trafficking compartments with alveolar epithelium," *Advanced Drug Delivery Reviews* 49 (2001): 281–300.

156. R. N. Alyaudtin et al., "Interaction of poly(butylcyanoacrylate) nanoparticles with the blood-brain barrier in vivo and in vitro," *Journal of Drug Targeting* 9, no. 3 (2001): 209–21.

157. Alexander Huw Arnall, *Future Technologies, Today's Choices.* Its secondary title is more enlightening: *Nanotechnology, Artificial Intelligence and Robotics; A technical, political and institutional map of emerging technologies* (London: Greenpeace Environmental Trust, July 2003): 36.

158. Vicki Colvin, "The potential environmental impact of engineered nanomaterials," *Nature Biotechnology* 21, no. 10 (October 2003), http://www.nature.com/cgi-af/Dyna Page.taf?file=/nbt/journal/v21/n10/full/nbt875.html (accessed March 9, 2004): 1168.

159. H. H. C. Chen et al, "Renal effects of water-soluble polyarylsulfonated C-60 in rats with an acute toxicity study," *Fullerene Science and Technology* 5 (1997): 1387–96, and H. H. C. Chen et al., "Acute and subacute toxicity study of water-soluble polyalkylsulfonated C-60 in rats," *Toxicological Pathology* 26 (1998): 1387–96.

160. Vicki Colvin, "The potential environmental impact of engineered nanomaterials," *Nature Biotechnology* 21, no. 10 (October 2003), http://www.nature.com/cgi-af/Dyna Page.taf?file=/nbt/journal/v21/n10/full/nbt875.html (accessed March 9, 2004): 1169.

161. David D. Allen, Joanna Koziara, Russell J. Mumper, Thomas J. Abbruscato, and Paul R. Lockman, "Novel nanoparticles demonstrate no adverse effects on blood-brain barrier baseline parameters in in vitro or in vivo preparations," 225th ACS National Meeting, New Orleans, LA, March 23–27, 2003, http://oasys2.confex.com/acs/225nm/tech program/P582270.htm (accessed December 1, 2003).

162. Ben Harder, "Conduit to the Brain: Particles Enter the Nervous System via the Nose," *Science News Online* (January 24, 2004): 54, http://www.phschool.com/science/science _news?articles/conduit_to_the_brain.html (accessed 2 June 2005).

163. Rick Weiss, "For science, nanotech poses big unknowns," *Washington Post*, February 1, 2004, p. A01, http://www.washingtonpost.com/ac2/wp-dyn/A1487-2004Jan31 (accessed February 4, 2004).

164. "Much ado about almost nothing," *Economist*, March 18, 2004, http://www .economist.com/science/displayStory.cfm?story_id=2521232 (accessed April 5, 2004).

165. Alex Kirby, "Tiny particles threaten brain," *BBC News*, January 8, 2004, http:// news.bbc.co.uk/go/pr/fr/-/2/hi/science/nature/3379759.stm (accessed January 22, 2004).

166. Eva Oberdörster et al., "Translocation of inhaled ultrafine particles to the brain," *Inhalation Toxicology* 16, part 6/7 (2004): 437–46.

167. Rick Weiss, "For science, nanotech poses big unknowns," *Washington Post*, Feb-

ruary 1, 2004, p. A01, http://www.washingtonpost.com/ac2/wp-dyn/A1487-2004Jan31 (accessed February 4, 2004).

168. Gunter Oberdörster, Z. Sharp, V. Atudorei et al., "Translocation of inhaled ultra-fine particles to the brain," *Inhalation Toxicology*, pre-print, 2004, http://www.mindfully .org/Heath/2004/Nonoparticles-On-Brain2004.htm (accessed June 3, 2005).

169. "Buckyballs cause brain damage in fish," *New Scientist.com news service*, March 29, 2004, http://www.newscientist.com/news/print.jsp?id=ns99994825 (accessed April 5, 2004).

170. Ibid.

171. Rick Weiss, "For science, nanotech poses big unknowns," *Washington Post*, February 1, 2004, p. A01, http://www.washingtonpost.com/ac2/wp-dyn/A1487-2004Jan31 (accessed February 4, 2004).

172. Eva Oberdörster, personal e-correspondence, June 23, 2004.

173. "Buckyballs cause brain damage in fish," *New Scientist.com news service*, March 29, 2004, http://www.newscientist.com/news/print.jsp?id=ns99994825 (accessed April 5, 2004).

174. Eric Berger, "Study raises concerns over buckyballs," *Houston Chronicle*, March 29, 2004, http://www.chron.com/cs/CDA/printstory.mpl/health/244472494 (accessed April 5, 2004).

175. Mark T. Sampson, "Type of buckyball shows to cause brain damage in fish," *NewsNanoApex*, March 29, 2004, http://news.nanoplex.com/modules.php?name=News &file=print&sid=4410 (accessed April 5, 2004).

176. Rick Weiss, "Nanoparticles toxic in aquatic habitat, study finds," *Washington Post*, March 29, 2004, p. A02, http://www.washingtonpost.com/ac2/wp-dyn/A31881-2004 Mar28?language=printer (accessed March 29, 2004).

177. Robert F. Service, "Nanotechnology Grows Up," *Science* 304 (June 18, 2004): 1733.

178. Eva Oberdörster, "Manufactured nanomaterials (Fullerene, C_{60}) induce oxidative stress in the brain of juvenile largemouth bass," *Environmental Health Perspectives* 112, no. 110 (July 2004): 1061.

179. Yoko Yamakoshi, Naoki Umezawa, Akemi Ryu et al., "Active oxygen species generated from photoexcited fullerene (C_{60}) as potential medicines," *Journal of the American Chemical Society* 125 (2003): 12803–809.

180. A. Derfus, W. Chan, and S. Bhatia, "Probing the cytotoxicity of semiconductor quantum dots," *Nano Letters* 4 (2004): 11–18.

181. Juliana Gruenwald, "Controversial study points to need for more federal research," *Small Times*, April 1, 2004, http://www.smalltimes.com/document_display.cfm ?document_id=7658 (accessed April 5, 2004).

182. Mark T. Sampson, "Type of buckyball shows to cause brain damage in fish," *NewsNanoApex*, March 29, 2004, http://news.nanoplex.com/modules.php?name=News &file=print&sid=4410 (accessed April 5, 2004).

183. R. Petkewich, "Buckyballs Batter Bacteria," *Science News-Environmental Science and Technology Online News*, May 4, 2005, http://pubs.acs.org/subscribe/journals/esthag-w/ 2005/may/science/rp_nanocrystals.html (accessed 6 May 2005).

184. Alexandra Goho, "Tiny Trouble: Nanoscale Materials Damage Fish Brains," *Science News Online* 165, no. 14 (April 3, 2004), http://www.sciencenew.org/scripts/printthis .asp?clip=%2Farticles%2F20040403%2Fclip%5Ffob1%2Easp (accessed June 2, 2005).

185. Barnaby J. Feder, "How big is nanotechnology's hazard," *International Herald Tribune*, March 29, 2004, http://www.iht.com/articles/512380.html (accessed April 5, 2004).

186. Juliana Gruenwald, "Controversial study points to need for more federal research," *Small Times*, April 1, 2004, http://www.smalltimes.com/document_display.cfm?document_id=7658 (accessed April 5, 2004).

187. Barnaby Feder, "Study raises concerns about carbon particles," *New York Times*, March 29, 2004, p. C5.

188. Eva Oberdörster, "Manufactured nanomaterials (Fullerene, C_{60}) induce oxidative stress in the brain of juvenile largemouth bass," *Environmental Health Perspectives* 112, no. 110 (July 2004): 1061.

189. Juliana Gruenwald, "Controversial study points to need for more federal research," *Small Times*, April 1, 2004, http://www.smalltimes.com/document_display.cfm?document_id=7658 (accessed April 5, 2004).

190. Robert F. Service, "Nanotechnology Grows Up," *Science* 304 (June 18, 2004): 1733.

191. Barnaby Feder, "Study raises concerns about carbon particles," *New York Times*, March 29, 2004, p. C5.

192. Ibid.

193. Liz Kalaugher, "Quantum dots could be toxic to cells," *Nanotechweb.org*, January 16, 2004, http://www.nanotechweb.org/articles/news/3/1/2/1 (accessed January 22, 2004).

194. Ibid.

195. CORDIS, "Nanotechnology: Opportunity or threat," *CORDIS*, June 17, 2003, http://www.godlikeproductions.com/news/item.php?keyid=46322 (accessed July 30, 2003).

196. Josh Wolfe, "Nanotech vs. environmentalists," *Forbes*, September 16, 2003, http://www.forbes.com/2003/09/16/cz_jw_0916soapbox.html (accessed October 17, 2003).

197. CORDIS, "Nanotechnology: Opportunity or Threat," *CORDIS*, June 17, 2003, http://www.godlikeproductions.com/news/item.php?keyid=46322 (accessed July 30, 2003).

198. Geoff Brumfiel, "Nanotechnology: A Little Knowledge," *Nature* 424, no. 6946 (July 17, 2003): 246.

199. Rick Weiss, "Some see solution, some potential dangers," *Washington Post*, February 10, 2004, http://newsday.com/news/printedition/health/ny-dsspdn3663472feb10,0,5417935,pri (accessed February 16, 2004).

200. Rick Weiss, "For science, nanotech poses big unknowns," *Washington Post*, February 1, 2004, p. A01, http://www.washingtonpost.com/ac2/wp-dyn/A1487-2004Jan31 (accessed February 4, 2004).

201. ETC Group, "No Small Matter! Nanotech Particles Penetrate Living Cells and Accumulate in Animal Organs," *Communiqué* 76 (May/June 2002), http://www.etcgroup.org/article.asp?newsid=356 (accessed May 21, 2003).

202. Vicki Colvin, "The potential environmental impact of engineered nanomaterials," *Nature Biotechnology* 21, no. 10 (October 2003), http://www.nature.com/cgi-af/DynaPage.taf?file=/nbt/journal/v21/n10/full/nbt875.html (accessed March 9, 2004): 1167.

203. Ibid.

204. Scott Mize, "Near-Term Commercial Opportunities in Nanotechnology," Comments made during a presentation at the foresight conference, October 10, 2002; ETC Group, *The Big Down: From Genomes to Atoms, Atomtech: Technologies Converging at the Nano-Scale*,

January 2003, http://www.etcgroup.org/documents/TheBigDown.pdf (accessed September 5, 2003), and p. 22.

205. Glenda Chui, "Who's afraid of nanotechnology," *Mercury News*, September 16, 2003, http://www.bayarea.com/mld/mercurynews/living/health/6779281.htm?template =content (accessed October 17, 2003).

206. The Royal Academy of Engineers and the Royal Society, *Nanoscience and Nanotechnologies: Opportunities and Uncertainties*, July 2004, http://www.nanotec.org/uk/final Report.htm (accessed August 11, 2004), p. 80.

207. Vicki Colvin, "The potential environmental impact of engineered nanomaterials," *Nature Biotechnology* 21 no. 10 (October 2003): 1166, http://www.nature.com/cgi-af/ DynaPage.taf?file=/nbt/journal/v21/n10/full/nbt875.html (accessed March 9, 2004).

208. A. Bridges, "Tiniest particles of matter don't behave, raising nanotech concerns," *USA Today*, April 2, 2004, http://www.usatoday.com/tech/news/2004-04-02-nano-flaw_x .htm (accessed April 5, 2004).

209. P. H. M. Hoet, A. Nemmar, and B. Nemery, "Health impact of nanomaterials," *Nature* 22, no. 1 (January 2004): 19, http://www.nature.com/cgi-taf/DynaPage.taf?file=/ nbt/journal/v22/n1/full/nbt0104-19.html (accessed January 12, 2004).

210. E. Hood, "Nanotechnology: Looking as we leap," *Environmental Health Perspectives* 112, no. 13 (September 2004): A740–49.

211. Alexandra Goho, "Buckyballs at Bat: Toxic nanomaterials get a tune-up," *Science News Online* 166, no. 14 (October 2, 2004): 211, http://www.sciencenews.org/articles/ 20041002/fob1.asp (accessed October 4, 2004).

212. T. Masciangioli and W. Zhang, "Environmental technologies at the nanoscale," *Environmental Science and Technology* (March 1, 2003): 102–108A.

213. J. Borchardt "Small science may clean a big problem," *Christian Science Monitor* February 10, 2005 e-mail communication from J. Moore, February 10, 2005.

214. "Iron nanoparticles may be effective in cleaning up carbon tetrachloride in contaminated groundwater," *NewsNanoApex*, January 14, 2005, http://news.nanoapex.com/ modules.php?name=News&file=print&sid=5444 (accessed January 28, 2005).

215. J. Uldrich, "Nanotech needs a hard sell, plus education," *Scientist* (August 30, 2004): 8.

216. J. Borchardt, "Small science may clean a big problem," *Christian Science Monitor*, February 10, 2005, e-mail communication from J. Moore, February 10, 2005.

217. See T. Masciangioli, and W. Zhang, "Environmental Technologies at the Nanoscale," *Environmental Science and Technology* (March 1, 2003): 102–108A.

218. Vicki Colvin, "Sustainability for Nanotechnology," *Scientist* (August 30, 2004): 26–27.

219. N. Stine and P. Lurie, "Small, but Dangerous," *Washington Post Online*, February 5, 2005: A18, http://www.washingtonpost.com/ac2/wp-dyn/A64962-2005Feb4?language =printer (accessed February 15, 2005).

220. E. Hood, "Nanotechnology: Looking as we leap," *Environmental Health Perspectives* 112, no. 13 (September, 2004): A740–49.

221. Gunter Oberdörster, Eva Oberdörster, and Jan Oberdörster, "Nanotoxicology: An emerging discipline evolving from studies of ultrafines," *Environmental Health Perspectives* (March 22, 2005): 30, http://ehp.niehs.nih.gov/members/2005/7339/7339.pdf (accessed June 4, 2005).

222. J. Gross, "New research raises questions about buckyballs and the environment," *EurekAlert!* May 9, 2005, http://www.eurekalert.org/pub_releases/2005-05/acs-nrr 050905.php (accessed May 9, 2005).

223. Gunter Oberdörster, Eva Oberdörster, and Jan Oberdörster, "Nanotoxicology: An emerging discipline evolving from studies of ultrafines," *Environmental Health Perspectives* (March 22, 2005): 30, http://ehp.niehs.nih.gov/members/2005/7339/7339.pdf (accessed June 4, 2005).

224. Vicki Colvin, "The potential environmental impact of engineered nanomaterials," *Nature Biotechnology* 21, no. 10 (October 2003), http://www.nature.com/cgi-af/ DynaPage.taf?file=/nbt/journal/v21/n10/full/nbt875.html (accessed March 9, 2004): 1166.

225. Wei Zhou, "Ethics of nanobiology at the frontline," *Santa Clara Computer and High Technology Law Review* 19 (2003): 487.

CHAPTER 10

1. W. J. Broad and J. Glanz, "Does science matter?" *New York Times*, November 11, 2003, p. D1.

2. Ibid.

3. Ibid.

4. Dan Gillmor, "Big breakthroughs can come in small packages," *SiliconValley.com*, February 16, 2002, http://www.siliconvalley.com/mld/siliconvalley/business/columnists/ 2685956.htm (accessed July 27, 2004).

5. National Science and Technology Council, Committee on Technology, Subcommittee on Nanoscale Science, Engineering, and Technology (NSET), *National Nanotechnology Initiative: Research and Development Supporting the Next Industrial Revolution, Supplement to the President's FY 2004 Budget* (Arlington, VA: National Nanotechnology Coordination Office, August 29, 2003), p. 2.

6. Ted Agres, "Nanotech ethics debated," *Scientist*, December 8, 2003, http:// www.biomedcentral.com/news/20031208/04 (accessed December 9, 2003).

7. C. Batt, Committee on Science in the House of Representatives, *H.R. 766, Nanotechnology Research and Development Act of 2003*, Hearings before the Committee on Science in the House of Representatives, March 19, 2003, p. 46.

8. A. Marty, Committee on Science in the House of Representatives, *H.R. 766, Nanotechnology Research and Development Act of 2003*, Hearings before the Committee on Science in the House of Representatives, March 19, 2003, p. 58.

9. Committee on Science in the House of Representatives, *The Societal Implications of Nanotechnology*, Hearings before the Committee on Science in the House of Representatives, April 9, 2003, p. 4.

10. National Academy of Sciences, Committee on Science in the House of Representatives, *H.R. 766, Nanotechnology Research and Development Act of 2003*, Hearings before the Committee on Science in the House of Representatives, March 19, 2003, p. 5.

11. S. Boehlert, Committee on Science in the House of Representatives, *The Societal Implications of Nanotechnology*, Hearings before the Committee on Science in the House of Representatives, April 9, 2003, p. 10.

12. National Academy of Sciences, *Preliminary Comments, Review of the National Nano-technology Initiative* (Washington, DC: National Academies Press, 2001), p. 7.

13. N. Smith, Committee on Science in the House of Representatives, *The Societal Implications of Nanotechnology*, Hearings before the Committee on Science in the House of Representatives, April 9, 2003, p. 13.

14. The National Nanotechnology Initiative (NNI, http://nano.gov) is a multi-agency effort within the US government that supports a broad program of federal nanoscale research in materials, physics, chemistry, and biology. It explicitly seeks to create opportunities for interdisciplinary work integrating these traditional disciplines. The NNI is balanced across five broad activities: fundamental research, grand challenges, centers and networks of excellence, research infrastructure, and societal/workforce implications.

15. The Twenty-First Century Nanotechnology Research and Development Act passed by the US Congress on November 18, 2003, and signed by President G. W. Bush on December 3, 2003.

16. Ted Agres, "U.S. Congress OKs nanotech bill," *Scientist*, November 24, 2003, http://www.biomedcentral.com/news/20031124/05 (accessed December 15, 2003).

17. Office of Technology Assessment at the German Parliament (TAB), *Summary of TAB working report No. 92*, July 2003, http://www.tab.fzk.de/en/projekt/zusammenfassung/ab92.htm (accessed July 11, 2004).

18. A. W. Sweeney, S. Seal, and P. Vaidyanathan, "The promises and perils of nanoscience and nanotechnology: Exploring emerging social and ethical issues," *Bulletin of Science, Technology, & Society* 23, no. 4 (August 2003): 242.

19. A. Leo, "Get ready for your nano future," *Technology Review*, May 4, 2001, http://www.technologyreview.com/articleswo_leo050401.asp (accessed October 5, 2003).

20. Julia Moore, "The Future Dances on a Pin's Head," *Woodrow Wilson International Center for Scholars: Knowledge in the Public Service*, November 26, 2002, http://wwics.si.edu/index.cfm?fuseaction=news.item&news_id=14638 (accessed July 27, 2004).

21. Vicki Colvin, Committee on Science in the House of Representatives, *The Societal Implications of Nanotechnology*, Hearings before the Committee on Science in the House of Representatives, April 9, 2003, p. 51.

22. Committee on Science in the House of Representatives, *The Societal Implications of Nanotechnology*, Hearings before the Committee on Science in the House of Representatives, April 9, 2003, p. 5.

23. Ben Wootliff, "Businesses begin to pay attention to high-profile nano health debate," *Small Times* (March/April 2004): 16.

24. Jack Mason, "As nanotech grows, leaders grapple with public fear and perception," *Small Times*, May 20, 2004, http://www.smalltimes.com/document_display.cfm?document_id=7926 (accessed July 11, 2004).

25. Ibid.

26. National Research Council, *Small Wonders, Endless Frontiers: A Review of the National Nanotechnology Initiative* (Washington, DC: National Academies Press, 2002), http://www.nap.edu/catalog/10395.html?onpi_newsdoc06102002 (accessed May 20, 2004).

27. Joe Kaplinsky, "Nanotechnology: A slippery debate," *Spiked Online*, May 7, 2003, http://131.104.232.9agnet/2003/5-2003/agnet_may_11.htm (accessed September 6, 2003).

28. Ibid.

29. "Who will control the nanobots," *Mail on Sunday*, April 27, 2003, p. 24.

30. Joe Kaplinsky, "Nanotechnology: A slippery debate," *Spiked Online*, May 7, 2003, http://131.104.232.9agnet/2003/5-2003/agnet_may_11.htm (accessed September 6, 2003).

31. "No small matter," *Red Herring*, December 16, 2002, http://www.redherring.com/Article.aspx?f=articles%2farchive%2finsider%2f2002%2f12%2f (accessed October 5, 2003).

32. P. Monaghan, "The humanities' new muse: Genomics," *Chronicle of Higher Education*, February 20, 2004, http://chronicle.com/temp/email.php?id=scjqtzo0kxzer26ofk1x4jsg3nfuwqk8 (accessed February 17, 2004).

33. J. F. Costello, Committee on Science in the House of Representatives, *The Societal Implications of Nanotechnology*, Hearings before the Committee on Science in the House of Representatives, 108 Cong. 1st Session, Serial No. 108–13, April 9, 2003, p. 14.

34. E. B. Johnson, Committee on Science in the House of Representatives, *The Societal Implications of Nanotechnology*, Hearings before the Committee on Science in the House of Representatives, April 9, 2003, p. 14.

35. R. Kurzweil, Committee on Science in the House of Representatives, *The Societal Implications of Nanotechnology*, Hearings before the Committee on Science, House of Representatives, April 9, 2003, p. 18.

36. Mihail C. Roco and William Sims Bainbridge, *Societal Implication of Nanoscience and Nanotechnology* (Arlington, VA: National Science Foundation, March 2001), p. 12.

37. Ibid., p. 2.

38. National Science and Technology Council, Committee on Technology, Subcommittee on Nanoscale Science, Engineering, and Technology (NSET), *National Nanotechnology Initiative: Research and Development Supporting the Next Industrial Revolution, Supplement to the President's FY 2004 Budget* (Arlington, VA: National Nanotechnology Coordination Office, August 29, 2003), p. 4.

39. J. Wolfe, "Nanotech vs. environmentalists," *Forbes*, September 16, 2003, http://www.forbes.com/2003/09/16/cz_jw_0916soapbox.html (accessed October 17, 2003).

40. Ibid.

41. Langdon Winner, Committee on Science in the House of Representatives, *The Societal Implications of Nanotechnology*, Hearings before the Committee on Science in the House of Representatives, April 9, 2003.

42. A. Mnyusiwalla, A. S. Daar, and P. Singer, "Mind the gap: Science and ethics in nanotechnology," *Nanotechnology* 14 (2003): R9.

43. Ibid., p. R10.

44. Ibid., p. R11.

45. Juliana Gruenwald, "Controversial study points to need for more federal research," *Small Times*, April 1, 2004, http://www.smalltimes.com/document_display.cfm?document_id=7658 (accessed April 5, 2004).

46. Vicki Colvin, "The potential environmental impact of engineered nanomaterials," *Nature Biotechnology* 21, no. 10 (October 2003), http://www.nature.com/cgi-af/DynaPage.taf?file=/nbt/journal/v21/n10/full/nbt875.html (accessed March 9, 2004): 1166.

47. Keay Davidson, "The promise and perils of the nanotech revolution," *SFGate.com*,

July 26, 2004, http://www.sfgate.com/cgi-bin/article.cgi?file=/c/a/2004/07/26/MNG 767SUKB1.DTL (accessed August 4, 2004).

48. Steve Crosby, "Study nano and the environment now, because what we can't see scares us," *Small Times* (January/February 2003): 4.

49. Vicki Colvin, "Responsible nanotechnology: Looking beyond the good news," *Eureka Alert: In Context*, November 2002, http://www.eurekalert.org/context.php?context =nano&show=essays&essaydate=1102 (accessed December 10, 2003).

50. Josh Wolfe, "Nanotech vs. environmentalists," *Forbes*, September 16, 2003, http:// www.forbes.com/2003/09/16/cz_jw_0916soapbox.html (accessed October 17, 2003).

51. Ibid.

52. Ibid.

53. Barnaby Feder, "As uses grow, tiny materials' safety is hard to pin down," *Silicon Investor*,November 3, 2003, http://www.siliconinvestor.com/stocktalk/msg.gsp?msgid= 19459234 (accessed December 9, 2003).

54. Ibid.

55. Ibid.

56. Vicki Colvin, "Responsible nanotechnology: Looking beyond the good news," *Eureka Alert: In Context*, November 2002, http://www.eurekalert.org/context.php?context =nano&show=essays&essaydate=1102 (accessed December 10, 2003).

57. Nick Smith, Committee on Science in the House of Representatives, *The Societal Implications of Nanotechnology*, Hearings before the Committee on Science in the House of Representatives, April 9, 2003, p. 13.

58. Ted Agres, "Nanotech ethics debated," *Scientist*, December 8, 2003, http:// www.biomedcentral.com/news/20031208/04 (accessed December 9, 2003).

59. Ibid.

60. Glenda Chui, "Who's Afraid of Nanotechnology," *Mercury News*, September 16, 2003, http://www.bayarea.com/mld/mercurynews/living/health/6783577.htm (accessed October 2, 2003).

61. Donald MacLeod, "Thinktank predicts nanotechnology backlash," *Guardian Unlimited*, February 13, 2003, http://education.guardian.co.uk/higher/research/story/ 0,9865,894755,00.html (accessed July 29, 2003).

62. Glenda Chui, "Who's Afraid of Nanotechnology," *Mercury News*, September 16, 2003, http://www.bayarea.com/mld/mercurynews/living/health/6783577.htm (accessed October 2, 2003).

63. Emmanuelle Schuler, "Perceptions of Risks and Nanotechnology," unpublished paper, p. 2.

64. "Nanotech advances need more safety screening-study," *ABC News*, June 14, 2005, http://abcnews.go.com/Technology/print?id=848241 (accessed June 15, 2005).

65. Doug Parr, foreword to *Future Technologies, Today's Choices*. Its secondary title is more enlightening: *Nanotechnology, Artificial Intelligence and Robotics; A technical, political and institutional map of emerging technologies* (London: Greenpeace Environmental Trust, July 2003): 5.

66. *Survey 2001* was sponsored by the National Geographic Society and the NSF. It was developed by Clemson University and had 2,196 questions, and some had to do with environmentalism. The survey used Likert scale items.

67. W. S. Bainbridge, "Sociocultural Meanings of Nanotechnology: Research Methodologies," an unpublished paper, *2nd Workshop on Societal Implications of Nanoscience and Nanotechnology* (Arlington, VA: NSF, December 3–5, 2003).

68. Similar to the National Nanotechnology User Network, the NNIN. SBE (Social, Behavioral, and Economic Sciences) must be incorporated into the NNIN (NSF 03-519, Proposal Solicitation, Proposal Content C. f.). Furthermore, the Director of the Societal Dimensions of Engineering, Science, and Technology Program, Division of Social and Economic Sciences, has outlined the standards of SBE participation in the NNIN. She maintains the management team must include a social scientist to coordinate specific research into social and ethical dimensions and the following proposal meets the requirements of the call.

69. A. H. Arnall, *Future Technologies, Today's Choices.* Its secondary title is more enlightening: *Nanotechnology, Artificial Intelligence and Robotics; A technical, political and institutional map of emerging technologies* (London: Greenpeace Environmental Trust, July 2003), p. 39.

70. The White House, "National Nanotechnology Initiative: Leading to the Next Industrial Revolution," January 21, 2002, http://clinton4.nara.gov/textonly/WH/New/html/20000121_4.html (accessed October 5, 2003).

71. Committee for the Review of the National Nanotechnology Initiative, National Research Council, *Small Wonders, Endless Frontiers: A Review of the National Nanotechnology Initiative* (Washington, DC: National Academies Press, 2002), p. 31.

72. Mihail C. Roco and William Sims Bainbridge, eds., *Societal Implication of Nanoscience and Nanotechnology* (Arlington, VA: National Science Foundation, March 2001).

73. Committee for the Review of the National Nanotechnology Initiative, National Research Council, *Small Wonders, Endless Frontiers: A Review of the National Nanotechnology Initiative* (Washington, DC: National Academies Press, 2002), p. 32.

74. A. Mnyusiwalla, A. S. Daar, and P. Singer, "Mind the gap: Science and ethics in nanotechnology," *Nanotechnology* 14 (2003): R10–11.

75. Committee for the Review of the National Nanotechnology Initiative, National Research Council, *Small Wonders, Endless Frontiers: A Review of the National Nanotechnology Initiative* (Washington, DC: National Academies Press, 2002), p. 48.

76. Ibid.

77. Ibid.

78. C. Peterson, Committee on Science in the House of Representatives, *The Societal Implications of Nanotechnology*, Hearings before the Committee on Science in the House of Representatives, April 9, 2003, p. 71.

79. Committee on Science, House of Representatives, *The Societal Implications of Nanotechnology*, Hearings before the Committee on Science, House of Representatives, 108 Cong. 1st Session, Serial No. 108–13, April 9, 2003, p. 5.

80. Committee for the Review of the National Nanotechnology Initiative, National Research Council, *Small Wonders, Endless Frontiers: A Review of the National Nanotechnology Initiative* (Washington, DC: National Academies Press, 2002), p. 49.

81. R. Colwell, "Welcoming remarks," Abstracts of the Workshop on Societal Implications of Nanoscience and Nanotechnology (December 3–5, 2003), p. 6f.

82. B. Feder, "As uses grow, tiny materials' safety is hard to pin down," November 3,

2003, http://www.siliconinvestor.com/stocktalk/msg.gsp?msgid=19459234 (accessed December 9, 2003).

83. V. Colvin, "Responsible nanotechnology: Looking beyond the good news," *Eureka Alert! In Context*, November 1, 2002, http://www.eurekaalert.org/context.php?context=nano &show=essays&essaydate=1102 (accessed October 5, 2003).

84. Josh Wolfe, "Nanotech vs. the Green Gang," *Forbes*, April 7, 2005, http://www.forbes .com/technology/sciences/2005/04/06/cz_jw_0406soapbox_inl.html, e-communication.

85. It is important at this point to distinguish between traditional cross-breeding of plants to produce new varieties and genetic manipulation between species, as in adding a gene from a flounder to a strawberry. While some have argued the GMO research, especially involving plants, has been going on for centuries, they avoid discussing the interspecies nature of some genetic research.

86. B. McKibben, *Enough: Staying Human in Engineered Age* (New York: Time Books, 2003).

87. Ibid., pp. 179–80.

88. Langdon Winner, Committee on Science in the House of Representatives, *The Societal Implications of Nanotechnology*, Hearings before the Committee on Science in the House of Representatives, April 9, 2003.

89. Ibid.

90. Mae-Wan Ho, "Nanotechnology, a Hard Pill to Swallow," http://www.i-sis .org.uk/nanotechnology.php (accessed October 5, 2003).

91. ETC Group, "Nanotech News in Living Colour: An Update on White Papers, Red Flags, Green Goo, Grey Goo (and Red Herrings)," *Communiqué* 85 (May/June 2004): 5.

92. Vicki Colvin, Committee on Science, House of Representatives, *The Societal Implications of Nanotechnology*, Hearings before the Committee on Science, House of Representatives, 108 Cong. 1st Session, Serial No. 108-13, April 9, 2003.

93. Vicki Colvin in Susan Morrissey, "Pacing the Way for Nanotech," *Chemical and Engineering News* (June 8, 2004) http://pubs.acs.org/cen/news/8223/8223earlygov.html (accessed June 15, 2004).

94. Langdon Winner, Committee on Science in the House of Representatives, *The Societal Implications of Nanotechnology*, Hearings before the Committee on Science in the House of Representatives, April 9, 2003, p. 56.

95. Carl Batt, Committee on Science in the House of Representatives, *H.R. 766, Nanotechnology Research and Development Act of 2003*, Hearings before the Committee on Science in the House of Representatives, March 19, 2003, p. 80.

96. V. Colvin, Committee on Science in the House of Representatives, *The Societal Implications of Nanotechnology*, Hearings before the Committee on Science in the House of Representatives, April 9, 2003.

97. Ibid.

98. Juliana Gruenwald, "D.C., nano union now put to the test," *Small Times* (January/February 2004): 9.

99. Candace Stuart, "Fertile ground: Small tech's growing clusters resemble biotech in its early life," *Small Times* (March/April 2004): 34.

100. NSF PR 03-89, "NSF Awards New Grants to Study Societal Implications of

Nanotechnology," August 25, 2003, http://www.nsf.gov/od/lpa/news/03/pr0389.htm (accessed June 26, 2004).

101. Lynne Zucker, e-mail correspondence, June 30, 2004.

102. Ibid.

103. Ibid.

104. NSF PR 03-89, "NSF Awards New Grants to Study Societal Implications of Nanotechnology," August 25, 2003, http://www.nsf.gov/od/lpa/news/03/pr0389.htm (accessed June 26, 2004).

105. Ibid.

106. "Follow the money," *Nanotech Report* 2, no. 9 (September 2003): 7.

107. *NSF Award Abstract—#0403847*, http://www.nsf.gov/awardsearch/showAward.do ?AwardNumber=0403847 (accessed July 23, 2004).

108. "Nanotechnology industry to address environmental and health effects of nanotech," *Business Wire*, July 7, 2003, http://ragingbull.lycos.com.mboard.boards.cgi?board -NANOTECH&read=1089 (accessed October 2, 2003).

109. Ibid.

110. Afterlife, much life aftermarket, refers to the implications associated with the life span of a product or process. While aftermarkets refer to the breadth of participation the produce may serve in maintaining the life span of its products, afterlife refers to the length of time a product continues to interact with the ecosystem into which it has been released.

111. "Nanotechnology industry to address environmental and health effects of nanotech," *Business Wire*, July 7, 2003, http://ragingbull.lycos.com.mboard.boards.cgi?board -NANOTECH&read=1089 (accessed October 2, 2003).

112. Doug Brown, "Changing of the guard," *Small Times* (January/February 2003): 14.

113. B. Goldfarb, "New task force focuses on nanotechnology health effects," *NanoBiotech News* (August 13, 2003): 3.

114. G. Chui, "Who's afraid of nanotechnology?" *Mercury News*, September 16, 2003, http://www.bayarea.com/mld/mercurynews/living/health/6779281.htm?template= content (accessed October 17, 2003).

115. Ibid.

116. Nancy Weil, "NanoBusiness Alliance starts safety task force," *Bio-IT World*, July 7, 2003, http://www.bio-itworld.com/news/070703_report2829.html?action=print (accessed on July 17, 2004).

117. Ibid.

118. "Nanotechnology Industry to Address Environmental and Health Effects of Nanotech," *Business Wire*, July 7, 2003, http://www.bio-itworld.com/news/070703 _report2829.html (accessed June 17, 2004).

119. Nancy Weil, "NanoBusiness Alliance starts safety task force," *Bio-IT World*, July 7, 2003, http://www.bio-itworld.com/news/070703_report2829.html?action=print (accessed on July 17, 2004).

120. "Nanotechnology Industry to Address Environmental and Health Effects of Nanotech," *Business Wire*, July 7, 2003, http://www.bio-itworld.com/news/070703_report 2829.html (accessed June 17, 2004).

121. Ibid.

122. Langdon Winner, Committee on Science in the House of Representatives, *The Societal Implications of Nanotechnology*, Hearings before the Committee on Science in the House of Representatives, April 9, 2003, p. 56.

123. V. Colvin, "Responsible nanotechnology: Looking beyond the good news," *Eureka Alert! In Context*, November 1, 2002, http://www.eurekaalert.org/context.php?context =nano&show=essays&essaydate=1102 (accessed October 5, 2003).

124. V. Colvin, Committee on Science in the House of Representatives, *The Societal Implications of Nanotechnology*, Hearings before the Committee on Science in the House of Representatives, April 9, 2003, p. 50.

125. Dana Rohrabacher, Committee on Science in the House of Representatives, *The Societal Implications of Nanotechnology*, Hearings before the Committee on Science in the House of Representatives, April 9, 2003, p. 76.

126. Ibid., cited by Douglas Brown, "Perception may be nano's biggest enemy, leaders tell Congress," *Small Times*, April 10, 2003, http://www.smalltimes.com/print_doc .cfm?doc_id=5809 (accessed April 24, 2003).

127. Howard Lovy, "Clash of the Nanotech Titans," *Howard Lovy's Nanobot*, December 1, 2003, http://nanobot.blogspot.com/2003_11_30_nanobot_archive.html (accessed December 2, 2003).

128. P. Monaghan, "The humanities' new muse: Genomics," *Chronicle of Higher Education*, February 20, 2004, http://chronicle.com/temp/email.php?id=scjqtzo0kxzer26 ofk1x4jsg3nfuwqk8 (accessed February 17, 2004).

129. Rick Weiss, "For science, nanotech poses big unknowns," *Washington Post*, February 1, 2004, p. A01, http://www.washingtonpost.com/ac2/wp-dyn/A1487-2004Jan31 (accessed February 4, 2004).

130. Langdon Winner, Committee on Science in the House of Representatives, *The Societal Implications of Nanotechnology*, Hearings before the Committee on Science in the House of Representatives, April 9, 2003, p. 78.

131. Robin Grove-White, Matthew Kearnes, Paul Miller et al., "Bio-to-Nano: Learning the Lessons, Interrogating the Comparison," a working paper by the Institute for Environment, Philosophy and Public Policy, Lancaster University and Demos, June 2004, p. 3.

132. C. Batt, Committee on Science in the House of Representatives, *H.R. 766, Nanotechnology Research and Development Act of 2003*, Hearings before the Committee on Science in the House of Representatives, March 19, 2003, p. 46.

133. B. Falconer, "Defense research agency seeks to create supersoldiers," *National Journal*, November 10, 2003, http://www.govexec.com/dailyfed/1103/111003nj1.htm (accessed May 2, 2004).

134. A. Leo, "Get ready for your nano future," *Technology Review*, May 4, 2001, http:// www.technologyreview.com/articleswo_leo050401.asp (accessed October 5, 2003).

135. Vicki Colvin, "Responsible nanotechnology: Looking beyond the good news," *Eureka Alert: InContext*, November 2002, http://www.eurekaalert.org/context.php?context =nano&show=essays&essaydate=1102 (accessed December 10, 2003).

CHAPTER 11

1. Rita Colwell, "Welcoming remarks," *Abstracts of the Workshop on Societal Implications of Nanoscience and Nanotechnology* (December 3–5, 2003): 6b–6c.

2. Ann Dowling, "Is nanotechnology the next GM," *New Scientist*, March 12, 2005, e-communication.

3. "Nanologue aims to boost nanotechnology discourse," *NanoTechWeb*, March 4, 2005, http://nanotechweb.org/articles/society/4/3/1/1 (accessed March 22, 2005).

4. "How safe is nanotechnology? Japan needs to find out soon," *Yomiuri Shimbun*, December 1, 2003, in *Small Times*, http://www.smalltimes.com/document_display.cfm?document_id=7021 (accessed December 9, 2003).

5. Ibid.

6. Roland Jackson, "Science and trust," *Independent*, February 16, 2005, e-communication.

7. Glenda Chui, "Who's afraid of nanotechnology?" *Mercury News*, September 16, 2003, http://www.bayarea.com/mld/mercurynews/living/health/6779281.htm?template=content (accessed October 17, 2003).

8. "Tough times ahead for nano," June 10, 2005, http://www.ferret.com.au/articles/97/0c030c97.asp (accessed June 16, 2005).

9. *Assessing Molecular and Atomic Scale Technologies (MAST)*, University of Texas at Austin, Office of Publications, 1989.

10. B. McKibben, *Enough: Staying Human in an Engineered Age* (New York: Times Books, 2003).

11. R. Birge et al., "The future of nanotechnology," *Wired* 3, no. 8 (August 1995): 58.

12. W. S. Bainbridge, "Public attitudes toward nanotechnology," *Journal of Nanoparticle Research* 4 (2002): 561–70.

13. The basic sample design applied in all member States is a multistage, random (probability) one. In each EU country, a number of sampling points was drawn with probability proportional to population size (for total coverage of the country) and to population density. In order to do this, the points were drawn systematically from each of the "administrative regional units," after stratification by individual unit and type of region. They thus represent the whole territory of the Member States according to the Eurostat-NUTS II and according to the distribution of the resident population of the respective EU nationalities in terms of metropolitan, urban, and rural areas. In each of the selected sampling points, a starting address was drawn at random. Further addresses were selected (every nth address) by standard random route procedures from the initial address. In each household, the respondent was drawn at random. All interviews were face-to-face in people's homes and in the appropriate national language.

14. BMRB Social Research, "Nanotechnology: Views of the general public, quantitative and qualitative research carried out as part of the nanotechnology study," *BMRB International Report 45101666*, 2004, http://www.nanotec.org.uk/Market%20Research.pdf (accessed June 25, 2004): 4.

15. "Outcome of the Open Consultation on the European Strategy for Nanotechnology," *Nanoforum*, December 2004, http://www.nanoforum.org/dateien/temp/nanosurvey6.pdf?20122004094532 (accessed July 18, 2005).

16. *ISTPP 2003 National Nanotechnology Survey* (College Station, TX: Institute for Science, Technology and Public Policy, George Bush School and Government and Public Service, Texas A&M University, June 16, 2004), unpublished.

17. Michael Cobb and Jane Macoubrie, "Public Perceptions about Nanotechnology: Risks, Benefits and Trust," 2004, Unpublished.

18. "U.S. Leadership in Nanoscience Should Be a Government Priority, Says Survey Respondents," *NewsNanoApex*, September 15, 2004, http://news.nanoapex.com/modules .php?name=News&fule=print&sid=5010 (accessed September 22, 2004).

19. Jane Macoubrie, "Informed Public Perceptions of Nanotechnology and Trust in Government," Project on Emerging Nanotechnologies, Woodrow Wilson International Center for Scholars, September 9, 2005, http://www.wilsoncenter.org/events/docs/ macoubriereport.pdf (accessed September 15, 2005).

20. Chul-joo Lee, Dietram A. Scheufele, and Bruce Lewenstein, "Public Attitudes toward Emerging Technologies: Examining the Interactive Effects of Cognitions and Affect on Public Attitudes toward Nanotechnology," *Science Communication*, preprint.

21. "University of Wisconsin–Madison study examines public attitudes on nanotechnology," *NanoTechWire*, August 30, 2005, http://nanotechwire.com/news.asp?nid=2266 &ntid=116&pg=1 (accessed September 15, 2005).

22. W. S. Bainbridge, "Sociocultural Meanings of Nanotechnology: Research Methodologies," an unpublished paper, 2nd Workshop on Societal Implications of Nanoscience and Nanotechnology (Arlington, VA: National Science Foundation, December 3–5, 2003).

23. Alexis de Tocqueville, *Democracy in America* (New York: Westvaco, 1938, 1999).

24. Max Weber, *The Protestant Ethic and the Spirit of Capitalism* (London: Unwin University Books, 1930, 1970).

25. John Dewey, *Democracy and Education* (New York: Macmillan, 1916).

26. Hannah Arendt, *The Origins of Totalitarianism* (New York: Harcourt, Brace, 1951).

27. Antonio Gramsci, *The Modern Prince and Other Writings* (London: Lawrence and Wishart, 1957).

28. Wayne Clark, *Activism in the Public Sphere: Exploring the Discourse of Political Participation* (Aldershot, UK: Ashgate, 2000), p. 43.

29. Carol Boggs, "Rise and Decline of the Public Sphere," *The End of Politics* (New York: Guilford Press, 2000), p. 108.

30. Jürgen Habermas, "Further Reflections on the Public Sphere," *Habermas and the Public Sphere*, ed. Craig Calhoun (Cambridge, MA: MIT Press, 1992), p. 433.

31. Charles Arthur Willard, "The Problem of the Public Sphere: Three Diagnoses," *Argument Theory and the Rhetoric of Assent*, ed. David C. Williams and Michael D. Hazen (Tuscaloosa: University of Alabama Press, 1990), p. 152.

32. Jurgen Habermas, *Habermas and the Public Sphere*, ed. Craig Calhoun (Cambridge, MA: MIT Press, 1992).

33. Walter Lippman, *The Phantom Public* (New Brunswick, NJ: Transaction Press, 1927, 1999).

34. John Dewey, "The Public and its Problems," *John Dewey: The Later Works, 1925–1953*, vol. 2, ed. Jo Ann Boydson (Carbondale: Southern Illinois University Press, 1981).

35. Walter Lippman, *The Phantom Public* (New Brunswick, NJ: Transaction Press, 1927, 1999).

36. Hannah Arendt, *The Human Condition*, 2nd ed. (Chicago: University of Chicago Press, 1958, 1998).

37. Jurgen Habermas, *The Structural Transformation of the Public Sphere: An Inquiry into a Category of Bourgeois Society*, trans. Thomas Burger (Cambridge, MA: MIT Press, 1963, 1989).

38. Oskar Negt and Alexander Kluge, *Public Sphere and Experience: Toward an Analysis of the Bourgeois and Proletarian Public Sphere*, trans. Peter Labanji, James Owen Daniel, and Assenka Oksilodd (Minneapolis: University of Minnesota Press, 1972, 1993).

39. Lynn Sanders, "Against Deliberation," *Political Theory* 25, no. 3 (June 1997): 4.

40. Seyla Benhabib, "The Embattled Public Sphere," *Reasoning Practically*, ed. Edna Ullmann-Margalit (New York: Oxford University Press, 2000), p. 170.

41. Jürgen Habermas, "Further Reflections on the Public Sphere," *Habermas and the Public Sphere*, ed. Craig Caluhoun (Cambridge, MA: MIT Press, 1992), p. 436.

42. Marc Lynch, "The Public Sphere Structure of International Politics," *State Interest and Public Spheres* (New York: Columbia University Press, 1999).

43. Wayne Clark, *Activism in the Public Sphere: Exploring the Discourse of Political Participation* (Aldershot, UK: Ashgate, 2000), p. 61.

44. John Dewey, "The Public and Its Problems," *John Dewey: The Latter Works, 1925–1953*, vol. 2: *1925–1927*, ed. Jo Ann Boydson (Carbondale: Southern Illinois University Press, 1927, 1981), p. 238.

45. Nathan Tinker, "Nano-Savvy Journalism," *Best of the NanoWeek* 2, no. 4 (April 20, 2005), http://www.voyle.net/Guest%20Writers/Darrell%20Brookstein/Darrell%20Brookstein%202005-001.htm (accessed July 18, 2005).

46. Ibid.

47. James Curran, "Rethinking the Media as a Public Sphere," *Communication and Citizenship: Journalism and the Public Sphere*, ed. Peter Dahlgren and Colin Sparks (London: Routledge, 1993), p. 32.

48. Ibid.

49. Ibid.

50. Ibid.

51. Nicholas Garnham, "The Media and the Public Sphere," *Habermas and the Public Sphere*, ed. Craig Calhoun (Cambridge, MA: MIT Press, 1992), p. 360.

52. Robert May, "Science and Society," *Science Year*, June 23, 2002, http://www.scienceyear.com/ about_sy/events/pdfs/Sci_SocActivities.pdf (accessed June 26, 2003).

53. Justin Sutcliffe, "MMR: More Scrutiny Please," *British Medical Journal* 326, no. 1272, (June 7, 2003), http://bmj.com/cgi/content/full/326/7401/1272.html (accessed June 26, 2003).

54. Jochen Pade et al., "Science on Television: Alternating between Elitism and Leveling," 1998, http://www.kommwiss.fu-berlin.de/~wissjour/studium/pcst98/Paper_pdf/pade.pdf (accessed June 26, 2003).

55. Michael Nitz and Sharon Jarvis, "Science in the News: The Potential Impact of Televised News Stories about Global Warming," a paper presented at the 5th International Conference on Public Communication of Sciences and Technology, Berlin, September,

1998, http://www.kommwiss.fu-berlin.de/~wissjour/studium/pcst98/Paper_pdf/nitz.pdf (accessed June 26, 2003).

56. Jochen Pade et al., "Science on Television: Alternating between Elitism and Leveling," 1998, http://www.kommwiss.fu-berlin.de/~wissjour/studium/pcst98/Paper_pdf/pade.pdf (accessed June 26, 2003).

57. Ibid.

58. Robert May, "Science and Society," *Science Year,* June 23, 2002, http://www.scienceyear.com/ about_sy/events/pdfs/Sci_SocActivities.pdf (accessed June 26, 2003).

59. "Nanoscience Technologies to develop media campaign," *Small Times,* May 20, 2005, http://www.smalltimes.com/document_display.cfm?document_id=9265 (accessed May 23, 2005).

60. Carol Boggs, "Rise and Decline of the Public Sphere," *The End of Politics* (New York: Guilford Press, 2000), p. 111.

61. G. W. F. Hegel, *Hegel's Philosophy of Right,* trans. T. M. Knox (Oxford: Clarendon Press, 1942, 1945), p. 207.

62. Andrew Scott, "Euro GM moratorium ends," *Scientist,* May 21, 2004, http://www.biomedcentral.com/news/20040521/03 (accessed May 22, 2004).

63. Ibid.

64. Glenda Chui, "Who's afraid of nanotechnology?" *Mercury News,* September 16, 2003, http://www.bayarea.com/mld/mercurynews/living/health/6779281.htm?template =content (accessed October 17, 2003).

65. Mary Ryan, *Women in Public: Between Banners and Ballots, 1825–1990* (Baltimore: Johns Hopkins Universtiy Press, 1990).

66. Nancy Fraser, "Rethinking the Public Sphere," *Habermas and the Public Sphere* (Cambridge, MA: MIT Press, 1992), p. 171.

67. Ibid., p. 126.

68. Moishe Postone, "Political Theory and Historical Analysis," *Habermas and the Public Sphere* (Cambridge, MA: MIT Press, 1992), p. 171.

69. Michael Warner, *Publics and Counterpublics* (New York: Zone Books, 2002), p. 91.

70. Jerome Davis, *Contemporary Social Movements* (New York: Century, 1930).

71. Mayer Zald and Roberta Ash, "Social Movement Organizations: Growth, Decay and Change," *Social Forces* 44, no. 3 (March 1966): 327–40.

72. "Some Subjects for Graduate Study Suggested by Members of the Department of Public Speaking of Cornell University," *Quarterly Journal of Speech 9* (April 1923): 147–53.

73. Doug McAdam and David Snow, *Social Movements: Readings on their Emergence, Mobilization, and Dynamics* (Los Angeles: Roxbury, 1997).

74. John A. Guidry, *Globalizations and the Public Sphere* (Ann Arbor: University of Michigan Press, 2000), p. 11.

75. Ibid., p. 12.

76. Jesse Lemisch, "A Movement Begins: The Washington Protests against IMF/World Bank," *New Politics* 8, no. 1 (new series), whole no. 29 (Summer 2000).

77. Renee Kjartan, "Coalition files petition against Frankenfoods," *Frankenfood* 46, (July/August 2000), http://www.speakeasy.org/wfp/46/frankenfoods.html (accessed June 26, 2003).

78. David Martosko, "Food fetish," *American Spectator on the Web*, 2003, http://tas.spectator.org/article.asp?art=8 (accessed June 26, 2003).

79. Friends of the Earth International, "Action Alerts: Playing with Hunger, *New FoEI GMO Report*," May 2003, 38, http://www.foei.org/publications/interlinkages/current.html (accessed June 26, 2003).

80. Reuters, "GMO crops here to stay or gone with the wind!" *Consumer Choice*, November 5, 2001, http://www.biotech-info.net/gone.html (accessed June 26, 2003).

81. Sue Mayer, "From genetic modification to nanotechnology: The dangers of sound science," *Science: Can we trust the experts?* ed. T. Gilland (London: Hodder and Stoughton, 2002), pp. 1–15, in Steven Wood, Richard Jones, and Alison Geldart, *The Social and Economic Challenges of Nanotechnology* (Swindon, UK: Economic and Social Research Council, 2003), http://www.esrc.ac.uk/esrccontent/DownloadDocs/Nanotechnology.pdf (accessed January 3, 2004), p. 30.

82. Glenn Harlan Reynolds, "Greenpeace and Nanotechnology," *Tech Central Station*, July 30, 2003, http://www.techcentralstation.com/073003B.html (accessed September 2, 2003).

83. "When nanopants attack," *10E20*, June 12, 2005, http://www.10e20webdesign.com/news/news_center_latest_technology_internet_news_12_june_05_When_Nanopants_Attack.htm (accessed July 18, 2005).

84. Vicki Colvin, "Responsible nanotechnology: Looking beyond the good news," *Eureka Alert! In Context*, November 1, 2002, http://www.eurekaalert.org/context.php?context=nano&show=essays&essaydate=1102 (accessed October 5, 2003).

85. Jim Lewis, "Regulating Nanotechnology," *Foresight Update* 51 (April 15, 2003): 4.

86. Vicki Colvin, "Responsible nanotechnology: Looking beyond the good news," *Eureka Alert! In Context*, November 1, 2002, http://www.eurekaalert.org/context.php?context=nano&show=essays&essaydate=1102 (accessed October 5, 2003).

87. David Berube in Ted Agres, "US Congress OKs nanotech bill," *Scientist*, November 24, 2003, http://www.biomedcentral.com/news/20031124/05 (accessed December 15, 2003).

88. "Trouble in Nanoland," *Economist*, December 5, 2002, http://www.economist.com/science/PrinterFriendly.cfm?Story_IS=1477445 (accessed October 5, 2003).

89. Joe Kaplinsky, "Nanotechnology: A slippery debate," *Spiked Online*, May 7, 2003, http://131.104.232.9agnet/2003/5-2003/agnet_may_11.htm (accessed September 6, 2003).

90. "Don't believe the hype," *Nature* 424, no. 6946 (July 17, 2003): 237.

91. Ibid.

92. "Nanobots," June 3, 2003, http://www.brightonlife.com/fastlike/articles.php?news_id=197 (accessed September 5, 2003).

93. "Don't believe the hype," *Nature* 424, no. 6946 (July 17, 2003): 237.

94. Katherine Viner, "'Luddites' we should not ignore," *Guardian*, September 29, 2000, http://www.zmag.org/luddites.htm (accessed July 9, 2004).

95. Ibid.

96. Wendy Barnaby, "National Forum for Science: Do We Trust Today's Scientists," March 2002, www.royalsoc.ac.uk/trackdoc.asp?id=556&pId=1988 (accessed July 18, 2005).

97. James Fishkin, *The Voice of the People* (New Haven, CT: Yale University Press, 1997).

98. "Austin Conference: A Microcosm of the Whole Country Changing Its Mind," *Current*, January 29, 1996, http://www.current.org/el602.html (accessed June 17, 2003).

99. Diana Claitor, "We listen and are willing to help each other," *Current*, January 29, 1996, http://www.current.org/el602.html (accessed June 17, 2003).

100. Jay Rosen, "As Democracy Goes, So Goes the Press," *What are Journalists For?* http://www.yale.edu/yup/chapters/078234chap.htm (accessed June 19, 2003).

101. Richard E. Sclove, "Town Meetings on Technology," *Technology Review*, July 1996, http://www.loka.org/pubs/techrev.htm (accessed June 17, 2003).

102. Ibid.

103. "NC State Professors to Discuss Public Perceptions of Nanotechnology at AAAS Conference," February 18, 2005, http://www.ncsu.edu/news/press_releases/05_02/045.htm (accessed July 18, 2005).

104. "Report of the Madison Area Citizen Consensus Conference on Nanotechnology," April 24, 2005, http://www.lafollette.wisc.edu/research/Nano/nanoreport42805.pdf (accessed July 18, 2005).

105. Jefferson Center, "Citizens Jury," http://www.jefferson-center.org/citizens_jury.htm. 2002 (accessed June 17, 2002).

106. "Last word: Now we're going public," *Guardian*, May 19, 2005, e-communication.

107. Glenn Reynolds, "NanoDynamism vs. NanoTimidity," *Tech Station*, August 4, 2004, http:// www.techcentralstation.com/080404B.html (accessed August 9, 2004).

BIBLIOGRAPHY

"$1 Billion for National Nanotechnology Initiative." 2004. http://www.azonano.com/news _old.asp?newsID=50 (accessed July 20, 2004).

"2003 Nanotech Product Guide." *Nanotech Report,* July 2003, p. 1.

"2nd Annual International Nanoscience Conference to Be Held in Grenoble, France." News.Nanoapex.com, May 18, 2004. http://news.nanoapex.com/modules.php?name =News&file=article&sid=4545 (accessed May 25, 2004).

Aeppli, Gabriel. "London Centre for Nanotechnology." Nanotec.org.uk, 2003. http://www .nanotec.org.uk/evidence/59aGabrielAeppli.htm (accessed March 30, 2004).

Agres, Ted. "Nanotech Ethics Debated." *Scientist*, December 8, 2003. http://www.biomed central.com/news/20031208/04 (accessed December 9, 2003).

———. "U.S. Congress OKs Nanotech Bill." *Scientist*, November 24, 2003. http://www .biomedcentral.com/news/20031124/05 (accessed December 15, 2003).

Alakeson, Vidhya. "Nanotechnology Call for Views." Nanotec.org.uk, 2003. http://www .nanotec.org.uk/evidence/62aForumForTheFuture.htm (accessed March 30, 2004).

———, and Tim Aldrich. "Weighed in the Nanoscale." *Green Futures*, January–February 2004. http://www.greenfutures.org.uk/features/default.asp?id=1723 (accessed February 12, 2004).

Albertario, Fabio, and Julian Snape. A solicited response to the European Commission Workshop. Nanotec.org.uk, 2003. http://www.nanotec.org.uk/evidence/79aFabioAl- bertario.htm (accessed March 30, 2004).

Allen, David D., Joanna Koziara, Russell J. Mumper, Thomas J. Abbruscato, and Paul R. Lockman. "Novel Nanoparticles Demonstrate No Adverse Effects on Blood-Brain Barrier Baseline Parameters in In Vitro or In Vivo Preparations." 225th ACS National Meeting, New Orleans, LA, March 23–27, 2003. http://oasys2.confex.com/acs/ 225nm/techprogram/P582270.htm (accessed December 1, 2003).

Altmann, Jürgen. "Military Uses of Nanotechnology—Risks and Proposals for Precautionary Action." Nanotec.org.uk, 2003. http://www.nanotec.org.uk/evidence/93b AltmannGobrudd.htm (accessed March 30, 2004).

Alyaudtin, R. N., A. Reichel, R. Lobenberg, P. Ramge, J. Kreuter, and D. J. Begley. "Interaction of Poly(butylcyanoacrylate) Nanoparticles with the Blood-Brain Barrier In Vivo and In Vitro." *Journal of Drug Targeting* 9, no. 3 (2001): 209–21.

American Society of Mechanical Engineers. *Monthly Report for the White House Office of Science and Technology Policy.* June 2003. http://www.asme.org/gric/fellows/reports/mackint_09-02.html (accessed June 10, 2003).

Arendt, Hannah. *The Human Condition.* 2nd ed. Chicago: University of Chicago Press, 1998. First published 1958.

———. *The Origins of Totalitarianism.* New York: Harcourt Brace, 1951.

Arnall, Alexander Huw. *Future Technologies, Today's Choices: Nanotechnology, Artificial Intelligence and Robotics; A Technical, Political and Institutional Map of Emerging Technologies.* London: Greenpeace Environmental Trust, 2003. http://www.greenpeace.org.uk/Multimedia Files/Live/FullReport/5886.pdf (accessed August 5, 2004).

Aronson, Neil. "Nanotech's Market Potential." *Boston Business Journal,* July 16, 2004. http://www.bizjournals.com/boston/stories/2004/07/19/editorial3.html (accessed July 26, 2004).

"Asian Governments Open Their Wallets for Nanotech." *Nanotech Report,* January 2003, p. 3.

"Asian Nanotech Fever Running Hot." *Nanotech Report,* January 2003, p. 2.

Associated Press. "Japanese Firm, State of New Mexico Sign Deal to Commercialize Technology." *Small Times.* http://www.smalltimes.com/print_doc.cfm?doc_id=8066 (accessed July 9, 2004).

"Austin Conference: A Microcosm of the Whole Country Changing Its Mind." *Current,* January 29, 1996. http://www.current.org/el602.html (accessed June 17, 2003).

Bainbridge, William Sims. "Sociocultural Meanings of Nanotechnology: Research Methodologies." Speaker, 2nd Workshop on Societal Implications of Nanoscience and Nanotechnology. Arlington, VA: NSF, December 3–5, 2003.

Baker, L. "Bush Continues NanoTech Funding Level." *High Tech Ceramic News,* February 2003. http://bcc.ecnext.com/coms2/summary_0002_001984_000000_000000_0002 _1 (accessed April 26, 2004).

Ball, Philip. "Natural Strategies for the Molecular Engineer." *Nanotechnology* 13 (2002): R15–28.

Balmer, Richard. "Hopes and Concerns of Nanotechnology." Nanotec.org.uk, 2003. http://www.nanotec.org.uk/evidence/90BalmerALDES.htm (accessed March 30, 2004).

Bates, D. "Health Indicates of the Adverse Effects of Air Pollution: The Question of Coherence." *Environmental Research* 59 (1992): 226–349.

Baum, Rudy. "Nanotechnology: Drexler and Smalley Make the Case for and against Molecular Assemblers." *Chemical and Engineering News,* December 1, 2003. http://pubs .acs.org/cen/coverstory/8148/8148counterpoint.html (accessed December 2, 2003).

Beardsley, Timothy. "Nanofuture: How Much Fun Would It Be to Live Forever?" *Scientific American,* January 1990, pp. 15–16.

Begley, Sharon. "Welcome to Liliput." *Newsweek*, April 15, 1991, pp. 60–61.

Benhabib, Seyla. "The Embattled Public Sphere." *Reasoning Practically.* Edited by Edna Ullmann-Margalit. New York: Oxford University Press, 2000.

Bennett, James C. "Funding Nanotechnology Ourselves." *Foresight Update*, July 1, 1993, p. 11.

———. "The Politics of Technology in the United States." *Prospects in Nanotechnology: Toward Molecular Manufacturing.* Edited by Markus Krummenacker and James Lewis. New York: Wiley, 1995.

Beradelli, Phil. "Interview: Marburger Defends R&D Policies." *United Press International*, June 23, 2004. http://www.upi.com/view.cfm?StoryID=20040622-07343506586r (accessed July 9, 2004).

Berg, Paul, D. Baltimore, H. W. Boyer, et al. Letter to the Editor. *Science* 185 (1974): 303.

———. "Study Raises Concerns over Buckyballs." *Houston Chronicle*, March 29, 2004. http://www.chron.com/cs/CDA/printstory.mpl/health/244472494 (accessed April 5, 2004).

Berger, Eric. "It's Going to Be a Giant Business." *Houston Chronicle*, March 4, 2004. http://www.chron.com/cs/CDA/printstory.mpl/business/2432521 (accessed March 18, 2004).

Bernard, Allen. "Zyvex: Building Nanoscale Machines with Microscopic Engines." *Nano-electronics Planet*, February 14, 2002. http://www.nanoelectronicsplanet.com/features/article/0,4028,6571_975231,00.html (accessed August 3, 2004).

"Beware of Nano-Pretenders." *Nanotech Report*, May 2002, p. 1.

"Beware the Nanobubble." *Nanotech Report*, June 2003, p. 3.

"Big Potential in TINY." *Nanotech Report*, March 2002, p. 1.

"Bingaman Calls for DARPA Head to Resign over Terrorism Futures Idea." *ABQ Journal*, July 29, 2003. http://www.abqjournal.com/news/metro/112110nm07-29-03.htm (accessed April 26, 2004).

Binks, Peter. "Questions Loom Large in Nanotech's Tiny World." *Age*, October 21, 2003. http://www.smh.com.au/articles/2003/10/20/1066631346170.html (accessed October 28, 2003).

"Biography of Philippe Van Nedervelde." Foresight.org, 2004. http://www.foresight.org/FI/VanNedervelde.html (accessed July 26, 2004).

"Blair Is the Enemy of the Greens." *Observer*, August 18, 2002. http://observer.guardian.co.uk/worldview/story/0,11581,776678,090.html (accessed July 21, 2003).

Blanpied, W. "Inventing US Science Policy." *Physics Today*, February 1998, pp. 34–40.

BMRB Social Research. *Nanotechnology: Views of the General Public, Quantitative and Qualitative Research Carried Out as Part of the Nanotechnology Study: BMRB International Report 45101666*, January 2004. http://www.nanotec.org.uk/Market%20Research.pdf (accessed June 25, 2004).

Boggs, Carol. "Rise and Decline of the Public Sphere." *The End of Politics.* New York: Guilford Press, 2000.

Boland, Michael. "Nanosys IPO Withdrawal; What Does It Mean for the Industry?" *Innovation World TechConnect*, August 6, 2004. http://web2.innovationworld.net/techconnect/ (accessed August 9, 2004).

Bond, Philip. "Converging Technologies and Competitiveness." In *Converging Technologies for Improving Human Performance: Nanotechnology, Biotechnology, Information Technology and Cognitive Science,* pp. 28–30. Edited by Mihail C. Roco and William Sims Bainbridge. Arlington, VA: National Science Foundation, 2000.

Boudreau, John. "Nanosys Pulls IPO." *Mercury News,* August 5, 2004. http://www.silicon valley.com/mld/siliconvalley/9324713.htm (accessed August 5, 2004).

Bounds, Jeff. "Zyvex Finding Nanotech to Be a Viable Business." *Dallas Business Journal,* July 9, 2004. http://www.bizjournals.com/dallas/stories/2004/07/12/newscolumn5 .html?jst=s3_rs_hl (accessed August 4, 2004).

Brazil, Rachel. "Nanotechnology—The Issues." Nanotec.org.uk, 2003. http://www.nanotec .org.uk/evidence/78aRSC.htm (accessed March 30, 2004).

Bridges, Andrew. "Tiniest Particles of Matter Don't Behave, Raising Nanotech Concerns." *USA Today,* April 2, 2004. http://www.usatoday.com/tech/news/2004-04-02-nano-flaw_x.htm (accessed April 5, 2004).

Briney, Andrew. "Hype, Hype, Hooray!" *InfoSecurity,* July 2003. http://infosecuritymag .techtarget.com/2003/jul/note.shtml (accessed July 23, 2004).

Broad, William J., and James Glanz. "Does Science Matter?" *New York Times,* November 11, 2003.

Broadhead, Lee-Anne, and Sean Howard. "The Heart of Darkness." *Resurgence,* November/December, 2003. http://resurgence.gn.apc.org/issues/broadhead221.htm (accessed November 4, 2003).

Brockman, John. "The Emerging Third Culture: Scientists Who Publish for General Audiences." *Whole Earth Review,* June 22, 1992, pp. 16–19.

———. *The Third Culture.* New York: Simon & Schuster, 1995.

Brother, Dallas. "The Road to Lilliput is Paved with Good Intentions: The Nanotechnology Conference Interviews, November, 1992." *MONDO 2000,* May 1992, p. 100.

Brown, Douglas. "Another Year, Another Threat to the Advanced Technology Program." *Small Times,* February 6, 2003. http://www.smalltimes.com/document_display.cfm? document_id=5456 (accessed June 23, 2004).

———. "Bush's Proposed Budget Makes Nanotechnology a Top Priority." *Small Times,* February 5, 2003. http://www.smalltimes.com/document_display.cfm?document_id =5449 (accessed April 26, 2004).

———. "Changing of the Guard." *Small Times,* January/February 2003, p. 14.

———. "Nano Litterbugs? Experts See Potential Pollution Problems." *Small Times,* March/April 2002. http://www.smalltimes.com/document_display.cfm?document_id =3266 (accessed July 30, 2003).

———. "Nanotech's Free Ride Is Over: Congress Will Want the Moon." *Small Times,* July/August 2003, p. 16.

———. "Nation Depends on More Money for Nano, Advocates Tell Senate." *Small Times,* 2002. http://www.smalltimes.com/document_display.cfm?&document_id=3829 (accessed on July 21, 2004).

———. "Perception May Be Nano's Biggest Enemy, Leaders Tell Congress." *Small Times,* April 10, 2003. http://www.smalltimes.com/print_doc.cfm?doc_id=5809 (accessed April 24, 2003).

Brown, Paul. "Printers Pulp Monsanto Edition of Ecologist." *Guardian,* September 29, 1998.

http://www.csdl.tamu.edu/FLORA/328Fall98/Reading14.html (accessed July 21, 2003).

Brubaker, Harold. "Nanotech Finds Fans in Industry." *Sun Herald*, April 30, 2004. http://www.sunherald.com/mld/thesunherald/business/8554757.htm (accessed May 3, 2004).

Brumfiel, Geoff. "A Little Knowledge." *Nature*, July 17, 2003, pp. 246–48.

Brunton, Michael. "Little Worries: Invasion of the Nanobots? Critics Warn of a 'Gray Goo' Future as Nanotechnology Enters the Marketplace." *Time Europe*, May 4, 2003. http://www.time.com/time/europe/magazine/article/0,13005,901030512-449458,00.html (accessed July 22, 2004).

Bryant Debra M. "Overview Bush Administration FY 2005 Budget Recommendations for Information Technology." Njtc.org, March 5, 2004. http://www.njtc.org/public policy/FY%202005%20IT%20Budget%20Rec.doc (accessed June 23, 2004).

"Buckyballs Cause Brain Damage in Fish." NewScientist.com, March 29, 2004. http://www.newscientist.com/news/print.jsp?id=ns99994825 (accessed April 5, 2004).

Burnell, Scott R. "Interview: Sen. Wyden Eyes Nanotech." *UPI Science News*, September 22, 2002. http://www.upi.com/view.cfm?StoryID=20020920-030036-4843r (accessed July 26, 2004).

Burnham, John. *How Superstition Won and Science Lost.* New Brunswick, NJ: Rutgers University Press, 1988.

"Bush and the Nano Pretenders." *Nanotech Report,* December 2003, p. 3.

Bush, George W. Address to a Joint Session of Congress and the American People. September 20, 2001. http://www.whitehouse.gov/news/releases/2001/09/20010920-8.html (accessed April 2, 2004).

Business Communications Company. Press release, February 3, 2003. http://www.bccresearch.com/editors/RGB-245R.html (accessed May 1, 2004).

"C60 Market Monitor of May 2004." NanoInvestorNews, May 16, 2004. http://www.nanoinvestornews.com/modules.php?name=News&file=article&sid=2792 (accessed May 20, 2004).

Calvey, Mark. "Jurvetson Pins Big Hopes on Tiny Nanomachines." *San Francisco Business Times*, July 16, 2001.

Cameron, David. "Walking Small." *Technology Review*, March 1, 2002. http://www.technologyreview.com/articles/print_version/cameron030102.asp (accessed July 31, 2003).

"Can Zac Save the Planet?" *Guardian*, November 7, 2000. http://www.guardian.co.uk/gmdebate/Story/0,2763,835010,00.html (accessed July 21, 2003).

Carbon Nanotechnologies, Inc. "Materials Safety Data Sheet." Cnanotech.com, April 25, 2002. http://www.cnanotech.com/download_files/MSDS%20for%20CNI%20SWNT.pdf (accessed August 3, 2004).

"Carbon Nanotechnologies, Inc. Announces the Allowance of a U.S. Patent for Coating Single-Walled Carbon Nanotubes." NanoInvestorNews, July 13, 2004. http://www.nanoinvestornews.com/modules.php?name=News&file=print&sid=3064 (accessed July 19, 2004).

Carroll, John. A solicited response to European Commission Workshop. Nanotec.org.uk, 2003. http://www.nanotec.org.uk/evidence/49JohnCarroll.htm (accessed March 30, 2004).

Cellucci, Thomas. "Fiscal Year 2005 NIST Budget: Views from Industry." U.S. House of Representatives' Committee on Science Environment, Technology, and Standards Subcommittee Hearing, April 28, 2004. http://www.zyvex.com/News/Testimony 042404_F.html (accessed on May 3, 2004).

Center for Disease Control. "Graphite, Synthetic." http://www.cdc.gov/niosh/pel88/ SYNGRAPH.html (accessed August 3, 2004).

Center for Responsible Nanotechnology. "Benefits…Risks…Solution…Urgency…Politics.…" http://www.crnano.org/index.html (accessed July 28, 2003).

———. "CRN and Students Team Up to Tackle Nanotechnology Issues." Press release, June 21, 2004. http://www.crnano.org/PR-Student%20Program.htm (accessed July 28, 2004).

———. "CRN Research: Current Results—Bootstrapping a Nanofactory." http://crnano .org/bootstrap.htm (accessed July 28, 2003).

———. "CRN Research: Current Results—Dangers of Molecular Nanotechnology." http://responsiblenanotechnology.org/dangers.htm (accessed July 28, 2003).

———. "CRN Research: Current Results—The Need for Early Development." http:// responsiblenanotechnology.org/early.htm (accessed July 28, 2003).

———. "CRN Research: Current Results—The Need for International Control." http:// responsiblenanotechnology.org/early.htm (accessed July 28, 2003).

———. "CRN Research: Current Results—The Need for International Development." http://responsiblenanotechnology.org/early.htm (accessed July 28, 2003).

———. "CRN Research: Current Results—No Simple Solutions." http://responsible nanotechnology.org/solutions.htm (accessed July 28, 2003).

———. "CRN Research: Current Results—Overview." http://responsiblenanotechnology .org/overview.htm (accessed July 28, 2003).

———. "CRN Research: Current Results—Possible Technical Restrictions." http:// responsiblenanotechnology.org/restrictions.htm (accessed July 28, 2003).

———. "CRN Research: Current Results—A Solution That Balances Many Interests." http://responsiblenanotechnology.org/early.htm (accessed July 28, 2003).

———. "Grey Goo Is a Small Issue." http://www.crnano.org/BD-Goo.htm (accessed December 15, 2003).

———. "Inside CRN." http://responsiblenanotechnology.org/inside.htm (accessed July 28, 2003).

———. "Invited Commentary on Royal Society Nanotechnology Workshop." December 2003. http://www.crnano.org/RSWorkshop1.htm (accessed December 8, 2003).

———. "Issues, Positions—and Urgency—in Nanotechnology Policy." http://responsible nanotechonlogy.org/positions.htm (accessed July 28, 2003).

———. "Managing Magic." http://responsiblenanotechnology.org/magic.htm (accessed July 28, 2003).

———. "Patchwork Regulation of Nanotech Could Be Grave Danger." February 15, 2003. http://responsiblenanotechnology.org/PR-patchwork.htm (accessed July 28, 2003).

———. "Published Papers." http://responsiblenanotechnology.org/papers.htm (accessed July 28, 2003).

———. "A Technical Commentary on Greenpeace's Nanotechnology Report." September 2003. http://crnano.org/Greenpeace.htm (accessed September 4, 2003).

———. "Transnational Regulation." April 2004. http://crnano.typepad.com/crnblog/2004/04/transnational_r.html (accessed July 28, 2004).

Cerf, Christopher, and Victor Navasky. *The Experts Speak: The Definitive Compendium of Authoritative Misinformation.* New York: Pantheon Books, 1984.

Chang, Kenneth. "IBM Creates a Tiny Circuit out of Carbon." *New York Times,* August 27, 2001.

———. "Smaller Computer Chips Built Using DNA as a Template." *New York Times,* November 21, 2003.

"Charles Fears Science Could Kill the Earth." *Floyd Report,* April 27, 2003. http://www.floyd report.com/view_article.php?lid=209 (accessed July 21, 2003).

Chen, H. H. C.. C. Yu, T. H. Ueng, S. Chen, B. J. Chen, K. J. Huang, and L.Y. Chiang. "Acute and Subacute Toxity Study of Water-soluble Polyalkylsulfonated C-60 in Rats." *Toxicological Pathology* 26 (1998): 1387–96.

Chen, H. H. C., B. J. Chen, L.Y. Chiang, C. C. Hong, C. T. Liang, T. H. Ueng, C. Yu. "Renal Effects of Water-soluble Polyarylsulfonated C-60 in Rats with an Acute Toxicity Study." *Fullerene Science and Technology* 5 (1997): 1387–96.

Chen, Hongda. "U.S. Department of Agriculture Research on Nanotechnology." *NNI: From Vision to Commercialization,* Washington, DC, March 31–April 2, 2004.

Choi, Charles. "Nano Bill Promises Real Results." TechNewsWorld, December 4, 2003. http://www.technewsworld.com/perl/story/32298.html (accessed December 9, 2003).

———. "Nanotubes, Buckyballs Surprise Investors." United Press International, May 19, 2004. http://www.upi.com/view.cfm?StoryID=20040518-011947-3488r (accessed May 25, 2004).

Chui, Glenda. "Who's Afraid of Nanotechnology?" *Mercury News,* September 16, 2003. http://www.bayarea.com/mld/mercurynews/living/health/6783577.htm (accessed October 2, 2003).

Cientifica. *The Nanotechnology Opportunity Report,* 2003. http://nanotechnow.com/nanotechnology-opportunity-report.htm (accessed April 26, 2004).

Claitor, Diana. "We Listen and Are Willing to Help Each Other." *Current,* January 29, 1996. http://www.current.org/el602.html (accessed June 17, 2003).

Clark, Wayne. *Activism in the Public Sphere: Exploring the Discourse of Political Participation.* Aldershot, UK: Ashgate, 2000.

Clinton, William J. "Remarks on Science and Technology Investments." California Institute of Technology, Pasadena, CA, January 21, 2000. http://www.columbia.edu/cu/osi/nanopotusspeech.html (accessed August 7, 2003).

Cobb, Michael, and Jane Macoubrie. "Public Perceptions about Nanotechnology: Risks, Benefits and Trust," Unpublished paper, North Carolina State University, 2004.

Cobb, Stewart. "International News." *Foresight Update,* October 30, 1990, pp. 4–5.

Collins, Graham. "Shamans of Small." *Scientific American,* September 2001, pp. 85–91.

Colvin, Vicki. "The Potential Environmental Impact of Engineered Nanomaterials." *Nature Biotechnology* 21, no. 10 (October 2003). http://www.nature.com/cgi-af/DynaPage.taf ?file=/nbt/journal/v21/n10/full/nbt875.html (accessed March 9, 2004).

———. "Responsible Nanotechnology: Looking Beyond the Good News." *Eureka Alert: In Context,* November 2002. http://www.eurekaalert.org/context.php?context=nano &show=essays&essaydate=1102 (accessed December 10, 2003).

Colwell, Rita. "Welcoming Remarks." Abstracts of the Workshop on Societal Implications of Nanoscience and Nanotechnology, Arlington, VA, December 3–5, 2003.

Commission of the European Communities. "Towards a European Strategy for Nanotechnology." 2004. ftp://ftp.cordis.lu/pub/nanotechnology/docs/nano_com_en.pdf (accessed July 13, 2004).

Committee for the Review of the National Nanotechnology Initiative, National Research Council. *Small Wonders, Endless Frontiers: A Review of the National Nanotechnology Initiative.* Washington, DC: National Academies Press, 2002.

"Companies to Watch." *Nanotech Report,* July 2004, 6.

"Congress OKs Grant for Nanotechnology." *Dallas Morning News,* June 22, 2004. http://wwww.dallasnews.com/cgi-bin/bi/gold_print.cgi (accessed June 30, 2004).

Cook, John. "Venture Capital: Nanotech May Take Years to Hit It Big." *Seattle Post-Intelligencer,* August 6, 2004. http://seattlepi.nwsource.com/venture/185099_vc06.html (accessed August 9, 2004).

Copeland, Phil. "Future Warrior Exhibits Super Powers." AFIS Defenselink, July 27, 2004. http://www.defenselink.mil/news/Jul2004/n07272004_2004072705.html (accessed July 30, 2004).

CORDIS. "Nanotechnology: Opportunity or Threat?" *CORDIS,* June 17, 2003. http://www.godlikeproductions.com/news/item.php?keyid=46322 (accessed July 30, 2003).

Correia, Antonio, and Pastora Martinex Samper. "Nanobiotechnology and the PHANTOMS Network." *Applied Nanoscience* 1, no. 1 (2004): 13–16.

Crosby, Steve. "Study Nano and the Environment Now, Because What We Can't See Scares Us." *Small Times,* January/February 2003, p. 4.

Crum, Rex. "Talking Nanotech with Mike Honda." CBS.Marketwatch.com, March 17, 2004. http://cbs.marketwatch.com/news/print_story.asp?print=1&guid={A637E67A-A5D7-41B- (accessed April 5, 2004).

Curran, James. "Rethinking the Media as a Public Sphere." In *Communication and Citizenship: Journalism and the Public Sphere.* Edited by Peter Dahlgren and Colin Sparks. London: Routledge, 1993.

Curtis, Wayne. "The Methuselah Report: Living to Be 120 Might Be Attainable, but Is It Desirable?" *AARP Bulletin Online,* July–August 2004. http://www.aarp.org/bulletin/yourhealth/Articles/a2004-07-07-methuselah.html (accessed July 12, 2004).

"DARPA Funds Collaborative Quantum Computer Center." *NanoDot,* February 14, 2001. http://nanodot.org/article.pl?sid=01/09/14/1313244 (accessed June 27, 2004).

Davey, Michael E. "RS20589: Manipulating Molecules: The National Nanotechnology Initiative." *CRS Report for Congress,* September 20, 2000. http://www.ncseonline.org/NLE/CRSreports/Sceince/st-48.cfm?&cfd=14971140&cftoken=12149058 (accessed July 19, 2004).

Davidson, Keay. "The Promise and Perils of the Nanotech Revolution." SFGate.com, July 26, 2004. http://www.sfgate.com/cgi-bin/article.cgi?file=/c/a/2004/07/26/MNG767SUKB1.DTL (accessed August 4, 2004).

Davis, Jerome. *Contemporary Social Movements.* New York: Century, 1930.

"Decoding Future Nanotech Investment Success." *Nanotech Report,* September 2002.

DeLong, J. Bradford. "Semi-daily Journal of Economist Brad DeLong: Fair and Balanced

Almost Every Day." December 3, 2003. http://www.j-bradford-delong.net.movable _type/2003_archives/002838.html (accessed December 8, 2003).

Derfus, Austin, Warren Chan, and Sangeeta Bhatia. "Probing the Cytotoxicity of Semiconductor Quantum Dots." *Nano Letters* 4 (2004): 11–18.

de Tocqueville, Alexis. *Democracy in America.* New York: Westvaco, 1999. First published in 1938.

Dewey, John. *Democracy and Education.* New York: Macmillan, 1916.

———. "The Public and Its Problems." *John Dewey: The Later Works, 1925–1953,* Volume 2. Edited by Jo Ann Boydson. Carbondale: Southern Illinois University Press, 1981.

Dinkelacker, Jamie. "Executive Director Notes." *Foresight Update,* July 15, 1992, p. 14.

———. "Interview: Jim Bennett." *Foresight Update,* July 15, 1992, pp. 12–13.

———. "Policy Watch." *Foresight Update,* April 15, 1994, pp. 8–9.

Dobson, Peter. "Nanotechnology." October 2003. http://www.nanotec.org.uk/evidence/ 58aPeterDobson.htm (accessed March 30, 2004).

"DOD Lacks Database to Determine Foreign Dependence, GAO Says." *Aerospace Daily,* April 6, 1994, p. 30.

Donaldson, K., V. Stone, A. Seaton, and W. MacNee. "Ambient Particle Inhalation and the Cardiovascular System: Potential Mechanisms." *Environmental Health Perspectives* 109 (2001): 523–27.

"Don't Believe the Hype." *Nature,* July 17, 2003, p. 237.

Dowie, Mark. "Brave New Tiny World." *California Magazine,* November 1988, pp. 148–49.

"Dr. Neal Lane to Return to Rice University." December 15, 2000. http://www.ostp.gov/ html/001215.html (accessed July 10, 2004).

Drexler, K. Eric. "Drexler Writes Smalley Open Letter on Assemblers." *Nanodot,* April 20, 2003. http://www.foresight.org/NanoRev/Letter.html (accessed May 15, 2003).

———. *Engines of Creation.* New York: Anchor Press, 1986.

———. *Nanosystems: Molecular Machinery, Manufacturing and Computation.* New York: Wiley-Interscience, 1992.

———. "Nanotechnology: From Feynman to Funding." *Bulletin of Science, Technology & Society* 24, no. 1 (February 2004): 21–27. http://www.metamodern.com/d/04/00/ FeynmanToFunding.pdf (accessed July 26, 2004).

———. "Visions of Nanotechnology: A World Divided." Invited Speaker, Imaging and Imagining Conference, University of South Carolina, Columbia, SC, March 3–7, 2004.

———, David Forrest, Robert Freitas, J. Storrs Halls, Neil Jacobstein, Thomas McKendree, Ralph Merkle, and Christine Peterson. "A Debate about Assemblers: Many Future Nanomachines; a Rebuttal to Whiteside's Assertion That Mechanical Molecular Assemblers Are Not Workable and Not a Concern." 2001. http://www.imm.org/Sci-AmDebate2/whitesides.html (accessed July 18, 2004).

Duncan, David Ewing. "The New Biochemphysicist: An Interview by David Ewing Duncan with George Whitesides." *Discover,* December 2003, pp. 22–24.

Dunford, Rosemary, Angela Salinaro, Lezhen Cai, Nick Serpone, Satoshi Horikoshi, Hisao Hidaka, and John Knowland. "Chemical Oxidation and DNA Damage Catalyzed by Inorganic Sunscreen Ingredients." *FEBS [Federation of European Biochemical Societies] Letters* 418 (1997): 1–2, 87–90.

"Earnings Season Hits the Nanosphere." *Nanotech Report,* November 2003, p. 4.

Economic and Social Research Council. "Our Mission." 2003. http://www.esrc.ac.uk/esrc-content/esrcgen/display.gen/mission.asp (accessed September 15, 2003).

——. "Who We Are." 2003. http://www.esrc.ac.uk/esrccontent/aboutesrc/aboutus.asp (accessed September 15, 2003).

Ehrrera, Stephan, and Lawrence Aragon. "Small Worlds: Nanotechnology Wins over Mainstream Venture Capitalists." *Red Herring,* December 18, 2001. http://www.redherring.com/Article.aspx?a=2975&hed=Small+Worlds (accessed August 3, 2004).

Eisenberg, Anne. "Benign Viruses Shine on the Silicon Assembly Line." *New York Times,* February 12, 2004.

"The End of Evolution." SMH.Com.Au, November 15, 2003. http://www.smh.com.au/articles/2003/11/14/1068674378878.html (accessed April 25, 2004).

ETC Group. "The Big Down: From Genomes to Atoms, Atomtech: Technologies Converging at the Nano-scale." January 2003. http://www.etcgroup.org/documents/TheBigDown.pdf (accessed September 5, 2003).

——. "Green Goo: Nanobiotechnology Comes Alive!" *Communique,* January/February 2003.

——. "The Little Bang Theory." *Communique,* February 6, 2003. http://www.etcgroup.org/article.asp?newsid=378 (accessed May 21, 2003).

——. "Nanoproducts." http://www.etcgroup.org/documents/nanoproducts_EPA.pdf (accessed July 20, 2004).

——. "Nanotech and the Precautionary Prince." *Genotype,* May 2, 2003. http://www.etcgroup.org/article.asp?newsid=397 (accessed May 21, 2003).

——. "Nanotech News in Living Colour: An Update on White Papers, Red Flags, Green Goo, Grey Goo (and Red Herrings)." *Communiqué,* May/June 2004.

——. "Nanotech Takes a Giant Step Down!" March 6, 2002. http://etcgroup.org/article.asp?newsid=305 (accessed May 21, 2003).

——. "Nanotech Un-gooed! Is the Grey/Green Goo Brouhaha the Industry's Second Blunder?" *Communiqué,* July/August 2003. http://www.etcgroup.org/article.asp?newsid=399 (accessed May 21, 2003).

——. "No Small Matter! Nanotech Particles Penetrate Living Cells and Accumulate in Animal Organs." *Communiqué,* July 23, 2002. http://www.etcgroup.org/article.asp?newsid=356 (accessed May 21, 2003).

——. "No Small Matter II: The Case for a Global Moratorium." *Occasional Paper Series,* April 2003.

——. "RAFI Becomes ETC Group." September 4, 2001. http://www.etcgroup.org/documents/news_rafietc.pdf (accessed July 28, 2004).

European Commission. "Third European Report on Science and Technology Indicators." *CORDIS,* March 2003. http://www.cordis.lu/indicators/third_report.htm (accessed April 26, 2004).

European Commission's Community Health and Consumer Protection Directorate. "Nanotechnologies: A Preliminary Risk Analysis on the Basis of a Workshop Organized in Brussels on March 1–2, 2004." http://europa.eu.int/comm/health/ph_risk/events_risk_en.htm (accessed June 20, 2004).

"European Free Alliance in the European Parliament." June 11, 2003. http://www.greens -ef.org/en/agenda/detail.php?id=1105&1g=en (accessed July 30, 2003).

"European Union Earmarks Billions to Prepare for a Nanofuture." *Small Times*, July 6, 2004. http://www.smalltimes.com/print_doc.cfm?doc_id=8144 (accessed July 12, 2004).

"EU-Government Funding for Nanotechnology." *Nanoinvestor News*, July 26, 2004. http:// www.nanoinvestornews.com/modules.php?name=Facts_Figures&op=sho&im =locfunding/eu (accessed July 26, 2004).

"EU-US-Japan Government Spending." *Nanoinvestor News*, July 27, 2004. http:// www.nanoinvestornews.com/modules.php?name=Facts_Figures&op=sho&im=fund _euusjapan9702 (accessed July 27, 2004).

"EU, CEOs Call for Nanotechnology Push." CNN.com, June 29, 2004. http://www .cnn.com/2004/TECH/biztech/06/29/nanotechnology.europe.reut/ (accessed August 4, 2004).

Evans, Barry. A solicited esponse to European Commission Workshop. Nanotec.org.uk, October 2003. http://www.nanotec.org.uk/evidence/56BarryEvans.htm (accessed March 30, 2004).

Fahey, Jonathan. "The Science of Small." *Forbes Global*, February 5, 2001. http://www .forbes.com/global/2001/0205/084.html (accessed August 3, 2004).

Falconer, Bruce. "Defense Research Agency Seeks to Create Supersoldiers." *National Journal*, November 10, 2003. http://www.govexec.com/dailyfed/1103/111003nj1.htm (accessed May 2, 2004).

Feder, Barnaby. "As Uses Grow, Tiny Materials' Safety Is Hard to Pin Down." *Silicon Investor*, November 3, 2003. http://www.siliconinvestor.com/stocktalk/msg.gsp?msgid =19459234 (accessed December 9, 2003).

———. "At IBM, a Tinier Transistor Outperforms Its Silicon Cousins." *New York Times*, May 20, 2002.

———. "Concerns That Nanotech Label Is Overused." *New York Times*, April 12, 2004. http://www.nytimes.com/2004/04/12/technology/12phone.html?ex=1397188800 &en=20db9db23c360970&ei=5007&partner=USERLAND (accessed May 2, 2004).

———. "Defense Department Expands Nanotechnology Research." *New York Times*, April 8, 2003. http://www.siliconvalley.com/mld/siliconvalley/news/5585217.htm (accessed August 5, 2004).

———. "How Big Is Nanotechnology's Hazard?" *International Herald Tribune*, March 29, 2004. http://www.iht.com/articles/512380.html (accessed April 5, 2004).

———. "It's a Tiny New World." *New York Times*, December 22, 2003.

———. "Study Raises Concerns about Carbon Particles." *New York Times*, March 29, 2004.

"FEI and Zyvex Sign Strategic Alliance Agreement." *NanoInvestorNews*, August 4, 2004. http://www.nanoinvestornews.com/modules.php?name=News&file=article &sid=3191 (accessed August 9, 2004).

Fenton, Gary. A solicited response to European Commission Workshop. Nanotec.org.uk, 2003. http://www.nanotec.org.uk/evidence/54GaryFenton.htm (accessed March 30, 2004).

Feynman, Richard. "Infinitesimal Machinery." *Journal of Microelectromechanical Systems* 2, no. 1 (March 1993): 4–14.

———. *The Meaning of It All: Thoughts of a Citizen-Scientist*. Reading, MA: Perseus Books, 1998.

———. *The Pleasure of Finding Things Out: The Best Short Works of Richard P. Feynman*. Edited by Jeffrey Robbins. Foreword by Freeman Dyson. Cambridge, MA: Perseus Books, 1999.

———. "There's Plenty of Room at the Bottom." *Engineering and Science* (1960): 22–36.

"First Feynman Prize in Nanotechnology Awarded: Caltech Chemistry Researcher Charles Musgrave Receives $5,000 Prize." *Foresight Update*, December 15, 1993, pp. 1–2.

"First Nanotechnology Development Company Formed and Seeking R&D Staff." *Foresight Update*, June 30, 1997. http://www.foresight.org/Updates/Updates29/update29.2.html (accessed July 26, 2004).

Fishbine, Glenn. *The Investors Guide to Nanotechnology and Micromachines*. New York: Wiley, 2002.

Fishkin, James. *The Voice of the People*. New Haven, CT: Yale University Press, 1997.

"Follow the Money." *Nanotech Report*, September 2003, p. 7.

"Forbes Ranks NanoBusiness Alliance Director as One of Nanotech Industry's Leading Power Brokers." *NanoTechWire.com*, March 18, 2004. http://nanotechwire.com/news.asp?nid=778&ntid=116&pg=2 (accessed June 25, 2004).

"Foresight Institute Offers $250,000 Feynman Grand Prize for Major Advanced in Molecular Nanotechnology." *Foresight Update*, April 15, 1996, pp. 1–2.

"Foresight Perspective & Policy." *Foresight Update*, June 1, 1995, p. 16.

"Foresight Publishes Guidelines for Development of Nanotechnology." *Foresight Update*, June 30, 2000. http://www.foresight.org/Updates/Updates41/update41.1.html (accessed July 26, 2004).

"Foresight Unveils Backlink Mediator to Provide Other Half of the Web." *Foresight Update*, December 15, 1997. http://www.foresight.org/Updates/Updates31/update31.3.html (accessed July 26, 2004).

"For Every Book You Buy Online, Amazon Donates to Foresight." *Foresight Update*, August 30, 1998. http://www.foresight.org/Updates/Updates34/update34.1.html (accessed July 26, 2004).

Forman, David. "Best Laid Plans." *Small Times*, July/August 2004, pp. 28–35.

———. "Industry Tries to Make Sense of Nanosys IPO Withdrawal." *Small Times*, August 5, 2004. http://www.smalltimes.com/print_doc.cfm?doc_id=8202 (accessed August 9, 2004).

———. "IP Storm Clouds Build on Horizon." *Small Times*, May/June 2004, pp. 21–24.

———. "Nano Takes to the Street: Second IPO Ready to Go." *Small Times*, July 21, 2004. http://www.smalltimes.com/document_display.cfm?section_id=51&document_id=8175 (accessed July 26, 2004).

———. "Nanotech Rides a Rising Tide." *Small Times*, March/April 2004, pp. 18–19.

———. "News Analysis: 2004 Finds 'Nano' Crossing into Speculative Space." *Small Times*, December 31, 2003. http://www.smalltimes.com/document_display.cfm?document_id=7155 (accessed January 6, 2004).

———. "VC Forecast: Slow, Steady, Boring." *Small Times*, November/December 2003, p. 8.

Foster, Richard, and Sarah Kaplan. *Creative Destruction*. New York: Doubleday, 2001.

"Fox Catches Prey." *IGN Insider*, July 26, 2002. http://filmforce.ign.com/articles/366/366010p1.html (accessed September 23, 2003).

Fraser, Bruce. "Interpretation of Novel Metaphors." *Metaphor and Thought*. 2nd ed. Edited by Andrew Ortony. Cambridge: Cambridge University Press, 1993.

Fraser, Nancy. "Rethinking the Public Sphere." In *Habermas and the Public Sphere*. Edited by Craig Calhoun. Cambridge, MA: MIT Press, 1992.

Fried, Jayne. "Investors See the Reality behind the Nano Hype—And Some Nice Pants." *Small Times*, March/April 2002.

Friends of the Earth International. "Action Alerts: Playing with Hunger, New FoEI GMO Report." FOEI.org, May 2003. http://www.foei.org/publications/interlinkages/current.html (accessed June 26, 2003).

Fubini, Bice. "Surface Reactivity in the Pathogenic Response to Particulates." *Environmental Health Perspectives* 105, supp. 5 (September 1997): 1013–20.

Fuller, Steve. *Philosophy, Rhetoric, and the End of Knowledge: The Coming of Science and Technology Studies.* Madison: University of Wisconsin Press, 1993.

"Future Nanotech IPOs—The A-list." *Nanotech Report*, June 2004, pp. 1–2.

Ganger, Gregory R., and David F. Nagle. "Enabling Dynamic Security Management of Networked Systems via Device-Embedded Security." December 2000. http://www.pdl.cmu.edu/PDL-FTP/Storage/CMU-CS-00-174.pdf (accessed August 19, 2003).

Gannett, Martin. A Solicited Response to European Commission Workshop. Nanotec.org.uk, 2003. http://www.nanotec.org.uk/evidence/85aMartinGannett.htm (accessed March 30, 2004).

Gardner, Elizabeth. "Brainy Food: Academia, Industry Sink Their Teeth into Edible Nano." *Small Times*, June 21, 2004, p. 4.

Garnham, Nicholas. "The Media and the Public Sphere." In *Habermas and the Public Sphere*. Edited by Craig Calhoun. Cambridge, MA: MIT Press, 1992.

Garreau, Joel. "From Internet Scientist, a Preview of Extinction." *Washington Post*, March 12, 2000. http://www2.gol.com/users/coynerhm/from_internet_scientist_htm (accessed July 28, 2004).

Gartner, Inc. "Hype Cycles Special Report." 2004. http://www4.gartner.com/research/special_reports/hype_cycle/hc_special_report.jsp (accessed July 23, 2004).

"Gartner Says No New Major IT Innovation before 2005." Press release, April 8, 2002. http://banners.noticiasdot.com/termometro/boletines/docs/ti/gartner/2002/gartner_Before_2005.pdf (accessed July 25, 2004).

Gasman, Lawrence. "Opportunities in the Emerging Nanostorage Market." *NanoMarkets White Paper*, July 2004.

———. "What Will the Future Nanotech Industry Look Like?" *NanoMarkets White Paper*, March 2004.

"Georgia Tech Researchers Use Lab Cultures to Control Robotic Device." *Science Daily*, April 28, 2003. http://www.sciencedaily.com/releases/2003/04/030428082503.htm (accessed May 13, 2003).

Gerard, Jasper. "Charles Gets in a Wee Tizz over Nanotechnology." *Sunday Times*, April 27, 2003, p. 17.

Gertner, Jon. "Proceed with Caution." *New York Times Magazine*, June 6, 2004, p. 32.

Gillmor, Dan. "Big Breakthroughs Can Come in Small Packages." SiliconValley.com, February 16, 2002. http://www.siliconvalley.com/mld/siliconvalley/business/columnists/2685956.htm (accessed July 27, 2004).

Gimzewski, James. A Solicited Response to European Commission Workshop. Nanotec

.org.uk, 2003. http://www.nanotec.org.uk/evidence/53aGSub.htm (accessed March 30, 2004).

Glassner, Barry. *The Culture of Fear: Why Americans Are Afraid of the Wrong Things; Crime, Drugs, Minorities, Teen Moms, Killer Kids, Mutant Microbes, Plane Crashes, Road Rage and So Much More.* New York: Basic Books, 1999.

Gleick, James. *Genius: The Life and Science of Richard Feynman.* New York: Vintage Books, 1992.

Goldfarb, Bruce. "New Task Force Focuses on Nanotechnology Health Effects." *NanoBiotech News,* August 13, 2003, p. 3.

Goldsmith, Zac. "Discomfort and Joy: Bill Joy Interview." *Ecologist,* September 22, 2000. http://www.theecologist.org/archive_article.html?article=188&category=84 (accessed March 22, 2004).

———. "The Seeds of Discord." *Sunday Telegraph,* March 2, 2002. http://millennium-debate.org/suntel3mar4.htm (accessed July 21, 2003).

———. "Time's up for the Global Economy." *Global Agenda Magazine,* 2003. http://www.globalagendamagazine.com/2003/zacgoldsmith.asp (accessed July 21, 2003).

Goldstein, Barbara. "Nanotechnology Is BIG at NIST." March 23, 2004. http://www.nist.gov/public_affairs/nanotech.htm (accessed July 14, 2004).

Gorman, Jessica. "Taming High-Tech Particles: Cautious Steps into the Nanotech Future." *Science News,* March 30, 2002. http://www.sciencenews.org/20020330/bob8.asp (accessed October 5, 2003).

Graham, Sarah. "Nanotech: It's Not Easy Being Green." ScientificAmerican.com, July 28, 2003. http://www.sciam.com/print_version.cfm?articleIID=00077C33-511E-1F20-B8E780A8418 (accessed March 18, 2004).

Gramsci, Antonio. *The Modern Prince and Other Writings.* London: Lawrence and Wishart, 1957.

Green, Katherine. "Zyvex Introduces S100 Nanomanipulator." *Canada NewsWire,* April 15, 2003.

Greenberg, Daniel S. *Science, Money, and Politics.* Chicago: University of Chicago Press, 2001.

Gruenwald, Juliana. "After the Bill: Bush's Budget a Bit Short." *Small Times,* March/April 2004, p. 10.

———. "Congress Decides to Study First and Regulate Later." *Small Times,* May/June 2004, p. 12.

———. "Controversial Study Points to Need for More Federal Research." *Small Times,* April 1, 2004. http://www.smalltimes.com/document_display.cfm?document_id=7658 (accessed April 5, 2004).

———. "D.C., Nano Union Now Put to the Test." *Small Times,* January/February 2004, p. 9.

———. "Despite House's Okay, Little Time Left to Pass Nanotech Bill This Year." *Small Times,* July 12, 2004. http://www.smalltimes.com/document_display.cfm?document_id=8159 (accessed July 19, 2004).

———. "Marburger Says Nano Regulators Ensure Health, Safety." *Small Times,* June 18, 2004. http://www.smalltimes.com/print_doc.cfm?doc+id=8075 (accessed July 9, 2004).

————. "PCAST Will Advise Bush on Nanotech." *Small Times*, July 20, 2004. http://www .smalltimes.com/document_display.cfm?document_id=8171 (accessed July 26, 2004).

————. "Researchers Discuss Safety Guidelines for Handling Nanomaterials." *Small Times*, May 19, 2004. http://www.smalltimes.com/print_doc.cfm?doc_id=7922 (accessed May 25, 2004).

————. "Senator Forming Caucus to Keep Nanotech Issues on Forefront." *Small Times*, April 8, 2004. http://www.smalltimes.com/print_doc.cfm?doc_id=7697 (April 13, 2004).

Guidry, John A. *Globalizations and Social Movements: Culture, Power, and the Traditional Public Sphere.* Ann Arbor: University of Michigan Press, 2000.

Gumbleton, M. "Caveolae as Potential Macromolecule Trafficking Compartments with Alveolar Epithelium." *Advanced Drug Delivery Reviews* 49 (2001): 281–300.

Habermas, Jurgen. *Habermas and the Public Sphere.* Edited by Craig Calhoun. Cambridge, MA: MIT Press, 1992.

————. *The Structural Transformation of the Public Sphere: An Inquiry into a Category of Bourgeois Society.* Translated by Thomas Burger. Cambridge, MA: MIT Press 1989. First published in 1963.

Hanson, Robin. A solicited response to European Commission Workshop. Nanotec.org.uk, 2003. http://www.nanotec.org.uk/evidence/73RobinHanson.htm (accessed March 30, 2004).

Hansson, P. "Nanotechnology: Prospects and Policies." *Futures*, October 1991, pp. 849–59.

Hapanowicz, Rick. "NanoCommerce 2003 Conference Overview Part II." News .NanoApex, December 15, 2003. http://news.nanoapex.com/modules.php?name =News&file=article&sid=4085 (accessed December 15, 2003).

"HARDtalk: Zac Goldsmith." *BBC Press Releases*, July 10, 2002. http://www.bbc.co.uk/ print/pressoffice/pressreleases/stories/2002/10_october/07/hardtalk (accessed July 26, 2003).

Harper, Tim. "Oral Evidence—Tim Harper, Cientifica." Nanotec.org.uk, 2003. http:// www.nanotec.org.uk/evidence/oralHarperTim.htm (accessed March 30, 2004).

"Harris & Harris Group Announced Closing of Follow-on Offering and Full Exercise of Overallotment Option." *NanoInvestorNews*, July 7, 2004. http://www.nanoinvestor news.com/modules.php?name=News&file=print&sid=3033 (accessed July 12, 2004).

"Harvard's George Whitesides on Nanotechnology: A Word, Not a Field." *Science Watch*, July/August 2002. http://www.sciencewatch.com/july-aug2002/sw_july-aug2002 _page3.htm (accessed June 30, 2004).

Haworth, W. Lance, and Ezio Andreta. European Commission—3rd Joint EC-NSF Workshop on Nanotechnology, January 31–February 1, 2002. *Nanotechnology Revolutionary Opportunities and Societal Implications.* Edited by Mihail Roco and Renzo Tomellini. Luxembourg: Office for Official Publications of the European Communities, 2002.

Healey, Peter. "Nanotechnology: Social, Ethical, Legal and Economic Issues—A STAGE Input." Nanotec.org.uk, 2003. http://www.nanotec.org.uk/evidence/69aPeterHealey .htm (accessed March 30, 2004).

Hegel, Georg W. F. *Hegel's Philosophy of Right.* Translated by T. M. Knox. Oxford: Clarendon Press, 1945. First published in 1942.

Helpman, Elhanan, ed. *General Purpose Technologies and Economic Growth*. Cambridge, MA: MIT Press, 1998.

Henderson, J. "Nanotechnology." *Small Times*, 2003.

Henderson, Tom. "Bayh-Dole Act of 1980 Opened the Doors for Researchers to Profit from Inventions." *Small Times*, July 15, 2001. http://www.smalltimes.com/document _display.cfm?document_id=1427 (accessed April 26, 2004).

Herrera, Stephen. "Fear Seems Not to Be a Factor in America Despite the Nanomonsters under the Bed." *Small Times*, January/February 2003, p. 45.

Hett, Annabelle. *Nanotechnology: Small Matter, Many Unknowns*. Zurich, Switzerland: Swiss Reinsurance, 2004.

Hett, Annabelle, ed. *Nanotechnology: Small Size—Large Impact*. Risk Dialogue Series. Zurich, Switzerland: Swiss RE Centre for Global Dialogue, 2005.

Hirsh, Lou. "Wireless Robots Work under a Microscope." *NewsFactor SciTech*, January 22, 2002. http://www.techextreme.com/perl/printer/15945/ (accessed July 31, 2003).

Ho, Mae-Wan. "Nanotechnology: A Hard Pill to Swallow." Institute of Science in Society. http://www.i-sis.org.uk/nanotechnology.php (accessed October 5, 2003).

Hoet, Peter H. M., Abderrahim Nemmar, and Benoit Nemery. "Health Impact of Nano-materials." *Nature*, January 2004. http://www.nature.com/cgi-taf/DynaPage.taf?file=/ nbt/journal/v22/n1/full/nbt0104-19.html (accessed January 12, 2004).

Hogan, Jenny. "DNA Robot Takes Its First Steps." *New Scientist*, May 8, 2004.

"Hot Debates: Is Techno-Surveillance Good? Will Lawsuits Throttle Nanotechnology." *Foresight Update*, August 30, 1998. http://www.foresight.org/Updates/Updates34/ update34.1.html (accessed July 26, 2004).

House of Commons—Committee on Science and Technology. *Science and Technology—Fifth Report*. March 22, 2004. http://www.publications.parliament.uk/pa/cm200304/cmselect /cmstech/56/5606.htm (accessed June 21, 2004).

"HRH the Prince of Wales: Menace in the Minutiae." *Independent*, July 11, 2004. http://argument.independent.co.uk/commentators/story,jsp?story=539977 (accessed July 12, 2004).

Huczko, Andrzej, Hubert Lange, Ewa Calko, Hanna Grubek-Jaworska, and Pawel Droszcz. "Physiological Testing of Carbon Nanotubes: Are They Asbestos-like?" *Fullerene Science and Technology* 9, no. 2 (2002): 251–54.

Hume, Claudia. "The Outer Limits of Miniaturization." *Chemical Specialties*, September 2000.

"IMM to Fund Molecular Manufacturing." *Foresight Update*, August 1, 1991, pp. 1, 5.

"In Nanotechnology, Josh Wolfe Is at the Door." *Red Herring*, January 28, 2002. http:// www.redherring.com/Article.aspx?a=4359 (accessed July 27, 2004).

In Realis. "A Critical Investor's Guide to Nanotechnology." *Inrealis.com*, February 2002. http://www.inrealis.com/nano.htm (accessed August 4, 2004).

Ingdahl, Waldemar. "Nitpicking Nanotechnology." *Tech Center Station*, January 24, 2004. http://www.techcentralstation.com/012704C.html (accessed April 18, 2004).

Institute for Biological Energy Alternatives. "IBEA Researchers Make Significant Advance in Methdology toward Goal of a Synthetic Genome," November 13, 2003. http:// www.bienergyalts.org/news.html (accessed June 25, 2004).

Institute of Nanotechnology. "Latest on Nanotechnology in the United States: MIT Wins

Army Contract." March 2003. http://www.nano.org.uk/thisweek71.htm (accessed June 11, 2003).

Interagency Working Group on Nanoscience, Engineering, and Technology. *Nanotechnology: Shaping the World Atom by Atom.* Washington, DC: GPO, 1999. http://www.wtec.org/loyola/nano/IWGN.Public.Brochure/ (accessed January 3, 2004).

———. *National Nanotechnology Initiative: Leading to the Next Industrial Revolution.* Washington, DC: NSTC/CT, February 2000.

"Interview: Nanotechnologist Ralph Merkle; Molecules and Computers—Modeling, Design and Visualization." *Foresight Update*, December 15, 1993, pp. 8–11.

"Interview with Bill Joy." *Weekly Edition: The Best of NPR News,* March 18, 2000. http://www.Lexis-Nexis.com/ (accessed October 2, 2003).

"Is Small Different? Not Necessarily, Say George Tech Researchers." NewsNanoApex, July 13, 2004. http://news.nanoapex.com/modules.php?name=News&file=article&sid=4731 (accessed July 26, 2004).

Jacobstein, Neil. "Nanotechnology Research and Development Sponsorship." In *Prospects in Nanotechnology: Toward Molecular Manufacturing.* Edited by M. Krummenacker and J. Lewis. New York: Wiley, 1995.

"Japan-Government Spending for Nanotechnology." *NanoInvestorNews,* July 27, 2004. http://www.nanoinvestornews.com/modules.php?name=Facts_Figures&op=sho&im=locfunding/japan_old (accessed July 27, 2004).

Jardim, Jesse. "Crowd Protests Opening of Molecular Research Lab." *Daily Californian Online,* January 30, 2004. http://www.dailycal.org/article.php?id=13950 (accessed April 16, 2004).

Jefferson Center. "Citizens Jury." 2002. http://www.jefferson-center.org/citizens_jury.htm (accessed June 17, 2002).

Jing, Fu. "Nanotech Needs Major Capital Injection." *China Daily,* May 20, 2004. http://www.chinadaily.com.cn/english/doc/12004-05/20/content_332367.htm (accessed May 25, 2004).

Jones, David. "Nanotechnology Bill Gets Bipartisan Support from Capital Hill." *Inside Energy,* May 12, 2003, p. 7. http://www.external.ameslab.gov/news/headlines/ie030512.pdf (accessed April 26, 2004).

Jones, Richard. "The Future of Nanotechnology." *NanoTechWeb,* August 2004. http://nanotechweb.org/articles/feature/3/8/1/1 (accessed August 9, 2004).

———. "FY 2005 National Science Foundation Budget Request." *AIP Bulletin of Science Policy News.* February 4, 2004. http://www.aip.org/fyi/2004/011.html (accessed on July 19, 2004).

Joy, Bill. "Why the Future Doesn't Need Us." *Wired,* April 2000. http://www.wired.com/wired/archive/8.04/joy.html (accessed March 22, 2004).

Kalaugher, Liz. "Quantum Dots Could Be Toxic to Cells." Nanotechweb.org, January 16, 2004. http://www.nanotechweb.org/articles/news/3/1/2/1 (accessed January 22, 2004).

Kallender, Paul. "Asia Pacific Governments Invest in Labs and Research Centers." *Small Times,* January/February 2004, p. 21.

———. "China and Taiwan Are Sinking Some Serious Cash into Nanotech." *Small Times,* November/December 2003, p. 10.

————. "Japan Boosts Nano Budget and Industrial Cooperation." *Small Times*, March/April 2004, p. 12.

————. "Japans Builds on its Nanopillars." *Small Times*, July/August 2003, p. 18.

————. "Japan's Nano Program Encourages Interdisciplinary Cooperation." *Small Times*, June 22, 2004. http://www.smalltimes.com/document_display.cfm?document_id =8079 (accessed July 10, 2004).

Kanellos, Michael. "Alternative Energies Are Looking Good Again." *CNET News.com*, July 12. http://news.com.com/Energy+heats+up+high+tech/2009-7337_3-5263772.html (accessed August 4, 2004).

————. "Nanosys Prices IPO, Looks Ahead." CNetNews.com, July 15, 2004. http:// news.com.com/Nanosys+prices+IPO,+looks+ahead/2100-1001_3-5271016.html (accessed July 29, 2004).

Kaplinsky, Joe. "Nanotechnology: A Slippery Debate." *Spiked Online*, May 7, 2003. http:// 131.104.232.9agnet/2003/5-2003/agnet_may_11.htm (accessed September 6, 2003).

Karoub, Jeff. "Ethics Center a Small Obstacle as Senate Nears Nano Bill Passage." *Small Times*, November 6, 2003. http://www.smalltimes.com/document_display.cfm ?document_id=6912 (accessed November 18, 2003).

————. "HP Official: Ignorance and Greed Could Spoil Nanotech's Credibility." *Small Times Magazine*, November 30, 2001. http://www.smalltimes.com/document_display .cfm?document_id=2655&keyword=karoub%20and%20hp&summary=1&startsum =1 (accessed September 16, 2003).

————. "Nano Tool Market Is No Small Change." *Small Times*, July/August 2003, p. 20.

————. "Nanotech Goes Commercial." *Small Times*, July/August 2003, 10.

————. "Q&A," *Small Times*, January/February 2004, p. 10.

Kary, Tiffany. "Is Small the Next Big Thing?" CNET News.com, February 11, 2002. http://news.com.com/2100-1001-833691.html (accessed July 27, 2004).

————. "Nanotech: More Science Than Fiction." CNET News.com, February 11, 2002. http://zdnet.com.com/2100-1103-833739.html (accessed May 3, 2004).

Kawai, Tomoji. "Nanotechnology in Japan: Present Situation and Future Outlook." Foreign Press Center—Japan, March 9, 2001. http://www.fpcj.jp/e/gyouji/br/2001/010309 .html (accessed June 13, 2003).

Keiper, Adam. "The Nanotechnology Revolution." *New Atlantis*, Summer 2003. http://www .thenewatlantis.com/archive/2/keiperprint.htm (accessed September 12, 2003).

Keiper, Adam, Yuval Levin, Christine Rosen, and Eric Brown. "The Nanotech Schism: High-Tech Pants or Molecular Revolution." *New Atlantis*, Winter 2004. http:// www.thenewatlantis.com/archive/4/soa/nanotechprint.htm (accessed April 5, 2004).

Kelly, Matt. "European Nanotech Depends Too Much on Government Handouts, Expert Says." *Small Times*, November 30, 2001. http://www.smalltimes.com/document _display.cfm?document_id=2656 (accessed April 26, 2004).

Kirby, Alex. "Tiny Particles Threaten Brain." BBC News, January 8, 2004. http://news.bbc .co.uk/go/pr/fr/-/2/hi/science/nature/3379759.stm (accessed January 22, 2004).

Kjartan, Renee. "Coalition Files Petition against Frankenfoods." Frankenfood, July/August 2000. http://www.speakeasy.org/wfp/46/frankenfoods.html (accessed June 26, 2003).

Knight, Will. "Military Robots to Get Swarm Intelligence." *New Scientist*, April 25, 2003.

http://www.newscientist.com/news/news.jsp?id=ns99993661 (accessed May 13, 2003).

Kordzik, Kelly "Prey Tell What Nano Is? Nope...." *Small Times,* January/February 2003, p. 8.

Kulinowski, Kristen. "Oral Evidence—Dr. Kristen Kulinowski, Center for Biological and Environmental Nanotechnology, Rice University, Houston, TX (USA)." Nanotec .org.uk, 2003. http://www.nanotec.org.uk/evidence/oralKulinowskiDrKristen.htm (accessed March 30, 2004).

Kurzweil, Ray. "The Drexler-Smalley Debate on Molecular Assembly." KurzweilAI.net, December 1, 2003, http://www.kruzweilai.net/articles/art0604.html (accessed December 2, 2003).

Kwuscha, K. "China Gaining Momentum in Nanotech R & D and Business." *Asia Pacific Nanotech Weekly,* November 3–5, 2002. http://www.nanoworld.jp/apnw/articles/ library/pdf/3.pdf (accessed April 26, 2004).

La Monica, Paul. "No-No for Nanotech." *CNNMoney,* August 5, 2004. http://money .cnn.com/2004/08/05/technology/techinvestor/lamonica/ (accessed August 9, 2004).

Laird, Burgess. "Policy Study Compares Nanotechnology Funding in the United States and Japan." *Foresight Update,* April 15, 1994, pp. 1–2.

Lall, Pavan. "Zyvex, UTD Collaborate on NASA Nano Contract." *Dallas Business Journal,* February 22, 2004. http://msnbc.msn.com/id/4332568/ (accessed February 23, 2004).

Lam, Chiu-Wing, John T. James, Richard McCluskey, and Robert L. Hunter. "Histopatho-logical Study of Single-Walled Carbon Nanotubes in Mice 7 and 90 Days after Instil-lation into the Lungs." 225th ACS National Meeting, New Orleans, LA, March 23–27, 2003. http://oasys2.confex.com/acs/225nm/techprogram/P595657.htm (accessed December 1, 2003).

———. "Pulmonary Toxicity of Single-Wall Carbon Nanotubes in Mice 7 and 90 Days After Intratracheal Instillation." *Toxicological Sciences* 77, no. 1 (2004): 126–34.

Lane, Neal. "Testimony before the Senate Appropriations Committee." Federal News Ser-vice, May 7, 1998. http://web.lexis-nexis.com/congroup/document (accessed Sep-tember 8, 2003).

Langberg, Mike. "Nanotech IPO May Lead to a New Bubble." *Mercury News,* June 18, 2004. http://www.mercurynews.com/mld/mercurynews/business/technology/8953611 .htm (accessed July 9, 2004).

———. "Nanotechnology Needs Vision." *Calgary Herald,* January 7, 1993.

LaPedus, Mark. "Agilent Labs Says 'Nano-stepper' Is Most Advanced MEMS Device." Silicon Strategics, March 14, 2002. http://www.siliconstrategies.com/article/printableArticle. jhtml;jessionid=RRRDILVKX1 (accessed July 31, 2003).

Lean, Geoffrey. "Hundreds of Firms Using Nanotech in Food." *Independent,* July 18, 2004. http://news.independent.co.uk/uk/environment/story.jsp?story=542140 (accessed July 19, 2004).

Leath, Audrey. "Advanced Technology Program: A Funding Retrospective." *American Insti-tute of Physics Bulletin of Science Policy News.* http://www.aip.org/fyi/2000/ fy00.080.htm (accessed June 23, 2004).

Leitl, Eugene. "Lord of the Dance." Transhumantech discussion group, May 25, 2001. http://groups.yahoo.com/group/transhumantech/message/7856 (accessed July 31, 2003).

Lemisch, Jesse. "A Movement Begins: The Washington Protests against IMF/World Bank." *New Politics* 8, no. 1, n.s. (Summer 2000).

Leo, Alan. "Get Ready for Your Nano Future." *Technology Review*, May 4, 2001. http://www.technologyreview.com/articleswo_leo050401.asp (accessed October 5, 2003).

Lessig, Lawrence. "Stamping Out Good Science." *Wired*, July 2004. http://www.wired.com/wired/archive/12.07/view.html?pg=5 (accessed July 9, 2004).

Lewis, Jim. "Regulating Nanotechnology." *Foresight Update* (April 15, 2003): 1, 4–8, 21.

Lin-Liu, Jen. "China, Emboldened by Breakthroughs, Sets Out to Become Nanotech Power." *Small Times*, December 18, 2001. http://www.smalltimes.com/section_display .cfm?section_id=51&summary=1&startpos=821 (accessed April 26, 2004).

Lippman, Walter. *The Phantom Public.* New Brunswick, NJ: Transaction Press, 1999. First published in 1927.

Lippmann, M. "Nature of Exposure to Chrysotile." *Annals of Occupational Hygiene* 38, no. 4 (1994): 459–67.

Lipsett, Althea. "Plenty More." *Guardian Unlimited*, June 22, 2004. http://education .guardian.co.uk/egweekly/story/0,5500,1244084,00.html (accessed July 9, 2004).

Lovy, Howard. "Business Has Redefined Nano in Its Own Image; Will That Vision Bring Us the Best of All Futures?" *Small Times*, January/February 2004, p. 14.

———. "Can Nano Create New Markets?" *NanoMarkets White Paper*, April 2004.

———. "Clash of the Nanotech Titans." *Howard Lovy's Nanobot*, December 1, 2003. http://nanobot.blogspot.com/2003_11_30_nanobot_archive.html (accessed December 2, 2003).

———. "The Greenpeace Report, Part II: NanoWars." *Howard Lovy's NanoBot*, July 25, 2003. http://nanobot.blogspot.com/2003_07_01_nanobot_archive.html#105905157013774 164 (accessed July 27, 2004).

———. "How to Fight Nanotech Misinformation in Two Easy Words: Honesty, Imagination." *Small Times*, May/June 2004, p. 20.

———. "Nano Re-created in Business's Image; Is This the Best of All Futures?" *Small Times*, January 23, 2004. http://www.smalltimes.com/document_display.cfm ?document_id=7279 (accessed July 11, 2004).

———. "Nanotechnology Has Reached a Crossroads Somewhere between Hypothesis and Hype." *Small Times*, July/August 2003, p. 6.

———. "A Sad Jab at the 'Bad Rad Lab.'" Biocritics.org. http://blogcritics.org/archives/ 2004/02/03/115423.php (accessed July 2, 2004).

———. "Societal Concerns vs. Scientific Accuracy." *Howard Lovy's Nanobot*, November 21, 2003. http://nanobot.blogspot.com/2003_11_16_nanobot_archive.html (accessed April 18, 2004).

———. "U.K. Recognizes Importance of Perception." *Howard Lovy's NanoBot*, September 30, 2003. http://nanobot.blogspot.com/2003_09_28_nanobot_archive.html (accessed October 15, 2003).

Lurie, Karen. "Smallest Robot." *ScienCentralNews*, July 19, 2004. http://www.sciencentral .com/articles/view/ohp3?type-article&article_id=218392303 (accessed July 19, 2004).

Lux Research. "About Lux Research." Lux Research Homepage, July 4, 2004. http://www.luxresearchinc.com/ (accessed July 11, 2004).

———. "Industry at a Glance." *San Franciso Chronicle*, February 1, 2004. http://sfgate

.com/cgi-bin/article.cgi?f=/chronicle/a/2004/02/01/BUGCJ4J1D51.DTL (accessed July 11, 2004).

Lynch, Marc. "The Public Sphere Structure of International Politics." *State Interest and Public Spheres.* New York: Columbia University Press, 1999.

MacLeod, Donald. "Thinktank Predicts Nanotechnology Backlash." *Guardian Unlimited,* February 13, 2003. http://education.guardian.co.uk/higher/research/story/0,9865, 894755,00.html (accessed July 29, 2003).

Mander, Jerry. *In the Absence of the Sacred: The Failure of Technology and the Survival of the Indian Nations.* San Francisco: Sierra Club Books, 1991.

Markoff, John. "Technologist Gives His Peers a Dark Warning." *New York Times,* March 13, 2000. http://www.nytimes.com/library/tech/00/03/biztech/articles/13joy.html (accessed July 27, 2004).

Martosko, David. "Food Fetish." *American Spectator,* 2003. http://tas.spectator.org/article.asp ?art=8 (accessed June 26, 2003).

Mason, Jack. "As Nanotech Grows, Leaders Grapple with Public Fear and Perception." *Small Times,* May 20, 2004. http://www.smalltimes.com/document_display.cfm ?document_id=7926 (accessed July 11, 2004).

———. "Melding of Nano, Bio, Info and Cogno Opens New Legal Horizons." *Small Times,* March 3, 2004. http://www.smalltimes.com/print_doc.cfmn?doc_id=7501 (accessed March 18, 2004).

———. "NanoBusiness Alliance Leader Steps Down, AtomWorks Chief Takes Over." *Small Times,* May 18, 2004. http://www.smalltimes.com/print_doc.cfm?doc_id=7915 (accessed May 18, 2004).

———. "Nanotechnology in the 21st Century." *Small Times Magazine,* 2003.

MAST, Policy Research Project on Anticipating Effects of New Technologies. *Assessing Molecular and Atomic Scale Technologies.* Austin, TX: Lyndon B. Johnson School of Public Affairs (University of Texas at Austin) Press, 1989.

Matthews, Christine M. "U.S. National Science Foundation: An Overview." *CRS Report for Congress.* July 1, 1998. http://www.ncseonline.org/NLE/CRSreports/Science/ st-6.cfm?&CFID=14969499&CFTOKEN=65710512 (accessed on July 20, 2004).

Mauboussin, Michael, and Kristen Batholdson. *Big Money in Thinking Small: Nanotechnology—What Investors Need to Know.* A Report by Credit Suisse/First Boston Equity Research, May 7, 2003. http://www.csfb.com/home/index/index.html (accessed July 15, 2002).

May, Robert. "Science and Society." Science Year, June 23, 2002. http://www.scienceyear.com/ about _sy/events/pdfs/Sci_SocActivities.pdf (accessed June 26, 2003).

Mayer, Sue. "From Genetic Modification to Nanotechnology: The Dangers of Sound Science." In *Science: Can We Trust the Experts?* Edited by T. Gilland. London: Hodder & Stoughton, 2002.

Maynard, Andrew, Paul Baron, Michael Foley, Anna Shvedova, Elena Kisin, and Vicent Castranova. "Exposure to Carbon Nanotube Material: Aerosol Release during the Handling of Unrefined Single-Walled Carbon Nanotube Material." *Journal of Toxicology and Environmental Health, Part A* 67 (2004): 87–107.

McAdam, Doug, and David Snow. *Social Movements: Readings on Their Emergence, Mobilization, and Dynamics.* Los Angeles: Roxbury, 1997.

McCarthy, Thomas. "Molecular Nanotechnology and the World System." January 9, 1996. http://www.bcf.usc.edu/~tmccarth/main.html (accessed October 23, 2003).

McCullagh, Declan. "House Earmarks Billions for Nanotech." CNET News.com, May 7, 2003. http://news.com.com/2100-1028_3-1000408.html?tag=prntfr (accessed July 14, 2004).

McGee, Patrick. "Nanotech, but Not in a Nanosecond." *Wired*, November 30, 2001. http://www.wired.com/news/technology/0,1282,48737,00.html?tw=wn_story_related (accessed July 27, 2004).

McKay, Nial. "A Baby Step for Nanotech." *Wired*, November 9, 1998. http://www.wired.com/news/technology/0,1282,16089,00.html (accessed August 3, 2004).

McKibben, Bill. *Enough: Staying Human in an Engineered Age*. New York: Time Books, 2003.

———. "More Than Enough." *Ecologist*, May 22, 2003. http://www.theecologist.org/archive_article.html?artilce=400 (accessed September 2, 2003).

Meakin, Ivan, and the Ministry for Education, Science, and Technology. *Nanotechnology in Japan—A Guide to Public Spending in FY2002*. 2002. Tokyo, Japan.

"Media Watch." *Foresight Update*, September 15, 1994, pp. 9–11.

Meeks, Ronald. "President's Budget Includes Modest Increase for R&D in FY 2004." *National Science Foundation InfoBrief*, October 2003.

"Megabucks for Nanotech." *Scientific American*, September 2001, p. 8.

Mehra, Jagdish. *The Beat of a Different Drum: The Life and Science of Richard Feynman*. Oxford: Clarendon Press, 1994.

Merel, Peter. "Security Fog." Sci.nanotech Discussion List, June 14, 1995. ftp://PLANCHET.RUTGERS.EDU (accessed October 7, 2003).

Merkle, Ralph. "A Response to Scientific American's New Story: Trends in Nanotechnology." Foresight.org, 1996. http://www.foresight.org/SciAmDebate/SciAmResponse.html (accessed July 21, 2004).

Merrill Lynch. "Comment: Nanotechnology—Introducing the Merrill Lynch Nanotech Index." ML.com, press release, April 8, 2004. http://www.ml.com/about/press_release/pdf/04012004_nano_index.pdf (accessed August 26, 2004).

———. "Merrill Lynch Creates *Nanotech Index* to Track Evolving Industry." ML.com, press release, April 1, 2004. http://www.ml.com/about/press_release/04012004-1_nanotech_index_pr.htm (accessed June 3, 2004).

"Merrill's Nano Index: Gilded List or Fool's Gold?" *Nanotech Report*, June 2004, p. 3.

Miller, John. "Beyond Biotechnology: FDA Regulation of Nanomedicine." *Columbia Science and Technology Law Review* 4 (2002/2003). http://lexis-nexis.com (accessed October 2, 2003).

Miller, Sonia. "Law in a New Frontier." Abstracts of the Workshop on Societal Implications of Nanoscience and Nanotechnology, Arlington, VA, December 3–5, 2003.

———. "A Matter of Scale." *New York Law Journal*, August 3, 2004.

Mills, Russell. "Steps toward Nanotechnology." *Foresight Update*, July 1, 1993, pp. 5–6.

Mize, Scott. "Near-Term Commercial Opportunities in Nanotechnology." 10th Foresight Conference on Molecular Nanotechnology, Bethesda, MD, October 10, 2002.

Mnyusiwalla, Anisa, Abdallah S. Daar, and Peter Singer. "Mind the Gap: Science and Ethics in Nanotechnology." *Nanotechnology* 14 (2003): R9–13.

Modzelewski, Mark. "The Greenpeace Report, Part II: NanoWars." *Howard Lovy's NanoBot*, July 25, 2003. http://nanobot.blogspot.com/2003_07_01_nanobot_archive.html #105905157013774164 (accessed July 27, 2004).

———. "Nanotech Industry Can Help Groundbreaking Bill Fulfill Its Promise through Collaboration, Public Policy." *Small Times*, January/February 2004, p. 12.

Mokhoff, Nicolas. "US Official Calls for Closer Cooperation on Nanotechnology." *EE Times*, December 9, 2003. http://www.eetimes.com/at/news/OEG20031209S0008 (accessed December 15, 2003).

Monaghan, Peter. "The Humanities' New Muse: Genomics." *Chronicle of Higher Education*, February 20, 2004. http://chronicle.com/temp/email.php?id=scjqtzo0kxzer26ofk1 x4jsg3nfuwqk8 (accessed February 17, 2004).

Mooney, Pat. "The ETC Century: Erosion, Technological Transformation and Corporate Concentration in the 21st Century." 1999. http://rafi.org/documents/other_etccentury.pdf (accessed October 2, 2003).

Moore, Julia. "The Future Dances on a Pin's Head." *Woodrow Wilson International Center for Scholars: Knowledge in the Public Service*, November 26, 2002. http://wwics.si.edu/ index.cfm?fuseaction=news.item&news_id=14638 (accessed July 27, 2004).

Moran, Ed. "Hype or Hope." Summer 2002. From *NNI: From Vision to Commercialization*, Washington D.C., March 31–April 2, 2004.

Morissey, Susan R. "Harnessing Nanotechnology." *Government and Policy* 82, no. 16 (April 14, 2004): 30–33. http://pubs.acs.org/cen/nlw/print/8216gov1html (accessed June 15, 2004).

———. "Paving the Way for Nanotech." *Chemical and Engineering News*, June 8, 2004. http:// pubs.acs.org/cen/news/8223/8223earlygov.html (accessed June 15, 2004).

Mosterín, Jesús. "Ethical Implication of Nanotechnology." *Nanotechnology: Revolutionary Opportunities and Societal Implications*. EC-NSF 3rd Joint Workshop on Nanotechnology, Lecce, Italy, January 31–February 1, 2002, pp. 91–94.

Moyer, Albert E. *A Scientist's Voice in American Culture: Simon Newcomb and the Rhetoric of the Scientific Methods*. Berkeley and Los Angeles: University of California Press, 1992.

"Much Ado about Almost Nothing." *Economist*, March 18, 2004. http://www.economist .com/science/displayStory.cfm?story_id=2521232 (accessed April 5, 2004).

Mullaney, Timothy, and Specer Ante. "Info Wars." *Business Week*, June 5, 2000, p. 107.

Mullen, Richard. "DOE Science 2004 Budget: A Few Winners." *New Technology Week*, 2003. http://www.kingpublishing.com/publications/ntw/ (accessed January 24, 2004).

"Nanobots." June 3, 2003. http://www.brightonlife.com/fastlike/articles.php?news_id=197 (accessed September 5, 2003).

NanoBusiness Alliance. "F. Mark Modzelewski." April 19, 2004. http://www.nanobusiness.org/ mark.html (accessed July 14, 2004).

———. "In the Spotlight: Jay Lindquist." *NanoBusiness News*, February 3, 2004. E-newsletter.

———. "In the Spotlight: Sherwood Boehlert." *NanoBusiness News*, November 24, 2003. E-newsletter.

———. "U.S. Senate Passes Nano Bill." *NanoBusiness News*, November 19, 2003. E-newsletter.

"NanoBusiness Alliance and Workingin-nanotechnology.com Join to Advance Nanotech Career Opportunities." *NanoInvestor News*, May 19, 2004. http://www.nano

investornews.com/modules.php?name=News&file=articles&sid=2908 (accessed May 25, 2004).

"NanoMaterial Leader Zyvex Introduces Carbon Nanotube-Based Additives for Polyurethanes." *NanoInvestor News*, May 4, 2004. http://www.nanoinvestor news.com/modules.php?name=News&file=print&sid=2743 (accessed May 10, 2004).

"Nano Name Game." *Nanotech Report*, February 2004, p. 1.

"Nanophase Announces Second Quarter 2004 Results." *NanoInvestor News*, July 22, 2004. http://www.nanoinvestornews.com/modules.php?name=News&file=article &sid=3102 (accessed July 26, 2004).

"Nano Stock Skepticism Spells Opportunity." *Nanotech Report*, October 2003, p. 3.

"Nanosys Awarded U.S. Defense Department Contract to Develop Flexible Solar Cells." *NanoInvestor News*, August 19, 2004. http://www.nanoinvestornews.com/ modules.php?name=News&file=article&sid=3274 (accessed August 25, 2004).

"Nanotech: Hype Yes, Bubble No." *Nanotech Report*, December 2002, p. 2.

"Nanotech Industry Review and Outlook." *Nanotech Report*, January 2003, p. 3.

"Nanotech Is Wall Street's Latest Love." *Seattle Post-Intelligencer*, July 13, 2004. http://seattle pi.nwsource.com/business/aptech_story.asp?category=1700&slug=Nanotechnology %20Hype (accessed July 19, 2004).

"Nanotechnology: Public Debate Takes Off." EurActiv.com Portal, July 31, 2003, http:// www.euractiv.com/cgi-bin/cgint.exe/1268169-86?14&1015=7&1004=1506022 (accessed October 17, 2003).

"Nanotechnology: Tiny Hope or Big Hype?" *CNN Technology*, March 15, 2004, p. 1.

"Nanotechnology in Brief." NanotechWeb.org, July 9, 2004. http://nanotechweb.org/ articles/news/3/7/15 (accessed July 10, 2004).

"Nanotechnology in Brief." Nanotechweb.org, July 23, 2004. http://nmanotechweb.org/ articles/news/3/7/15 (accessed July 24, 2004).

"Nanotechnology Industry, an Estimated $961 Million for FY 2004." *NanoInvestorNews*, July 15, 2004. http://www.nanoinvestornews.com/modules.php?name=News&file-Print &sid=3-72 (accessed July 19, 2004).

"Nanotechnology Industry to Address Environmental and Health Effects of Nanotech." *Business Wire*, July 7, 2003. http://ragingbull.lycos.com.mboard.boards.cgi?board -NANOTECH&read=1089 (accessed October 2, 2003).

Nanotechnology Process Foundry. "Strategy of Nanotechnology Research and Development." January 26, 2004. http://www.sanken.osaka-u.ac.jp/labs/foundry/en/project .html (accessed August 5, 2004).

"Nanotechnology's Homeland Security Potential to Be Explored." *Space Daily*, December 11, 2003. http://www.spacedaily.com/news/nanotech-03zzr.html (accessed August 5, 2004).

"Nanotechnology's Power Brokers." *Nanotech Report*, March 2003, pp. 1–3.

"Nanotechnology under the Spotlight." *NewsWales*, September 8, 2004. http://www .newswales.co.uk/?section=Business&F=1&id=7245 (accessed August 16, 2004).

"Nanotech on the Front Lines." *Nanotech Report*, November 2002, p. 1.

"Nanotech Patent Wars." *Nanotech Report*, October 2003, p. 1.

"Nanotech Relief for G.I. Joe." *Nanotech Report*, April 2004, p. 3.

"Nanotech's New Apollo Mission: Energy." *Nanotech Report,* June 2003, p. 1.

"Nanotech's Power Elite: 2004." *Nanotech Report,* March 2004, p. 2.

National Academy of Sciences. *Preliminary Comments, Review of the National Nanotechnology Initiative.* Washington, DC: National Academies Press, 2001.

National Consumer Council. A solicited response to European Commission Workshop. Nanotec.org.uk, 2003. http://www.nanotec.org.uk/evidence/45NNCSubl.htm (accessed March 30, 2004).

National Institute of Mental Health. "Technology Transfer Legislation Summary." August 2000. http://intramural.nimh.nih.gov/techtran/legislation.htm (accessed June 10, 2003).

National Institute of Standards and Technology. "Advanced Technology Program: A Progress Report on the Impacts of an Industry-Government Technology Partnership." 1996. http://www.atp.nist.gov/atp/repcong/repcong.htm (accessed June 23, 2004).

———. "ATP's Project Portfolio in Nanotechnology." http://www.nist.gov/public_affairs/atp_nanotech.htm (accessed June 23, 2004).

National Nanotechnology Initiative Grand Challenge Workshop. Nanoscience Research for Energy Needs, March 16–18, 2004.

National Research Council. *Small Wonders, Endless Frontiers: A Review of the National Nanotechnology Initiative.* Washington, DC: National Academies Press, 2002. http://www.nap.edu/catalog/10395.html?onpi_newsdoc06102002 (accessed May 20, 2004).

National Science Foundation. "NSF Awards New Grants to Study Societal Implications of Nanotechnology." August 25, 2003. http://www.nsf.gov/od/lpa/news/03/pr0389.htm (accessed June 26, 2004).

———."NSF Award Abstract—#0403847." http://www.nsf.gov/awardsearch/showAward.do?AwardNumber=0403847 (accessed July 23, 2004).

National Science Foundation and U.S. Department of Commerce. *Converging Technologies for Improving Human Performance: Nanotechnology, Biotechnology, Information Technology and Cognitive Science.* Edited by M. C. Roco and W. S. Bainbridge. http://www.wtec.org/ConvergingTechnologies/ (accessed August 12, 2004).

National Science and Technology Council, Committee on Technology, Subcommittee on Nanoscale Science, Engineering, and Technology (NSET). *National Nanotechnology Initiative: Research and Development Supporting the Next Industrial Revolution, Supplement to the President's FY 2004 Budget.* Arlington, VA: National Nanotechnology Coordination Office, August 29, 2003.

Nazrozov, Lev. "Will Drexler Save the West from Nano Annihilation?" Newsmax, December 19, 2003. http://216.26.163.62/2003/lev12_19.html (accessed December 29, 2003).

Negt, Oskar, and Alexander Kluge. *Public Sphere and Experience: Toward an Analysis of the Bourgeois and Proletarian Public Sphere.* Translated by Peter Labanji, James Owen Daniel, and Assenka Oksilodd. Minneapolis: University of Minnesota Press, 1993. First published in 1972.

Nelson, Max, and Calvin Shipbaugh. *The Potential of Nanotechnology for Molecular Manufacturing.* Santa Monica, CA: Rand, 1995.

Nelson, Nancy M. "Technology: Metacomputers, Nanotechnology, Electronic Publishing." *Information Today,* April 1991, p. 13.

Nemmar, Abderrahim, M. Hoet, B. Vanquickenbone, , D. Dinsdale, M. Thomeeer, M. Hoylaerts, H. Vanbilloen, L. Mortelmans, and B. Nemery. "Passage of Inhaled Particle into the Blood Circulation of Humans." *Circulation* 105, no. 411 (2002). http://circ.aha journals.org/cgi/content/full/105/4/411 (accessed February 11, 2004).

Nemmar, Abderrahim, M. Hoylaerts, Peter Hoet, J. Vermylen, and B. Nemery. "Size Effect of Intratracheally Instilled Particles on Pulmonary Inflammation and Vascular Thrombosis." *Toxicology and Applied Pharmacology* 1 (Janaury 2003): 38–45.

Nemmar, Abderrahim, M. Hoylaerts, Peter Hoet, D. Dinsdale, T. Smith, H. Xu, J. Vermylen, and B. Nemery. "Ultrafine Particles Affect Experimental Thrombosis in an In Vivo Hamster Model." *American Journal of Respiratory and Critical Care Medicine* 166 (2002): 998–1004. http://ajrccm.atsjournals.org/cgi/content/full/166/7/998 (accessed June 10, 2004).

"New Advisers Named." *Foresight Update,* February 15, 1993, p. 11.

"New Faces at IMM." *Foresight Update,* July 15, 1992, p. 15.

"New NanoBusiness Alliance Task Force Launches to Search for 'Real Answers' and Dispel Claims Based on 'Science Fiction.'" *Business Wire,* July 7, 2003. http://www.businesswire. come/cgi-bin/cb_headline.cgi?&story_file=bw.070703/231885425 (accessed June 17, 2004).

"New World Is Born." *Nanotech Report,* March 2002, p. 2.

"NIH 'Soft Landing' Turns Hard in 2005." The AAAS R&D Funding Update on R&D in the FY 2005 NIH Budget, February 20, 2004. http://www.aaas.org/spp/rd/nih05p .htm (accessed on July 21, 2004).

"NIST Launches Advanced Measurement Laboratory to Support Nanotech." http:// www.linuxelectronics.com/article.php?story=20040622083043387 (accessed on June 30, 2004).

Nitz, Michael, and Sharon Jarvis. "Science in the News: The Potential Impact of Televised News Stories about Global Warming." 5th International Conference on Public Communication of Sciences and Technology, Berlin, September 1998. http://www .kommwiss.fu-berlin.de/~wissjour/studium/pcst98/Paper_pdf/nitz.pdf (accessed June 26, 2003).

"NNUN Abstracts 2002." http://nsf.cornell.edu/nnun/2002NNUNinfo.pdf (accessed July 13, 2004).

"Nobel Winner Smalley Responds to Drexler's Challenge: Fails to Defend National Nanotech Policy." *Foresight Institute Press Center,* December 1, 2003. http://www.foresight .org/press.html (accessed December 2, 2003).

"No Small Matter." *Red Herring,* December 16, 2002. http://www.redherring.com/ Article.aspx?f=articles%2farchive%2finsider%2f2002%2f12%2f (accessed October 5, 2003).

Novartis Foundation. A solicited response to European Commission Workshop. Nanotec .org.uk, 2003. http://www.nanotec.org.uk/evidence/86DerekChadwick.htm (accessed March 30, 2004).

"NWNR Wins Award for Editorial Excellence!" *Nanotech Report,* June 2003, p. 1.

Oberdörster, Eva. "Manufactured Nanomaterials (Fullerence, C_{60}) Induce Oxidative Stress in the Brain of Juvenile Largemouth Bass." *Environmental Health Perspectives* 112, no. 110 (July 2004): 1058–62.

Oberdörster, E., Z. Sharp, V. Atudorei, A. Elder, R. Gelein, W. Kreyling, and C. Cox.

"Translocation of Inhaled Ultrafine Particles to the Brain." *Inhalation Toxicology* 16, part 6/7 (2004): 437–46.

Oberdöster, Günter. "Effects and Fate of Inhaled Ultrafine Particles." 225th ACS National Meeting, New Orleans, LA, March 23–27, 2003. http://oasys2.confex.com/acs/225nm/techprogram/P598970.htm (accessed December 1, 2003).

———. "Emerging Concepts in Nanoparticle Toxicology." EC's Community Health and Consumer Protection Directorate. *Nanotechnologies: A Preliminary Risk Analysis on the Basis of a Workshop Organized in Brussels on March 1–2, 2004.* http://europa.eu.int/comm/health/ph_risk/events_risk_en.htm (accessed June 20, 2004).

———. "Pulmonary Effect of Inhaled Ultrafine Particles." *International Archives of Occupational and Environmental Health* 74 (2001): 1–8.

Oberdörster, Gunter, J. Ferin, and B. Lehnert. "Correlation between Particle Size, In Vivo Particle Persistence, and Lung Injury." *Environmental Health Perspectives* 102 (1994): 173–79.

Oberdörster, Gunter, Eva Oberdörster, and Jan Oberdörster. "Nanotoxicology: An Emerging Discipline Evolving from Studies of Ultrafines." *Environmental Health Perspectives* (March 22, 2005): 30. http://ehp.niehs.nih.gov/members/2005/7339/7339.pdf (accessed June 4, 2005).

Oberdörster, Gunter, Z. Sharp, V. Atudorei, A. Elder, R. Gelein, A. Lunts, W. Kreyland, and C. Cox. "Extrapulmonary Translocation of Ultrafine Carbon Particles Following Whole-Body Inhalation Exposure of Rats." *Journal of Toxicology and Envrionmental Health* 65, part A (2002): 1531–43.

Office of Basic Energy Sciences, Office of Science. *Nanoscale Science, Engineering and Technology in the Department of Energy: Research Directions and Nanoscale Science Research Centers,* March 2004.

The Office of Management and Budget, The Executive Office of the President. "National Science Foundation." 2004. http://www.whitehouse.gov/omb/budget/fy2005/nsf.html (accessed July 19, 2004).

Office of Technology Assessment at the German Parliament (TAB), Summary of TAB Working Report No. 92, July 2003. http://www.tab.fzk.de/en/projekt/zusammenfassung/ab92.htm (accessed July 11, 2004).

Oger, Genevieve. "European Union Turns to Research Projects in Hopes of Building up Its Sagging Economy." *Small Times,* January/February 2004, p. 20.

———. "French Academy Urges Government to Invest in Small Tech." *Small Times,* April 30, 2004. http://www.smalltimes.com/document_display.cfm?document_id=7787 (accessed May 3, 2004).

———. "Government Panel Displeased with British Support for Nano." *Small Times,* May/June 2004, p. 12.

Oliver, Jonathan. "Nightmare of the Grey Goo." *Mail,* April 27, 2003. http://static.highbeam.com/t/themailonsundaylondonengland/april272003/nightmareofthegreygootoscientiststhebillionsspento/ (accessed August 27, 2004).

"On the Record: Nanotechnology—Unlocking the Smallest Secrets." *San Francisco Chronicle,* February 1, 2004. http://sfgate.com/cgi-bin/article.cgi?file=/chronicle/archive/2004/02/01/BUGCJ4J2761.DTL&type=business (accessed July 11, 2004).

"Opposition to Nanotechnology." *New York Times*, August 19, 2002. http://www.maine science.org/news/2002/08h_nanotech.html (accessed July 30, 2003).

"Overview of ATP." September 10, 2003. http://www.atp.nist.gov/atp/overview.htm (accessed January 3, 2004).

Pade, Jochen, Klaus Scluepmann, Frachbereich Physik, and Carl von Ossietzky. "Science on Television: Alternating between Elitism and Leveling." 1998. http://www.kommwiss .fu-berlin.de/~wissjour/studium/pcst98/Paper_pdf/pade.pdf (accessed June 26, 2003).

"Parables, Predictions, and Wisdom of Crowds." Forbes/Wolfe Blog, May 31, 2004. http:// www.forbeswolfe.com/ (accessed June 15, 2004).

Paris, Arnaud. "Micro- and Nanosystems in Biology: From Innovative Applications to Market." *Applied Nanoscience* 1, no. 1 (2004): 3–6.

Parkinson, Stuart. "Nanotechnology." Nanotec.org.uk, July 10, 2003. http://www.nanotec .org.uk/evidence/76aScientistsforGlobalResponsibility.htm (accessed March 30, 2004).

Partnership in Nanotechnology. From NSF Grantees Conference, Arlington, VA, January 29–30, 2001.

Pescovitz, David. "Nano's Got the Ways and Memes for a Viral Assault on Pop Culture." *Small Times*, January/February 2004, p. 56.

———. "Plenty of G's and a Ton of Gee-Whiz." *Small Times*, January/February 2004, p. 30.

Peterson, Christine. "Inside Foresight: Paradigm Shift in Progress." *Foresight Update*, June 30, 1997. http://www.foresight.org/Updates/Updates29/update29.2.html (accessed July 26, 2004).

———. "Nanotechnology: Evolution of a Concept." *Journal of the British Interplanetary Society*, October 1, 1992, pp. 395–400.

———. "Societal Implications of Nanotechnology." *Federal Document Clearing House*, April 9, 2003. http://web.lexis_nexis.com/congcomp/document?_m=b3afd9dc2b8d55 cdb425cc4c2fd2c2b (accessed April 24, 2003).

———. "Taking Action: Making a Difference in Nanotechnology." *Foresight Update*, December 15, 1993, pp. 14–15.

———. "Thank You." *Foresight Update*, May 30, 1998. http://www.foresight.org/Updates/ Updates33/update33.5.html (accessed July 26, 2004).

Pethoukoukis, James M. "Drexler Speaks!—Or at Least Writes!" *U.S. News*, September 4, 2003. http://www.usnews.com/usnews/nycu/tech/nextnews/archive/next030904 .htm (accessed September 12, 2003).

———. "Turning Green over Nanotech." *US News*, July 31, 2003. http://www.usnews .com/usnews/nycu/tech/nextnews/archive/next030731.htm (accessed August 7, 2003).

Phelps, Lewis M. "Progress toward Molecular Manufacturing Detailed by Foresight Conference Speakers." *Foresight Update*, November 30, 1995, pp. 1–3.

Phibbs, Pat. "Nanotechnology: Nonprofit Institute to Work with Industry, Organizations to Develop Voluntary Standards." *Daily Environment*, June 17, 2004.

Phoenix, Chris. "Letters." *CENEAR* 82: 6–8. http://pubs.acs.org/isubscribe/journals/ cen/82/i04/print/8204lett.html (accessed February 4, 2004).

———. "Of Chemistry, Nanobots, and Policy." *Center for Responsible Nanotechnology*, December 2003. http://crnano.org/Debate.htm (accessed December 2, 2003).

Phoenix, Chris, and Eric Drexler. "Safe Exponential Manufacturing." *Nanotechnology* 15 (2004): 869–72.

Phoenix, Chris, and Mike Treder. "Applying the Precautionary Principle to Nanotechnology." Center for Responsible Nanotechnology, January 2003. http://responsiblenano technology.org/precautionary.htm (accessed July 28, 2003).

————. "Three Systems of Ethics: A Proposed Application for Effective Administration of Molecular Nanotechnology." Center for Responsible Nanotechnology, October 2003. http://crnano.org/systems.htm (accessed October 22, 2003).

Pincock, Stephen. "US Leading World Science." *Scientist,* July 15, 2004. http://www .biomedcentral.com/news/20040715/03 (accessed July 22, 2004).

Poggi, Mark, Jeffrey S. Boyles, Lawrence A. Bottomley, Andrew W. McFarland, Jonathan S. Colton, Cattien V. Nguyen, Ramsey M. Stevens, and Peter T. Lillehei. "Measuring the Compression of a Carbon Nanospring." *NanoLetters* 4, no. 6. (2004): 1009–16.

Porteus, Liza. "Psyching Up Kids for Tech Could Help U.S. Jobs." Fox News, July 5, 2004. http://www.foxnews.com/story/0,2933,123562,00.html (accessed July 9, 2004).

"Portfolio Plays in Nanotech." *Nanotech Report,* April 2003, p. 2.

Postone, Moishe. "Political Theory and Historical Analysis." *Habermas and the Public Sphere.* Cambridge, MA: MIT Press, 1992.

Postrel, Virginia. "Joy, to the World." Reasononline, June 2000. http://reason.com/0006/ co.vp.joy.shtml (accessed July 27, 2004).

"President Bush's FY 2005 Budget Request: How Does Science Fare?" February 2004. http://www.chemistry.org/portal/a/c/s/1/feature_pol.html?DOC=policymakers %5Cpol_capcon_feb04.html (accessed on June 23, 2004).

"President's Council of Advisors on Science and Technology." PCAST Home Page, June 22, 2004. http://www.ostp.gov/PCAST/pcast.html (accessed July 11, 2004).

Prince, Marcelo. "Nanotech Start-Ups: Wall St.'s New Fad." DelawareOnline.com, July 18, 2004. http://www.delawareonline.com/newsjournal/business/2004/07/18nanotech startup.html (accessed August 4, 2004).

Proceedings of the Joint NSF-NIST Conference on Nanoparticles: Synthesis, Processing into Functional Nanostructures, and Characterization. Arlington, VA, May 12–13, 1997.

Proceedings of the Joint NSF-NIST Conference on Ultrafine Particle Engineering. Arlington, VA, May 25–27, 1994.

"Rage against the Machines." *SciFi,* December 2003, p. 59.

"Ralph Merkle, a Leading MNT Theorist, Joins Zyvex." *Foresight Update,* December 30, 1999. http://www.foresight.org/Updates/Updates39/update39.2.html (accessed July 26, 2004).

Rama, Michelle. "Nanotech: Mini-Science, Mega-Potential." Boston.com, July 25, 2004. http://www.boston.com/business/technology/articles/2004/07/25/nanotech_mini_ science_mega_potential/ (accessed August 4, 2004).

Ratner, Daniel, and Mark Ratner. *Nanotechnology and Homeland Security.* Upper Saddle River, NJ: Pearson Education, 2004.

"Recent Events: Extropy Institute Conference." *Foresight Update,* September 1, 1997. http:// www.foresight.org/Updates/Updates30/update30.5.html (accessed July 26, 2004).

Regalado, Antonio. "Greenpeace Warns of Pollutants Derived from Nanotechnology." *Wall*

Street Journal, June 25, 2003. http://www.mindfully.org/Technology/2003/Pollutants-From-Nanotechnology25ju03.htm (accessed July 30, 2003).

Regis, Ed. *Nano: The Emerging Science of Nanotechnology: Remaking the World—Molecule by Molecule*. New York: Little, Brown, 1995.

Reid, David. "Nanotechnology Pioneer Slays *Grey Goo* Myths." EurekaAlert!, June 8, 2004. http://www.eurekaalert.org/pub_releases/2004-06/iop-nps060704.ph (accessed July 9, 2004).

Reuters. "GMO Crops Here to Stay or Gone with the Wind?" *Consumer Choice*, November 5, 2001. http://www.biotech-info.net/gone.html (accessed June 26, 2003).

————. "Nanosys Files to Go Public." April 22, 2004. http://www.reuters.com/finance NewsArticle.jhtml?type=bondsNews&storyID=4914432 (accessed May 3, 2004).

————. "Prince Charles Warns of Dangers of the Breakthrough Science of Nanotechnology." July 10, 2004. http://nanotechwire.com/news.asp?nid=925&ntid=&pg=i (accessed July 11, 2004).

Reynolds, Glenn. "Forward to the Future: Nanotechnology and Regulatory Policy." November 2002. http://www.pacificresearch.org/pub/sab/techno/forward_to_nanotech .pdf (accessed October 3, 2003).

————. "Greenpeace and Nanotechnology." TechCentralStation, July 30, 2003. http://www.techcentralstation.com/073003B.html (accessed September 2, 2003).

————. "Inside Foresight: Greenpeace and Nanotechnology." *Foresight Update*, August 31, 2003. http://www.foresight.org/Updates/Updates52/update52.2.html (accessed July 26, 2004).

————. "NanoDynamism vs. NanoTimidity." TechCentralStation, August 4, 2004. http://www.techcentralstation.com/080404B.html (accessed August 9, 2004).

————. "A Tale of Two Nanotechs." TechCentralStation.com, January 28, 2004. http://www.techcentralstation.com/012804A.html (accessed July 11, 2004).

Reynolds, Melanie. "UK Government Rules Out Nano Fab Plans." *Electronics Weekly*, June 30, 2004. http://nanotechwire.com/news.asp?nid=900&ntid=&pg=2 (accessed July 9, 2004).

Rickerby, David. "Societal and Policy Aspects of the Introduction of Nanotechnology in Healthcare." Nanotech.org.uk, 2003. http://www.nanotec.org.uk/evidence/47b SocietalAndPolicyAspects.htm (accessed March 30, 2004).

"Rick Smalley's Top 5 Nano-Energy Solutions." *Nanotech Report*, June 2003, p. 2.

Rincon, Paul. "Nanotech Guru Turns Back on Goo." BBC News, June 9, 2004. http://newsvote.bbc.co.uk/mpapps/pagetools/print/news.bbc.co.uk/1/hi/sci/tech/3788673 ,stm (accessed June 10, 2004).

"Robert A. Freitas Jr., Author of Nanomedicine, Joins Zyvex." *Foresight Update*, March 31, 2000. http://www.foresight.org/Updates/Updates40/update40.2.html (accessed July 26, 2004).

Roco, Mihail. "Crichton Preys on Fear, but Real Nano Man's a Fan." *Small Times*, January/February 2003, p. 7.

————. "National Nanotechnology Investment in the FY 2005 Budget Request." In *AAAS Report on U.S. R&D in FY 2005*, Washington, DC, March 2004.

————. "The U.S. National Nanotechnology Initiative after Three Years (2001–2003): Setting New Goals for a Responsible Nanotechnology." *Abstracts of the Workshop on Societal Implications of Nanoscience and Nanotechnology*, December 3–5, 2003.

————. "Nanoscale Science and Engineering: Unifying and Transforming Tools." *AIChE Journal* 50, no. 5 (May 2004): 890–97.

————, and William Sims Bainbridge, eds. *Societal Implications of Nanoscience and Nanotechnology*. New York: Kluwer, 2001.

Roman, Cristina. "It's Ours to Lose—An Analysis of EU Nanotechnology Funding and the Sixth Framework Program." http://www.nanoeurope.org/docs/European%20 Nanotech%20Funding.pdf (accessed April 26, 2004).

Rosen, Jay. "As Democracy Goes, So Goes the Press." *What are Journalists For?* March 1999. http://yalepress.yale.edu/YupBooks/viewbook.asp?isbn=0300078234 (accessed June 19, 2003).

Roy, Rustum. "Giga Science and Society." *Materials Today* 5, no. 12 (2002): 72. http:// www.materialstoday.com/pdfs_5_12/opinion.pdf (accessed October 4, 2003).

Royal Society. "Royal Society Working Group on Best Practice in Communicating the Results of New Scientific Research to the Public." October 31, 2003. http://www .royalsoc.ac.uk/news/comm/htm (accessed November 24, 2003).

————. "Summary of Evidence from Civil Society Groups at a Meeting with the Nanotechnology Working Group on 30 October 2003." Nanotec.org.uk. http://www.nano tec.org.uk/CivilSocietyGroups.pdf (accessed April 18, 2004).

————. "Nanotechnology and Nanoscience—Progress Report." Nanotec.org.uk, September 2, 2003. http://www.nanotec.org.uk/ReportSep03.htm (accessed August 2, 2004).

————. "Nanotechnology: Views of Scientists and Engineers." December 5, 2003. http://www.nano tec.org.uk/workshopOct03.htm (accessed June 25, 2004).

———— and Royal Academy of Engineers. *Nanoscience and Nanotechnologies: Opportunities and Uncertainties.* July 29, 2004. http://www.nanotec.org.uk/finalReport.htm (accessed July 30, 2004).

Roylance, Frank D. "Tiny Tubes and Big Dreams." *Baltimore Sun*, May 10, 2004. http:// www.lexis-nexis.com/ (accessed July 26, 2004).

Ryan, John. "Nanotechnology: Separating Fact, Fiction, Hype and Hope." PhysicsWeb, August 2004. http://physicsweb.org/article/world/17/8/2 (accessed August 9, 2004).

Ryan, Mary. *Women in Public: Between Banners and Ballots, 1825–1990*. Baltimore: Johns Hopkins University Press, 1990.

Salleh, Anna. "Nanotech Risk Put on Insurance Agenda." *News in Science*, June 7, 2004. http://www.abc.net.au/science/news/stories/s1124943.htm (accessed July 26, 2004).

Sampson, Mark T. "Type of Buckyball Shows to Cause Brain Damage in Fish." News NanoApex, March 29, 2004. http://news.nanoplex.com/modules.php?name=News &file=print&sid=4410 (accessed April 5, 2004).

Sanders, Lynn. "Against Deliberation." *Political Theory* 25, no. 3 (June 1997): 4.

Schehr, David. "Emerging Technologies and Innovation in Financial Services: Deciphering Opportunity from Irrelevance." http://www.nicsa.org/Archives/download/TF04 _Schehr.ppt (accessed July 26, 2004).

Schuler, Emmanuelle. "Perceptions of Risks and Nanotechnology." http://www.ifs .tu-darmstadt.de/phil/Schuler.pdf (accessed July 29, 2004).

Schulz, William. "Crafting a National Nanotechnology Effort." *Chemical and Engineering News*, October 16, 2000. http://pubs.acs.org/cen/nanotechnology/7842/print/7842 government.html (accessed on June 15, 2004).

————. "Nanotechnology under the Scope." *Chemical and Engineering News*, December 4, 2002. http://pubs.acs.org/cen/today/dec4.html (accessed October 5, 2003).

Schwartz, J. "Air Pollution and Daily Mortality: A Review and Meta-Analysis." *Envrionmental Research* 64 (1994): 36–52.

"Science Scope." *Science* 287, no. 5457 (February 25, 2000). http://www.sciencemag.org/content/vol287/issue5457/s-scope.shtml.

Sclove, Richard E. "Town Meetings on Technology." *Technology Review*, July 1996. http://www.loka.org/pubs/techrev.htm (accessed June 17, 2003).

Scott, Andrew. "Euro GM Moratorium Ends." *Scientist*, May 21, 2004. http://www.biomed central.com/news/20040521/03 (accessed May 22, 2004).

Seaton, A., W. MacNee, K. Donaldson, and D. Godden. "Particular Air Pollution and Acute Health Effects." *Lancet*, January 21, 1995.

Seeman, Nadrian. "Nanotech and DNA." *Scientific American*, June 2004, pp. 65–75.

Selin, Cynthia. "Expectations in the Emergence of Nanotechnology." September 25, 2002. http://web.cbs.dk/departments/mpp/forskerskolen/calendar/Selin.pdf (accessed May 2, 2004).

"Senate Bill: $103 Million for Oregon Defense Work." Bend.com New Sources, June 23, 2004. http://www.bend.com/news/ar_view%5E3Far_id%5E3D16350.htm (accessed July 10, 2004).

Service, Robert F. "Nanotechnology Grows Up." *Science* 304 (June 18, 2004): 1732–34.

Shibata, Yuichi. "How Safe Is Nanotechnology? Japan Needs to Find Out Soon." *Small Times*, December 1, 2003. http://www.smalltimes.com/document_display.cfm ?document_id=7021 (accessed December 9, 2003).

Shropshire, Philip. "The Big Letdown." Betterhumans.com, March 3, 2003. http://www .betterhumans.com/Features/Columns/Red_Hour_Orgy/column.aspx?articleID =2003-03-02-4 (accessed July 27, 2004).

Shvedova, Anne, Vincent Castronova, Elena Kisin, Diane Schwegler-Berry, Asley Murray, Vadim Gandelsman, Andrew Maynard, and Paul Baron. "Exposure to Carbon Nanotubes Material: Assessment of Nanotube Cytotoxicity Using Human Keratinocyte Cells." *Journal of Toxicology and Environmental Health* 66, part A (2003): 1909–26.

Siegel, Richard. Executive Summary of *Nanostructure Science and Technology: A Worldwide Study.* Loyola College, MD: WTEC, September 1999.

Sienko, Tanya C., and Asian Technology Information Program. "Overview of Nanotechnology Efforts in Japan." Tokyo, Japan, 1996.

"SI: StockTalk: Miscellaneous (Technology): TINew York (Nasdaq)—Small-Technology Venture Capital, Inc." Siliconinvestor.com, April 2, 2002. http://www.siliconinvestor .com/stocktalk/subject.gsp?subjectid=52700 (accessed June 27, 2004).

Singer, M., and D. Soll. "Guidelines for DNA Hybrid Molecules." *Science* 181 (1973): 1114.

Slee, Tom. "Semi-Daily Journal of Economist Brad DeLong: Fair and Balanced Almost Every Day." December 3, 2003. http://www.j-bradford-delong.net.movable_type/ 2003_archives/002838.html (accessed December 8, 2003).

Smalley, Richard. "Nanotechnology: The State of Nano-Science and Its Prospects for the Next Decade." Hearings before the Subcommittee on Basic Research of the House Committee on Science. 104th Cong. June 22, 1999. http://www.house.gov/science/ smalley_062299.htm (accessed April 26, 2004).

————. "Of Chemistry, Love, and Nanobots." *Scientific American*, September 16, 2001, pp. 76–77.

"Small Science, Big Questions." SNH.com.au, November 27, 2003. http://www.smh.com .au/articles/2003/11/16/10768917669906.html (accessed November 25, 2003).

"*Small Times* Magazine: 2003 Innovator of the Year: Finalists." *Small Times*, November/ December 2003, p. 33.

Smith, Jeremy, and Tom Wakeford. "Who's in Control?" *Ecologist*, May 22, 2003. http:// www.theecologist.org/archive_article.html?article=405 (accessed September 2, 2003).

Smith, Simon. "Dismissing Drexler Is Bad for Business." Betterhumans.com, February 26, 2004. http://www.betterhumans.com/Print/article.aspx?articleID=2004-02-26-1 (accessed March 2, 2004).

Smith, Tobin. "National Science Foundation in the FY 2004 Budget." *AAAS Report XXVIII: Research & Development FY 2004*. http://www.aaas.org/spp/rd/04pch7.htm (accessed on July 20, 2004).

Snow, Charles P. *The Two Cultures and a Second Look*. New York: Oxford University Press, 1963.

"Some Subjects for Graduate Study Suggested by Members of the Department of Public Speaking of Cornell University." *Quarterly Journal of Speech* 9 (April 1923): 147–53.

"So Where Is the Hero Who Can Save the Planet Before It's Too Late?" *BIT@E: Monthly Bulletin on Armenian Information Technology Market*, May 2003, p. 24.

Stix, Gary. "Little Big Science." *Scientific American*, September 16, 2001. http://www.ruf .rice.edu/~rau/phys600/stix.htm (accessed May 2, 2004).

————. "Razing the Tollbooths: A Call for Restricting Patents on Basic Biomedical Research." *Scientific American*, April 2003. http://www.sciam.com/article.cfm?articleID =00092FBB-6BDC-1E61-A98A809EC5880105 (accessed April 26, 2004).

————. "Waiting for Breakthroughs." *Scientific American*, April 1996, pp. 94–99.

Strauss, Stephen. "A Far-Fetched Theory That Won't Come Unstuck." GlobeandMail.com, July 24, 2004. http://www.theglobeandmail.com/servlet/ArticleNews/TPStory/ LAC/20040724/GOO24/TPScience/ (accessed July 26, 2004).

Stuart, Candace. "Fertile Ground: Small Tech's Growing Clusters Resemble Biotech in Its Early Life." *Small Times*, March/April 2004, p. 34.

————. "He Has a Nobel, a Company and Washington's Ear." *Small Times*, November/ December 2003, p. 37.

————. "Small Tech Provides Some Answers to a Nation Questioning Its Safety." *Small Times*, September 11, 2002. http://www.smalltimes.com/document_display.cfm ?section_id=45&document_id=4586 (accessed August 5, 2004).

————. "States Should Look beyond Borders for True Nature of Small Tech Clusters." *Small Times*, March/April 2004, p. 7.

"Summer Story Stocks." *Nanotech Report*, July 2003, p. 3.

Sutcliffe, Justin. "MMR: More Scrutiny Please." *British Medical Journal* 326, no. 1272 (June 7, 2003). http://bmj.com/cgi/content/full/326/7401/1272.html (accessed June 26, 2003).

Sweeney, Aldrin E., S. Seal, and P. Vaidyanathan. "The Promises and Perils of Nanoscience and Nanotechnology: Exploring Emerging Social and Ethical Issues." *Bulletin of Science, Technology, & Society* 23, no. 4 (August 2003): 236–45.

Swinbanks, David. "MITI Aims at Small Research." *Nature* 349 (February 7, 1991): 449.

Sykes, Christopher. *No Ordinary Genius: The Illustrated Richard Feynman.* New York: Norton, 1994.

Takenaka, S., E. Karg, C. Roth, A. Schulz, U. Heinzmann, P. Schramel, and J. Heyder. "Pulmonary and Systemic Distribution of Inhaled Intrafine Silver Particles in Rats." *Environmental Health Perspectives Supplements* 109, no. 4 (August 2001): 547–51.

Taneguchi, Norio. "On the Basic Concept of Nanotechnology." Proceedings of the International Conference on Production Engineering, Tokyo, 1974.

Taylor, John M. "New Dimensions for Manufacturing: A UK Strategy for Nanotechnology." http://www.oft.osd.mil/library/library_files/document_134_nanotechnology report.pdf (accessed June 21, 2004).

Terra, Richard. "National Nanotechnology Initiative in FY2001 Budget." *Foresight Update,* March 31, 2000. http://www.foresight.org/Updates/Update40/Update40.1.html (accessed on July 19, 2004).

"Thinking Small: Mark Modzelewski." *Nanotech Report,* December 2002, p. 5.

"Thinking Small: Mike Roco." *Nanotech Report,* July 2002, p. 5.

"Thinking Small: Phillip J. Bond." *Nanotech Report,* July 2003, p. 51.

Thomas, Jim. "Future Perfect?" *Ecologist,* May 22, 2003. http://www.theecologist.org/archive.html?article=398 (accessed September 2, 2003).

Tickner, Joel, Carolyn Raffensperger, and Nancy Myers. *The Precautionary Principle in Action: A Handbook.* 1st edition. http://biotech-info.net/handbook.pdf (accessed August 17, 2003).

"Tiny Science Is Lost on UK Public." BBC News, March 15, 2004. http://newsvote.bbc.co.uk/mpapps/pagetools/print/news.bbc.co.uk/2/hi/science/nature/3513 (accessed March 18, 2004).

Toth-Fejel, Tihamer. "Grey Goo Begone." Sci.nanotech Discussion List, April 26, 1995. ftp://PLANCHET.RUTGERS.EDU (accessed October 6, 2003).

"Trouble in Nanoland." *Economist,* December 5, 2002. http://www.economist.com/science/PrinterFriendly.cfm?Story_IS=1477445 (accessed October 5, 2003).

"UK Organic Picnic Slams GM Food." BBC News, July 25, 1999. http://news.bbc.co.uk/1/hi/uk/403352.stm (accessed July 21, 2003).

Uldrich, Jack, and Deb Newberry. *The Next Big Thing Is Really Small: How Nanotechnology Will Change the Future of Your Business.* New York: Crown Business, 2003.

United States Army Logistics Management College. "Thinking Small: Technologies That Can Reduce Logistics Demand." January 2000. http://www.almc.army.mil/alog/issues/MarApr00/MS523.htm (accessed June 13, 2003).

"USA-Government Funding for Nanotechnology." *NanoInvestor News.* http://www.nanoinvestornews.com/modules.php?name=Facts_Figures&op=sho&im=locfunding/usa (accessed August 4, 2004).

U.S. Congress. House of Representatives—Committee on Science. *H.R. 766, Nanotechnology Research and Development Act of 2003: Hearings before the Committee on Science.* 108th Cong., 1st Sess. Serial No. 108-13. March 19, 2003.

———. *The Societal Implications of Nanotechnology: Hearings before the Committee on Science. House of Representatives.* 108 Cong., 1st Sess. Serial No. 108-13. April 9, 2003.

————. Opening Statement: *Hearing on Nano Consequences*. April 9, 2003. http://www.house.gov/science/hearings/full03/apr09/boehlert.htm (accessed January 7, 2004).

Van, Jon. "Nanotechnology Gets $3.7 Billion Federal Backing." November 26, 2003. http://www.menafn.com/qn_news_story.asp?StoryId=Cp8we0eicveiTtKfotlrfq0G (accessed December 2, 2003).

Viner, Katherine. "'Luddites' We Should Not Ignore." *Guardian*, September 29, 2000. http://www.zmag.org/luddites.htm (accessed July 9, 2004).

Von Hippel, Arthur R. *Molecular Science and Molecular Engineering*. New York: MIT Press, 1959.

Wainwright, Phil. "Weekly Review: Web Services Journey Will Be Long." ASPNews, October 29, 2002. http://www.aspnews.com/analysis/analyst_cols/print.php/11275_1490151 (accessed July 23, 2004).

Wakefield, Julie. "Science's Political Bulldog." *Scientific American*, May 2004, pp. 50–52.

Wang, Brian. A solicited response to European Commission Workshop. Nanotec.org.uk, 2003. http://www.nanotec.org.uk/evidence/83BrianWang.htm (accessed March 30, 2004).

Warheit, David B., Brett R. Lawrence, Kenneth L. Reed, and Thomas R. Webb. "Comparative Pulmonary Toxicity Assessment of Single-Wall Carbon Nanotubes in Rats." *Toxicological Sciences* 77, no. 1 (2004): 117–25.

————. "Pulmonary-Toxicity Screening Studies with Single-Wall Carbon Nanotubes." The 225th ACS National Meeting, New Orleans, LA, March 23–27, 2003. http://oasys2.confex.com/acs/225nm/techprogram/P595346.htm (accessed December 1, 2003).

Warner, Michael. "Publics and Counter Publics." In *Habermas and the Public Sphere*. Edited by Craig Calhoun. Cambridge, MA: MIT Press, 1992.

Warwick University—Nanosystems Group. A solicited response to European Commission Workshop. Nanotec.org.uk, 2003. http://www.nanotec.org.uk/evidence/89WarwickUniversity.htm (accessed March 30, 2004).

Weber, Max. *The Protestant Ethic and the Spirit of Capitalism*. London: Unwin University Books, 1970. First published in 1930.

Weil, Nancy. "NanoBusiness Alliance Starts Safety Task Force." *Bio-IT World*, July 7, 2003. http://www.bio-itworld.com/news/070703_report2829.html?action=print (accessed July 17, 2004).

Weisman, Jonathan. "2006 Cuts in Domestic Spending on Table." *Washington Post*, May 27, 2004.

Weiss, Rick. "For Science, Nanotech Poses Big Unknowns." *Washington Post*, February 1, 2004. http://www.washingtonpost.com/ac2/wp-dyn/A1487-2004Jan31 (accessed February 4, 2004).

————. "Language of Science Lags behind Nanotech." *Washington Post*, May 17, 2004. http://Lexis-Nexis.com (accessed July 15, 2004).

————. "Nanoparticles Toxic in Aquatic Habitat, Study Finds." *Washington Post*, March 29, 2004. http://www.washingtonpost.com/ac2/wp-dyn/A31881-2004Mar28?language=printer (accessed March 29, 2004).

————. "Some See Solution, Some Potential Dangers." *Washington Post*, February 10, 2004.

http://newsday.com/news/printedition/health/ny-dsspdn3663472feb10,0,5417935
,pri (accessed February 16, 2004).

Wendman, Mark. "Letters." *CENEAR* 82 (2004): 6–8. http://pubs.acs.org/isubscribe/journals/
cen/82/i04/print/8204lett.html (accessed February 4, 2004).

"What Is the Foresight Institute?" *Foresight Background* (1987): 5–6.

"What Will the Future Nanotech Industry Look Like?" *NanoMarkets White Paper*, March
2004. http://www.nanomarkets.net/docs/nanowp3-04.pdf (accessed April 16, 2004).

Wherrett, Carl, and John Yelovich. "Commercializing Nanotechnology." *Motley Fool*, March
2, 2004. http://www.fool.com/Server/FooltPrint.sdp?File=/news/commentary/2-4/
commentary040 (accessed March 18, 2004).

———. "Nano's Banner Company?" *Motley Fool*, August 4, 2004. http://www.fool.com/
News/mft/2004/mft04080307.htm (accessed August 9, 2004).

———. "Is Nanotechnology for Real?" *Motley Fool*, July 23, 2004. http://www.fool.com/
news/commentary/2004/commentary04072305.htm (accessed July 26, 2004).

White House. "National Nanotechnology Initiative: Leading to the Next Industrial Revolution." January 21, 2000. http://clinton4.nara.gov/textonly/WH/New/html/
20000121_4.html (accessed October 5, 2003).

Whitesides, George M. "Nanotechnology: Art of the Possible." *Technology Review* 101, no. 6
(November/December 1998): 85–87.

———. "The Once and Future Nanomachine." *Scientific American*, September 2001, pp.
78–83.

"Who Will Control the Nanobots?" *Mail on Sunday*, April 27, 2003, p. 24.

Wiggins, Jason. A solicited response to European Commission Workshop. Nanotec.org.uk,
2003. http://www.nanotec.org.uk/evidence/65aJasonWiggins.htm (accessed March
30, 2004).

Willard, Charles Arthur. "The Problem of the Public Sphere: Three Diagnoses." In *Argument Theory and the Rhetoric of Assent*. Edited by David C. Williams and Michael D.
Hazen. Tuscaloosa: University of Alabama Press, 1990.

"Winstead Agrees Smaller Is Better—Comments on Passing of Nanotechnology Act."
NanoInvestor News, December 14, 2003. http://www.nanoinvestornews.com/modules.
php?name=News&file=articles&sid=2126 (accessed December 15, 2003).

Witze, Alexandra. "Science's Big Unknown: Downside to Downsizing." CentralDaily.com,
June 23, 2003. http://www.centraldaily.com/mld/centredaily/news/6150971.1.htm
(accessed December 9, 2003).

Wolfe, Josh. "Nanomaterials: A Chemical Reaction." *Nanotech Report,* January 2004, p. 1.

———. "Nanotech Insider." *Nanotech Weekly Newsletter*, August 6, 2004. E-subscription.

———. "Nanotech vs. Environmentalists." *Forbes*, September 16, 2003. http://www
.forbes.com/2003/09/16/cz_jw_0916soapbox.html (accessed October 17, 2003).

Wong, Eugene. "Nano-Scale Science and Technology: Opportunities for the Twenty First
Century." June 22, 1999. http://www.house.gov/science/wong_062299.htm (accessed
April 26, 2004).

Wood, Steven, and Alison Geldart. *The Social and Economic Challenges of Nanotechnology.*
Swindon, UK: Economic and Social Research Council, 2003. http://www.esrc.ac.uk/
esrccontent/DownloadDocs/Nanotechnology.pdf (accessed January 3, 2004).

Wood, Steven, and Richard Jones. "Report of Oral Evidence Session with Professor

Stephen Wood (SW) and Professor Richard Jones (RJ)." Nanotec.org.uk, 2003. http:// www.nanotec.org.uk/evidence/oralJones&Wood.htm (accessed March 30, 2004).

Wootliff, Ben. "Businesses Begin to Pay Attention to High-Profile Nano Health Debate." *Small Times,* March/April 2004, p. 16.

World Technology Evaluation Center. *R&D Status and Trends in Nanoparticles, Nanostructured Materials, and Nanodevices in the United States.* Proceedings of the WTEC Workshop May 8–9, 1997.

Wyden, Ron. "Wyden Chairs First Senate Nanotechnology Hearing: 'Small Science' Could Change the Way Americans Live, Work, Treat Disease." September 17, 2002. http:// wyden.senate.gov/media/2002/09172002_nanotech.html (accessed July 26, 2004).

"Young, Gifted and Zac." *Observer,* April 13, 2003. http://observer.guardian.co.uk/ magazine/story/0,11913,935831,00.html (accessed July 21, 2003).

Zald, Mayer, and Roberta Ash. "Social Movement Organizations: Growth, Decay and Change." *Social Forces* 44, no. 3 (March 1966): 327–40.

Zhang Yihong, Theodore C. Lee, Benedicte Guillemin, Ming-Chih Yu, and William N. Rom. "Enhances IL-1, and Tumor Necrosis Factor-Release and Messenger RNA Expression in Macrophages from Idiophatic Pulmonary Fibrosis and after Asbestos Exposure." *Journal of Immunology* 150, no. 9 (May 1, 1993): 4188–96.

Zhou, Wei. "Ethics of Nanobiology at the Frontline." *Santa Clara Computer and High Technology Law Review* 19, 2003. http://Lexis-Nexis.com (accessed August 27, 2004).

Zucker, Lynne, Michael R. Darby, Roy Doumani, Jonathan Furner, and Evelyn Hu. "NanoBank: A National Resource Now under Construction." http://www.sscnet.ucla .edu/soc/faculty/zucker/nanobank/NanoBank.pdf (accessed June 30, 2004).

"Zyvex Forms Two MEMS Development Partnerships." *Foresight Update,* April 1, 2001. http://www.foresight.org/Updates/Updates44/update44.5.html (accessed July 26, 2004).

"Zyvex Hires North American Sales Manager." *NanoInvestorNews,* July 8, 2004. http://www .nanoinvestornews.com/modules.php?name=News&file=print&sid=3-34 (accessed July 12, 2004).

INDEX

Abraxane, 202
Accelrys Software, 186
Acrongenomics, 203, 416
Acta, 210, 421
AcuFlex, 210
Advanced Battery, 209
Advanced Magnetics, 204
Advanced Measurement Laboratory, 100
Advanced Micro Devices, 196
Advanced Nanotechnology, 205, 255
Advanced Technology Program (ATP), 6, 25, 91, 99–101, 104, 160, 199
AFM (atomic force microscope), 187, 198, 221, 338
Agriculture, Department of (Department of Agriculture, USDA), 119, 127, 164, 193, 240
Aguavia, 191
AIDS (acquired immune deficiency syndrome; see also HIV), 130, 202, 299
Alivisatos, Paul, 232
Allen, David, 296
Allen, George (Sen., R-VA), 24, 82, 84, 91, 93, 132, 148
Alliance for Nanotechnology in Cancer, 201

Altair Nanotechnologies, 192, 209, 219, 312
Altana Chemie, 189
Alzheimer's Disease, 202, 206, 297
American Chemical Society, 174, 297
American Pharmaceutical Partners, 202
Androsorb, 201
Angel Network (NbA) 241
anti-GMO movement 351, 355–356, 360
antiglobalization movement, 77, 348, 351–352, 355–356, 360
anti-nanotechnology movement, 27, 312, 352
ANTS (autonomous nanotechnology swarms), 120
AO-DVD (Articulated Optical—Digital Versatile Disc), 197
ApNano Materials, 192
Applied Microstructures, 192
Applied Nanotech, 198, 205
Argonide Nanomaterials, 315
Argonne National Laboratory, 113, 238, 301
Arizona State University, 262, 323
Arnall, Alexander Huw, 174, 271–272, 286

Armstrong, Amanda, 39
asbestos (asbestosis), 43, 166, 182, 246, 275,
    289–291, 295, 299, 326
Asensio and Co., 32
assembler(s), 16, 24, 31, 38–39, 65, 67–68,
    70–71, 129, 226, 236, 255, 257–258,
    265, 277, 279, 338
Association of Liberal Democrat Engi-
    neers and Scientists (ALDES), 168
AtomWorks, 242

Babolet, 210
backlash, 31, 43, 72, 166, 186, 230, 258, 271,
    310, 316, 321, 327, 354
Bainbridge, W. S., 316, 338
Baird, Davis, 13, 27, 323–325
BASF, 219
Basulin, 201
Bayer, 219
Bayh-Dole Act, 107, 109
Begbroke Science Park (UK), 168
Bennett, Jim, 247, 249, 251–253, 255
Bethe, Hans, 50
Big Down, The, 78, 245, 264–265, 280, 290
Big Money in Thinking Small: Nanotech-
    nology—What Investors Need to Know,
    176
biochip(s), 118, 204, 221
BioMEMS, 118
biomimetics, 219
Bionic-Nano-Atomtech, 265
BioRyx Platform, 187
Blair, Tony, 79, 245, 342, 353
Bock, Larry, 218
Boehlert, Sherwood (Rep., R-NY), 40, 104,
    132, 135, 307
Boggs, Carol, 347
Bond, Philip, 93, 96, 178
bootstrapping, 322, 330
BrachySil, 204
Brazil, 137
British Market Research Bureau (BMRB),
    169
Brookhaven National Lab, 113, 207

buckyball(s), 284–286, 297–300, 320
Bucky USA, 284
Bush, George W. (Pres., R), 37, 82, 84–85,
    90–91, 93, 103, 111, 120, 126–127,
    130–132, 135, 179, 305, 319, 322, 342
Business Communications Co., 213–214

CalTech (California Technological Uni-
    versity), 24, 47, 50–51, 53, 61, 66, 71,
    87, 129, 250, 254
California at Santa Barbara, University of
    (UCSB), 130, 323
California NanoSystems Initiative, 323
Cambridge Display Technology (CDT),
    199, 231, 234
Carbolex, 284
Carbon Nanotechnologies (CNI), 24, 69,
    115, 150, 190, 209, 284, 312, 315
Carbon Nanotech Research Institute, 177,
    190
carbon nanotubes (single-, many-, and
    multiwalled, SWNT), 43, 58, 69, 92,
    150–151, 161, 164, 166, 187, 190–191,
    193, 195–199, 202, 206, 208–209, 211,
    213, 219, 238, 246, 270, 276, 283–295,
    299, 309, 312–313, 333
Catalytic Solutions, 188
Cavendish Kinetics, 197
CBRE (chemical, biological, radiological,
    and explosive), 114, 161
cellectronics, 206
CellTracks Analyzer II, 234
Cellucci, Thomas (*see also* Zyvex), 32,
    100–101
Center for Advanced Engineering Fibers
    and Films, 208
Center for Biological and Environmental
    Nanotechnology (CBEN), 11, 87, 167–
    168, 281–282, 287, 300, 314–315, 326
Center for Nanotechnology in Society
    (CNS), 13, 133–135, 296, 307,
    322–323, 325
Center for Neutron Research (NCNR),
    100, 164

Centers of Cancer Nanotechnology Excellence, 201

Center of Environmental Kinetics Analysis (CEKA), 205

Center for Constitutional Issues in Technology (CCIT), 248, 254–255

CFC(s) (chlorofluorocarbons), 77

Chemical Abstract Service Registry (CAS), 173

Chen, H. H. C., 296

Chernobyl, 43

Chevron Texaco Technology Ventures, 36, 222

Chicken Little, 44

China (People's Republic of China, PRC), 25, 43, 109, 135–137, 144–145, 148, 151–152, 190, 198, 220, 228, 284

China Nano 2005, 151

China Syndrome, The, 43

Chinese Academy of Science, 151

chromium-6, 205

Cientifica, 14, 168, 172, 190

citizen-consumer(s), 44, 46, 134, 304, 310, 335, 338, 340, 356

citizens' jury(ies) (*see also* NanoJury, UK), 261, 271, 273, 358

Citizen's School (South Carolina Citizen's School on Nanotechnology), 325

Clinton, Hillary Rodham (Sen., D-NY), 240

Clinton, William (Bill, Pres., D), 24, 50, 57, 61, 85–87, 89–90, 97, 99, 102, 110, 127, 130, 157, 240, 276, 281, 305

CMOS (complementary metal oxide semiconductor), 195

colloidosomes, 194

Colvin, Vicki, 41, 175, 275, 281, 284–285, 287–288, 293, 295–298, 300, 308, 314, 316, 321, 353–354

Colwell, Rita, 97, 99, 129, 316, 336–337

Combidex, 204

CombiFil Nano, 191

Commerce, Department of, 24–25, 37, 90–91, 99–100, 102, 129

Committee on Science and Technology, Science and Technology Committee (UK) 143, 170–171

Commonwealth Scientific and Industrial Research Organization (CSIRO), 182

Community Research and Development Information Service (CORDIS), 14, 172–173

consensus conference(s) ,172, 357–358

Convergence, 15, 21, 23, 38, 81, 130, 168–169, 201, 265–266, 272, 277, 311, 324

Cornell Nanofabrication Facility, 131

counterpublic(s), 348–352, 356, 359

Center for Responsible Nanotechnology (CRN), 12, 26, 63, 71, 169, 178, 246, 256–263, 266, 299, 323

C-Sixty (CSixty), 69

creative destruction, 23, 236

Credit Swisse (Credit Suisse/First Boston), 14, 17, 25, 155, 176, 242, 277

Crichton, Michael (*see also Prey*), 20, 41–43, 47, 270, 321, 345, 347

Crosby, Steve, 42, 281

crossbar latch, 196

CyberDisplay, 199

Cyclics, 189

Dag Hammarskjöld Foundation, 264

daphnia (water fleas; *see also* Oberdörster, Eva), 297

DARPA (Defense Advanced Research Projects Agency), 25, 113–114, 116–118, 232

Day After Tomorrow, The, 43

DDT (dichlorodiphenyltrichloroethane), 77, 275

Defense, Department of, 25, 105, 112–114, 116–117, 119, 127–129, 133, 156, 160–161, 163–164, 307, 318, 372, 387–389

deliberative polling, 332, 356

Deloitte and Touche, 224

Dendritic NanoTechnologies, 201

Department of Energy Laboratory Act, 110
Department of Trade and Industry (DTI, UK), 79, 170–171
Detz, Cliff, 36
Dewey, John, 341–342, 344
Diamondoid, 71
Dinkelacker, Jamie, 249
disruptive technologies, 171, 217
Distributed Sensing Smart Dust (Smart Dust), 200
DNA microchip, 222
Donaldson, Ken, 238, 279, 296
dot-com (dot-com bubble), 33, 103, 179, 215, 225, 236
Dow Automotive, 191
Dow Chemical, 187, 196, 199
Dow Corning, 115
Dow Jones Industrial Index, 177, 221
Dowling, Ann, 336
Draper, Fisher, and Jurvetson, 26, 225, 236–237
Drexler, K. Eric, 11–12, 17, 23–24, 26, 31, 38–40, 46–47, 50, 52, 54–68, 70–75, 128–129, 166–167, 169, 176, 218, 226, 234, 236–237, 2410242, 245, 247–248, 251–252, 254–257, 261, 277, 306, 338
Dunford, R., 293
Dupont, 69, 115, 197, 219, 221, 231–232, 292, 312
D-Wave, 237
dynamic force microscopy, 221

Easton's Sports, 210
Ecology Coatings, 192
Economic and Social Research Council, the (ESRC), 40, 165, 168, 336
Eco-Tru, 203
EHS (Environmental Health and Safety), 26, 279, 288, 301–302, 316
Eigler, Don, 30, 252
Elan Corp., 201
electron-beam lithography, 198
Embedded Security Management (ESM), 259

eMembrane, 191, 205
Emergency Filtration Products, 203
EMPA (the Swiss Federal Laboratories for Materials Testing and Research), 336
Ener1, 209
Engines of Creation, 11, 23, 40, 47, 55, 250, 347
Environmental Protection Agency (EPA) 14, 77, 127, 132, 163–164, 280, 286–287, 298, 307, 312, 319, 326
EnvironSystems, 203
e-paper, 195
Estrasorb, 201
ETC Group (Action Group on Erosion, Technology and Concentration), 12, 23, 26, 38, 42, 74, 76, 79, 193, 213, 245–247, 263, 268, 272, 276, 280–281, 283, 290, 299–300, 310, 313, 321
Ethical, Legal, and Social Issues in Science (ELSI), 308, 310–311, 330
Eurobarometer, 338
EuropaBio, 348
European NanoBusiness Alliance (ENA), 14, 140
European Union (EU), 14, 25, 101, 136–142, 148, 159, 165, 171–173, 251, 263, 328, 338–339
Evident Technologies, 192, 203, 209
Evolution Capital, 177
Evolution (nanotech golf club shaft), 210
Ewing's sarcoma, 204
EXIST (Existenzgründungen aus Hochschulen),138
Experimental Program to Stimulate Competitive Research (EPSCOR), 133
Export Council (US), 85
Extropy Institute (extropians), 236, 256
Exxon, 219

Fan, John, 32
FAST-ACT, 118
fat finger(s) (*see also* sticky fingers), 71
Feeney, JoAnne, 228
FEI Comp., 187

Feynman Prize, 250

Feynman, Richard, 23–24, 47, 49–54, 60–61, 64–65, 128–129

FibrMark Gessner, 191

Fiorito, Silvana, 246, 294

Fishbine, Glenn, 14, 109, 136

Fishkin, James, 356, 359,

fish study (*see also* Oberdörster, Eva), 92, 298

Flamel Technologies, 201

Flash memory, 196–197, 278

flexible photovoltaics (*see also* photovoltaics), 210, 228

FlexICs, 237

Fluorophores, 222

Food and Drug Administration (FDA), 182, 202–203, 234, 283, 311

Foley & Lardner, 326

Foresight Institute, 12, 26, 38, 54–56, 72, 236, 242, 246–147, 250, 255–256

Foresight Nanotech Institute, 255–256

Foresight Update (*see also* FU), 247

Forman, David, 14, 225–226, 230, 233–234

Fortner, J. D., 287

Fortune 500, 221, 240

Forum for the Future (UK), 168, 336

Foundation for Scientific and industrial Research at the Norwegian Institute of Technology, the (SINTEF), 194

FRAM (ferroelectric random access memory), 197

Frankenfood, 40, 353

Frankenstein, 41

Fraser, Nancy, 349

Freitas, Robert, 38, 254

Friesland Foods, 193

Front Edge Technologies, 209

FTM Consulting, Inc., 195

fuel cell(s), 110–113, 137, 168, 189, 207–210, 219, 221, 243, 277, 284, 287

Fujitsu, 195, 199

Fullerene International, 150

Future Force Warrior System, 115

FutureGen, 208

Garreau, Joel, 75

Gartner Hype Cycle, 35, 231

GE Advanced Materials, 191

GE (General Electric), 191, 221

General Motors, 219,221

Genesis Air, 192

genetically modified organisms (foods) (GMO, GM), 10, 27, 76, 163, 167, 169, 186, 199, 169, 186, 245, 262, 264, 266, 270–271, 275, 301, 310, 316–317, 320–322, 325, 328, 339, 348, 351, 353–356, 358, 360

Georgia Tech (Georgia Technological University), 196, 201, 204–205

Germany, 137–138, 159, 171, 191, 336

Gingrich, Newt, 90, 99, 123, 240

Gleick, James, 50, 52

Good Fellow Group (China), 152

Goldsmith, Zac, 24, 73, 76–80, 245, 247, 269

golden rice, 353

GolinHarris, 338

gray goo (grey goo), 24, 38–41, 56, 62–63, 65, 67–68, 73–75, 78, 174, 176, 260, 270, 272, 275–276

GreatCell (Switzerland), 209

green goo, 38

Green Millennium, 192

Greenpeace Environmental Trust (Greenpeace), 26, 174, 246–247, 261–263, 265, 270–273, 283, 286–287, 290, 299, 316, 358

Green Party, 79, 350

Green Plus, 210

Green-shield, 211

Guangzhou Yorkpoint, (China) 264

Habermas, Jurgen, 333, 341–344, 346, 348–349

halo-regime, 33

Hall, J. Storrs, 254

Hamlett, Patrick W., 358

Harris and Harris Group (TINY), 26, 211, 215, 228, 233, 238–239

Harvard University, 24, 65–66, 130, 161–
    162, 195, 203, 232, 238, 306, 323–324,
    373
Health and Science University (Oregon), 205
Hebrew University of Jerusalem, 232
HEITF (Health and Environmental Issues
    Task Force), 27, 241, 325–329
Helmut Kaiser Consultancy, 191, 194, 288
Hett, Annabelle, 175, 181, 268, 289
herpes, 201
Hewlett-Packard, 36, 56, 91, 108, 137, 196,
    219, 221, 222, 236
High Technology Research and Develop-
    ment Program (China, 863 Program),
    151
Hitachi, 197, 221
HIV (human immuno-virus; see also AIDS),
    201–202, 235
Hoet, Peter H. M., 300
Hollings, Fritz (Sen., R-SC), 103–104
Homeland Security, Department of
    (DHS), 75, 119–120, 130, 156,
    163–164, 359
Honda, Mike (Rep., D-CA), 36, 104, 132
Hosokawa Powder Technology Research,
    205
Hubs Initiative (NbA), 241
Huczko, Andrjez, 246, 289, 290
Human Genome Project, 308
hydrogen storage carbon nanotubes pro-
    gram (China), 151
hydroxypatite crystals, 202
hype buying, 224
hype curve, 179, 231
HYPERMEC, 210
Hysitron, 187

IBM, 30, 187, 196, 198, 219, 221, 312
Iljin, 284
Imago Scientific, 237
Immunicon, 230–231, 234
India, 137, 353
Industrial Nanotech, 189
Infineon, 196

Infocast, 298
Inmat, 210
In-Q-Tel, 232
In Realis, 91, 176–177
Institute for Microelectronics Technology
    (Russia), 196
Institute for Molecular Manufacturing
    (IMM), 251–255
Institute for Soldier Nanotechnologies
    (ISN), 69, 115, 188
Institute of Advanced Industrial Science
    and Technology (AIST, Japan),
    146–147, 149, 190
Institute of Physical and Chemical
    Research (RIKEN, Japan), 146–147
Intel, 135, 196–197, 219, 221, 231–232
intellectual property (IP), 69, 109, 134,
    150, 177, 219–220, 227, 230, 232–233,
    261, 264, 266–267, 272
intelligent wood, 193
Interagency Working Group on
    Nanoscience, Engineering and Tech-
    nology (IWGN), 87, 125, 157
International Center for Young Scientists
    (ICYS), 150
International School for Advanced Studies
    (SISSA/ISAS), 206
International Union of Pure and Applied
    Chemistry, 173
Interdisciplinarians, 46
Introgen Therapeutics, 204
IPO (initial public offering), 179, 209, 228,
    230–234, 238
Itochu, 149
Israel, 137, 190–191
ISTPP (Institute for Science, Technology
    and Public Policy—Texas A&M Uni-
    versity), 338

Jackson-Lee, Sheila (Rep., R-TX), 133
Jacobstein, Neil, 252
Japan, 25, 69, 87, 135–137, 140, 144–152,
    156, 159, 171, 190, 205, 220, 253, 284,
    336

Jenkins, Mark, 294
Jet Propulsion Lab (JPL), 51
Johns Manville, 191
Johnson and Johnson, 201
Johnson Matthey FuelCells, 209
Joint Center for Bioethics (University of
    Toronto), 318, 337
Jones, Richard, 168
Joy, Bill, 24, 39, 47, 73–76, 260
J. P. Morgan Partners, 225
JR Nanotech, 194
Justice, Department of (DOJ), 127,
    163–164,
Jurassic Park, 20, 40, 42
juvenile largemouth bass (*see also* Eva
    Oberdörster), 297
Jurvetson, Steve, 26, 55, 222, 225, 229, 234,
    236–237, 239–240, 243

Kalil, Thomas, 25, 89–90, 93, 125, 126, 281
Keiper, Adam, 23, 63
Keithley Instruments, 187–188
King Phamaceuticals, 201
Kleiner Perkins Caufield & Byers, 227
Knowledge and Distributed Intelligence,
    (KDI) 86
Kopin, 32, 189, 199
Kordzik, Kelly, 42
Kostat, 69
k to gray, 94
KRI (Japan), 210
Kulinowski, Kristin, 11, 168, 282, 300
Kurzweil, Ray, 71, 236, 251, 311
Kvamme, E. Floyd, 134
KX Industries, 191

LabNow, 202
lab-on-a-chip, 219, 222
Lam, Chiu–Wing, 286, 291–292
Lancome, 204
Landemann, J. H., 295
Lane, Neal, 24, 50, 61, 85–87, 89, 93, 125,
    129, 157–158
Lawrence Berkeley National Lab, 113, 204

leading force, 13, 248, 261
Lecoanet, Helene, 286
limited molecular nanotechnology
    (LMNT), 169, 262
Lin, Patrick (UCSB), 323
Lippman, Walter, 342, 347
Liquidia Technologies, 201
lithium-ion (lithium) batteries, 190,
    208–209
L'Oreal, 204
Lovy, Howard, 58, 70, 72, 89, 93, 214,
    218–219, 270, 329
Luddite (neo-Luddite), 75, 270, 356
Lumera, 230, 234
Luna Innovation, 84
Luna nanoWorks, 189–190
Lux Capital, 12, 26, 128, 213, 235–237, 241,
    243, 326
Lytitek, 207

Macoubrie, Jane, 339
Maebius, Stephen 326
magnetic-resonance force imaging micro-
    scope (MRFM), 187
magnetic-resonance imaging (MRI), 187,
    204, 278
Manufacturing Technology Competitive-
    ness Act (MTCA), 104
Marburger, John, 134–135, 144, 163, 269,
    298
MAST (Assessing Molecular and Atomic
    Scale Technologies), 337
Materials and Electrochemical Research,
    150
Material Safety Data Sheet(s), 182
Matsishita Electric Works, 197
Matsushita Electric Industrial, 149
May, Robert, 346
McCain, John (Sen., R-AZ), 63
mechanosynthesis, 31, 38, 57, 59, 72, 254,
    262, 267
Mercedes Benz, 191
Merck, 219
Merkle, Ralph, 56, 217, 248, 250, 252–253

Merrill Lynch, 214, 225, 228–229, 235

MetaMateria, 192

micellar nanoparticle drug-delivery platform, 201

Michigan State University, 27, 325

Micro-Air Vehicles (MAV) Project, 116

microarray(s), 187, 203

Micro-Electro-Mechanical Systems (MEMS), 21, 33, 62, 118–119, 137, 214, 237

microfluidics, 203

Microsoft, 196, 236, 261

Millipede, 196, 221

Milunovich, Steven, 229

Minding the Gap, 258

Ministry for Education, Science and Technology (MEXT), 145, 149

Ministry of Economy, Trade and Industry (METI), 145, 149

Ministry of Science and Technology (China), 151

Minsky, Marvin, 55–56, 251

Mintz, Levin, Cohn, Ferris, Glovsky, and Popeo, PC, 225

MIT (Massachusetts Institute of Technology), 55, 66, 69, 115–118, 187, 232, 238

Mitsubishi, 147, 150, 283

Mitsui, 147, 177, 284

Modzelewski, Mark, 26, 59, 63, 128, 135, 191, 219, 228, 239–242, 271, 326, 328

Molecular Diamond Technologies, 222

molecular electronics, 221

Molecular Foundry (UC Berkeley), 113, 232, 354

Molecular Imprints, 219, 231, 233

molecular nanotechnologies (MNT), 38

Monsanto, 77, 353

Montemagno, Carlo D., 254

Monteiro-Riviere, Nancy, 293

Mooney, Pat, 26, 30, 263–269, 276, 299

Moore, Julia, 308

Moore's Law, 195

Moran, Ed, 224

movement(s), 12, 27, 47, 49, 77, 80–82, 85, 116, 179, 211, 241, 247–248, 262, 307, 312, 328, 342–343, 350–353, 355–356, 360

MRAM (magnetic random access memory), 196–197

Mulhall, Douglas, 257

multiwall nanotube supercapacitor film, 207

Murday, James, 60

Murdock, Sean, 26, 186, 242, 302

Nanergy, 192, 210

nanoaluminum, 189

NanoBank (UCLA), 323–324

nanobiology, 148, 222

nanobiotechnology, 102, 134

nanoblock, 257, 259–260

nanobot(s), 19–21, 37, 39–42, 59, 61–62, 67, 70–72, 74–75, 120, 169, 206, 235, 245, 256, 262, 265, 267, 272, 277–278, 306

NanoBreeze, 210

Nanobubble, 179, 233

nanobumps, 202

NanoBusiness Alliance (NbA), 12, 26–27, 59, 61–63, 93, 123, 132, 177, 213, 239–242, 247, 263, 302, 309, 326–329, 354

NanoBusiness Development Group (NBDG), 240

Nano-C, 190

nano-C_{60}, 287

nanocables, 191

Nanocarblab (Russia), 284

Nano-Care, 32

nanocarpet, 190

nanocatalyst(s), 187, 210

NanoCeram Superfilter, 191

nanochip(s), 243

NanoChromics, 199

NanoClarity, 214

nanocoating(s), 219

nanocomposite(s), 100, 163, 186, 191–192, 210, 219, 221, 243

NanoConnection to Society NanoIndicator Series (*see also* Harvard), 323
NanoCoolers, 237
NanoCrystal, 201
nanocrystal(s), 188, 200, 202–203, 205–206, 222, 243, 283, 298
NanoCures, 201
Nanodesu, 210
Nanodivide, 262, 272
NanoDynamics, 93, 192, 233
nanoelectronics, 102, 114, 117, 127, 221, 277
nano-EM, 188
nano-emissive display, 199
nano-emulsion(s), 222
nano-enabled products, 186, 192
NanoEnvironBank (*see also* Harvard), 323
NanoEthicsBank (*see also* Harvard), 323
Nanoethics Group (*see also* California at Santa Barbara, University of), 323
nanofactory(ies), 39, 40, 71, 169, 255, 257–262
Nanofilm, 192, 233
Nanofur, 118
Nanogate, 210
Nanogen, 135, 179–180
NanoHorizons, 199, 209, 211
nanohype, 27, 30, 34, 37, 57, 61, 153
nanoimprint lithography, 195
NanoInk, 69
Nanoinstruments, 187
nanointermediates, 186
Nano-JETA platform, 203
NanoJury, UK, 273, 358
NanoKinetix, 210
Nanoledge (France), 284
Nanolithography, 54, 60, 188
NanoLogix, 118, 203
Nanologue, 336
NanoLub, 192
nanomachine(s), 21, 23, 39, 66–68, 72, 129, 265, 268, 277
nanomagnetics (nanomagnets), 100, 127, 221

nanomanufacturing, 91, 269
Nanomanufacturing Investment Partnership, 104
NanoMarkets, 195, 197–198, 200, 205
Nanomat, 172, 219
NanoMatrix, 150
Nanomedicine, 38, 206, 222, 254
nanomeme, 168
nanometal oxide(s), 137
Nanometrics, 180, 219
nanometrology, 102
Nanomix, 199, 206, 227, 233
NanoNed, 138
nanopants, 35, 177
Nanoparticle Synthesis and Processing Initiative, 124
Nanopharma, 238
Nanophase (NANX), 113, 189, 219, 230, 234, 238, 283, 299, 312
nanophotonics, 127, 138, 243, 330
NanoPierce Technologies, 32
nanopigment(s), 221
Nanopoint, 187
nanopolies, 266
nanopore(s), 191
nanopowders (nano-powders), 189, 218, 268, 283
NanoOpto, 227, 233, 237–238
nanopretenders, 180, 218
NanoProprietary, 180
NanoProtect, 118
NanoQuest, 40
Nanorex, 256
Nanoscale, Materials 281
Nanoscale Science and Engineering Center (NSEC), 87, 133–134, 307, 322–323, 325, 332
Nanoscience and Nanotechnologies: Opportunities and Uncertainties, 175
NanoScience Technologies, 347
Nano Sci Tech Industrial Park (China), 152
nanosensor(s), 115–116, 187, 193–194, 200, 202

NanoSig, 327
NanoSign, 210
Nanosolar, 209
NanoSonic, 191
Nanospectra Biosciences, 69, 204
Nanosphere, 150
nanosphere, 180, 231
Nanostellar, 186, 188
nanostocks, 225
Nanosys, 14, 26, 118, 197, 199, 209, 218, 228, 230–233, 238
Nanosystems, 56, 60, 252, 256
Nanoscience and Microtechnologies Institute (OR), 84
nanoshell(s), 178, 204
NanoTech Center (Albany State University, NY), 84
Nanotech Company, The, 345
Nanotechnologies: A Preliminary Risk Analysis on the Basis of a Workshop Organized in Brussels on 1 –2 March 2004, 173
Nanotechnology: Shaping the World Atom by Atom, 156
Nanotechnology: Small Matter, Many Unknowns, 181
Nanotechnology Victoria, 182
Nanotechnology: Views of the General Public, 169
Nanotech Report (Forbes/Wolfe Nanotech Report), 12, 14, 21, 93, 155, 178–180, 235–236, 238
Nantero, 197, 227, 233, 237–238
NanoTex (Nano-Tex) ,32, 93, 150, 177, 227, 231, 234
Nano-Touch, 32
nanotoxicity, 79, 175, 268–269, 321
nanotoxicology ,26, 156, 168, 275, 279–281, 297, 302, 315
Nanotronics, 267
Nanotube Power, 210
NanoTwin, 210
Nanovax, 201
nanoweapon(s), 260
nanowire(s), 193, 195–196, 200, 202–203, 207, 209, 214, 221, 243

nanoyarn, 190
Nantero, 197, 227, 233, 237–238
Nara Institute of Science and Technology, 149
Nansulate, 189
National Academy of Science (NAS), 26, 75, 104, 257, 306, 309, 318–319
National Advanced Interdisciplinary Research Laboratory (NAIR), 146
National Cancer Institute (NCI), 200–201
National Consortium of Materials, Science, and Technology (INSTM), 206
National Consumer Council, the (UK), 168
National Institute of Advanced Industrial Science and Technology (AIST), 146–147, 149
National Institute of Environmental Health Sciences, 280–281, 354
National Institute of Materials Science (NIMS), 149–150
National Institute of Occupational Safety and Health (NIOSH; *see also* OSHA), 164, 269
National Institute of Standards and Technology (NIST), 25, 91, 99–100, 101–103, 109, 128, 132, 160, 163–164, 318
National Institutes of Health (NIH), 94, 96–98, 127, 160–161, 163–164, 173, 200–201, 318
National Nanofabrication Users Network (NNUN), 25, 124–125, 127, 130, 135, 160, 170, 324
National Nanotechnology Advisory Panel, 134
National Nanotechnology Coordination Office (NNCO) (*see also* Teague, Clayton), 89, 127, 159, 314
National Nanotechnology Infrastructure Network (NNIN), 7, 25, 127, 130–131, 133, 135, 160, 307, 317, 322, 324, 330
National Nanotechnology Initiative (NNI), 7, 15, 16, 25, 33, 35, 50, 54,

57–58, 60–61, 63–64, 75, 83–84, 86–87, 89–91, 96–99, 101, 104, 106–107, 109–110, 112–114, 116, 119–121, 125–130, 132, 136–140, 145–146, 151, 157–164, 177, 218–219, 241, 252–255, 257, 298, 306–307, 309, 311, 313–314, 317–318, 332–335

National Nanotechnology Initiative and Its Implementation Plan, 161

National Nanotechnology Initiative: Leading to the Next Industrial Revolution, 157

National Nanotechnology Initiative: The Initiative and Its Implementation Plan, 158

National Nanotechnology Initiative: Research and Development Supporting the Next Industrial Revolution, 163

National Science and Technology Council (NSTC,) 87, 90, 125–127, 136, 163, 306–307, 312

National Science Foundation (NSF), 11, 13, 15–17, 24–26, 28, 34–35, 50, 57, 60, 67–68, 84–87, 90, 94–99, 101, 109, 112, 114, 117, 121, 124–125, 127, 129, 131–133, 160–165, 176–177, 185, 213, 219, 257, 260, 276, 307–308, 316–320, 322–323, 330, 332, 334, 336–337, 358–359

National Toxicology Program, 280

NaturalNano, 189

Naval Research Laboratory, 199

nCode, 210

Nemmar, A., 282, 290

Neo-EpCAM cancer detection kit, 203

neo-Marxism (neo-Marxist), 76, 269

Netherlands, the, 87, 137

Network for Computational Nanotechnology, (NCN) 127, 164,

New Dimensions for Manfacturing: A UK Strategy for Nanotechnology (Taylor Report), 170–171

NG-DVD (Nano-Grating-DVD),197

NGen Partners, 213

NGOs (nongovernmental organizations), 26, 28, 30, 47, 126, 167–168, 316, 328

Nokia, 142

Nomadics, 115

Nordan, Matthew, 186, 188, 209, 220, 223, 226, 229

Norel Optronics, 199

North Carolina Citizens Technology Forum, 358

Northwest Venture Partners, 234

Novation Environmental Technologies, 191

NuCelle, 210

Nucryst Pharmaceuticals, 203

nutraceuticals, 194

NVE, 197

Oberdörster, Eva, 92, 297–298, 302–303

Oberdörster, Gunter, 282, 290, 292, 296, 302–303

Objective Force Warrior, 115

Occupational Safety and Health Administration (OSHA; *see also* NIOSH), 312

Office of Basic Energy Sciences (BES), 109–110, 112, 165

Office of Science (DOE), 109–110, 112

Office of Science and Technology Policy (OSTP), 85–86, 125–126, 163, 298

OilFresh, 195

Oil of Olay, 189

Olum, Paul, 50

Optisol UV absorber, 204

Optiva, 225, 233, 238

Optoelectronics, 219, 221, 231

organic light emitting diode (OLED) displays, 199

overhype, 30, 32 37

Oxonica, 204, 210

Pacific Northwest National Lab, 205

Pakistan, 137

Parkinson's Disease, 202, 206

Parr, Doug, 174, 271–273, 316

participatory democracy, 356

Particle Sciences, 205

patent(s), 69, 107, 151, 220, 223, 225,

230–231, 259–260, 264, 266–267, 272, 355
Patent Trademark Office (USPTO), 107
PCBs (polychlorinated biphenyls), 285–286
perception management, 93, 309, 312, 320, 328, 332
Peterson, Christine, 12, 26, 54, 59, 247–249, 251, 318
PHANTOMS Network, 143
pharmaceutical(s), 34, 66–67, 88, 156, 167, 176, 185–187, 201, 220, 222, 279, 294, 296, 356
Phoenix, Chris, 26, 38, 71–72, 169, 256–258, 260, 262–263
photovoltaics, 118, 207, 209–210, 228
picks and shovels, 186, 235,
Plenitude, 204, 210
Polyflon, 205
PolyFuel, 210
PolyMetallix, 205
Polytetrafluoroethylene, 292
post-9/11 world, 46, 118
posthumanism, 311
precautionary principle, 168–169, 183, 260–262, 272, 283, 301–302
President's Council of Advisers on Science and Technology (PCAST), 126–127, 134
Prey (*see also* Crichton, Michael), 20, 40–43, 84, 181, 270, 322, 345, 347
Prince Charles, HRH, 24, 39, 73, 76, 78–80, 245, 264, 283, 313
PRINT (Particle Replication In Nonwetting Templates), 201
prolepsis, 55
pSilvida, 204
public(s), 256, 258, 265, 271–272, 307–308, 310, 312–316, 319, 321–323, 325–328, 330, 333–337, 339–344, 346–360
public intellectual(s), 46
public sphere, 24, 27, 43–44, 134, 271–272, 313, 335,–360
Punk Ziegel & Co., 214, 228–229
PVNanofilm, 192

Quantum Architecture Research Center, 117
quantum computer(s) (computing), 198
Quantum Dot, 203
quantum dot(s) (*see also* semiconducting nanocrystals), 161, 192, 198, 201, 203–204, 243, 280, 297, 299
QuantumSphere, 188–189, 210
Quick Start, 141

RAFI (Rural Advancement Foundation International), 26, 253–254, 264
Raytheon, 115
Regional Development Agency (RDA), 171
Regis, Ed, 56, 64
Rejeski, David, 210
Renault, 283
Reynolds, Glenn, 58–59, 241, 251, 270–271, 354, 359
Rice University (*see also* CBEN), 11, 14, 24, 41, 65, 69, 85–87, 161–162, 168, 175, 205, 281–282, 285–288, 293–294, 297, 300, 302, 308, 314–316, 326
Roco, Mihail (Mike), 12, 15, 25, 28, 42, 47, 60, 62–63, 87–90, 93, 97–98, 105, 119, 125, 132, 164, 178, 311–312, 317, 321, 322, 348
Rohm and Haas, 189
Rohrabacher, Dana (Rep., R-CA), 329
Rossette (Cyrpus), 284
Roy, Rustum, 33–34
Royal Academy of Engineering, the, 167, 169, 301
Royal Society, the, 77, 167–169, 175, 264, 273, 285, 301, 336
Russia, 109, 187, 196, 284

Sainsbury, Lord, 79
Samsung Group (Samsung), 196, 199
Sayes, Christie, 286, 297
Science and Technology Basic Plan (Japan), 145
Scientific Committee on Cosmetic Products and Non-Food Products Intended for Consumers (UK), 182

Seagate, 221

Second Sight Medical Products, 206

Seeman, Nadrian C., 250

SEIN (Societal and Ethical Implications of Nanotechnology), 26, 27, 35, 50, 54, 57, 68, 75, 87, 92, 93, 133, 156, 162, 164, 166, 168, 254, 270, 276–277, 270, 302–304, 306–307, 309–314, 317–326, 328–337, 348, 360

self-assembly, 55, 59, 63, 131, 193, 211, 221, 265

self-replicators, 38

Senszoko odor-eating refrigerators, 210

sexually transmitted diseases (STDs), 201

Shalleck, Alan, 214

Shand, Hope, 264, 268, 300

Shanghai Institute of Nuclear Research, 152

Showa Denko, 284

silicon-carbide ceramics, 189

Silicon Storage Technology, 197

Singer, Peter, 38–40, 313–314, 321, 337

Sixth Framework Programme (EU), 139–141, 165

Skypharma, 202

Smalley, Richard, 17, 24, 47, 52, 58, 63, 65–73, 106, 111, 129, 162, 167, 209, 235, 237, 245, 262, 266, 338, 374–376

Small Business Innovation Research (SBIR), 104, 162, 318

Small Business Technology Transfer (STTR), 104, 162

Small Times, 11–12, 14, 35, 42, 149, 225–226, 231, 281

Small Wonders, Endless Frontiers, 162, 309

Snow, C. P., 45, 199, 319, 351

Social and Economic Challenges of Nanotechnology, The, 165

solar cell(s) (*see also* photovoltaics), 74, 111, 118, 137, 158, 209–210, 232, 328

Solicore, 237

Somitomo, 150

Sonny Oh, 195

South Carolina, University of, 17, 27–28, 57, 64, 164, 323–325, 332

South Korea (Republic of Korea), 95, 137

Spence, Bill, 40

Spintronic Science and Applications Center, 198

Spire, 206

Spitzer, Eliot, 32

SRAM (static random access memory), 196

STAGE (Science, Technology and Governance in Europe), 168

Starfire Systems, 189

Starpharma, 201

start-up(s), 26, 69, 94, 103–104, 108, 126, 133, 138, 140, 143, 147, 152, 176–177, 179–180, 187, 202, 211, 213, 219, 220, 222–228, 230, 233, 235, 237–240, 242, 255, 280, 315, 324

Stealth CNT baseball bats, 120

sticky finger(s) (*see also* fat fingers), 66–67, 70

STI (Australia), 209

Stix, Gary, 61–62, 252

STM (scanning tunneling microscope), 41, 54, 188, 338

STMicroelectronics, 142

STS (Science and Technology Studies), 13, 308, 313, 424

Stupp, Samuel, 15, 28, 202, 204, 309

Sumitomo, 69

Sun Microsystems, 73, 227, 229, 249

Sun Nanotech, 284

Sunsense sunscreen, 210

superhumanism, 41

superthermites, 189

Surface Logix, 66

Swiss Reinsurance Company (Swiss RE), 17, 25, 93, 155, 175–176, 181–183, 268

Syngento, 348

Taiwan, 137, 211

Takenaka, S., 291

Tan, M., 295

Taneguchi, Norio, 23, 54

Takara Bio, 150

Taylor, John M., 170–171

TCE (trichloroethylene), 205
Teague, Clayton, 89, 314
Technical Instruments, 219
Technology Transfer Commercialization Act, 107
terminator gene (*see also* Monsanto), 353
Terrawatt Challenge, 111
Texas Instruments, 196, 221
Texas Nanotechnology Initiative, 42
Thailand, 137
Thalidomide, 77–78, 275
Thayer, Ann, 14
third-culture intellectuals (new-public intellectuals), 46–47
Thomas Swan and Co., 190
Thompson, Paul, 27, 308, 325, 333
THONG (Topless Humans Organized for Natural Genetics), 263, 354,
Three Mile Island, 43
Tice, Donn, 32, 231,
Tinker, Nathan, 63, 239, 242, 345
Tinkle, S. J., 295
tipping point, 93, 223, 225
titanium dioxide (titanium oxide), 182, 192, 205, 209, 211, 295, 293, 299
Tokyo Institute of Technology, 149
Tomson, Mason, 285–286
Toray Industries, 150
Toshiba, 196, 210
Toyota, 210, 283
Trane, 205
transdisciplinary, 29
transhumanism (transhumanists), 28, 236, 256–257, 311
transnational(s) (transnational corporations), 8, 26, 211, 219–221, 262, 267, 308, 349, 352, 356
Treder, Michael, 26, 63, 256–258, 260–263
trimetaspheres, 189
Triple Innova (Germany), 336
Triton Systems, 115
Tsukuba Science City, 146
TVA Productions, 347
Twentieth Century Fox (*see also Prey*), 20

Twenty-first Century Nanotechnology Research and Development (R&D) Act, 33, 37, 59, 61–63, 82–83, 128, 131, 179, 218, 254, 262, 307, 329, 334

ultrafine particles (UFPs), 282, 288, 290–292, 296
ultra-nanocrystalline diamond films, 206
Ultratech, 187
United Kingdom (UK), 25, 40, 76, 79, 137–138, 143–144, 165, 167–168, 170–171, 269, 273, 299–300, 310, 336, 358
universal assembler(s), 31, 38
US Global Nanospace, 31

Veeco, 187, 205
venture capital (VC), 103, 141, 211, 215, 218, 223–228, 233–240, 243, 253, 309, 315, 326, 336
Venture One, 225
Vision 2020 Future Warrior System, 115
von Ehr, James, 12, 36, 326
VS Drive, 210

Warheit, David B., 289, 292
Warwick Nanosystems Group, 168
West, Jennifer, 288, 293–294,
Wet Dreams All-Natural Sunscreen, 205
Whitesides, George, 66–68 245, 306
Wiesner, Mark, 286, 288, 293, 300
Wilson, 210
Winstead, Sechrest and Minick, 42
Williams, R. Stanley, 35, 108
Wolfe, Josh, 12, 14, 21, 25–26, 42–43, 55, 62, 69, 93–94, 114, 130, 150, 155, 178–180, 201, 211, 219, 227, 229, 232, 234–239, 241, 243, 312, 315, 319, 326
Wood, Stephen, 165, 168
Woodrow Wilson International Center for Scholars, 41, 210, 308, 339
Wyden, Ron (Sen., D-OR), 24, 33, 82–85, 93, 97, 126, 132

Yamakoshi, Yoko, 297

Zatasizer Nano, 205
Zettacore, 197, 237
Zhang, Yihong, 290
Zhongke Nano Engineering Center
    (China), 152

ZinClear, 205
Zucker, Lynne, 27, 324
Zylon, 284
Zyvex, 12, 20, 32, 36, 62, 100, 219, 233, 326